METHODS IN MOLECULAR BIOLOGY™

For further volumes:
http://www.springer.com/series/7651

Gene Regulatory Networks

Methods and Protocols

Edited by

Bart Deplancke

Inst. Bioengineering, École Polytechnique Fédérale de Lausanne, Lausanne, Switzerland

Nele Gheldof

Faculté de Biologie det Médecine, Centre Intégratif de Génomique, Université de Lausanne, Bâtiment Génopode, Lausanne, Switzerland

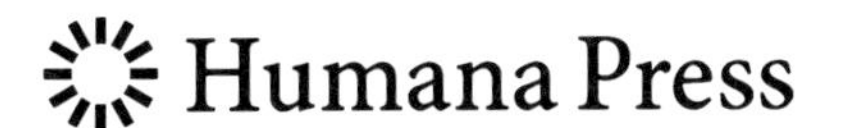

Editors
Bart Deplancke, Ph.D
Inst. Bioengineering
École Polytechnique Fédérale de Lausanne
Lausanne 1015, Switzerland
bart.deplancke@epfl.ch

Nele Gheldof
Faculté de Biologie det Médecine
Centre Intégratif de Génomique
Université de Lausanne
Bâtiment Génopode
Lausanne 1015, Switzerland
nele.gheldof@unil.ch

ISSN 1064-3745 e-ISSN 1940-6029
ISBN 978-1-61779-291-5 e-ISBN 978-1-61779-292-2
DOI 10.1007/978-1-61779-292-2
Springer New York Dordrecht Heidelberg London

Library of Congress Control Number: 2011936383

Printed on acid-free paper

Humana Press is part of Springer Science+Business Media (www.springer.com)

Preface

Gene regulatory networks (GRNs) play a vital role in organismal development and function by controlling gene expression, and deregulation of these networks is often implicated in disease because of inappropriate gene expression or silencing. GRNs depict the dynamic interactions between genomic and regulatory state components. The genomic components comprise genes and their associated *cis*-regulatory elements. The regulatory state components consist primarily of transcriptional complexes that bind the latter elements as well as noncoding RNAs such as microRNAs that control gene expression via posttranscriptional mechanisms. With the availability of complete genome sequences, several novel approaches have recently been developed which promise to significantly enhance our ability to identify either the genomic or regulatory state components, or the interactions between these two. In this volume, we briefly detail how each of these approaches contributes to a more comprehensive understanding of the composition and function of GRNs, and provide a comprehensive protocol on how to implement them in the laboratory.

In a first section, we focus on approaches aiming to identify and characterize regulatory state components. First, a protocol is provided on how to computationally identify genes coding for transcription factors in prokaryotic and eukaryotic genomes (Chapter 1). Once identified, then the function of these DNA-binding proteins needs to be determined in terms of where and when these proteins are active (Chapter 2), which DNA-binding sites they recognize (Chapters 3–6), which proteins they interact with (Chapter 7), and by which posttranscriptional or posttranslational mechanisms (Chapter 8) or signaling events (Chapter 9) their activity is controlled. We conclude this first section by detailing a protocol on how to generate small RNA deep-sequencing libraries to examine the function of RNA-based regulatory components (Chapter 10).

In a second section, we present approaches to identify genomic components. In Chapter 11, a detailed protocol is presented on how to identify transcription start sites and thus gene promoters by cap analysis gene expression or CAGE coupled to high-throughput sequencing. Chapter 12 explains how to identify DNaseI-hypersensitive sites, which may connote a regulatory function, in high-throughput but sensitive fashion via multiplex ligation-dependent probe amplification or MPLA. Finally, Chapter 13 details how to perform a chromosome conformation capture (4C)-seq assay. Such an assay reveals long-range physical interactions between DNA elements such as a promoter and an enhancer, thus allowing the identification of active regulatory elements.

In the next section, we discuss methods that map interactions between regulatory state and genomic components. There are two groups of methods. The first are transcription factor-centered and aim to identify genomic targets for one transcription factor of interest at a time. The best known representative of such group of methods is chromatin immunoprecipitation or ChIP, and Chapters 14 and 15 detail how to perform genome-wide ChIP using (multiplexed) high-throughput sequencing in *Drosophila* embryos and mammalian cells, respectively. Chapter 16 provides a guide on how to computationally process and analyze ChIP-seq data. ChIP has the advantage of being able to detect in vivo interactions. Nevertheless, in vitro or in silico transcription factor-centered methods are still useful by allowing the detection of protein–DNA interactions that can happen irrespective of whether they are relevant in vivo or not. It is well-accepted that ChIP often fails to detect interactions that occur in a small number of cells or in a small time frame. Moreover, ChIP requires specific antibodies for the protein of interest, which are still not available for many transcription factors. In this regard, there will always be room for methods complementary to ChIP and two such methods, one in vitro and one in silico, are featured in Chapters 17 and 18, respectively. A second group of protein–DNA interaction detection methods are gene-centered since they aim to identify the proteins that can bind to a specific DNA element. In Chapter 19, several proteomic methodologies are presented which allow the identification of transcription factors or even entire transcriptional complexes binding to delineated DNA-binding sites or regulatory elements. Chapter 20 details how to conduct high-throughput yeast one-hybrid assays to rapidly identify TFs binding to promoters or enhancers of interest.

In a final section, we present two computational approaches. The first (Chapter 21) explains how to visualize the observed protein–DNA interactions in intuitive fashion with the aim of deducing the structural and dynamic properties of the underlying GRNs. The second provides an in-depth overview of how to qualitatively model different aspects of increasingly large regulatory networks such as steady state behavior, stochasticity, and gene perturbation experiments.

In summary, this volume aims to provide a comprehensive and timely toolkit to study GRNs from the point of data generation to processing, visualization, and modeling. The featured approaches thereby illustrate how the high rate of novel assay development and the constantly increasing sensitivity, throughput, and miniaturization of existing methods will enable us to address many questions related to the structural and dynamic properties of GRNs.

Lausanne, Switzerland

Bart Deplancke
Nele Gheldof

Contents

Contributors

STEIN AERTS • *Laboratory of Computational Biology, Center for Human Genetics, K. U. Leuven, Leuven, Belgium*
ZEYNEP KALENDER ATAK • *Laboratory of Computational Biology, Center for Human Genetics, K. U. Leuven, Leuven, Belgium*
MUKESH BANSAL • *Joint Centers for Systems Biology, Columbia University, New York, NY, USA*
EUGENE BEREZIKOV • *InteRNA Technologies BV, Utrecht, The Netherlands; Hubrecht Institute, Utrecht, The Netherlands*
MARK D. BIGGIN • *Genomics Division, Lawrence Berkeley National Laboratory, Berkeley, CA, USA*
ANNA BOTVINIK • *Research Group 'Gene Expression', Max-Planck-Institute of Experimental Medicine, Goettingen, Germany*
SHAWN M. BURGESS • *National Human Genome Research Institute, National Institutes of Health, Bethesda, MD, USA*
ANDREA CALIFANO • *Joint Centers for Systems Biology, Department of Biomedical Informatics, Institute of Cancer Genetics and Herbert Irving Comprehensive Cancer Center, Columbia University, New York, NY, USA*
PIERO CARNINCI • *Riken Omics Science Center, Riken Yokohama Institute, Yokohama, Kanagawa, Japan*
PIL JOONG CHUNG • *Genomics Genetics Institute, GreenGene Bio Tech Inc., Yongin, Kyonggi-Do, South Korea*
HANNAH L. CRAIG • *Faculty of Biological Sciences, Institute of Integrative and Comparative Biology, The University of Leeds, Leeds, UK*
GIOVANNI DE MICHELI • *School of Computer and Communication Sciences, École Polytechnique Fédérale de Lausanne (EPFL), Lausanne, Switzerland*
BART DEPLANCKE • *Laboratory of Systems Biology and Genetics, École Polytechnique Fédérale de Lausanne (EPFL), Lausanne, Switzerland*
ALEXANDER DÜNKLER • *Department of Biology, Institute of Molecular Genetics and Cell Biology, Ulm University, Ulm, Germany*
STEFANIE EGGERS • *Molecular Development Laboratory, Murdoch Childrens Research Institute, Royal Children's Hospital, Melbourne, VIC, Australia*
HUIYUN FENG • *Faculty of Biological Sciences, Institute of Integrative and Comparative Biology, The University of Leeds, Leeds, UK*
JEAN-DANIEL FEUZ • *Laboratory of Systems Biology and Genetics, École Polytechnique Fédérale de Lausanne (EPFL), Lausanne, Switzerland*
EILEEN E.M. FURLONG • *Genome Biology Unit, European Molecular Biology Laboratory, Heidelberg, Germany*
ABHISHEK GARG • *Department of Systems Biology, Harvard Medical School, Boston, MA, USA*

MARCEL GEERTZ • *Department of Molecular Biology, University of Geneva, Geneva, Switzerland*
YAD GHAVI-HELM • *Genome Biology Unit, European Molecular Biology Laboratory, Heidelberg, Germany*
NELE GHELDOF • *Center for Integrative Genomics, University of Lausanne, Le Génopode, Quartier UNIL-Sorge, Lausanne, Switzerland*
WILLEMIJN M. GOMMANS • *InteRNA Technologies BV, Utrecht, The Netherlands*
KORNEEL HENS • *Laboratory of Systems Biology and Genetics, École Polytechnique Fédérale de Lausanne (EPFL), Lausanne, Switzerland*
CARL HERRMANN • *TAGC Inserm U928, Marseille, France; Université de la Mediterranée, Marseille, France*
IAN A. HOPE • *Faculty of Biological Sciences, Institute of Integrative and Comparative Biology, The University of Leeds, Leeds, UK*
HARRY W. JARRETT • *Department of Chemistry, University of Texas San Antonio, San Antonio, TX, USA*
NILS JOHNSSON • *Department of Biology, Institute of Molecular Genetics and Cell Biology, Ulm University, Ulm, Germany*
SACHI KATO • *Riken Omics Science Center, Riken Yokohama Institute, Yokohama, Kanagawa, Japan*
MIN-JEONG KIM • *Genomics Genetics Institute, GreenGene Bio Tech Inc., Yongin, Kyonggi-Do, South Korea*
TAE-HOON KIM • *Genomics Genetics Institute, GreenGene Bio Tech Inc., Yongin, Kyonggi-Do, South Korea*
YEON-KI KIM • *Genomics Genetics Institute, GreenGene Bio Tech Inc., Yongin, Kyonggi-Do, South Korea*
TAE-HO LEE • *Genomics Genetics Institute, GreenGene Bio Tech Inc., Yongin, Kyonggi-Do, South Korea*
MARION LELEU • *School of Life Sciences and Bioinformatics and Biostatistics Core Facility, Swiss Institute of Bioinformatics, Ecole Polytechnique Fédérale (EPFL), Lausanne, Switzerland*
WILLIAM J.R. LONGABAUGH • *Institute for Systems Biology, Seattle, WA, USA*
NICHOLAS M. LUSCOMBE • *EMBL-European Bioinformatics Institute, Wellcome Trust Genome Campus, Cambridge, UK; EMBL-Heidelberg Gene Expression Unit, Heidelberg, Germany*
SEBASTIAN J. MAERKL • *Institute of Bioengineering, École Polytechnique Fédérale de Lausanne (EPFL), Lausanne, Switzerland*
KARTIK MOHANRAM • *Departments of Electrical and Computer Engineering and Computer Science, Rice University, Houston, TX, USA*
JUDITH MÜLLER • *Department of Biology, Institute of Molecular Genetics and Cell Biology, Ulm University, Ulm, Germany*
MITSUYOSHI MURATA • *Riken Omics Science Center, Riken Yokohama Institute, Yokohama, Kanagawa, Japan*
FELIX NAEF • *Laboratory of Computational Systems Biology, School of Life Sciences, École Polytechnique Fédérale de Lausanne (EPFL), Lausanne, Switzerland*
BAEK HIE NAHM • *Division of Bioscience and Bioinformatics, MyongJi Unviersity, Yongin, Kyonggi-Do, South Korea*

DAAN NOORDERMEER • *School of Life Sciences, Ecole Polytechnique Fédérale (EPFL), Lausanne, Switzerland*
MARCUS B. NOYES • *Lewis-Sigler Institute of Integrative Genomics, Princeton University, Princeton, NJ, USA*
NOBUO OGAWA • *Genomics Division, Lawrence Berkeley National Laboratory, Berkeley, CA, USA*
THOMAS OHNESORG • *Molecular Development Laboratory, Murdoch Childrens Research Institute, Royal Children's Hospital, Melbourne, VIC, Australia*
DELPHINE POTIER • *TAGC Inserm U928 and Université de la Mediterranée, Marseille, France; IBDML CNRS UMR 6216 and Université de la Mediterranée, Marseille, France*
SUNIL KUMAR RAGHAV • *Laboratory of Systems Biology and Genetics, École Polytechnique Fédérale de Lausanne (EPFL), Lausanne, Switzerland*
ALEXANDRE REYMOND • *Center for Integrative Genomics, University of Lausanne, Le Génopode, Quartier UNIL-Sorge, Lausanne, Switzerland*
SYLVIE ROCKEL • *Institute of Bioengineering, École Polytechnique Fédérale de Lausanne (EPFL), Lausanne, Switzerland*
MORITZ J. ROSSNER • *Research Group 'Gene Expression', Max-Planck-Institute of Experimental Medicine, Goettingen, Germany*
JACQUES ROUGEMONT • *Bioinformatics and Biostatistics Core Facility, Swiss Institute of Bioinformatics, Ecole Polytechnique Fédérale (EPFL), Lausanne, Switzerland*
MARINA NAVAL SANCHEZ • *Laboratory of Computational Biology, Center for Human Genetics, K. U. Leuven, Leuven, Belgium*
HAZUKI TAKAHASHI • *Riken Omics Science Center, Riken Yokohama Institute, Yokohama, Kanagawa, Japan*
SARAH A. TEICHMANN • *MRC Laboratory of Molecular Biology, Cambridge, UK*
JUAN M. VAQUERIZAS • *EMBL-European Bioinformatics Institute, Cambridge, UK*
STEFAN J. WHITE • *Centre for Reproduction and Development, Monash Institute of Medical Research, Clayton, VIC, Australia*
IOANNIS XENARIOS • *Swiss Institute of Bioinformatics, University of Lausanne, Lausanne, Switzerland*
JIZHOU YAN • *Institute for Marine Biosystem and Neurosciences, Department of Hydrobiology, Shanghai Ocean University, College of Fisheries and Life Sciences, Lingang New City, Shanghai, China*

Part I

Regulatory State Components

Chapter 1

How Do You Find Transcription Factors? Computational Approaches to Compile and Annotate Repertoires of Regulators for Any Genome

Juan M. Vaquerizas, Sarah A. Teichmann, and Nicholas M. Luscombe

Abstract

Transcription factors (TFs) play an important role in regulating gene expression. The availability of complete genome sequences and associated functional genomic data offer excellent opportunities to understand the transcriptional regulatory system of an entire organism. To do so, however, it is essential to compile a reliable dataset of regulatory components. Here, we review computational methods and publicly accessible resources that help identify TF-coding genes in prokaryotic and eukaryotic genomes. Since the regulatory functions of most TFs remain unknown, we also discuss approaches for combining diverse genomic datasets that will help elucidate their chromosomal organisation, expression, and evolutionary conservation. These analysis methods provide a solid foundation for further investigations of the transcriptional regulatory system.

Key words: Transcription factor, Genomics, TF, Transcriptional regulation, Evolution, Gene expression, Genome organisation

1. Introduction

Transcriptional regulation is one of the most fundamental mechanisms for controlling the amount of protein produced by cells under different environmental conditions and developmental stages (1, 2). In eukaryotes, transcription can be modulated at many different levels, from the assembly of the core transcriptional machinery to the architecture and intranuclear localisation of chromosomes (3). A vast array of proteins, including RNA polymerases, histones, histone modifiers, transcription factors, and co-factors, are involved in maintaining the accuracy and specificity of the regulatory process.

Bart Deplancke and Nele Gheldof (eds.), *Gene Regulatory Networks: Methods and Protocols*,
Methods in Molecular Biology, vol. 786, DOI 10.1007/978-1-61779-292-2_1, © Springer Science+Business Media, LLC 2012

Among these, DNA-binding transcription factors (TFs) play a central role, as they are responsible for directing transcription initiation to specific gene promoters based on their sequence-recognition abilities (4). The importance of TFs is highlighted by the amount of research devoted to understanding how they function, ranging from the basis for DNA-sequence recognition to their regulatory function in particular cell types (5, 6).

The availability of fully sequenced genomes and the development of high-throughput experimental techniques over the past decade have expanded our capacity to explore regulatory systems on a whole-organism scale. Using these data, computational studies have characterised repertoires of TFs for organisms across all phylogenetic groups including bacteria, fungi, and plants (see Table 1). Further, integration of these repertoires with functional and evolutionary data has led to insights about basic organisational properties of these regulatory systems. For example, our recent analysis of the human TF repertoire revealed a two-tier organisation of tissue-specific and ubiquitous expression patterns, and a step-wise pattern of evolutionary conservation (7). Since the regulatory functions of many TFs remain unknown, these types of general analyses provide a good starting point for further detailed molecular and computational studies.

In this review, we introduce several online resources describing TF repertoires in different eukaryotic genomes. We also describe possible computational strategies for identifying such a repertoire and discuss approaches to annotate newly identified TF-coding genes with additional information, using our recent publication of the human TF repertoire as an example (7). It is worth noting here that we restrict our definition of TFs as proteins that recognise DNA in a sequence-specific manner, but neither display enzymatic activity nor form part of the core transcriptional machinery. The tutorial refers to our recently published study of human TFs for demonstration purposes (Fig. 1), but the analysis could be extended to any other organism with a sequenced genome.

2. Materials

2.1. Identification of the TF Repertoire

1. Genome sequences and associated annotations. The sequence, gene assembly, and protein-sequence annotation for the human genome, as well as many others, can be obtained from Ensembl (http://www.ensembl.org) (8). Programmatic access to the database is also available through an Application Programming Interface (API).

Table 1
Summary of available online resources for TF annotation

Resource	Organism	References	Link
Prokaryotic			
RegulonDB	*E. coli*	(66)	http://regulondb.ccg.unam.mx/
DBTBS	*B. subtilis*	(53)	http://dbtbs.hgc.jp/
Moreno-Campuzano et al.	*B. subtilis*	(56)	http://www.biomedcentral.com/1471-2164/7/147/additional/
CoryneRegNet	Corynebacteria	(51)	http://www.coryneregnet.de/
cTFbase	Cyanobacteria	(52)	http://cegwz.com/
PRODORIC	Bacteria	(63)	http://prodoric.tu-bs.de
RegTransBase	Bacteria	(65)	http://regtransbase.lbl.gov
ArchaeaTF	Archaea	(49)	http://bioinformatics.zj.cn/archaeatf/
BacTregulators	Prokaryotes	(50)	http://www.bactregulators.org/
GTOP_TF	Prokaryotes	(70)	http://spock.genes.nig.ac.jp/~gtop_tf/index2.html
Eukaryotic			
Vaquerizas et al.	*H. sapiens*	(7)	http://www.valleyofpigs.org/humantfs
Messina et al.	*H. sapiens*	(48)	http://genome.cshlp.org/content/14/10b/2041/suppl/DC1
TFcat	*H. sapiens/ M. musculus*	(68)	http://www.tfcat.ca/
TFCONES	*H. sapiens/ M. musculus/ T. rubripes*	(69)	http://tfcones.fugu-sg.org
Gray et al.	*M. musculus*	(59)	http://www.sciencemag.org/cgi/content/full/306/5705/2255/DC1
TFdb	*M. musculus*	(46)	http://genome.gsc.riken.jp/TFdb/
FlyTF	*D. melanogaster*	(54)	http://flytf.org/
EDGEdb	*C. elegans*	(45)	http://edgedb.umassmed.edu/
RARTF	*A. thaliana*	(64)	http://rarge.gsc.riken.jp/rartf/
AtTFDB	*A. thaliana*	(47)	http://arabidopsis.med.ohio-state.edu/AtTFDB/
SoyDB	*G. max*	(67)	http://casp.rnet.missouri.edu/soydb/
TOBFAC	*N. tabacum*	(71)	http://compsysbio.achs.virginia.edu/tobfac/
wDBTF	*T. aestivum*	(74)	http://wwwappli.nantes.inra.fr:8180/wDBFT/
ITFP	Mammals	(57)	http://itfp.biosino.org/itfp/
FTFD	Fungi	(55)	http://ftfd.snu.ac.kr/
PlanTAPDB	Plants	(60)	http://www.cosmoss.org/bm/plantapdb/
PlantTFDB	Plants	(61)	http://planttfdb.cbi.pku.edu.cn/
PlnTFDB	Plants	(62)	http://plntfdb.bio.uni-potsdam.de/v2.0/
JASPAR	Eukaryotes	(58)	http://jaspar.cgb.ki.se/
TRANSFAC	Eukaryotes	(72)	http://www.gene-regulation.com/pub/databases.html/
TrSDB	Eukaryotes	(73)	http://bioinf.uab.es/cgi-bin/trsdb/trsdb.pl/
Cross-kingdom			
DBD	–	(13)	http://www.transcriptionfactor.org/

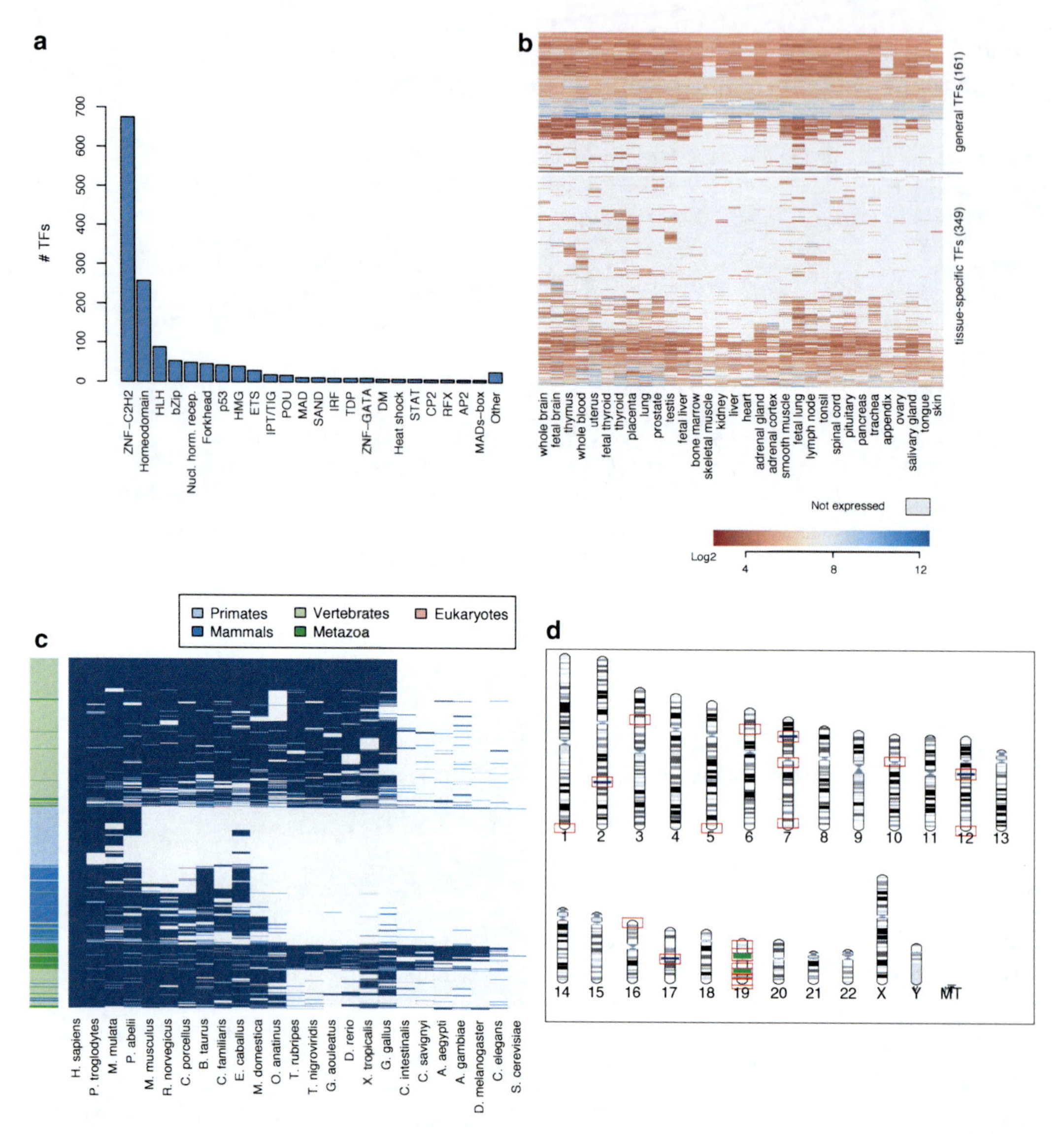

Fig. 1. Computational annotation of sequence-specific DNA binding transcriptional factors. (**a**) Human TFs classified according to their DNA-binding domain composition. Families with less than five members are classified as "other". (**b**) Heat-map representation of transcription factor expression (*rows*) across 32 human tissues and organs (*columns*). Cells are coloured according to the expression level (*dark blue* for high expression, *dark red* for low expression). General (ubiquitous) and tissue-specific TFs are grouped according to their expression values using hierarchical clustering. Expression levels below the threshold of detection are shown as white cells. (**c**) Evolutionary conservation of human transcription factors across 24 eukaryotic genomes. Transcription factors (*rows*) and species (*columns*) are clustered based on the presence (*blue cell*) or absence (*white cell*) of orthologous genes. The coloured bar on the left indicates the level of conservation for each TF depending whether these are primate-specific (*light blue*), mammalian-specific (*dark blue*), vertebrate-specific (*light green*), metazoa-specific (*dark green*) or present in all examined species (*pink*). (**d**) Chromosomal location of transcription factor clusters in the human genome. The positions of 23 chromosomal regions with high density of TFs are *highlighted with red boxes*. The Hox clusters in chromosomes 2, 7, 12, and 17 are depicted in *blue*. Zn-finger clusters in chromosome 19 are depicted in *light green*. First published in [Nature Reviews Genetics, 10(4), 2009, doi:10.1038/nrg2538. © Nature Publishing Group, a division of Macmillan Publishers Limited].

2. Protein sequences representing DNA-binding domains (DBDs). A high-confidence dataset of 347 InterPro hidden Markov models (InterPro release 18) (9) of DBDs can be obtained from (http://www.valleyofpigs.org/humantfs). Complete datasets of protein domains can also be obtained from Pfam (http://pfam.sanger.ac.uk) (10), SUPERFAMILY (http://supfam.cs.bris.ac.uk/SUPERFAMILY_1.73) (11), or PROSITE (http://www.expasy.ch/prosite) (12). The DBD database (http://www.transcriptionfactor.org) also provides hidden Markov models for DBDs from SUPERFAMILY and Pfam (13, 14).
3. Sequence search software. The InterProScan software can be downloaded from (http://www.ebi.ac.uk/Tools/InterProScan) (15).

2.2. Assessing the Coverage of the TF Repertoire

1. Gene-function annotations. Gene Ontology annotations for genes can be downloaded from (http://www.geneontology.org) (16).

2.3. Assigning Regulatory Functions to the TF Repertoire

1. GO functional enrichments. Enrichment of biological functions can be calculated using g:Profiler (http://biit.cs.ut.ee/gprofiler) (17).
2. Literature citations. Journal abstracts citing the genes above can be obtained from the PubMed database (http://www.ncbi.nlm.nih.gov/pubmed) (18). It is possible to query PubMed automatically, but please read the database documentation as uncontrolled querying may lead to restricted access.

2.4. Measuring the Expression of the TF Repertoire

1. Gene-expression data. Datasets can be downloaded from ArrayExpress or GEO. Here we will use the GNF SymAtlas dataset measuring expression across 79 human tissues and cell lines (http://biogps.gnf.org) (19). A GCRMA-normalised version of the dataset is available from ArrayExpress (20) with accession number E-TABM-145.
2. Bioinformatics analysis software. The R statistical software package and the BioConductor bioinformatics suite can be obtained from (http://www.r-project.org) and (http://www.bioconductor.org) (21).

2.5. Examining the Evolutionary Conservation of the TF Repertoire

1. Orthologue predictions. These can be obtained from the Ensembl Compara database (http://www.ensembl.org) for most major eukaryotic genomes (22).

2.6. Assessing the Genomic Organisation of the TF Repertoire

1. Genomic coordinates. Coordinates describing the genomic location of genes can be accessed in Ensembl (see Subheading 2.1).

3. Methods

3.1. Identification of the TF Repertoire

Potential TF-coding genes can be determined by several computational approaches. A common method is to use pair-wise sequence-alignment algorithms such as BLAST (23) to identify homologues of known TFs. A more sensitive approach is to search for genes containing known DBDs using profile-based methods such as InterProScan, HMMER, and PSI-BLAST (24). Resources like InterPro, Pfam, and SUPERFAMILY provide curated hidden Markov models describing the amino-acid sequences for groups of conserved protein-sequence regions and domains. Among these are models representing well-known DBDs for all kingdoms, which we have identified and listed in the dataset from Subheading 2.1.

For most well-annotated genomes, searches can be performed directly against a reference set of protein sequences that are provided by resources such as UniProt (http://www.uniprot.org) (25). For genomes that are still unannotated, searches may have to be performed on the underlying nucleotide sequence itself.

The HMM search will return a set of genes coding for potential TFs containing a DBD. Some DBDs and their sequence models, however, can be promiscuous and produce false-positive hits to non-TF proteins that nonetheless bind DNA. The HMM search will also miss TFs that lack a conventional DBD (26), so generating false negatives. Therefore, we recommend that users refine the dataset to exclude non-TF hits and include known non-standard TFs; this refinement will usually consist of a combination of literature-based curation and inspection of the domain organisation of the protein.

In our own analysis of the human genome, we obtained 1,960 initial hits from a search of 347 DBD models against all human protein sequences. We manually curated this dataset by filtering out 596 probable false positives and included 27 known TFs that were missed. The final high-confidence dataset contained 1,391 TF-coding genes (Fig. 1a) (7).

Below is a step-by-step protocol to identify the TF repertoire for a given genome, using human as an example:

1. Download a list of manually curated DBDs (http://www.valleyofpigs.org/humantfs). Universal prokaryote/eukaryote DBD lists can also be downloaded from DBD (http://dbd.mrc-lmb.cam.ac.uk/DBD/index.cgi?Download).
2. Download and install InterProScan (http://www.ebi.ac.uk/Tools/InterProScan).
3. Download a unique reference set of human proteins from UniProt (http://www.uniprot.org).

4. Run InterProScan on a set of human proteins with default parameters.
5. Examine InterProScan results and filter proteins with hits to a DBD.
6. (Optional) Manually curate the list of proteins obtained in step 5 using the Ensembl gene identifier and online resources such as NCBI's Entrez (http://www.ncbi.nlm.nih.gov/sites/entrez?db=gene), GeneCards (http://www.genecards.org) (27), and Uniprot. Also examine the DBD and partner domain composition (i.e. non-DBDs) to remove genes that are unlikely to be TFs (e.g. enzymes, transmembrane proteins, etc.).
7. Finally, augment this list with any known TFs that were missed.

3.2. Assessing the Coverage of the TF Repertoire

Once TFs have been identified, we recommend automated benchmarks against other datasets to gauge the coverage of the repertoire. A common approach is to compare against a reference dataset such as genes annotated as TFs in the Gene Ontology database (GO, molecular function annotation). Note that the reference dataset is not necessarily a true gold standard – indeed the aim is to create a much more comprehensive list of TFs than any other – and these benchmarks will necessarily be estimates. The coverage is calculated as the proportion of entries in the reference dataset that are included in the repertoire.

In our own analysis, we used a list of 62 human and 207 mouse experimentally validated TFs as our reference dataset. The estimated coverage of the repertoire was 85–95% (7).

Below is a step-by-step protocol to estimate the coverage of the TF repertoire, using human as an example:

1. Extract a reference dataset of genes from GO (http://www.geneontology.org) annotated with the terms "transcription factor activity" and "DNA-binding", restricting to experimentally derived evidence (Inferred from Direct Assay, IDA; Inferred from Mutant Phenotype, IMP).
2. Estimate the coverage by calculating the fraction of the reference dataset that is included in your TF repertoire.
3. (Optional) In case the GO annotation is sparse in the organism of interest, it is possible to repeat the analysis using orthologues from a better-annotated genome such as the mouse. Multiple estimates obtained across species will ensure a robust measure of coverage. Alternatively, it is also possible to compare against carefully annotated TF repertoires in other genomes, such as worm or fly.

3.3. Assigning Regulatory Functions to the TF Repertoire

It is possible to survey the cellular and biological processes that are known to be regulated by the TF repertoire by using resources such as GO, Entrez, or PubMed.

In our analysis, we found that the most common GO regulatory functions were for developmental (263 TFs) and cellular processes (221 TFs) (7).

Below is a step-by-step protocol to annotate regulatory functions to TFs:

1. Extract the GO annotations (biological process category) for your repertoire. Depending on stringency, it is possible to restrict annotations to experimentally derived evidence as before (IDA, IMP; very strict) or relax them to include annotations inferred from automatic computational assignments (Inferred from Electronic Annoatation; IEA; very lenient).
2. Upload the TF repertoire to g:Profiler (http://biit.cs.ut.ee/gprofiler) to assess the biological processes that are statistically enriched.
3. (Optional) Query PubMed to access research articles that cite members of the TF repertoire. Note that gene names often resemble common scientific terms, and so careful filtering of hits is required.

3.4. Structural Classification of the TF Repertoire

TFs are most commonly classified by the identity of the DBD. This classification has proved to be very useful in tracing the evolutionary origins of TF families, and understanding how they recognise and bind specific DNA sequences (28). Furthermore, the identity of the DBD itself may indicate the regulatory function of the TF; for example, homeodomains are most commonly found in developmental regulators.

Most importantly, investigating the relative frequency of different TF families across species can bring interesting observations regarding their evolutionary history. For instance, 80% of the human repertoire comprises C_2H_2-Zn fingers, homeodomains, and helix-loop-helix proteins (7). However, this distribution varies substantially between species, as there are lineage-specific expansions such as the nuclear hormone receptor family in worms. Similarly, a recent survey of TFs across the three kingdoms of life described a number of kingdom-specific DBDs (29, 30).

Below is a step-by-step protocol to classify TFs by their DBD:

1. Download a list of TF families and associated InterPro IDs (http://www.valleyofpigs.org/humantfs). It is possible to reconstruct similar groups of TF families using the parent–child relationships between InterPro domains (http://www.ebi.ac.uk/interpro).
2. Calculate the proportions of distinct TF families in the repertoire and compare it against those of other species (see Table 1).

3.5. Measuring Expression of the TF Repertoire

Large-scale gene expression datasets allow us to assess patterns of TF usage across different cell types and conditions (7, 31–33). Two major types of information can be inferred from such data: (1) expression of a TF in specific circumstances is indicative of its function as a regulator in the condition and, therefore, is a good candidate for further investigation; (2) global expression pattern of the TF repertoire might reveal important organisational principles of the regulatory system, such as the hierarchy of control between TFs.

In performing this analysis, it is important to note that many TFs are post-translationally regulated through covalent modifications, subcellular localisation and oligomerisation; therefore, their expression alone may not be indicative of regulatory activity. It is also important to remember that most eukaryotic TFs function in combination with other regulators; therefore, appreciation of combinatorial regulation is important to understand TF functionality (34).

There are large numbers of array-based, and more recently increasing numbers of sequencing-based (RNA-Seq), transcriptomic datasets that can be used to examine tissue-, condition-, and disease-specific TF-expression. The GNF SymAtlas measuring expression across 79 human tissues and cell lines is one of the most widely used datasets (19). More recently, a compendium of over 6,000 array hybridisations has been published, which covers most publicly available datasets generated using the Affymetrix U133a GeneChip (35). One of the limitations of microarrays is that many genes, including TFs, are not represented in the array design. The application of high-throughput sequencing circumvents this problem as does not require prior knowledge of the transcripts that will be present. Additionally RNA-Seq is considered to provide more sensitive and accurate quantification of expression levels (RNA-Seq) (36). Although there are still few datasets available and analysis methods are still being developed, the use of RNA-Seq should increase rapidly in the coming years.

In our analysis, we used a subset of the GNF SymAtlas to examine TF expression across 32 healthy tissue samples. 873 TF-coding genes were present on the Affymetrix HG-U133a GeneChip. We detected expression of 510 TFs in at least one of the tissue types. In general, we found that TFs are expressed at lower levels than non-TFs. In addition, we observed general expression of 161 TFs across most tissues in the dataset, whereas 349 displayed tissue-specific expression (Fig. 1b) (7).

Below is a step-by-step protocol to assess TF expression using the GNF SymAtlas dataset:

1. Download the raw .CEL microarray files for the dataset from the SymAtlas website (http://biogps.gnf.org/downloads). Alternatively, the processed dataset is available from ArrayExpress (accession number E-TABM-145).

2. Download and install the R statistical software (http://www.r-project.org) and the BioConductor bioinformatics analysis suite (http://www.bioconductor.org).
3. Load the .CEL files into R and perform a quality check of the raw data using arrayQualityMetrics package (http://www.bioconductor.org/packages/2.6/bioc/html/arrayQualityMetrics.html) (37).
4. Pre-process and normalise the data using GCRMA (GeneChip Robust Multi-array Analysis) as implemented in Bioconductor. This will output values representing $\log_2$ expression levels for each probe set across all arrays. Alternative normalisation methods can also be used.
5. Download the mapping between probe sets and Ensembl gene IDs using the biomaRt package within BioConductor (http://www.bioconductor.org/packages/2.6/bioc/html/biomaRt.html) (38) or from the Ensembl website (http://www.ensembl.org).
6. Compare $\log_2$ expression values between TF and non-TF-coding genes. Statistical significance for the comparison can be obtained using a Wilcoxon or *t*-test available within R.
7. Determine the "presence" or "absence" (i.e. expressed or not expressed status) for each probe set using the PANP package (http://www.bioconductor.org/packages/2.6/bioc/html/panp.html) or MAS5.0 (http://www.bioconductor.org/packages/2.6/bioc/html/affy.html).
8. Evaluate the number of TF and non-TF-coding genes expressed in each tissue using the presence and absence calls.
9. Calculate tissue-specific expression of genes using for each gene the SpeCond package (http://www.bioconductor.org/packages/2.6/bioc/html/SpeCond.html).
10. Apply hierarchical clustering (available through the *hclust* function in R) to the expression levels and tissue-specific expression values, and represent the output graphically in a heat map.

3.6. Examining the Evolutionary Conservation of the TF Repertoire

The evolutionary history of TF-coding genes can reveal useful information about their regulatory functions and genomic organisation. For example, in our recent study of the human TF repertoire, we showed how regulators could be separated into five distinct patterns of evolutionary conservation: those present only in primates, predominantly in mammals, vertebrates, metazoans, and finally all eukaryotes (7).

There are many ways to determine the evolutionary conservation of a gene, ranging from orthologue-finding using reciprocal best-matching BLAST hits to detailed phylogenetic methods. There are also online resources that provide automatically calculated phylogenetic relationships such as Ensembl GeneTrees (22), TreeFam (39), and Inparanoid (40). We recommend manual inspection of automated results, if it is important to obtain accurate information about particular genes. A major challenge in examining the evolution of TFs is that they undergo a great deal of lineage-specific expansion; as a result it is often difficult to discriminate between orthologous and paralogous relationships.

In our analysis, we observed dramatic, step-wise increases in the size of the TF repertoire at key points during the evolution of the human lineage (Fig. 1c) (7). Interestingly, different classes of TF families expanded at these stages: for example, C_2H_2-Zn fingers expanded rapidly in vertebrates, whereas helix-loop-helix proteins originated in metazoan organisms and have not expanded significantly since.

Below is a step-by-step description of a possible procedure for identifying orthologues using Ensembl Compara GeneTrees. Similar analyses are possible using orthologue descriptions obtained from other sources. Note that this method highlights the evolutionary history of TFs in the genome of interest only and does not include TF-coding genes that have arisen in other lineages.

1. Use the Ensembl API to obtain Ensembl Compara GeneTrees for the TF repertoire.
2. Create matrices representing the evolutionary conservation of the TF repertoire: rows represent TFs, columns represent organisms, and intersecting cells describe the presence or absence of an orthologue.
3. Visualise the conservation of the TF repertoire by applying a clustering algorithm to the matrix and displaying the results in a heat map.
4. Classify TFs by their pattern of conservation. Human TFs were grouped according to their conservation in specific groups of genomes: for example, to define the five phylogenetic groups (primate-specific, mammalian-specific, vertebrate-specific, metazoan-specific, and eukaryotic), we examined the presence of orthologues in a step-wise manner. For each TF, (1) flag it as eukaryotic if they are present in any single-celled organism such as yeast; (2) for those not present in (1), classify as metazoan if present in two or more metazoan genomes; (3) continue for vertebrates, mammals, and non-human primates.
5. Assess when different TF families appeared in the human lineage by calculating the proportion of DBD types represented among the primate-specific, mammalian-specific, vertebrate-specific, metazoan-specific, and all eukaryotic groups of TFs.

3.7. Assessing the Genomic Organisation of the TF Repertoire

The genomic location of a gene is closely linked to its evolutionary history. As a result, genes with similar functions are often located in clusters at particular loci. Among human TFs, this is exemplified by the HOX clusters, which are found on chromosomes 2, 7, 12, and 17, and the organisation of these genes is extremely well conserved through much of the human lineage. The clustering of HOX genes is important functionally also, since it impacts on their expression; displacement of genes from their original locations results in mis-expression and eventually developmental defects (41).

The coordinated expression of gene clusters probably arises from a combination of influencing factors, including shared enhancer elements and the effects of local chromatin structure. Further with the recent availability of high-resolution datasets such as Hi-C and ChIA-PET (42, 43), there is current renewed interest in the spatial arrangement of chromosomes within the nucleus, and the impact this has on transcriptional regulation.

In our study of the human TF repertoire, we found that chromosome 19 is particularly enriched for TF-coding genes compared with others; moreover, we identified 15 new clusters of potentially evolutionarily related TF-coding genes (Fig. 1d) (7). Many of these clusters reside in sub-telomeric and centromeric regions of chromosomes, suggesting that they may experience rapid turnover during evolution compared with other gene types.

Below is a step-by-step description to examine the organisation of TF-coding genes in the human genome:

1. Download the genomic coordinate of TF-coding genes (including chromosome, strand, start and end coordinates) from Ensembl (http://www.ensembl.org).
2. For each chromosome, count the number of TF- and non-TF-coding genes in a 500 kb sliding window using a step size of 100 kb.
3. For each window, construct a 2×2 contingency table of the numbers of TF and non-TF-coding genes present inside and outside the window.
4. Use a Fisher's exact test as implemented in the R statistical software (http://www.r-project.org) to calculate the significance of the TF enrichment for each window. As we are only interested in over-representations of TFs a one-side test can be used.
5. Correct the resulting p-values for multiple-testing using FDR as implemented in R (44) and identify windows that are significantly enriched for TFs using a cutoff of $p < 0.05$.
6. To define the TF clusters, merge overlapping windows with significant p-values.

4. Conclusions and Further Directions

Transcriptional regulation in eukaryotes is an area of research that attracts a great deal of interest owing to its importance to understanding how cells function. In addition to defining the TF repertoire or a genome, it is possible to integrate large-scale datasets to describe their potential regulatory functions and trace their evolutionary histories. By identifying further datasets and methods of data integration, it should be possible to perform further investigations such as examining the combinatorial usage of TFs, and the regulatory interaction with microRNAs. Such analyses will provide a valuable starting point for more detailed characterisation of individual TFs.

References

1. Simon, I., Barnett, J., Hannett, N., Harbison, C. T., Rinaldi, N. J., Volkert, T. L., Wyrick, J. J., Zeitlinger, J., Gifford, D. K., Jaakkola, T. S., and Young, R. A. (2001) Serial regulation of transcriptional regulators in the yeast cell cycle, *Cell* ***106***, 697–708.
2. Bain, G., Maandag, E. C., Izon, D. J., Amsen, D., Kruisbeek, A. M., Weintraub, B. C., Krop, I., Schlissel, M. S., Feeney, A. J., and van Roon, M. (1994) E2A proteins are required for proper B cell development and initiation of immunoglobulin gene rearrangements, *Cell 79*, 885–892.
3. Lemon, B., and Tjian, R. (2000) Orchestrated response: a symphony of transcription factors for gene control, *Genes Dev 14*, 2551–2569.
4. Mitchell, P. J., and Tjian, R. (1989) Transcriptional regulation in mammalian cells by sequence-specific DNA binding proteins, *Science 245*, 371–378.
5. Levine, M., and Tjian, R. (2003) Transcription regulation and animal diversity, *Nature 424*, 147–151.
6. Badis, G., Berger, M. F., Philippakis, A. A., Talukder, S., Gehrke, A. R., Jaeger, S. A., Chan, E. T., Metzler, G., Vedenko, A., Chen, X., Kuznetsov, H., Wang, C., Coburn, D., Newburger, D. E., Morris, Q., Hughes, T. R., and Bulyk, M. L. (2009) Diversity and complexity in DNA recognition by transcription factors, *Science 324*, 1720–1723.
7. Vaquerizas, J. M., Kummerfeld, S. K., Teichmann, S. A., and Luscombe, N. M. (2009) A census of human transcription factors: function, expression and evolution, *Nat. Rev. Genet 10*, 252–263.
8. Flicek, P., Aken, B. L., Ballester, B., Beal, K., Bragin, E., Brent, S., Chen, Y., Clapham, P., Coates, G., Fairley, S., Fitzgerald, S., Fernandez-Banet, J., Gordon, L., Gräf, S., Haider, S., Hammond, M., Howe, K., Jenkinson, A., Johnson, N., Kähäri, A., Keefe, D., Keenan, S., Kinsella, R., Kokocinski, F., Koscielny, G., Kulesha, E., Lawson, D., Longden, I., Massingham, T., McLaren, W., Megy, K., Overduin, B., Pritchard, B., Rios, D., Ruffier, M., Schuster, M., Slater, G., Smedley, D., Spudich, G., Tang, Y. A., Trevanion, S., Vilella, A., Vogel, J., White, S., Wilder, S. P., Zadissa, A., Birney, E., Cunningham, F., Dunham, I., Durbin, R., Fernández-Suarez, X. M., Herrero, J., Hubbard, T. J. P., Parker, A., Proctor, G., Smith, J., and Searle, S. M. J. (2010) Ensembl's 10th year, *Nucleic Acids Res 38*, D557–562.
9. Hunter, S., Apweiler, R., Attwood, T. K., Bairoch, A., Bateman, A., Binns, D., Bork, P., Das, U., Daugherty, L., Duquenne, L., Finn, R. D., Gough, J., Haft, D., Hulo, N., Kahn, D., Kelly, E., Laugraud, A., Letunic, I., Lonsdale, D., Lopez, R., Madera, M., Maslen, J., McAnulla, C., McDowall, J., Mistry, J., Mitchell, A., Mulder, N., Natale, D., Orengo, C., Quinn, A. F., Selengut, J. D., Sigrist, C. J. A., Thimma, M., Thomas, P. D., Valentin, F., Wilson, D., Wu, C. H., and Yeats, C. (2009) InterPro: the integrative protein signature database, *Nucleic Acids Res 37*, D211–215.
10. Finn, R. D., Mistry, J., Tate, J., Coggill, P., Heger, A., Pollington, J. E., Gavin, O. L., Gunasekaran, P., Ceric, G., Forslund, K., Holm, L., Sonnhammer, E. L. L., Eddy, S. R., and Bateman, A. (2010) The Pfam protein families database, *Nucleic Acids Res 38*, D211–222.
11. Wilson, D., Pethica, R., Zhou, Y., Talbot, C., Vogel, C., Madera, M., Chothia, C., and Gough, J. (2009) SUPERFAMILY – sophisticated comparative genomics, data mining, visualization and phylogeny, *Nucleic Acids Res 37*, D380–386.
12. Sigrist, C. J. A., Cerutti, L., de Castro, E., Langendijk-Genevaux, P. S., Bulliard, V., Bairoch, A., and Hulo, N. (2010) PROSITE, a protein domain database for functional characterization and annotation, *Nucleic Acids Res 38*, D161–166.
13. Kummerfeld, S. K., and Teichmann, S. A. (2006) DBD: a transcription factor prediction database, *Nucleic Acids Res* 34, D74–81.
14. Wilson, D., Charoensawan, V., Kummerfeld, S. K., and Teichmann, S. A. (2008) DBD – taxonomically

broad transcription factor predictions: new content and functionality, *Nucleic Acids Res 36*, D88–92.

15. Mulder, N., and Apweiler, R. (2007) InterPro and InterProScan: tools for protein sequence classification and comparison, *Methods Mol. Biol 396*, 59–70.
16. Ashburner, M., Ball, C. A., Blake, J. A., Botstein, D., Butler, H., Cherry, J. M., Davis, A. P., Dolinski, K., Dwight, S. S., Eppig, J. T., Harris, M. A., Hill, D. P., Issel-Tarver, L., Kasarskis, A., Lewis, S., Matese, J. C., Richardson, J. E., Ringwald, M., Rubin, G. M., and Sherlock, G. (2000) Gene ontology: tool for the unification of biology. The Gene Ontology Consortium, *Nat. Genet 25*, 25–29.
17. Reimand, J., Kull, M., Peterson, H., Hansen, J., and Vilo, J. (2007) g:Profiler – a web-based toolset for functional profiling of gene lists from large-scale experiments, *Nucleic Acids Res 35*, W193–200.
18. Sayers, E. W., Barrett, T., Benson, D. A., Bolton, E., Bryant, S. H., Canese, K., Chetvernin, V., Church, D. M., Dicuccio, M., Federhen, S., Feolo, M., Geer, L. Y., Helmberg, W., Kapustin, Y., Landsman, D., Lipman, D. J., Lu, Z., Madden, T. L., Madej, T., Maglott, D. R., Marchler-Bauer, A., Miller, V., Mizrachi, I., Ostell, J., Panchenko, A., Pruitt, K. D., Schuler, G. D., Sequeira, E., Sherry, S. T., Shumway, M., Sirotkin, K., Slotta, D., Souvorov, A., Starchenko, G., Tatusova, T. A., Wagner, L., Wang, Y., John Wilbur, W., Yaschenko, E., and Ye, J. (2010) Database resources of the National Center for Biotechnology Information, *Nucleic Acids Res 38*, D5–16.
19. Su, A. I., Wiltshire, T., Batalov, S., Lapp, H., Ching, K. A., Block, D., Zhang, J., Soden, R., Hayakawa, M., Kreiman, G., Cooke, M. P., Walker, J. R., and Hogenesch, J. B. (2004) A gene atlas of the mouse and human protein-encoding transcriptomes, *Proc. Natl. Acad. Sci. USA 101*, 6062–6067.
20. Parkinson, H., Kapushesky, M., Kolesnikov, N., Rustici, G., Shojatalab, M., Abeygunawardena, N., Berube, H., Dylag, M., Emam, I., Farne, A., Holloway, E., Lukk, M., Malone, J., Mani, R., Pilicheva, E., Rayner, T. F., Rezwan, F., Sharma, A., Williams, E., Bradley, X. Z., Adamusiak, T., Brandizi, M., Burdett, T., Coulson, R., Krestyaninova, M., Kurnosov, P., Maguire, E., Neogi, S. G., Rocca-Serra, P., Sansone, S., Sklyar, N., Zhao, M., Sarkans, U., and Brazma, A. (2009) ArrayExpress update – from an archive of functional genomics experiments to the atlas of gene expression, *Nucleic Acids Res 37*, D868–872.
21. Gentleman, R. C., Carey, V. J., Bates, D. M., Bolstad, B., Dettling, M., Dudoit, S., Ellis, B., Gautier, L., Ge, Y., Gentry, J., Hornik, K., Hothorn, T., Huber, W., Iacus, S., Irizarry, R., Leisch, F., Li, C., Maechler, M., Rossini, A. J., Sawitzki, G., Smith, C., Smyth, G., Tierney, L., Yang, J. Y. H., and Zhang, J. (2004) Bioconductor: open software development for computational biology and bioinformatics, *Genome Biol* 5, R80.
22. Vilella, A. J., Severin, J., Ureta-Vidal, A., Heng, L., Durbin, R., and Birney, E. (2009) EnsemblCompara GeneTrees: Complete, duplication-aware phylogenetic trees in vertebrates, *Genome Res 19*, 327–335.
23. Altschul, S. F., Gish, W., Miller, W., Myers, E. W., and Lipman, D. J. (1990) Basic local alignment search tool, *J. Mol. Biol 215*, 403–410.
24. Altschul, S. F., Madden, T. L., Schäffer, A. A., Zhang, J., Zhang, Z., Miller, W., and Lipman, D. J. (1997) Gapped BLAST and PSI-BLAST: a new generation of protein database search programs, *Nucleic Acids Res 25*, 3389–3402.
25. (2010) The Universal Protein Resource (UniProt) in 2010, *Nucleic Acids Res 38*, D142–148.
26. Hu, S., Xie, Z., Onishi, A., Yu, X., Jiang, L., Lin, J., Rho, H., Woodard, C., Wang, H., Jeong, J., Long, S., He, X., Wade, H., Blackshaw, S., Qian, J., and Zhu, H. (2009) Profiling the human protein-DNA interactome reveals ERK2 as a transcriptional repressor of interferon signaling, *Cell 139*, 610–622.
27. Safran, M., Dalah, I., Alexander, J., Rosen, N., Iny Stein, T., Shmoish, M., Nativ, N., Bahir, I., Doniger, T., Krug, H., Sirota-Madi, A., Olender, T., Golan, Y., Stelzer, G., Harel, A., and Lancet, D. (2010) GeneCards Version 3: the human gene integrator, *Database (Oxford)* 2010, baq020.
28. Luscombe, N. M., Austin, S. E., Berman, H. M., and Thornton, J. M. (2000) An overview of the structures of protein-DNA complexes, *Genome Biol 1*, REVIEWS001.
29. Charoensawan, V., Wilson, D., and Teichmann, S. A. (2010) Genomic repertoires of DNA-binding transcription factors across the tree of life, *Nucleic Acids Res.*
30. Charoensawan, V., Wilson, D., and Teichmann, S. A. (2010) Lineage-specific expansion of DNA-binding transcription factor families, *Trends Genet 26*, 388–393.
31. Zaslaver, A., Mayo, A. E., Rosenberg, R., Bashkin, P., Sberro, H., Tsalyuk, M., Surette, M. G., and Alon, U. (2004) Just-in-time transcription program in metabolic pathways, *Nat. Genet 36*, 486–491.

32. Freilich, S., Massingham, T., Bhattacharyya, S., Ponsting, H., Lyons, P. A., Freeman, T. C., and Thornton, J. M. (2005) Relationship between the tissue-specificity of mouse gene expression and the evolutionary origin and function of the proteins, *Genome Biol* *6*, R56.
33. Luscombe, N. M., Babu, M. M., Yu, H., Snyder, M., Teichmann, S. A., and Gerstein, M. (2004) Genomic analysis of regulatory network dynamics reveals large topological changes, *Nature* ***431***, 308–312.
34. Ravasi, T., Suzuki, H., Cannistraci, C. V., Katayama, S., Bajic, V. B., Tan, K., Akalin, A., Schmeier, S., Kanamori-Katayama, M., Bertin, N., Carninci, P., Daub, C. O., Forrest, A. R. R., Gough, J., Grimmond, S., Han, J., Hashimoto, T., Hide, W., Hofmann, O., Kamburov, A., Kaur, M., Kawaji, H., Kubosaki, A., Lassmann, T., van Nimwegen, E., MacPherson, C. R., Ogawa, C., Radovanovic, A., Schwartz, A., Teasdale, R. D., Tegnér, J., Lenhard, B., Teichmann, S. A., Arakawa, T., Ninomiya, N., Murakami, K., Tagami, M., Fukuda, S., Imamura, K., Kai, C., Ishihara, R., Kitazume, Y., Kawai, J., Hume, D. A., Ideker, T., and Hayashizaki, Y. (2010) An atlas of combinatorial transcriptional regulation in mouse and man, *Cell* ***140***, 744–752.
35. Lukk, M., Kapushesky, M., Nikkilä, J., Parkinson, H., Goncalves, A., Huber, W., Ukkonen, E., and Brazma, A. (2010) A global map of human gene expression, *Nat. Biotechnol* *28*, 322–324.
36. Wang, Z., Gerstein, M., and Snyder, M. (2009) RNA-Seq: a revolutionary tool for transcriptomics, *Nat. Rev. Genet* ***10***, 57–63.
37. Kauffmann, A., Gentleman, R., and Huber, W. (2009) arrayQualityMetrics – a bioconductor package for quality assessment of microarray data, *Bioinformatics* *25*, 415–416.
38. Durinck, S., Moreau, Y., Kasprzyk, A., Davis, S., De Moor, B., Brazma, A., and Huber, W. (2005) BioMart and Bioconductor: a powerful link between biological databases and microarray data analysis, *Bioinformatics* *21*, 3439–3440.
39. Ruan, J., Li, H., Chen, Z., Coghlan, A., Coin, L. J. M., Guo, Y., Hériché, J., Hu, Y., Kristiansen, K., Li, R., Liu, T., Moses, A., Qin, J., Vang, S., Vilella, A. J., Ureta-Vidal, A., Bolund, L., Wang, J., and Durbin, R. (2008) TreeFam: 2008 Update, *Nucleic Acids Res* *36*, D735–740.
40. Ostlund, G., Schmitt, T., Forslund, K., Köstler, T., Messina, D. N., Roopra, S., Frings, O., and Sonnhammer, E. L. L. (2010) InParanoid 7: new algorithms and tools for eukaryotic orthology analysis, *Nucleic Acids Res* *38*, D196–203.
41. Garcia-Fernàndez, J. (2005) The genesis and evolution of homeobox gene clusters, *Nat. Rev. Genet* *6*, 881–892.
42. Lieberman-Aiden, E., van Berkum, N. L., Williams, L., Imakaev, M., Ragoczy, T., Telling, A., Amit, I., Lajoie, B. R., Sabo, P. J., Dorschner, M. O., Sandstrom, R., Bernstein, B., Bender, M. A., Groudine, M., Gnirke, A., Stamatoyannopoulos, J., Mirny, L. A., Lander, E. S., and Dekker, J. (2009) Comprehensive mapping of long-range interactions reveals folding principles of the human genome, *Science* *326*, 289–293.
43. Fullwood, M. J., Liu, M. H., Pan, Y. F., Liu, J., Xu, H., Mohamed, Y. B., Orlov, Y. L., Velkov, S., Ho, A., Mei, P. H., Chew, E. G. Y., Huang, P. Y. H., Welboren, W., Han, Y., Ooi, H. S., Ariyaratne, P. N., Vega, V. B., Luo, Y., Tan, P. Y., Choy, P. Y., Wansa, K. D. S. A., Zhao, B., Lim, K. S., Leow, S. C., Yow, J. S., Joseph, R., Li, H., Desai, K. V., Thomsen, J. S., Lee, Y. K., Karuturi, R. K. M., Herve, T., Bourque, G., Stunnenberg, H. G., Ruan, X., Cacheux-Rataboul, V., Sung, W., Liu, E. T., Wei, C., Cheung, E., and Ruan, Y. (2009) An oestrogen-receptor-alpha-bound human chromatin interactome, *Nature* ***462***, 58–64.
44. Benjamini, Y., and Hochberg, Y. (1995) Controlling the false discovery rate – a practical and powerful approach to multiple testing, *Journal of the Royal Statistical Society Series B-Methodological* *57*, 289–300.
45. Reece-Hoyes, J. S., Deplancke, B., Shingles, J., Grove, C. A., Hope, I. A., and Walhout, A. J. M. (2005) A compendium of *Caenorhabditis elegans* regulatory transcription factors: a resource for mapping transcription regulatory networks, *Genome Biol* *6*, R110.
46. Kanamori, M., Konno, H., Osato, N., Kawai, J., Hayashizaki, Y., and Suzuki, H. (2004) A genome-wide and nonredundant mouse transcription factor database, *Biochem. Biophys. Res. Commun* *322*, 787–793.
47. Palaniswamy, S. K., James, S., Sun, H., Lamb, R. S., Davuluri, R. V., and Grotewold, E. (2006) AGRIS and AtRegNet. A platform to link cis-regulatory elements and transcription factors into regulatory networks, *Plant Physiol* ***140***, 818–829.
48. Messina, D. N., Glasscock, J., Gish, W., and Lovett, M. (2004) An ORFeome-based analysis of human transcription factor genes and the construction of a microarray to interrogate their expression, *Genome Res* ***14***, 2041–2047.
49. Wu, J., Wang, S., Bai, J., Shi, L., Li, D., Xu, Z., Niu, Y., Lu, J., and Bao, Q. (2008) ArchaeaTF: an integrated database of putative transcription factors in Archaea, *Genomics* ***91***, 102–107.

50. Martínez-Bueno, M., Molina-Henares, A. J., Pareja, E., Ramos, J. L., and Tobes, R. (2004) BacTregulators: a database of transcriptional regulators in bacteria and archaea, *Bioinformatics* ***20***, 2787–2791.
51. Baumbach, J., Brinkrolf, K., Czaja, L. F., Rahmann, S., and Tauch, A. (2006) CoryneRegNet: an ontology-based data warehouse of corynebacterial transcription factors and regulatory networks, *BMC Genomics* *7*, 24.
52. Wu, J., Zhao, F., Wang, S., Deng, G., Wang, J., Bai, J., Lu, J., Qu, J., and Bao, Q. (2007) cTFbase: a database for comparative genomics of transcription factors in cyanobacteria, *BMC Genomics* ***8***, 104.
53. Sierro, N., Makita, Y., de Hoon, M., and Nakai, K. (2008) DBTBS: a database of transcriptional regulation in *Bacillus subtilis* containing upstream intergenic conservation information, *Nucleic Acids Res* ***36***, D93–96.
54. Pfreundt, U., James, D. P., Tweedie, S., Wilson, D., Teichmann, S. A., and Adryan, B. (2010) FlyTF: improved annotation and enhanced functionality of the Drosophila transcription factor database, *Nucleic Acids Res* ***38***, D443–447.
55. Park, J., Park, J., Jang, S., Kim, S., Kong, S., Choi, J., Ahn, K., Kim, J., Lee, S., Kim, S., Park, B., Jung, K., Kim, S., Kang, S., and Lee, Y. (2008) FTFD: an informatics pipeline supporting phylogenomic analysis of fungal transcription factors, *Bioinformatics* ***24***, 1024–1025.
56. Moreno-Campuzano, S., Janga, S. C., and Pérez-Rueda, E. (2006) Identification and analysis of DNA-binding transcription factors in *Bacillus subtilis* and other Firmicutes – a genomic approach, *BMC Genomics* *7*, 147.
57. Zheng, G., Tu, K., Yang, Q., Xiong, Y., Wei, C., Xie, L., Zhu, Y., and Li, Y. (2008) ITFP: an integrated platform of mammalian transcription factors, *Bioinformatics* ***24***, 2416–2417.
58. Portales-Casamar, E., Thongjuea, S., Kwon, A. T., Arenillas, D., Zhao, X., Valen, E., Yusuf, D., Lenhard, B., Wasserman, W. W., and Sandelin, A. (2010) JASPAR 2010: the greatly expanded open-access database of transcription factor binding profiles, *Nucleic Acids Res* ***38***, D105–110.
59. Gray, P. A., Fu, H., Luo, P., Zhao, Q., Yu, J., Ferrari, A., Tenzen, T., Yuk, D., Tsung, E. F., Cai, Z., Alberta, J. A., Cheng, L., Liu, Y., Stenman, J. M., Valerius, M. T., Billings, N., Kim, H. A., Greenberg, M. E., McMahon, A. P., Rowitch, D. H., Stiles, C. D., and Ma, Q. (2004) Mouse brain organization revealed through direct genome-scale TF expression analysis, *Science* ***306***, 2255–2257.
60. Richardt, S., Lang, D., Reski, R., Frank, W., and Rensing, S. A. (2007) PlanTAPDB, a phylogeny-based resource of plant transcription-associated proteins, *Plant Physiol* ***143***, 1452–1466.
61. Guo, A., Chen, X., Gao, G., Zhang, H., Zhu, Q., Liu, X., Zhong, Y., Gu, X., He, K., and Luo, J. (2008) PlantTFDB: a comprehensive plant transcription factor database, *Nucleic Acids Res* ***36***, D966–969.
62. Pérez-Rodríguez, P., Riaño-Pachón, D. M., Corrêa, L. G. G., Rensing, S. A., Kersten, B., and Mueller-Roeber, B. (2010) PlnTFDB: updated content and new features of the plant transcription factor database, *Nucleic Acids Res* ***38***, D822–827.
63. Grote, A., Klein, J., Retter, I., Haddad, I., Behling, S., Bunk, B., Biegler, I., Yarmolinetz, S., Jahn, D., and Münch, R. (2009) PRODORIC (release 2009): a database and tool platform for the analysis of gene regulation in prokaryotes, *Nucleic Acids Res* ***37***, D61–65.
64. Iida, K., Seki, M., Sakurai, T., Satou, M., Akiyama, K., Toyoda, T., Konagaya, A., and Shinozaki, K. (2005) RARTF: database and tools for complete sets of Arabidopsis transcription factors, *DNA Res* ***12***, 247–256.
65. Kazakov, A. E., Cipriano, M. J., Novichkov, P. S., Minovitsky, S., Vinogradov, D. V., Arkin, A., Mironov, A. A., Gelfand, M. S., and Dubchak, I. (2007) RegTransBase – a database of regulatory sequences and interactions in a wide range of prokaryotic genomes, *Nucleic Acids Res* ***35***, D407–412.
66. Gama-Castro, S., Jiménez-Jacinto, V., Peralta-Gil, M., Santos-Zavaleta, A., Peñaloza-Spinola, M. I., Contreras-Moreira, B., Segura-Salazar, J., Muñiz-Rascado, L., Martínez-Flores, I., Salgado, H., Bonavides-Martínez, C., Abreu-Goodger, C., Rodríguez-Penagos, C., Miranda-Ríos, J., Morett, E., Merino, E., Huerta, A. M., Treviño-Quintanilla, L., and Collado-Vides, J. (2008) RegulonDB (version 6.0): gene regulation model of *Escherichia coli* K-12 beyond transcription, active (experimental) annotated promoters and Textpresso navigation, *Nucleic Acids Res* ***36***, D120–124.
67. Wang, Z., Libault, M., Joshi, T., Valliyodan, B., Nguyen, H. T., Xu, D., Stacey, G., and Cheng, J. (2010) SoyDB: a knowledge database of soybean transcription factors, *BMC Plant Biol* ***10***, 14.
68. Fulton, D. L., Sundararajan, S., Badis, G., Hughes, T. R., Wasserman, W. W., Roach, J. C., and Sladek, R. (2009) TFCat: the curated catalog of mouse and human transcription factors, *Genome Biol* ***10***, R29.

69. Lee, A. P., Yang, Y., Brenner, S., and Venkatesh, B. (2007) TFCONES: a database of vertebrate transcription factor-encoding genes and their associated conserved noncoding elements, *BMC Genomics 8*, 441.

70. Fukuchi, S., Homma, K., Sakamoto, S., Sugawara, H., Tateno, Y., Gojobori, T., and Nishikawa, K. (2009) The GTOP database in 2009: updated content and novel features to expand and deepen insights into protein structures and functions, *Nucleic Acids Res 37*, D333–337.

71. Rushton, P. J., Bokowiec, M. T., Laudeman, T. W., Brannock, J. F., Chen, X., and Timko, M. P. (2008) TOBFAC: the database of tobacco transcription factors, *BMC Bioinformatics 9*, 53.

72. Matys, V., Kel-Margoulis, O. V., Fricke, E., Liebich, I., Land, S., Barre-Dirrie, A., Reuter, I., Chekmenev, D., Krull, M., Hornischer, K., Voss, N., Stegmaier, P., Lewicki-Potapov, B., Saxel, H., Kel, A. E., and Wingender, E. (2006) TRANSFAC and its module TRANSCompel: transcriptional gene regulation in eukaryotes, *Nucleic Acids Res 34*, D108–110.

73. Hermoso, A., Aguilar, D., Aviles, F. X., and Querol, E. (2004) TrSDB: a proteome database of transcription factors, *Nucleic Acids Res 32*, D171–173.

74. Romeuf, I., Tessier, D., Dardevet, M., Branlard, G., Charmet, G., and Ravel, C. (2010) wDBTF: an integrated database resource for studying wheat transcription factor families, *BMC Genomics 11*, 185.

Chapter 2

Expression Pattern Analysis of Regulatory Transcription Factors in *Caenorhabditis elegans*

Huiyun Feng, Hannah L. Craig, and Ian A. Hope

Abstract

Expression pattern data are fundamental to understanding transcriptional regulatory networks and the biological significance of such networks. For *Caenorhabditis elegans*, expression pattern analysis of transcription factor genes, with cellular resolution, typically involves generation of transcription factor gene/reporter gene fusions. This is followed by the creation of *C. elegans* strains transgenic for, and determination of expression patterns driven by, these fusions. Physiologically relevant regulatory relationships between transcription factors are both inferred from their expression patterns, in combination with protein–DNA interaction data, and evidenced from alterations of expression patterns when networks are disturbed.

Key words: Transcription factor, Gateway, Recombineering, Reporters, *Caenorhabditis elegans*

1. Introduction

Differential gene expression of metazoan genomes is regulated at both transcriptional and post-transcriptional levels by a complex and highly organized network of *cis*-acting elements and *trans*-acting factors (1). At the transcriptional level, spatial and temporal gene expression is controlled through the action of regulatory transcription factors that bind, sequence specifically, to their cognate *cis*-acting elements. The combinational nature of gene control by transcription factors results in exquisitely fine-tuned gene expression patterns and levels.

While techniques such as yeast one hybrid and ChIP analyses can provide important information regarding potential interactions between regulatory factors and DNA, expression pattern analysis

Bart Deplancke and Nele Gheldof (eds.), *Gene Regulatory Networks: Methods and Protocols*,
Methods in Molecular Biology, vol. 786, DOI 10.1007/978-1-61779-292-2_2, © Springer Science+Business Media, LLC 2012

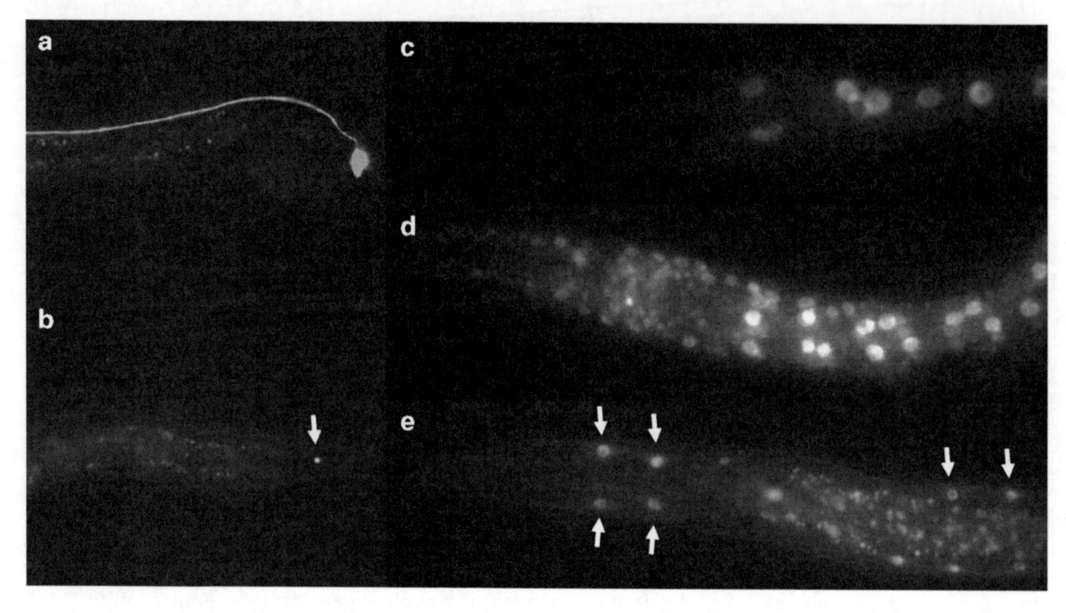

Fig. 1. Examples of *C. elegans* transcription factor gene expression patterns for reporter gene fusions generated by Gateway cloning and λ-red mediated recombineering. A promoter GFP transcriptional fusion was constructed by Gateway cloning and introduced into *C. elegans* by ballistic transformation (**a**) and an N-terminal GFP translational fusion was constructed by recombineering and introduced by micro-injection (**b**), for the homeodomain transcription factor gene *C02F12.10*. Expression is in the same nerve cell, DVC, but the GFP fusion protein (**b**) is nuclear localized (*arrow*), while the free GFP (**a**) is expressed at higher levels and is unlocalized, revealing the nerve cell axon projection along the nerve cord. GFP tagging of alternative isoforms of the basic leucine zipper transcription factor C27D6.4 by fosmid recombineering yielded different expression patterns (**c**–**e**) in *C. elegans* transgenics generated by microinjection. The C27D6.4::GFP fusion for transcript c gave expression in the intestinal nuclei only (**c**), the fusion for transcript a gave broad expression in many cells types (**d**) and the fusion for transcript b gave a short window of expression in the nuclei of the seam cells (*arrows*) forming the lateral hypodermis (**e**). *C. elegans* strains: UL2650 (**a**), UL3015 (**b**), UL3809 (**c**), UL3706 (**d**), U3709 (**e**). Images were captured at 400× magnification with anterior to the left.

is critical in linking the genome sequence information with the developmental events that are observable anatomically, in both spatial and temporal scales. Anatomic expression patterns, revealed by reporter gene fusions (Fig. 1), can provide dynamic, in vivo information on where and when genes are expressed and which genes are expressed with the same cellular location. This allows network mapping from sets of co-expressed genes (2).

The nematode *C. elegans* is a particularly good multicellular model for gene expression pattern analysis, with its simple and transparent anatomy, short life cycle, small and sequenced genome, invariant and completely characterized cell lineage, and fully mapped neuronal circuit (3–5). Moreover, there is a broad array of systems-level resources and high-throughput approaches available for expression pattern analysis in this species, which facilitates the mapping and annotation of transcription regulatory networks (6–8).

This chapter focuses on the methods and recent updates for expression pattern analysis in *C. elegans*. Protocols that are currently and commonly used by *C. elegans* research laboratories, to construct reporter gene fusions, to generate *C. elegans* strains transgenic for these fusions, and to analyze expression patterns in live animals, are presented.

1.1. Construction of Reporter Gene Fusions

1.1.1. Gateway Recombination Cloning

The Gateway technology is a universal cloning method based on the site-specific recombination properties of bacteriophage lambda (9, 10). It enables rapid and highly efficient parallel transfer of genome-wide sets of DNA sequences into multiple vector systems for functional analysis and protein expression (10).

Recombination between short, site-specific, attachment (*att*) sites is catalyzed by a mixture of enzymes that bind DNA sequence specifically. BP Clonase enzyme mix catalyzes the BP reaction: recombination of an *attB* substrate (an *attB*-containing PCR product or a linearized *attB*-containing plasmid) with an *attP* substrate (donor vector) to create *attL* and/or *attR*-containing products. LR Clonase enzyme mix catalyzes the LR reaction: recombination of an *attL* substrate (in one plasmid) with an *attR* substrate (in a destination vector) to create *attB* and/or *attP*-containing products. Appropriate positioning of bacterial genes in the plasmid vectors allows strong selection for the desired recombinant product upon bacterial transformation. Remarkably, sequence variation in the core of the *att* sites is tolerated with recombination restricted to *att* sites with the same core sequence, allowing a set of distinct *att* cloning sites to be developed. This then confers directionality of cloning and makes complex cloning schemes, involving multiple fragments, simple and efficient. Distinct upstream, protein-encoding, and downstream genetic elements can be brought together in a single LR reaction that results in the simultaneous and precisely ordered cloning of multiple fragments into a destination vector.

In a typical expression pattern analysis, two types of fusions, promoter::reporter transcriptional fusions and promoter::ORF::reporter translational fusions might be constructed. The *C. elegans* ORFeome and Promoterome resources include more than 12,000 ORFs and ~6,000 promoters cloned as Gateway entry clones (6, 7) that can be used to create these fusions, in a single LR reaction. Of the approximately 1,000 potential transcription factor genes in the *C. elegans* genome, 652 are represented in the ORFeome and 640 in the Promoterome (11, 12). However, promoter or ORF entry clones can also be created easily *ab initio* for a gene not represented in these resources (for a strategy overview see Fig. 2). Presented here are protocols routinely used in our laboratory to generate reporter fusions through Gateway cloning. For more detailed procedures and sequences of donor and destination vectors, see Invitrogen Gateway manuals (http://www.invitrogen.com).

1.1.2. λ-Red Mediated Recombineering

Gateway cloning has advantages over the traditional one-gene-at-a-time, restriction enzyme-based cloning strategy, especially for

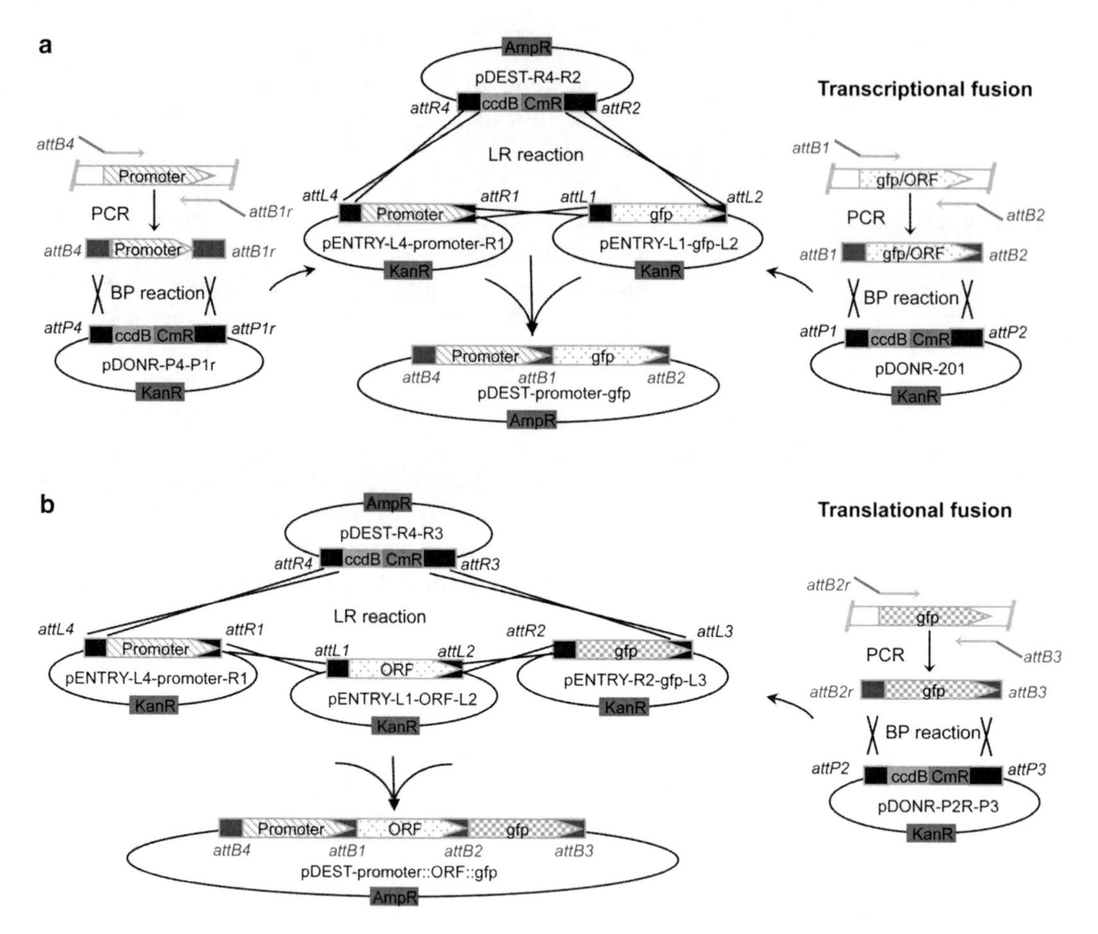

Fig. 2. Schematic representation of procedures for constructing *C. elegans* reporter gene fusions by Gateway cloning. *att*B flanked PCR products are amplified from genomic DNA for the promoter, from cDNA clones for the protein-coding ORF, and from plasmids for the reporter gene. Entry clones are then generated by a BP reaction between *att*B flanked PCR products and *att*P containing donor vectors. A two-fragment LR reaction is then performed to recombine pENTRY-L4-promoter-R1 and pENTRY-L1-*gfp*-L2 with the destination vector pDEST-R4-R2 to generate the promoter::*gfp* transcriptional fusion (**a**). A three-fragment LR reaction is performed to recombine pENTRY-L4-promoter-R1 (from (**a**)), pENTRY-L1-ORF-L2 (from (**a**)), and pENTRY-R2-*gfp*-L3 with the destination vector pDEST-R4-R3 to generate the promoter::ORF::*gfp* translational fusion (**b**).

studying large numbers of genes. However, technically it limits the sizes of fragments easily assayable to a few kilobase pairs due to the increased frequency of errors introduced by PCR and the decreased efficiency of both PCR and Gateway recombination reactions, as the fragment sizes increase. Biologically, using this approach means assuming that the regulatory elements for a *C. elegans* gene are located in the few kilobase pairs of DNA upstream from the predicted translational start codon, and any potential *cis*-acting elements outside of this assayed promoter region would be missed. Although the compact nature of the *C. elegans* genome means for the majority of genes these promoter regions appear to include all key transcriptional regulatory elements, an approach including the broader genomic environment may be preferable.

Recombineering technology, based on homologous recombination mediated by the bacteriophage λ Red system, has been developed to generate reporter gene fusions within large (~40 kb) genomic DNA fragments (13). For almost all *C. elegans* genes, such a genomic environment is likely to include all transcriptional regulatory elements. Such fusions can also retain elements involved in post-transcriptional regulation and signal sequences that direct intracellular localization of the endogenous protein.

The essence of the recombineering strategy, to insert fluorescent protein reporter genes seamlessly into any desired site of a *C. elegans* genomic DNA fosmid clone, involves two steps (Fig. 3). It also allows multiple alterations to be made in the same fosmid; for example two or more different reporters can be inserted at distinct sites (14). First, a positive/negative selection cassette, flanked by the first and last 50 nucleotides of the reporter gene, is inserted at a precise position by λ Red mediated homologous recombination. Recombination is directed by two further 50 nucleotide end regions matching the target in the fosmid, with positive selection for the cassette. The selection cassette is subsequently replaced by homologous recombination between the 50 nucleotide ends of the reporter gene and the cassette inserted during the first step, with negative selection against the cassette. The positive/negative selection cassette contains two genes, *rpsL* and *TetA(C)* (together referred to as RT), which confer streptomycin

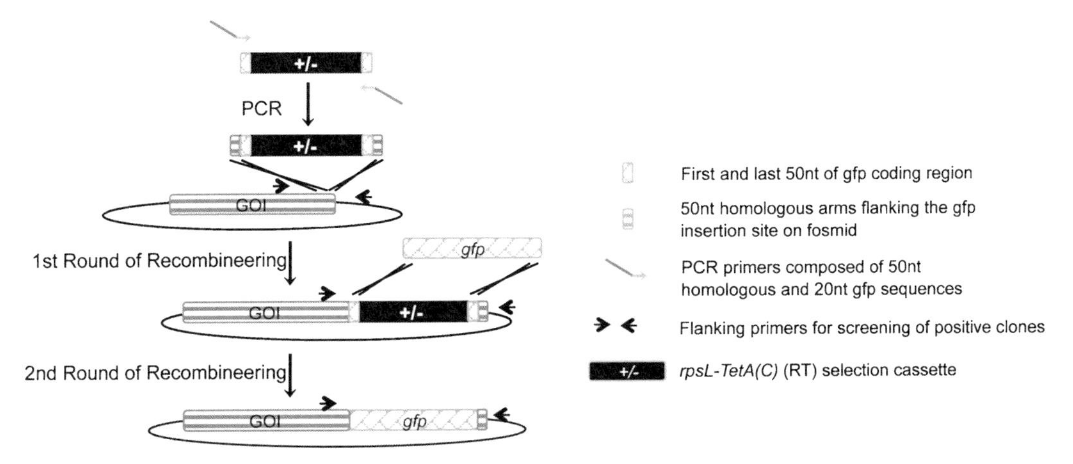

Fig. 3. Schematic showing introduction of a *gfp* reporter gene into a *C. elegans* fosmid genomic DNA clone by recombineering. A PCR template consisting of an *rpsL-TetA*(C) cassette, allowing both positive and negative selection, and the first and last 50 nucleotides of the *gfp* coding region is utilized. A fosmid specific selection cassette is generated by PCR amplification of this template with a pair of primers consisting of a 50-nucleotide target gene specific section plus a 20-nucleotide section matching the *gfp* ends of the template. The selection cassette is inserted into the fosmid by homologous recombination and positive selection for tetracycline resistance. The *gfp* reporter gene fragment is prepared by restriction enzyme digestion from a plasmid. Homologous recombination between the *gfp* ends in the second recombineering step replaces the selection cassette with the *gfp* coding sequence, through negative selection for loss of streptomycin sensitivity. At each step, selected clones are screened by colony PCR using the flanking primers.

sensitivity and tetracycline resistance, the negative and positive selection markers, respectively (15).

This technique can also be adapted to seamlessly alter fosmid clones in other ways. For instance, point mutations can be engineered or larger regions of DNA can be deleted. To facilitate targeted mutation in this manner, the RT cassette first replaces the region of DNA to be altered. A fragment containing this region plus the desired alteration is made by PCR and then used to replace the RT cassette. In combination with fluorescent protein reporter gene insertion, this provides a highly flexible method for investigation of the regulatory elements and functional domains associated with a particular gene.

Previously, two strains of *E. coli* were used. Recombination was performed in the recombineering-competent *E. coli* strain EL350. EL350 contains a chromosomally integrated defective λ prophage including the Red *gam*, *bet*, and *exo* genes in their natural context. Expression of *gam*, *bet*, and *exo* genes is under the control of a native *PL* promoter, which is tightly regulated by the temperature-sensitive cI857 repressor. Inactivating the repressor, by brief temperature shift to 42°C, permits high-level, coordinated expression of *gam*, *bet*, and *exo* that together mediate homologous recombination between the recipient target gene and an introduced double-stranded linear donor DNA. *Caenorhabditis elegans* fosmid clones were constructed in the copy number inducible pCC1FOS vector and maintained in the EPI300 *E. coli* strain (16). Addition of arabinose to an EPI300 culture induces the expression of the *trfA* gene, the product of which initiates replication from the high copy *oriV* origin of replication, allowing a high yield of fosmid DNA to be reached. Fosmid DNA isolated from EPI300 was transferred into EL350 to perform recombineering, and the recombineered fosmid was then transferred back to EPI300 for DNA isolation after copy number induction. The need for this final transfer back to EPI300 has been avoided by retrofitting EL350 with the P_{BAD}-driven *trfA* gene to generate the strain MW005 (17). This allows recombineering and transient induction of copy number in a single strain, reducing the time required.

1.2. Transformation of C. elegans with Reporter Gene Fusions

Two different techniques, microinjection and ballistic transformation, have been used in generating *C. elegans* strains transgenic for reporter gene fusions (18, 19). Compared to construction of the fusion genes, which can be performed in a high-throughput manner, *C. elegans* transformation, by either approach, has been the rate-limiting step for determination of expression patterns. Nevertheless, thousands of strains, including those from several large-scale projects assaying reporter gene fusions, have been created by these techniques (12, 20).

Microinjection involves direct injection of DNA into the distal arm of the gonad of adult hermaphrodites of *C. elegans*, one animal at a time. The distal arm of the gonad contains many germ cell nuclei in a common cytoplasm; thus, DNA from a single injection is incorporated into multiple fertilized eggs when oocytes mature and separate from the syncytium. Injected DNAs are rapidly concatemerized in the gonad, to form large extrachromosomal arrays that are replicated and delivered to daughter cells during mitotic and meiotic cell divisions with rates that vary between independent transgenic strains. Concatemerization means that reporter gene fusions and transformation marker genes become linked together in an extrachromosomal array even if not on the same DNA molecule prior to injection. Microinjection is a relatively simple and flexible technique, once mastered, as it requires minimal preparation. Individual adult *C. elegans* hermaphrodites are immobilized on a dried pad of agarose under oil to allow insertion of a glass microneedle containing DNA. The gonad of the dried down animal is filled with the DNA solution before rehydration and recovery. Progeny of the injected animal are screened to identify those that are transgenic and allowed to reproduce, with screening for the transgenic phenotype at successive generations to establish a transgenic strain.

In ballistic transformation, DNAs are bound onto gold particles which are then propelled at high speed into a packed *C. elegans* population using a biolistic bombardment instrument or "helium gun". This technique yields transformed strains with chromosomally integrated, low copy number transgenes at high frequency, in addition to strains with extrachromosomal arrays. However, the shooting is not specifically directed at the germ line and involves scattering the projectiles at a very large number of individuals with a specific genetic background, which makes this approach less flexible than injection. The forces involved may also result in shearing of large DNA molecules such as fosmids. Transformants are selected from the rescue of the uncoordinated and dauer defective phenotype of DP38 *[unc-119(ed3)]* mutants by a wild-type copy of the *unc-119* gene. The inability of *unc-119(ed3)* worms to enter the dauer life stage means they die without access to food allowing strong selection for the rare transformants generated from the large population that is bombarded. The protocol described below is based on personal communications from Rob Andrews, Jane Shingles, and John Reece-Hoyes as well as published protocols (21–23) and Wormbase.

1.3. Expression Pattern Analysis

With transgenic strains established, large numbers of animals can be generated, indefinitely, with which to characterize the expression pattern driven by a reporter gene fusion. The expression pattern can potentially be characterized with cellular resolution, or even subcellular resolution for fusion proteins, through all developmental stages of the life cycle, including the alternative dauer stage, in both the male and hermaphrodite sex, in different environments and in different genetic backgrounds. The ease of determination depends on the complexity of the expression pattern for a particular fusion, even in an animal with so few cells; distinguishing cells in the pre-elongation embryo or in the nervous system is not trivial. The matter is further complicated by the extrachromosomal arrays being randomly lost from or silenced in some cell lineages. The extent of such mosaicism varies between strains. Occasionally rare extra sporadic expression appears to be generated in an individual as if the promoter of the fusion gene has initiated transcription in a spontaneous and irrelevant fashion. Such complications mean that any single individual of a non-chromosomally integrated transgenic strain is unlikely to provide a complete expression pattern and the complete expression pattern needs to be inferred from observations of large numbers of animals.

2. Materials

2.1. Construction of Reporter Gene Fusions by Gateway Recombination Cloning (see Note 1)

2.1.1. PCR Amplification of Promoter, Protein-Coding ORF, and Reporter Gene Coding Fragments

1. *Caenorhabditis elegans* genomic DNA or a *C. elegans* genomic DNA clone containing the promoter to be assayed. Clones covering a target gene can be identified in Wormbase (http://www.wormbase.org). Fosmid genomic DNA clones and Promoterome plasmids are conveniently available from Geneservice (http://www.geneservice.co.uk/products/clones/Celegans_Fos.jsp and http://www.geneservice.co.uk/products/clones/Celegans_Prom.jsp).
2. cDNA or plasmid DNA containing the coding region or ORF of the gene to be studied. ORFeome plasmids are also available from Geneservice (http://www.geneservice.co.uk/products/cdna/Celegans_ORF.jsp).
3. Plasmid DNA containing the coding region of a reporter gene (*gfp*, *rfp*, *cfp*, etc.) available from Addgene (http://www.addgene.org/Andrew_Fire).
4. Oligonucleotide primers, composed of the appropriate *attB* sites plus promoter/ORF/reporter specific complementarity (Table 1). Purify primers by de-salting. Make primer stock solutions in ddH_2O at 200 μM concentration and dilute to 10 μM for PCR.
5. High-fidelity proof-reading DNA polymerase such as Platinum® Taq DNA Polymerase High Fidelity (Invitrogen) and the accompanying 10× PCR buffer and 50 mM $MgSO_4$.

Table 1
Primers for PCR amplification of substrates for BP reactions

Purpose	Primer direction	*att* site	*att* site containing region – 5′ end	Gene specific complementarity – 3′ end	Notes
Promoter amplification	Forward	B4	GGGGACAACTTTGTATAGAAAA GTTGNN	Promoter region – upstream end	
	Reverse	B1r	GGG GAC TGC TTT TTT GTA CAA ACT TGN	Promoter region – downstream end	a, b
Reporter amplification (transcriptional fusions)	Forward	B1	G GGG ACA AGT TTG TAC AAA AAA GCA GGC TNN	Reporter coding region – upstream end	b
	Reverse	B2	GGG GAC CAC TTT GTA CAA GAA AGC TGG GTN	Reporter coding region – downstream end	b, c
ORF amplification	Forward	B1	G GGG ACA AGT TTG TAC AAA AAA GCA GGC TNN	Protein coding region – upstream end	b
	Reverse	B2	GGG GAC CAC TTT GTA CAA GAA AGC TGG GTN	Protein coding region – downstream end	b, d
Reporter amplification (translational fusions)	Forward	B2r	G GGG ACA GCT TTC TTG TAC AAA GTG GNN	Reporter coding region – upstream end	b
	Reverse	B3	GGG GAC AAC TTT GTA TAA TAA AGT TGN	Reporter coding region – downstream end	b, c

[a] Start from and include nucleotides complementary to the translational initiation codon of the coding sequence in the reverse primer
[b] The reading frame of the ORF or reporter being amplified should match the indicated translational reading frame of the *att* site containing region
[c] Include the stop codon of the reporter
[d] Give consideration to inclusion or omission of a stop codon

6. Nucleotide mix. dATP, dTTP, dGTP, and dCTP; 10 mM each.
7. 1% agarose gel for DNA analysis.
8. PCR purification kit (Qiagen).

2.1.2. Preparation of Chemically Competent E. coli

1. 200 ml LB medium: 1% tryptone, 0.5% yeast extract, 1% NaCl (w/v) in 1-L conical flask, autoclaved.
2. 0.1 M $CaCl_2$, autoclaved and stored at 4°C.
3. 0.1 M $CaCl_2$ with 15% glycerol, autoclaved and stored at 4°C.

2.1.3. BP Recombination Reaction

1. Purified *att*B4-promoter-*att*B1r, and either *att*B1-ORF-*att*B2 and *att*B2r-reporter-*att*B3 or *att*B1-reporter-*att*B2 PCR products.
2. Plasmid vectors (Invitrogen), pDONR-P4-P1R and pDONR-201, plus pDONR-P2R-P3 if required, all dissolved in ddH_2O at 150 ng/μl.
3. Gateway BP Clonase II Enzyme Mix and 2 μg/μl Proteinase K solution.
4. 1×TE buffer: 10 mM Tris–HCl, pH 8.0, 1 mM EDTA, pH 8.0.
5. High-efficiency chemically competent DH5α bacteria, stored at –80°C.
6. LB medium and LB agar plates (LB medium plus 1.5% w/v agar) supplemented with 50 μg/ml kanamycin.

2.1.4. LR Recombination Reaction

1. Purified pENTRY-L4-promoter-R1, and either pENTRY-L1-ORF-L2 and pENTRY-R2-reporter-L3 or pENTRY-L1-reporter-L2 DNA, at 50–150 ng/μl.
2. Invitrogen pDEST-R4-R2 vector or pDEST-DD03 vector which contains the *Caenorhabditis briggsae unc-119(+)* marker for *C. elegans* transformation or pDEST-R4-R3, all dissolved in ddH_2O at 150 ng/μl.
3. Gateway LR Clonase II Enzyme Mix and 2 μg/μl Proteinase K solution.
4. 1×TE buffer, pH 8.0.
5. High-efficiency chemically competent DH5α bacteria, stored at –80°C.
6. LB medium and LB agar plates supplemented with 100 μg/ml ampicillin.

2.1.5. Verifying Positive Clones from BP and LR Reactions

1. Oligonucleotide primers complementary to sequences flanking the *att* sites of each vector (Table 2).
2. Standard Taq DNA polymerase, such as BioTaq Red DNA Polymerase (BIOLINE), and the accompanying 10× buffer, 50 mM $MgCl_2$ and nucleotide mix, containing 10 mM each of dATP, dTTP, dGTP, and dCTP.

Table 2
Primers to check products of recombination reactions

Primer direction	Sequence 5′–3′	Vector appropriate for
M13 forward M13 reverse	GTAAAACGACGGCCAG CAGGAAACAGCTATGAC	pDONR-P4-P1R or pDONR-P2R-P3
Forward Reverse	TCGCGTTAAACGCTAGCATGGATCTC GTAACATCAGAGATTTTGAGACAC	pDONR-201

2.2. Construction of Reporter Gene Fusions by λ-Red Mediated Recombineering

2.2.1. Fosmid Clones and Fosmid DNA Isolation

1. EPI300 bacterial strain containing *C. elegans* genomic DNA fosmid for the target gene. Clones covering a target gene can be identified in Wormbase (http://www.wormbase.org) and are available from Geneservice (http://www.geneservice.co.uk/products/clones/Celegans_Fos.jsp).
2. 10% L-arabinose stock solution for copy number induction of fosmid.
3. Plasmid Miniprep DNA purification kit (Qiagen).
4. LB agar plates with 12.5 μg/ml chloramphenicol.
5. LB medium.

2.2.2. Generation of the Fosmid-Specific Positive/Negative Selection Cassette and the Reporter Gene

1. Cloned DNA containing the *rpsL-TetA(C)* selection cassette flanked by the first and last 50 nucleotides of the reporter gene. It is important to keep the RT template in a low copy number vector. A fosmid for which the first step has been completed is ideal but prepare this template DNA from a non-induced culture. Maintaining the template at high copy number often results in loss of tetracycline resistance.
2. 70 nucleotides primers, PAGE purified. The forward primer, 5′ to 3′, contains 50 nucleotides matching genomic DNA upstream of the insertion site followed by 20 nucleotides matching the start of the reporter gene. The reverse primer, 5′ to 3′, contains 50 nucleotides matching genomic DNA downstream of the insertion site (on the opposite strand to the forward primer) followed by 20 nucleotides matching the reverse complement of the end of the reporter gene.
3. Platinum® Taq DNA Polymerase High Fidelity (Invitrogen) and accompanying 10× buffer and 50 mM $MgSO_4$, dATP/dCTP/dGTP/dTTP mix (each 10 mM), and 1% agarose gel for analysis.
4. Plasmid (e.g. pUL#SB94) containing the full-length reporter gene coding region with appropriate flanking restriction enzyme sites to allow release of the reporter gene.
5. PCR purification and gel extraction kits (Qiagen).

2.2.3. Preparation of Electrocompetent E. coli

1. Fresh MW005 colonies on agar plates.
2. LB medium, autoclaved.
3. SOB[–Mg] medium: 2% tryptone, 0.5% yeast extract, 0.06% NaCl, 0.05% KCl, autoclaved.
4. 10% glycerol, autoclaved, stored at 4°C.
5. ddH_2O, autoclaved, stored at 4°C.

2.2.4. Two-Step Counter-Selection Recombineering

1. Purified DNA fragments, either PCR-amplified cassette or restriction enzyme liberated reporter.
2. Electrocompetent MW005 bacteria.
3. Electroporation cuvettes, 0.1 cm gap, 100 μl capacity.
4. SOB[–Mg] medium, autoclaved.
5. LB medium and LB agar plates with 12.5 μg/ml chloramphenicol and 5 μg/ml tetracycline.
6. NSLB (LB minus NaCl) agar plates with 12.5 μg/ml chloramphenicol and 500 μg/ml streptomycin.
7. Short (~20 nucleotide) target specific primers matching genomic DNA, ~200 bp away from and for both sides of the insertion site.
8. Standard Taq DNA polymerase plus accompanying 50 mM $MgCl_2$, 10× buffer, dNTP mix (10 mM each of dATP, dCTP, dGTP, and dTTP), and 1% agarose gel for analysis.

2.3. Transformation of C. elegans by Microinjection

2.3.1. Equipment

1. Inverted DIC microscope. The microscope (e.g. Zeiss Axiovert) should have 5× and 40× DIC objectives and a free-sliding stage with centred rotation.
2. Micromanipulator. The manipulator (e.g. Narashige MO-202) should provide fine mobility in the *x*, *y*, and *z* axes, and allow easy alteration of needle angle and position.
3. Pressurized injection system with needle holder. The injection needle is placed in a holder with a tight-seal collar, which is then attached by plastic tubing to a regulated pressure source. A valve in the plastic tubing allows the flow pressure to be switched on and off. The pressure regulator is attached to a nitrogen cylinder.
4. Needle puller. The needle puller should provide needles fine enough to penetrate the hermaphrodite cuticle. The Narashige PD-5 uses a heated filament and pulls needles horizontally.

2.3.2. Agarose Pads

Many agarose pads can be made in one go and stored indefinitely at room temperature in a coverslip box.

1. Make a 2% agarose solution by heating in a microwave until the agarose is dissolved.

2. Lay out glass cover slips (22 × 40 mm) on a flat surface.
3. Using wide-mouth glass Pasteur pipette place a drop (~50 μl) of hot agarose onto the centre of each coverslip and quickly place a second coverslip on the drop, tapping gently to flatten.
4. After the agarose is solidified, slide the coverslips apart and bake the pads in an oven at 80°C for 1 h or leave them on the bench at room temperature overnight.

2.3.3. Reagents for Microinjection

1. Microinjection needles. Needles are pulled from borosilicate glass capillaries with an internal filament (e.g. Harvard Apparatus, Kent, UK: GC100F-10, 1.0 mm × 0.58 mm × 100 mm). The needles should be pulled to taper evenly to a sharp but open tip, and this will depend on fine-tuning of the needle puller. Experimentation may be necessary to pull needles of a shape that provides an appropriate degree of flexibility and sharpness. Needles can be retained with a strip of modelling clay in a large petri dish to prevent damage to the fine tips. Wetted tissue within the petri dish maintains a humid environment to reduce evaporation of the small volume of DNA solution in loaded needles.
2. Heavy colourless liquid paraffin (BDH).
3. M9 buffer: 3 g KH_2PO_4, 6 g Na_2HPO_4, 5 g NaCl in ddH_2O, and 1 ml 1 M $MgSO_4$, per litre, autoclaved.
4. *Caenorhabditis elegans* for microinjection. Healthy, uncontaminated, well-fed, young to middle-aged gravid hermaphrodite adults are needed.
5. DNA-loading pipettes. Place a standard glass capillary (e.g. 1.15 × 1.55 × 75 mm micro haematocrit tubes from Brand, Wertheim, Germany) over a Bunsen flame until soft. Remove from the flame and quickly pull out to draw out the capillary to a diameter that will fit within the injection needle capillaries. Snap to give two loading pipettes.
6. DNA solution to be injected. DNA purified with standard kits (e.g. Qiagen) in ddH_2O is appropriate for injection. DNA is injected at a total concentration of 100 μg/ml, with the plasmid bearing the transformation marker gene at 50–95 μg/ml, mixed with an assay plasmid at 20–50 μg/ml or assay fosmid at a lower concentration, 5–20 μg/ml.
7. NGM (nematode growth medium) agar plates seeded with OP50 bacteria.

2.3.4. Single-Hermaphrodite PCR Reagents

1. 10× Taq polymerase buffer can be used as 10× lysis buffer: 500 mM KCl, 100 mM Tris–HCl pH 8.3, and 15 mM $MgCl_2$. Aliquot and store at −20°C. Dilute to 1× in ddH_2O with addition of Proteinase K before use.

2. 10 mg/ml Proteinase K in Proteinase K storage buffer: 5 ml glycerol, 0.1 ml 1 M Tris–HCl pH 7.5, 29 mg $CaCl_2$, and 4.9 ml ddH_2O, autoclaved. Aliquot and store at −20°C.
3. Taq polymerase (e.g. Biotaq Red) with accompanying 10× buffer, 50 mM $MgCl_2$, and dNTP mix.
4. Appropriate primers (10 μM) for specific PCR detection of the transgenic DNA.

2.4. Transformation of C. elegans by Microprojectile Bombardment

2.4.1. Liquid Worm Culture

1. *E. coli* strain HB101.
2. LB medium, autoclaved.
3. M9 buffer, autoclaved.
4. 5 cm NGM agar plates seeded with OP50 *E. coli.*
5. Bleach (5% sodium hypochlorite) and 5 M NaOH.
6. S-basal medium supplemented with 50 μg/ml nystatin and 50 μg/ml streptomycin.
7. Refrigerated centrifuge with rotor for 250-ml or 500-ml centrifuge bottles.

2.4.2. DNA and Gold Particle Preparation

1. 7 μg of each transforming DNA.
2. Restriction endonucleases for linearization of DNA.
3. Gold particles (0.3–3 μm, Chempur, Germany).
4. Sterile 50% glycerol.
5. 1.5-ml tubes (Treff).
6. 2.5 M $CaCl_2$ and 0.1 M spermidine.

2.4.3. Bombardment

1. BioRad PDS-1000/He with Hepta adapter, which requires the following accessories: Rupture discs (1,350 psi), Macrocarriers, and Stopping screens.
2. 9 cm NGM plates seeded with OP50 *E. coli.*

2.5. Expression Pattern Analysis

2.5.1. Compound Microscope with DIC Optics, Equipped for Epifluorescence, and Image Capture System

Epifluorescent microscopy is fundamental to analysis of fluorescent reporter protein expression patterns but for *C. elegans* the epifluorescent microscope should be equipped with DIC optics. Appropriate filter sets for specific excitation and emission wavelengths are needed for reporter proteins with different fluorescent properties. The image capture system needs to generate high-quality images with low signal intensity. The range of epifluorescent microscopes and digital cameras that are appropriate to this purpose, however, is beyond the scope of this chapter.

2.5.2. Reagents

1. Agarose.
2. M9 buffer, autoclaved.
3. Plain microscope slides, eight-well multitest microscope slides (MP Bio) and coverslips.

4. Silicone oil (Sigma, melting point bath oil).
5. Levamisole (50 mM) or sodium azide (50 mM) in M9 buffer.

3. Methods

3.1. Construction of Reporter Gene Fusions by Gateway Recombination Cloning

3.1.1. PCR Amplification of Promoter, Protein Encoding ORF, and Reporter Encoding DNA Fragments with attB-Flanked Primers

The B4-promoter-B1r, B1-ORF-B2, B1-reporter-B2, and B2r-reporter-B3 DNA fragments are PCR-amplified using primers including the appropriate *attB*-site. B4-promoter-B1r and B1-reporter-B2 DNA fragments are used to generate a promoter::reporter transcriptional fusion, whereas B4-promoter-B1r, B1-ORF-B2 and B2r-reporter-B3 DNA fragments are used to generate a promoter::ORF::reporter translational fusion.

1. Set up a standard 30 μl PCR with high-fidelity proof-reading DNA polymerase for amplification of each fragment. This should contain 5–10 ng of the template, 1 μl of dNTP mix, primers at 0.5 μM each, 2 mM $MgSO_4$, and 1 U of the polymerase.

   Use the PCR program: 1 step of 2 min at 94°C, followed by 30 cycles of 30 s at 94°C, 45 s at 55°C, and 2 min at 68°C, with then 1 final step of 10 min at 68°C. The length of the repeated 68°C elongation step should be adjusted according to the size of fragment using 1 min for each kilobase of DNA.
2. Check the PCR products by electrophoresis, purify with the PCR purification kit, and elute the DNA in ddH_2O at 50–150 ng/μl.

3.1.2. Preparation of Chemically Competent E. coli DH5α

While commercially available competent cells could be used, the following protocol provides competent cells suitable for transformation of Gateway recombination reaction products.

1. Pick a single colony from a freshly streaked plate into 3 ml of LB medium and grow overnight at 37°C with shaking.
2. Use 2 ml of this culture to inoculate 200 ml LB medium in a 1-L conical flask and incubate at 37°C with shaking until the OD600 reaches 0.5–0.6. Perform all subsequent procedures on ice and chill solutions to ice-coldness.
3. Split the 200 ml culture into four 50-ml Falcon tubes and chill on ice for 15 min. Centrifuge at 4°C for 10 min at 3,300×*g*. Discard the supernatant and re-suspend the cell pellet very gently in 2 ml 0.1 M $CaCl_2$. Add more 0.1 M $CaCl_2$ to a 30 ml final volume in each tube.
4. Leave the cell suspension on ice for 15 min and centrifuge again as above. Gently re-suspend the pellet in 10 ml 0.1 M $CaCl_2$.

5. Pool the cells together into one tube. Centrifuge again as above. Discard the supernatant and gently re-suspend the pellet in 4 ml 0.1 M $CaCl_2$ containing 15% glycerol.
6. Store 50-μl aliquots at −80°C.

3.1.3. BP Recombination Reaction to Generate Entry Clones (see Note 2)

Perform the BP recombination reaction between each *attB*-flanked DNA fragment and the appropriate *attP*-containing donor vector to generate an entry clone.

1. Add the following components into 1.5-ml eppendorf tubes on ice:

PCR product (20–50 fmoles)	1–7 μl
Plasmid vector, 150 ng/μl	1 μl
1×TE buffer, pH 8.0	to 8 μl

To generate pENTRY-L4-promoter-R1, use B4-promoter-B1r PCR product and pDONR-P4-P1R vector.

To generate pENTRY-L1-reporter-L2, use B1-reporter-B2 PCR product and pDONR-201 vector.

To generate pENTRY-L1-ORF -L2, use B1-ORF-B2 PCR product and pDONR-201 vector.

To generate pENTRY-R2-reporter-L3, use B2r-reporter-B3 PCR product and pDONR-P2R-P3 vector.

2. Vortex BP Clonase II enzyme mix briefly. Add 2 μl to the components above and mix well by vortexing briefly twice. Incubate the reaction at 25°C overnight.
3. Add 1 μl of 2 μg/μl Proteinase K solution and incubate at 37°C for 10 min. Put the tubes on ice.
4. Chill the reactions and thaw an aliquot of competent *E. coli* DH5α cells for each on ice. Mix 5 μl of each reaction with an aliquot of cells and keep on ice for 20 min. Heat-shock the mixture in a 42°C water bath for 90 s and put back on ice. Add 500 μl LB medium and transfer the cells to a sterile culture tube. Incubate with shaking for 1 h. Plate 100 μl per LB agar plate containing 50 μg/ml kanamycin and incubate overnight at 37°C.

3.1.4. Quick Screen for Positive Clones from BP Reaction by Colony PCR

From a successful BP reaction, tens to hundreds of colonies can be obtained on kanamycin selective plates and they all should contain the desired recombinational fusion. However, to check for positive colonies, one may re-streak several colonies and perform colony PCR. Alternatively one may pick a few colonies for overnight mini-culture, plasmid DNA preparation and digestion with suitable restriction enzymes or sequencing to check the fusion.

The same *attB*-flanked primers as used for PCR amplification of the promoter, ORF and reporter fragments (Table 1) can be used for the colony PCR. Alternatively, a set of standard primers (Table 2) complementary to regions flanking the *ccdB* cassette can be used for all products from a particular vector.

Prepare a PCR master mix of 15 μl reaction per colony containing 0.6 U standard Taq DNA polymerase, 2 mM $MgCl_2$, 0.5 μl dNTP mix, and 0.5 μM each primer. Aliquot into individual tubes before adding a small scraping of bacteria from each colony and mixing well. Use the following PCR program: 1 step of 2 min at 94°C, followed by 30 cycles of 30 s at 94°C, 45 s at 55°C, and 2 min at 72°C, with then 1 final step of 10 min at 72°C. The length of the repeated 72°C elongation step should be adjusted according to the size of fragment using 1 min for each kilobase of DNA. Check the PCR products by electrophoresis.

3.1.5. Verify the Sequences of Entry Clones

Entry clone inserts should be verified by sequencing before proceeding to the next step, especially for entry clones of ORF and reporter coding genes. The same primers used for colony PCR as above can be used for sequencing.

3.1.6. LR Recombination Reaction to Generate the Promoter::Reporter and Promoter::ORF::Reporter Expression Fusions

Perform LR recombination reaction between multiple entry clones and either pDEST-R4R2 and pDEST-DD03 or pDEST-R4R3 vectors to generate promoter::reporter or promoter::ORF::reporter expression fusions, respectively.

To set up the LR reaction:

1. Add the following components into a 1.5-ml eppendorf tube on ice:

Entry clone plasmid DNAs (10 fmoles each)	1–7 μl
(pENTRY-L4-promoter-R1/pENTRY-L1-reporter-L2/pENTRY-L1-ORF-L2/pENTRY-R2-reporter-L3)	
Vector plasmid DNA (20 fmoles)	1 μl
(pDEST-R4R2 or pDEST-DD03 or pDEST-R4R3)	
1×TE buffer, pH 8.0	to 8 μl

2. Vortex LR Clonase II enzyme mix briefly. Add 2 μl to the components above and mix well by vortexing.

   Incubate reaction at 25°C overnight.
3. Add 1 μl of 2 μg/μl Proteinase K solution and incubate at 37°C for 10 min. Put the tubes on ice.
4. Transform 5 μl of the reaction into competent *E. coli* DH5α cells as before and select for ampicillin-resistant clones on LB/ampicillin plates.

3.1.7. Verify the Expression Clones by Colony PCR and Digestion with Restriction Endonucleases

As with the entry clones, positive colonies can be quickly screened by colony PCR. DNA can also be isolated from the positive clones for verification by restriction enzyme digestion or sequencing.

3.2. Construction of Reporter Gene Fusions by λ-Red Mediated Recombineering

3.2.1. Selection of Fosmid Clones, Copy-Number Induction, and DNA Isolation

Streak EPI300 containing the fosmid clone of interest onto LB/chloramphenicol agar and grow overnight at 37°C. Pick a single colony and inoculate 10 ml LB medium supplemented with 12.5 μg/ml chloramphenicol and 0.01% L-arabinose in a 50-ml Falcon tube. Incubate this culture with vigorous shaking overnight (16 h max) at 37°C. Isolate fosmid DNA using the Miniprep DNA purification kit and elute in 20–40 μl ddH$_2$O.

3.2.2. Generation of the Fosmid-Specific Recombineering Cassette and Reporter Gene Fragment

To generate the fosmid specific RT cassette, perform a PCR on the reporter sequences-flanked RT cassette template using the 70 nucleotide primers and clean up the PCR products using a PCR purification kit.

1. Set up a standard 30 μl PCR with high-fidelity proof-reading DNA polymerase for amplification of each fragment. The reaction should contain 1–10 ng of the template, 1 μl of dNTP mix (10 mM each nucleotide), primers at 0.5 μM each, 2 mM MgSO$_4$, and 1 U of the polymerase. 1–2 μl of a 50 μl fosmid DNA template preparation purified from a 3 ml overnight uninduced culture is typical. Use the following PCR program: 1 step of 2 min at 94°C, followed by 30 cycles of 30 s at 94°C, 45 s at 55°C, and 2.5 min at 68°C, then one final step of 10 min at 68°C.
2. Check the PCR product by agarose gel electrophoresis and purify the products with the PCR purification kit. Elute the purified product in ddH$_2$O at 100–200 ng/μl.
3. The reporter gene DNA fragment required for the second step should be released from the vector by restriction enzyme digestion, subjected to agarose gel electrophoresis and purified using the gel extraction kit. Elute the reporter DNA at 100–200 ng/μl in ddH$_2$O.

3.2.3. Preparation of Electrocompetent E. coli

A large-scale preparation of electrocompetent *E. coli* MW005 cells can be generated as a stock for receipt of numerous fosmid clones from the EPI300 cells in which they are provided. A small-scale preparation is used to make MW005 cells containing specific fosmids competent for transformation with linear DNA fragments in the recombineering steps.

Large-Scale Preparation for Transformation of Original Fosmid DNAs

1. Prepare a pre-culture by inoculating a single colony of *E. coli* MW005 in 2 ml LB medium and shaking at 32°C overnight.
2. Inoculate 1 ml of the pre-culture into 1 L of LB medium and incubate at 32°C with shaking until the OD600 reaches ~0.6. Chill on ice for 30 min.

3. Collect cells by centrifuging at 4°C, 3,300×*g* for 15 min. Discard the supernatant and re-suspend the pellets thoroughly but very gently in 20 ml of ice-cold sterile ddH_2O before adding more to a total volume of 500 ml. Centrifuge again as above and re-suspend the pellet in 250 ml of ice-cold sterile ddH_2O. Centrifuge again and re-suspend the cell pellet in 25 ml ice-cold 10% glycerol. Centrifuge again and discard supernatant leaving ~2 ml total volume in which the cells are re-suspended. Distribute in 20-μl aliquots and store at –80°C.

Preparation of Competent Fosmid-Containing MW005 Cells for Recombineering

1. Prepare a pre-culture by inoculating a single colony in 2 ml LB medium with appropriate antibiotic (12.5 μg/ml chloramphenicol for fosmid only, 5 μg/ml tetracyline for fosmid containing the RT cassette) and shaking at 32°C overnight.
2. Inoculate 20 ml SOB medium plus appropriate antibiotic with 200 μl pre-culture in a 50 ml Falcon tube and incubate at 32°C with shaking until OD600 reaches ~0.6.
3. Split the culture equally between two 50 ml Falcon tubes. Keep one tube of culture on ice. This is the non-induced control for recombineering. Heat-shock the other tube of culture in a 42°C water bath with shaking at 150 rpm for 20 min then chill on ice for 20 min.
4. Centrifuge both tubes at 4°C, 3,300×*g*, and re-suspend each cell pellet very gently in 10 ml ice-cold 10% glycerol. Centrifuge again and re-suspend in 5 ml 10% glycerol. Centrifuge again and discard supernatant leaving a final volume of ~100 μl in which to re-suspend each pellet. It is best that the cells are used immediately. Cells can be stored at –80°C for future use but recombineering efficiency may be reduced.

3.2.4. Electroporation of Fosmid DNA into Electrocompetent Cells

Fosmid DNA isolated from EPI300 cells for each target gene is transformed into a recombineering-competent *E.coli* strain MW005 by electroporation.

1. Thaw an aliquot of electrocompetent cells on ice.
2. Chill fosmid DNA and electoporation cuvettes on ice.
3. Gently mix 1 μl of fosmid DNA with 20 μl competent cells, on ice. Transfer the mixture to a cuvette using a narrow pipette tip, ensuring the cells are in contact with both electrodes and devoid of air bubbles.
4. For a 0.1 cm gap cuvette, set the parameters to 1.8 kv, 200 ohms, 25 μF and electroporate. A time constant of four or more generally gives efficient uptake of DNA.
5. Add 1 ml room-temperature SOB medium immediately to the cuvette and transfer the cells to a sterile culture tube. Incubate the cells for 1–3 h at 32°C with shaking then plate 50–100 μl on LB agar plates supplemented with 12.5 μg/ml chloramphenicol and grow overnight at 32°C.

3.2.5. First Recombineering Step to Insert the RT Cassette

1. On ice, mix 150–250 ng purified PCR-generated gene-specific RT cassette DNA with 100 μl aliquots of the non-induced control or induced electrocompetent MW005 cells, containing the target fosmid, and electroporate as above.
2. Add 1 ml of SOB to each cuvette, transfer to a culture tube and incubate at 32°C, with shaking at 220 rpm, for 2–3 h to allow the cells to recover. Plate 50 μl recovered cells onto LB agar plates containing 12.5 μg/ml chloramphenicol and 5 μg/ml tetracycline. Incubate at 32°C for 36–48 h.
3. No colonies are expected on tetracycline plates for non-induced cells. The number of colonies for induced cells differs from one reaction to another as efficiency can vary dramatically depending on the specific fosmid and target site. Streak four to eight discrete positive colonies from the induced plate for each fusion onto a fresh LB agar plate containing chloramphenicol and tetracycline and grow overnight at 32°C. Perform a colony PCR as described above with flanking primers to confirm insertion of the cassette.
4. Grow a 2 ml overnight culture of positive clones in LB plus tetracyline, at 32°C, with shaking at 220 rpm. Use 1 ml to inoculate a culture for preparation of electrocompetent cells for the second recombineering step.

3.2.6. Second Recombineering Step to Replace the RT Cassette with the Reporter Gene

1. Make the MW005 bacteria, containing the fosmid with the RT cassette inserted at the desired position, electrocompetent with Red function-induced and non-induced cultures as per Subheading "Preparation of competent fosmid-containing MW005 cells for recombineering" above.
2. Transform the cells by electroporation with 150–250 ng of the purified reporter gene DNA fragment. After cells are recovered, spread 50 μl aliquots of recovered cells for non-induced and induced cells, respectively, onto NSLB/agar plates containing chloramphenicol and streptomycin. Incubate at 32°C for 36–48 h.
3. For each gene, re-streak eight discrete colonies from the induced plates and screen by colony PCR for replacement of the RT cassette by the reporter gene. Mutations affecting the *rpsL* gene, in cells that have not undergone the recombination of the second recombineering step, can lead to a background of streptomycin-resistant colonies of varying severity. Such mutations can mean it is necessary to repeat the second recombineering step with another independent colony from the first recombineering step.
4. Once MW005 cells containing the desired end-product fosmid have been identified, inoculate 10 ml LB plus chloramphenicol and L-arabinose with a single colony and incubate

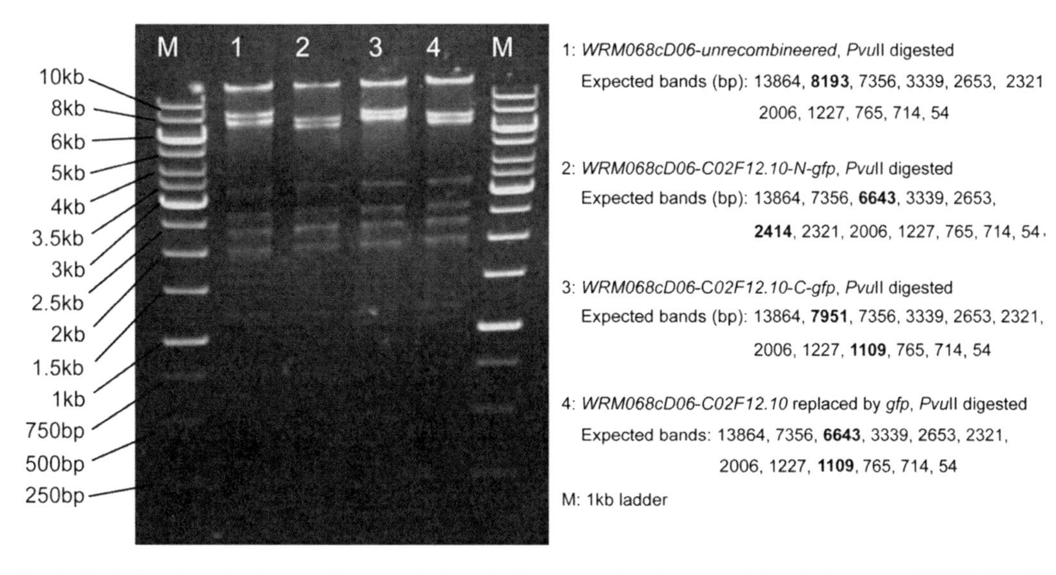

Fig. 4. *PvuII* digestion of *gfp* recombineering fusions for the homeodomain transcription factor gene *C02F12.10* in the fosmid clone *WRM068cD06*. Three *gfp* fusions were generated for *C02F12.10*: an N-terminal fusion with *gfp* inserted after the *C02F12.10* translation initiation codon, a C-terminal fusion with *gfp* inserted before the *C02F12.10* stop codon, and a third fusion with the *C02F12.10* coding region replaced by *gfp*. The gel shows the alteration of *PvuII* digestion fragments of the fosmid from before to after recombineering. Fragments altered by recombineering are in *bold*, while the other fragments remain in all four fosmids.

overnight (16 h max) at 32°C with shaking before fosmid DNA purification.

5. Verify the recombineered fosmid DNA structure by restriction enzyme digestion (e.g. Fig. 4). It may also be desirable to sequence the region that has undergone recombination, as rare errors in the sequence of the long oligonucleotides can be incorporated.

3.3. Transformation of C. elegans by Microinjection

3.3.1. Microinjection (see Note 3)

1. To prevent blocking of the very fine needle tip by contaminating particles, the reporter and transformation marker DNA preparations are microcentrifuged at 12,000 × *g* for 30 min prior to and after mixing.
2. Attach the DNA-loading pipette to a tube that will allow pressure to be applied. Place the tip of the DNA-loading pipette into the DNA mixture, taking care not to touch the bottom or sides of the tube, and allow 1–2 μl to enter by capillary action.
3. Insert the pipette tip into the injection needle through the wide end and expel the DNA solution onto the needle's internal surface near the tip. Capillary action will take the DNA to the tip of the needle.

4. Place a loaded needle into the needle holder and mount on the micromanipulator. Place a pad on the microscope stage. Position the needle so that the tip is in the centre of the microscope's field of view using the 5× objective. Shift to the 40× objective and apply the pressure (15–20 lb/in^2). The flow of liquid should be slow and smooth, but may need starting by gently touching the tip against the pad. If still no liquid is expelled, replace the needle.
5. Put a drop of liquid paraffin onto the pad. Under a dissection microscope, pick 1–5 (depending on expertise) adult *C. elegans* hermaphrodites from a bacteria-free region of the NGM plate, into the drop. Minimize transfer of bacteria. The animals should fall through the liquid paraffin and stick automatically to the pad over a few minutes as you transfer the preparation to the high-power microscope.
6. Place the pad, with worms on, on the stage of the injection microscope, securing with a small piece of tape and move into the field of view with the 5× objective. Rotate the stage so that the worm is at a 20° angle to the needle and shift to the 40× objective. Bring the needle tip into the same focal plane as the gonad arm to be injected and then up against the cuticle so as to cause an indentation. A tap of the microscope should cause the needle to pierce the cuticle and enter the gonad with the needle tip in the centre of the syncytium.
7. Apply the pressure. Allow the gonad to fill with the DNA solution, but do not overfill. Remove the needle from the worm quickly, with the pressure left on, to prevent blocking of the needle. Turn off the pressure. Both arms of the gonad of a single individual can be injected.
8. When all individuals on a pad have been injected, add two to three drops of M9 buffer onto the liquid paraffin. The M9 buffer will drop through the liquid paraffin to the pad surface and rehydrate the injected animals.
9. Place the pad on the inside surface of the lid of an inverted NGM agar plate to provide a humid environment to minimize evaporation. The animals should lift off the pad and start to thrash in the M9 buffer. Leave to recover for 10–60 min before carefully transferring each worm individually to a NGM plate seeded with OP50 bacteria using a glass pipette. Recovery rate varies so it is worth transferring worms that are not moving. Place injected worms at 20°C.

3.3.2. After Microinjection

All transgenic progeny of injected animals should be picked to fresh seeded NGM plates, up to five individuals per plate, 3–4 days after injection. The most common co-transformation marker is *rol-6 (su1006),* on the plasmid pRF4, which confers a readily recognized dominant roller phenotype. Most of these initial transgenic progeny do not transmit the transgenes, but transgenic lines can be established by transferring transgenic individuals at successive generations for those that do. Retain only one line per injected individual to ensure independence. Typically at least three independent lines are established for each reporter gene fusion to identify aberrant reporter expression resulting from inappropriate joining of DNA elements during array formation, although this is very rarely encountered.

Co-injection of plasmids usually results in both being incorporated into the transgenic extrachromosomal array. With reporter gene fusions made in fosmids, incorporation of the fosmid into the array with the transformation marker plasmid is less reliable. Therefore, if no reporter expression is observed, it is worth confirming that the reporter gene fusion is present in the transgenic lines established.

1. Mix 2 μl 10 mg/ml Proteinase K per 100 μl 1× lysis buffer.
2. Transfer a single adult hermaphrodite to a 6-μl aliquot in the bottom of a PCR tube and freeze at −80°C for ≥20 min.
3. In a PCR machine, incubate at 65°C for 1 h followed by 95°C for 15 min. The lysis is ready for use as PCR template immediately or can be stored either at 4°C for a short time or at −20°C for a longer time.
4. Prepare a master mix, 12.5 μl per PCR: 8.8 μl ddH_2O, 1.5 μl 10× PCR buffer, 0.6 μl of 50 mM $MgCl_2$, 0.5 μl dNTPs (10 mM each nucleotide), 0.5 μl of each primer, and 0.6 μl of Taq polymerase (1 Unit/μl).
5. Aliquot 12.5 μl into individual tubes before adding 2.5 μl of lysed extract. Use the PCR program: 1 step of 5 min at 94°C, followed by 30 cycles of 30 s at 94°C, 45 s at 55°C, and 2 min at 72°C, then one final step of 10 min at 72°C. The length of the repeated 72°C elongation step should be adjusted according to the size of fragment using 1 min for each kilobase of DNA.
6. Check PCR products by agarose gel electrophoresis.

3.4. Transformation of *C. elegans* by Microprojectile Bombardment

3.4.1. Large-Scale Liquid Culture of *C. elegans*

1. To maximize the proportion of the population subjected to bombardment that are adults it is worth synchronizing the nematode culture. A small population is generated first on NGM plates at 20°C. Five 5-cm OP50 seeded NGM plates are inoculated with approximately 50 larvae per plate. After 8–9 days, just before the bacteria is exhausted, the nematodes are washed from the plates using M9 buffer and collected in a 15-ml Falcon tube.
2. The volume is adjusted to 7 ml before addition of 2 ml bleach (5% sodium hypochlorite) and 1 ml 5 M NaOH. The tube is vortexed for a few sec, repeating every 2 min for 10 min.

The nematodes are centrifuged for 30 s at 1,300 × *g* and then washed five times with 5 ml sterile water. The final egg pellet is re-suspended in 100 ml S-basal plus antibiotics. Incubation overnight at 20°C allows the eggs to hatch to obtain arrested L1s.

3. 500 ml LB medium in a 2-L flask is inoculated with colonies of HB101 bacteria and incubated in a rotating incubator at 37°C for 24 h. Following centrifugation at 4°C, 3,300 × *g* force for 20 min the bacteria pellet is re-suspended in an equal volume (approximately 6 ml) of M9 buffer.
4. 3 ml of the HB101 bacterial suspension is added to the arrested L1s. The culture is then grown at 20°C, with shaking at 200 rpm. After approximately 4 days, the cultures contain a large number of young adult worms.
5. The grown cultures are allowed to settle for 10 min in 50 ml glass tubes at room temperature. This first pellet contains predominantly older nematodes suitable for bombardment. 1 ml of nematode pellet is sufficient for one hepta bombardment. The supernatant is removed to a fresh tube and incubated on ice for 20 min to form another pellet. This second pellet, containing young larval stages, is then used to inoculate 100 ml of S-basal medium plus bacteria and antibiotics as before, which is then re-incubated. In 3–4 days, this culture is ready to harvest again.
6. After the first 4-day culture, the number of nematodes present may be still quite low, and a second round of incubation may be required before there are sufficient animals to bombard. In this case, the two pellets are combined into fresh medium with new bacteria. A large population, for repeated bombardments of a series of DNAs, can be maintained by this cyclic culturing. The cultures must not be allowed to starve, and addition of extra bacteria between sedimentations is sometimes required. Such cyclic growth of nematode populations risks contamination with foreign microorganisms, in which case the cultures are bleached and the eggs collected as above. The cultures are maintained in the cyclic system for no more than 6 weeks, as transformation efficiency drops off eventually. When a reduction in efficiency is observed (e.g. less than five lines are generated per hepta bombardment), then a totally new population is generated from a fresh frozen aliquot of *C. elegans*. It takes about 8 weeks from frozen aliquot to first bombardment.

3.4.2. DNA Preparation and Coating of Gold Particles

1. Approximately 7 μg DNA is required for each plasmid for a hepta bombardment. Plasmid DNA is prepared from overnight bacterial cultures using a plasmid mini-preparation kit and is linearized, by restriction enzyme digestion, prior to bombardment.

2. Add 60 mg gold particles to 2 ml 70% ethanol, vortex for 5 min, and then soak for 15 min. Microfuge briefly, discard the supernatant, and wash the particles three times with sterile water. Re-suspend the final pellet in 1 ml of 50% sterile glycerol. The final gold particle suspension is 60 mg/ml and can be stored for up to 2 months at 4°C.
3. Linearized DNA can be precipitated directly onto the gold particles from the restriction enzyme digestion without further purification, but this should be performed on the day of bombardment. Use Treff tubes to reduce DNA loss from adherence to the tube. 30 μl (7 μg) DNA is added drop-wise, mixing between drops, to 70 μl of gold suspension. 300 μl of 2.5 M $CaCl_2$ and 112 μl 0.1 M spermidine are also added drop-wise and then vortexed for 5 min. Particles, pelleted at 2,500 × *g* for 5 s, are then washed in 800 μl 70% ethanol before re-suspension in 70 μl of 100% ethanol. Vortex until ready to use. It is important to add the reagents slowly and to vortex between each addition to prevent the gold beads clumping. The final preparation should look like a homogeneous brown liquid.

3.4.3. Bombardment

This part of the protocol is carried out in a sterile hood. The biolistic gun is situated in the flow hood and routinely cleaned with ethanol.

1. Pipette 10 μl of DNA-coated gold beads suspended in ethanol over the centre of each macrocarrier in the hepta macrocarrier holder. Allow the macrocarrier to dry; the beads then appear white.
2. Just before bombardment, 1 ml of young adult nematodes is distributed evenly over the seven hepta target spots on a 9 cm NGM plate. 1 ml of packed adult nematodes contains approximately 500,000–1,000,000 individuals.
3. Refer to the Biorad manual for the biolistic gun for more details. The vacuum is set to 26 in.Hg. The target shelf is placed in the second position from the bottom. Connect the vacuum tubing to a vacuum source. Open the valve on the helium cylinder and check that helium pressure is near or above 2,200 psi. Place a rupture disc soaked in isopropanol in the retaining cap and tighten. Place the hepta stopping screen and macrocarrier holder in the appropriate positions in the chamber. Place an uncovered NGM plate with worms on the target shelf in the bombardment chamber and close the door.
4. Evacuate the chamber to 26 in.Hg and press the fire button until the disc ruptures. Release the vacuum and remove the bombarded plate. Unscrew the retaining cap and discard the ruptured disc. Turn off the vacuum and close the helium cylinder valve, ensuring no pressure is left in the line.

Turn the bombardment machine off and clean the bombardment chamber and macrocarrier holder with 70% ethanol.

3.4.4. Post-bombardment

1. 1 ml of M9 buffer is spread over the surface of each bombarded plate and the nematodes are left to recover for 1–2 h at room temperature. The plates are then washed with 4 ml of M9 buffer and 0.5 ml of this is placed directly onto seven separate OP50 seeded 9 cm NGM plates, leaving the remaining small amount of liquid and nematodes on the original bombardment plate. All the eight plates are incubated at 20°C.
2. After 3–4 weeks, successful bombardments will result in the presence of transformants on the plates. Transformants were able to form dauers to survive starvation and have wild-type locomotion. From each of the large plates containing transformants, four individuals are each transferred to a separate 5-cm OP50 seeded NGM plate. Seven days later all small plates are assessed for transmission stability of the transgenic line. Within each set of four small plates, the one with the highest level of transmission is chosen and the remaining three discarded. The maximum number of lines generated per plasmid bombardment is, therefore, eight.

3.5. Expression Pattern Analysis

3.5.1. Initial Examination

For initial assessment of new transgenic lines, examine reporter expression in a mixed stage population in a rapid mount. Wash a population from an NGM agar plate with M9 buffer. Place in a 1.5-ml tube and centrifuge 30 s at 2,500 × *g*. Pipette 4 μl from the loose pellet and 1 μl of 50 mM sodium azide or 50 mM levamisole, as anaesthetic, into each well of an eight-well microscope slide. Apply a coverslip carefully. Examine on a high-power fluorescence microscope.

3.5.2. Closer Examination

To minimize movement for more careful observation and to capture images of the expression pattern, transgenic animals should be mounted with anaesthetic on an agarose pad on a microscope slide, under a coverslip. Place a drop of molten 2% agarose in ddH_2O and on a plain glass microscope slide, parallel to and between two other slides thickened with two layers of tape. Immediately place another slide over the drop, at right angles to the other slides, to spread the agarose into a thin disc, 1–2 cm in diameter. Wait a minute or so for the agarose to set and then remove the top slide by sliding laterally.

Place 5 μl of 10 mM sodium azide or 10 mM levamisole in M9 buffer on the agarose pad. Pick up worms individually from a culture plate, transferring as little additional bacteria as possible, and place in the drop of buffer. Initially, the animals thrash vigorously but eventually become paralyzed by the anaesthetic. Alternatively, for a larger number of animals, wash a population from a NGM agar plate with M9 buffer. Place in a 1.5-ml tube and centrifuge 30 s at 2,500×*g*. Pipette 4 μl from the loose pellet and 1 μl of 50 mM sodium azide or 50 mM levamisole, onto the agarose pad. Carefully place a cover slip on top, avoiding air bubbles. The mount can be sealed with a few drops of silicone oil, allowed to fill the space between coverslip and slide, to eliminate evaporation.

3.5.3. Expression Pattern Characterization

Fluorescent protein distributions can be examined by conventional epifluorescence microscopy using 20×, 40×, 63×, and 100× objectives. DIC optics allows visualization of the nuclei of any of the cells present in *C. elegans*, to which the distribution of the fluorescent protein reporter can be related. However, expression pattern characterization usually depends upon a detailed knowledge of *C. elegans* anatomy. Fortunately, WormAtlas (http://www.wormatlas.org) provides a detailed description of *C. elegans* anatomy integrated with images that are particularly convenient for comparison to direct observations of an expression pattern. This resource can prove of value even to experienced *C. elegans* biologists for some expression patterns. WormBook (http://www.wormbook.org) includes a valuable chapter on Nomarski/DIC imaging of *C. elegans* (24).

Comparison with previously determined gene expression patterns can be useful for confirming interpretations. Such data can be found in WormBase (http://www.wormbase.org), as well as WormAtlas and web sites of individual laboratories. Two expression patterns can be compared directly by using reporters of different colour fluorescence. An *rfp* reporter fusion for which the expression pattern is known can be co-injected with novel *gfp* fusions, injected into existing *gfp* expressing lines or *rfp* and *gfp* expressing lines can be crossed to give co-expressing lines to reveal co-localization.

Confocal imaging systems can prove useful for interpreting subcellular distributions of reporter fusion proteins and have been utilized in steps taken to automate analysis of expression patterns. For example, a system has been developed which permits automatic analysis of reporter expression with cellular resolution continuously during embryogenesis, up to the 350 cell stage (25). This automation also allows quantitative measurement of expression levels from which subtle regulatory relationships can be interpreted (26, 27).

4. Notes

1. The starting pDONR and pDEST vectors contain the *ccdB* gene between the two *attP* sites for negative selection and, therefore, before the *ccdB* gene is replaced in the Gateway recombination reaction, can only be maintained in *E. coli* strain

DB3.1. This bacterial strain must be used if more of these vector DNAs needs to be prepared.

2. The weight in nanogram of 1 fmole of a DNA fragment equals the length of the DNA in base pairs multiplied by 6.6×10^{-4}.

3. The injection technique requires practice as the procedure described in Subheading 3.3.1 should only take a few minutes; the animals will become desiccated and die if left on the agarose pad for too long. Success of microinjection is determined by multiple factors. Animals need to stick well to the pads, and both the state of individuals to be injected and the quality of pads are critical. Populations to be injected should be free of contamination and well-fed, without starvation at a recent generation. Bleaching the population and growing on fresh NGM plates may improve efficiency. Adjusting the thickness of the agarose pad can alter the rate of desiccation. The needle tip needs to be sharp enough to pierce the cuticle without causing fatal damage, but also open enough so that DNA can flow steadily. Efficiency is also greatly influenced by the quality of the DNA solution, which needs to be free of particles that may block the needle. Plasmids and fosmids must be at high enough concentration for extrachromosomal array formation but not so as toxic. Fosmids are generally less efficient than plasmids. Failure to create stable transmitting lines for a particular fosmid after several attempts may indicate that the fosmid includes a gene that is lethal to *C. elegans* when on an array; reducing the concentration of the fosmid in the injection mixture may help.

4. While reporter gene fusion technology is convenient for determining gene expression patterns in *C. elegans*, data generated may not necessarily reflect the expression of the endogenous gene. Clearly, short DNA fragments containing the promoter, as used in transcriptional reporter gene fusions, may lack *cis*-acting regulatory elements important to the endogenous gene's expression. Translational reporter gene fusions constructed by recombineering fosmids are quite likely to include all the *cis*-acting elements for a target gene given the compact nature of the *C. elegans* genome. Nevertheless, such reporter gene insertion could disrupt regulatory interactions, and the rate of synthesis and degradation of a reporter fusion protein is unlikely to reflect precisely those of the endogenous gene products. Alternative transcripts also complicate the use of reporter gene technology. If the expression pattern of the endogenous gene is critical, then independent confirmation, from for example immunofluorescence microscopy with specific antibodies, is

recommended. Despite these concerns, reporter gene fusions do give reproducible data that report reliably on transcriptional expression properties of the DNA fragment assayed that will be vital for revealing the physiological relevance in transcription factor regulatory networks.

Acknowledgments

The authors thank Drs. Jane Shingles and John Reece-Hoyes for training in bombardment. Some protocols presented here were adapted from Invitrogen Gateway manuals, Wormbase, Wormatlas, and Wormbook, which are all greatly appreciated. This work was supported by BBSRC grant BB/E008038/1 and Wellcome Trust Grant 082603/B/07/Z.

References

1. Hobert, O. (2008) Gene regulation by transcription factors and microRNAs. *Science* **319**, 1785–1786.
2. Wenick, A. S. and Hobert, O. (2004) Genomic *cis*-regulatory architecture and *trans*-acting regulators of a single interneuron-specific gene battery in *C. elegans*. *Dev. Cell* **6**, 757–770.
3. Sulston, J. E., Schierenberg, E., White, J. G., and Thomson, J. N. (1983) The embryonic cell lineage of the nematode *Caenorhabditis elegans*. *Dev. Biol.* **100**, 64–119.
4. Sulston, J. E. (1977) Postembryonic cell lineages of the nematode *Caenorhabditis elegans*. *Dev. Biol.* **56**, 110–156.
5. The *C. elegans* Sequencing Consortium (1998) Genome sequence of the nematode *C. elegans*: A platform for investigating biology. *Science* **282**, 2012–2018.
6. Dupuy, D., Li, Q.-R., Deplancke, B., Boxem, M., Hao, T., Lamesch, P., et al. (2004) A first version of the *Caenorhabditis elegans* promoterome. *Genome Res.* **14**(10b), 2169–2175.
7. Lamesch, P., Milstein, S., Hao, T., Rosenberg, J., Li, N., Sequerra, R., et al. (2004) *C. elegans* ORFeome Version 3.1: increasing the coverage of ORFeome resources with improved gene predictions. *Genome Res.* **14**(10b), 2064–2069.
8. Luan, C.-H., Qiu, S., Finley, J. B., Carson, M., Gray, R. J., Huang, W., et al. (2004) High-Throughput Expression of *C. elegans* Proteins. *Genome Res.* **14**(10b), 2102–2110.
9. Invitrogen Life Technologies. MultiSite Gateway three-fragment vector construction kit – using Gateway technology to simultaneously clone multiple DNA fragments. Available from: www.invitrogen.com.
10. Hartley, J. L., Temple, G. F. and Brasch, M. A. (2000) DNA cloning using *in vitro* site-specific recombination. *Genome Res.* **10**(11), 1788–1795.
11. Reece-Hoyes, J. S., Deplancke, B., Shingles, J., Grove, C. A., Hope, I. A., and Walhout, A. J. M. (2005) A compendium of *Caenorhabditis elegans* regulatory transcription factors: a resource for mapping transcription regulatory networks. *Genome Biol.* **6**, R110.
12. Reece-Hoyes, J. S., Shingles, J., Dupuy, D., Grove, C. A., Walhout, A. J., Vidal, M., et al. (2007) Insight into transcription factor gene duplication from *Caenorhabditis elegans* Promoterome-driven expression patterns. *BMC Genomics* **8**, 27.
13. Dolphin, C. T. and Hope, I. A. (2006) *Caenorhabditis elegans* reporter fusion genes generated by seamless modification of large genomic DNA clones. *Nucleic Acids Res.* **34**(9), e72.
14. Bamps, S., Wirtz, J., Savory, F. R., Lake, D., and Hope, I. A. (2009) The *Caenorhabditis elegans* sirtuin gene, *sir-2.1*, is widely expressed and induced upon caloric restriction. *Mech. Ageing & Dev.* **130**, 762–770.
15. Stavropoulos, T. A. and Strathdee, C. A. (2001) Synergy between *tetA* and *rpsL* provides high-stringency positive and negative selection in bacterial artificial chromosome vectors. *Genomics* **72**(1), 99–104.

16. Epicenter Biotechnologies. CopyControl™ Fosmid Library Production Kit. Available from: www.epibio.com.
17. Westenberg, M., Bamps, S., Soedling, H., Hope, I. A., and Dolphin, C. T. (2010) *Escherichia coli* MW005: lambda Red-mediated recombineering and copy-number induction of *oriV*-equipped constructs in a single host. *BMC Biotechnology* **10**, 27.
18. Mello, C. C., Kramer, J. M., Dan, S., and Ambros, V. (1991) Efficient gene transfer in *C. elegans*: extrachromosomal maintenance and integration of transforming sequences. *The EMBO Journal* **10**(12), 3959–3970.
19. Praitis, V., Casey, E., Collar, D., and Austin, J. (2001) Creation of low-copy integrated transgenic lines in *Caenorhabditis elegans*. *Genetics* **157**(3), 1217–1226.
20. Hunt-Newbury, R., Viveiros, R., Johnsen, R., Mah, A., Anastas, D., Fang, L., et al. (2007) High-throughput *in vivo* analysis of gene expression in *Caenorhabditis elegans*. *PLoS Biology* **5**(9), e237.
21. Berezikov, E., Bargmann, C. I. and Plasterk, R. H. A. (2004) Homologous gene targeting in *Caenorhabditis elegans* by biolistic transformation. *Nucleic Acids Res.* **32**(4), e40.
22. Wilm, T., Demel, P., Koop H.-U., Schnabel H., and Schnabel, R. (1999) Ballistic transformation of *Caenorhabditis elegans*. *Gene* **229**, 31–35.
23. Jackstadt, P., Wilm, T. P., Horst, Z., and Gerd, H. (1999) Transformation of nematodes *via* ballistic DNA transfer *Mol. Biochem. Parasitology* **103**(2), 261–266.
24. Yochem, J., *Nomarski images for learning the anatomy, with tips for mosaic analysis*, in *WormBook*, The *C. elegans* research community, Editor. 2006. (http://www.wormbook.org).
25. Bao, Z., John I. Murray, Thomas Boyle, Siew Loon Ooi, Matthew J. Sandel, and Waterston, R. H. (2006) Automated cell lineage tracing in *Caenorhabditis elegans*. *Proc. Natl. Acad. Sci. USA*. **103**, 2707–2712.
26. Murray, J. I., Bao, Z., Boyle, T., and Waterston, R. H. (2006) The lineaging of fluorescently labeled *Caenorhabditis elegans* embryos with StarryNite and AceTree. *Nat. Protoc.* **1**, 1468–1476.
27. Boyle, T. J., Bao, Z., Murray, J. I., Araya, C. L., Waterston, R. H., and (2006). (2006) AceTree: a tool for visual analysis of *Caenorhabditis elegans* embryogenesis. *BMC Bioinformatics* 7, 275.

Chapter 3

High-Throughput SELEX Determination of DNA Sequences Bound by Transcription Factors In Vitro

Nobuo Ogawa and Mark D. Biggin

Abstract

SELEX (systematic evolution of ligands by exponential enrichment) was created 20 years ago as a method to enrich small populations of bound DNAs from a random sequence pool by PCR amplification. It provides a powerful way to determine the in vitro binding specificities of DNA-binding proteins such as transcription factors. Here, we present a robust version of the SELEX protocol for high-throughput analysis. Protein-bound beads prepared from insoluble recombinant 6× HIS-tagged transcription factor protein are used in a simple pull-down assay. To allow efficient determination of the enriched DNA sequences, bound oligonucleotides are concatenated, allowing approximately 1,000 oligonucleotides to be sequenced from one 96-well format plate. Successive rounds of SELEX data are statistically useful for understanding the full range of moderate affinity and high-affinity binding sites.

Key words: Transcription factor, SELEX, DNA-binding sequence, In vitro assay

1. Introduction

Mapping functional transcriptional *cis*-regulatory sequences in the nonprotein coding regions of the genome is one of the biggest challenges in the postgenome sequencing era. While *cis*-regulatory modules have been identified by a range of molecular genetic and biochemical approaches for more than 25 years (1) over the last decade, high-throughput in vivo crosslinking assays (ChIP-chip and ChIP-seq) have allowed maps of nuclear protein bound regions to be identified genome wide (e.g., refs. 2–4). Although not all of these in vivo crosslinked regions appear to contain functional regulatory elements (3, 5), their identification is a critical step toward understanding the functional structure of the genome. Yet another approach to identify functional elements has been the use

Bart Deplancke and Nele Gheldof (eds.), *Gene Regulatory Networks: Methods and Protocols*,
Methods in Molecular Biology, vol. 786, DOI 10.1007/978-1-61779-292-2_3, © Springer Science+Business Media, LLC 2012

of computer predictions that score statistically significant local clusters of recognition sites for multiple factors (6). To be optimally effective, however, this predictive approach requires detailed data on the full spectrum of high and moderate affinity DNA sequences recognized by each transcription factor in a system.

SELEX (systematic evolution of ligands by exponential enrichment) involves the binding of protein to a mixture of oligonucleotides containing, in the first round, a random set of DNA sequences of typically 16–24 bp in length, flanked by primers that allow PCR amplification (Fig. 1) (7). Oligonucleotides bound by a transcription factor are then PCR amplified and used as input in a further DNA-binding reaction, and so on. After multiple rounds of selection, the oligonucleotides are sequenced to determine which sequences are preferentially bound by the transcription factor (Fig. 2). As a part of the Berkeley Drosophila Transcription Network Project (http://bdtnp.lbl.gov/Fly-Net/), we have developed a robust version of this method to more accurately determine the range and affinities of DNA sequences bound in vitro by *Drosophila melanogaster* transcription factors.

Our protocol is based on work by Roulet et al. (8), who concatenated bound oligonucleotides prior to DNA sequencing to allow more oligonucleotides to be sequenced at a reasonable cost (Figs. 1 and 2). Our variant of this method allows proteins produced from a bacterial expression system that are insoluble to be used, a great advantage as many transcription factors are not soluble when expressed this way. In our method, insoluble 6X HIS affinity tagged

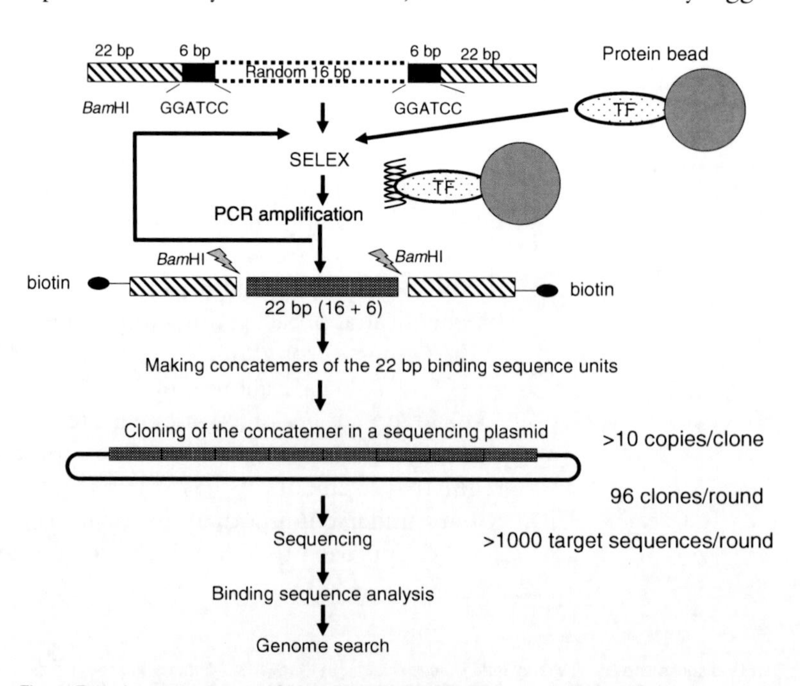

Fig. 1. Experimental scheme of high-throughput SELEX.

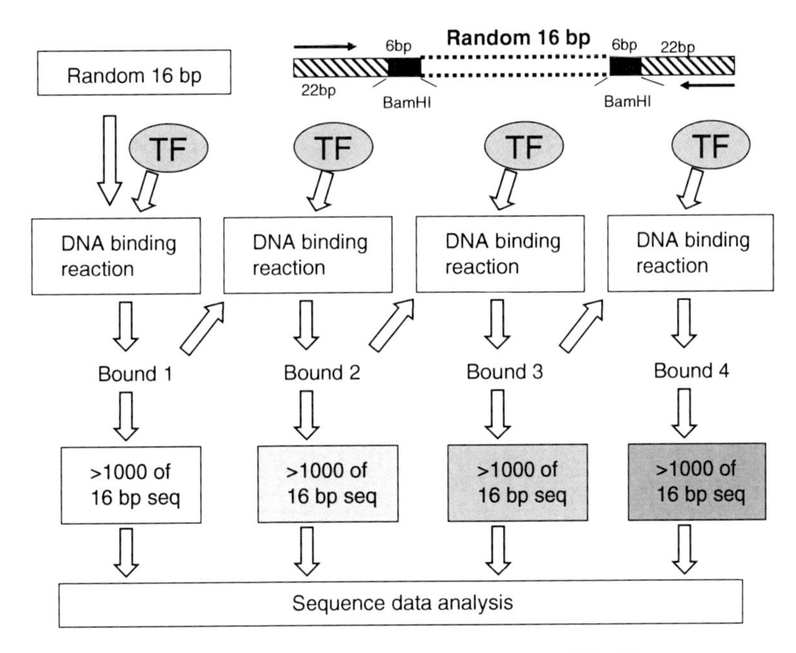

Fig. 2. Enrichment of binding sequences in successive rounds of SELEX.

proteins are renatured only after they are bound to an affinity resin bead, which stabilizes insoluble proteins. In addition, our variant of the original protocol allows sequences of approximately 1,000 oligonucleotides to be obtained from each round of SELEX, a sufficient depth of data to allow use of sophisticated statistical methods that provide more accurate models of factors DNA-binding specificity than previously available.

2. Materials

2.1. Preparation of Renatured Protein Bound to Bead

1. Buffer P: 50 mM sodium phosphate buffer (Na_2HPO_4) (pH 7.0), 300 mM NaCl, 10% glycerol, 10 μM $ZnSO_4$, 0.1% NP-40.
2. Buffer P (pH 9.0): 50 mM Na_2HPO_4 (do not adjust pH, final pH ~9.0), 0.3 M NaCl, 10% glycerol, 10 μM $ZnSO_4$. Add EDTA-free proteinase inhibitor tablet.
3. Buffer PB: Buffer P containing 0.1 mg/ml BSA.
4. CAPS buffer containing 8 M urea: 50 mM CAPS buffer (pH 10.0), 300 mM NaCl, 8 M urea.
5. TALON Metal Affinity Resins was obtained from Clontech (see Note 1).
6. EDTA-free proteinase inhibitor tablet "complete, Mini, EDTA-free Protease Inhibitor Cocktail" (Roche).

2.2. Preparation of Double-Stranded Random oligoDNA

1. Random sequence oligoDNA for SELEX (see Note 2).

   Random72: GGATTTGCTGGTGCAGTACAGT-GGAT-CC-$(N)_{16}$-GGATCC-TTAGGAGCTTGAAATCGAGCAG

   Random72R: TCCATCGCTTCTGTATGACGCA-AGATCT $(N)_{16}$AGATCTGTCCTAACCGACTCCGTTGATT

   Random72HR: TCCATCGCTTCTGTATGACGCA-AAGCTT $(N)_{16}$AAGCTTGTCCTAACCGACTCCGTTGATT

2. Forward and Reverse primers for PCR amplification of the random sequence oligoDNA.

   Random72 Forward: GGATTTGCTGGTGCAGTACA

   Random72 Reverse: CTGCTCGATTTCAAGCTCCT

   Random72R Forward (common Forward primer to Random72HR): TCCATCGCTTCTGTATGACG

   Random72R Reverse (common Reverse primer to Random72HR): AATCAACGGAGTCGGTTAGG

3. LoTE: 3 mM Tris–HCl (pH 7.5), 0.2 mM EDTA.
4. *Taq* DNA polymerase and 10× PCR buffer (New England BioLabs).
5. QIAquick Gel Extraction kit (Qiagen).

2.3. SELEX, DNA-Protein Binding Reaction

1. 5× DNA-binding buffer without NaCl: 50 mM Tris–HCl (pH 7.5), 25% glycerol, 37 mM $MgCl_2$.
2. Washing buffer: 10 mM Tris–HCl (pH 7.5), 5 mM NaCl, 7.4 mM $MgCl_2$, 5% glycerol, 0.1% NP-40.
3. poly(dIdC) (GE Healthcare): Make 1 μg/μl poly(dIdC) in LoTE solution. Sonicate the solution to shear the DNA into ~500 bp lengths. Warm the solution at 65°C for 15 min, followed by slow cooling down to room temperature. Store at –20°C.
4. 2 mg/ml BSA: diluted from 10 mg/ml BSA obtained from New England BioLabs.
5. Elution buffer: 10 mM Tris–HCl (pH 7.5), 500 mM NaCl, 7.4 mM $MgCl_2$, 5% glycerol, 0.1% NP-40.
6. Co-precipitant: Co-Precipitant Pink (Bioline).
7. SYBR kit for Q-PCR (ABI).

2.4. Isolation, Cloning, and Sequencing of Large Numbers of Binding Sequence Units

1. Forward and Reverse 5′-Biotin primers

   Random72 Forward: Biotin-GGATTTGCTGGTGCAGTACA

   Random72 Reverse: Biotin-CTGCTCGATTTCAAGCTCCT

   Random72R Forward (common Forward primer to Random72HR): Biotin-TCCATCGCTTCTGTATGACG

   Random72R Reverse (common Reverse primer to Random72HR): Biotin-AATCAACGGAGTCGGTTAGG

2. *Bam*HI, *Bgl*II, and *Hin*dIII restriction enzymes, and their 10× reaction buffers (New England BioLabs).
3. Streptavidin Sepharose (GE Healthcare).
4. Ultrafree centrifugal filter unit (Ultrafree-MC, HV Filter, 0.45 μm, Millipore).
5. T4 DNA Ligase and 10× Ligation buffer (New England BioLabs).
6. T4 DNA polymerase, DNA I polymerase Klenow fragment and their 10× reaction buffers (New England BioLabs).
7. *Sma*I-digested pUC19 (20 ng/μl).
8. Competent cells (e.g., Invitrogen, ElectroMAX DH10B).
9. M13 Universal primer: GTTTTCCCAGTCACGAC.
10. M13 Reverse primer: AACAGCTATGACCATG.

3. Methods

3.1. Preparation of Renatured Protein Bound to Bead

Recombinant 6× HIS-tagged protein is commonly used to obtain purified protein for studying protein activity (9). This approach often produces large amounts of recombinant protein in bacterial cells, but the proteins are often nonfunctional due to insolubility in aqueous solution. We have found that the vast majority (more than 90%) of the *Drosophila* transcription factors we have tested (more than 30) are insoluble. Insoluble protein can be solubilized by denaturing with 8 M urea or 6 M guanidine and the resulting solubilized 6× HIS-tagged protein can be purified by Ni-column chromatography. Detail methods of bacterial expression and purification of recombinant 6× HIS-tagged protein from bacterial cells are available from commercial instruction books (e.g., Qiagen and Invitrogen) depending upon vector and host strain used. Therefore, our procedure in this text starts from the step after denatured and purified 6× HIS-tagged protein by Ni-column in 8 M urea solution.

1. Prepare and purify recombinant 6× HIS-tagged protein in 2–5 ml of Buffer P containing 8 M urea.
2. Change the buffer of the 6× HIS-tagged protein to 50 mM CAPS buffer containing 8 M urea at room temperature by using a desalting column or dialysis against 1,000-fold volume of buffer.
3. Measure protein concentration by Bradford or Lowry method, and adjust concentration to 1 mg/ml.
4. Take 3.5 ml of protein solution and add 13 μl of β-mercaptoethanol.

5. While stirring, add 3.3 ml of the protein solution drop-by-drop into 30 ml of Buffer P (pH 9.0) (this will give a final urea concentration of 0.8 M).
6. Incubate the diluted protein solution for 15 min at room temperature.
7. Adjust solution pH to 7 by the addition of 1 M H_3PO_4.
8. Let the solution stand overnight at 4°C.
9. Centrifuge for 30 min at 10,000×*g* at 4°C and transfer the supernatant to a fresh tube. Most proteins should be still soluble at this step.
10. Make 1 ml aliquots and freeze them at –70°C.
11. Thaw the frozen 6× HIS-tagged protein solution on ice and mix using a rocking shaker for 10 min at 4°C.
12. Mix an amount of protein solution defined using a pilot test run (see Note 3) with 250 μl of TALON Metal Affinity Resin equilibrated with Buffer PB containing 0.8 M urea in a 1.5 ml tube. When the protein solution used is less than 250 μl, add the required amount of Buffer PB containing 0.8 M urea to adjust the total buffer plus protein solution to 250 μl. Rock the beads suspension using a rocking shaker for 2 h at 4°C.
13. Spin at 500×*g* for 1 min. Remove much of the supernatant, leaving 250 μl supernatant with the beads.
14. Add 250 μl of Buffer PB (no urea) to the resin suspension, and rock for 15 min at 4°C. Spin the tube at 500×*g* for 1 min and remove the 250 μl of supernatant.
15. Repeat this step-wise dilution of urea five times.
16. At the last step, remove as much supernatant as possible and add 250 μl of Buffer PB. This is the 50% (v/v) protein bead suspension to be used in the DNA-binding reactions below.

3.2. Preparation of Double-Stranded Random OligoDNA

1. Run a one cycle PCR with a Random72 oligoDNA (one of the three designs, see Note 4) and its Reverse primer. The Reverse primer should be in eightfold excess over the Random72 oligoDNA. Typical reaction mixture and cycle are as follows: 4 μl of 10× PCR buffer, 4.8 μl of 2.5 mM dNTP mix, 1 μl of 10 μM Random72 oligoDNA, 8 μl of 10 μM Random72 Reverse primer, 1 μl *Taq* DNA polymerase, 21.2 μl water in a total of 40 μl. Run one PCR cycle for 1 min at 94°C, 5 min at 64°C (60°C for "R" design), and 9 min at 72°C.

   Multicycle PCR is not recommended due to the loss of a random population by PCR bias.
2. Apply 1 μl out of 40 μl PCR on to a 15% polyacrylamide (PAGE) gel and on lanes of either side of the single-strand Random72 oligoDNA and the Reverse primer. Perform electrophoresis to confirm the formation of the double-stranded

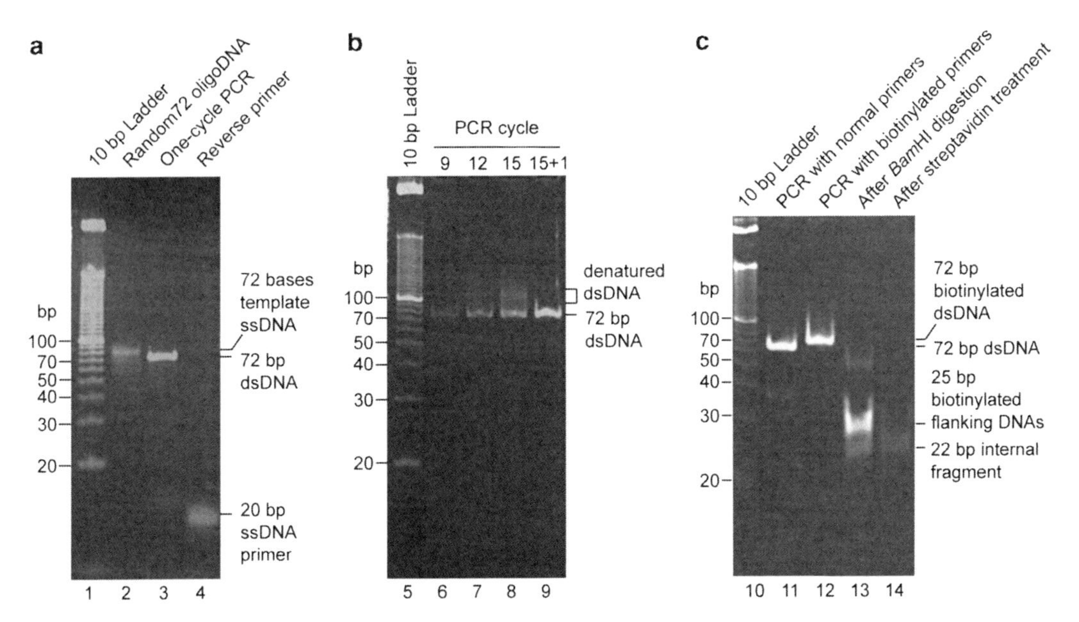

Fig. 3. Analysis of double-stranded (ds) and single-stranded (ss) DNAs by 15% polyacrylamide gel electrophoresis. (**a**) Double-stranded random oligoDNA (*lane 3*) was prepared by one cycle PCR from a single-strand template DNA (*lane 2*) and a single-stranded Reverse primer (*lane 4*). (**b**) Further cycles of PCR produce incomplete dsDNA for the random oligoDNA or SELEX-bound DNA (*lanes 6–8*). Following such a series of second round PCR cycles with a single cycle reaction converts the incomplete dsDNA to the completed dsDNA (*lane 9*). (**c**) PCR with biotinylated primers creates a biotinylated 72 bp dsDNA (*lane 12*) which can be distinguished from its nonbiotinylated dsDNA (*lane 11*) by differences in migration on PAGE. Restriction enzyme (*Bam*HI in this case) separates the 22 bp internal fragment containing the transcription factor bound sequence from the flanking biotinylated DNAs, which can be removed by treatment with streptavidin beads (*lanes 13* and *14*).

DNA (dsDNA), which will appear as a band in addition to the single-strand template and the primer DNAs (Fig. 3a).

3. Purify the dsDNA with a QIAquick Gel Extraction kit using LoTE as the elution buffer. More than 10 ng of dsDNA should be obtained from the PCR. Adjust the DNA concentration to 10 ng/μl. Keep the DNA solution at 4°C. It should not be frozen or heated up to avoid possible denaturation.

3.3. SELEX, DNA-Protein Binding Reaction

1. In 10 μl, make the DNA-binding premixture solution using 10 ng of random dsDNA and an amount of poly(dIdC) determined in a pilot study (see Note 4). 10 ng of the random 72 bp dsDNA theoretically contains approximately 100 copies of each of all possible 16-mer sequences. A typical DNA-binding premixture solution can be made as follows: 2 μl of 5× DNA-binding buffer without NaCl, 0.5 μl of 1 M NaCl, 0.5 μl of 2 mg/ml BSA, 0.5 μl of 2% NP-40, 1 μl of 1 μg/μl poly(dIdC), 1 μl of Random72 dsDNA (10 ng/μl) or SELEX PCR DNA (10 ng/μl), and 4.5 μl of water in a total of 10 μl.

2. Mix the 10 μl of the premixture solution with 10 μl of 50% (resin volume) protein resin suspension equilibrated with the same buffer as the binding reaction (minus any Random72 dsDNA) in a 1.5 ml reaction tube. Incubate the reaction by standing at room temperature for 20 min with occasional tapping.
3. Add 1 ml of ice-cold Washing buffer into the reaction. Shake gently several times. Spin the reaction tube for 30 s at 500×*g* and remove the supernatant as much as possible.
4. Repeat this wash two times more, and rock the tube for 3 min at the third washing before centrifugation.
5. Add 300 μl of Elution buffer to the resin and rock the tube for 5 min at room temperature.
6. Spin the tube for 30 s at 500×*g* and transfer supernatant to a fresh tube. Repeat the elution with an additional 100 μl of Elution buffer. Combine the first (300 μl) and the second (100 μl) elutions.
7. Extract the resulting 400 μl of elute with 400 μl of phenol/chloroform/isoamylalcohol (50:49:1).
8. Precipitate the DNA by Ethanol-precipitation with co-precipitant, and dry the pellet.
9. Dissolve the DNA pellets in 10 μl of LoTE.
10. Amplify the eluted DNA (bound fraction) by two rounds of PCR using 1 μl out of the 10 μl as template. For the first round PCR use: 2 μl of 10× PCR buffer, 2 μl of 2.5 mM dNTP mix, 1 μl of SELEX DNA (total 10 μl), 2 μl of 10 μM Forward primer, 2 μl of 10 μM Reverse primer, 1 μl of *Taq* DNA polymerase, and 10 μl of water in total of 20 μl. Run 20 PCR cycles for 30 s at 94°C, 1 min at 68°C (62°C for "R" design), and 1 min at 72°C, followed by 7 min at 72°C (see Note 5).
11. For the second round of PCR, add 20 μl of the following supplement mix: 2 μl of 10× PCR buffer, 2 μl of 2.5 mM dNTP mix, 4 μl of 10 μM Forward primer, 4 μl of 10 μM Reverse primer, 1 μl of *Taq* DNA polymerase, and 6 μl of water. Run one PCR cycle for 5 min at 94°C, 5 min at 68°C (62°C for "R" design), and 9 min at 72°C.
12. Purify the PCR-amplified DNAs by QiaQuick Gel Extraction kit, and elute the DNA in 30 μl of LoTE.
13. Check the DNA by 15% PAGE. If the second round PCR is incomplete, a smear above the band (incomplete dsDNA) will be visible on the PAGE in addition of the correct dsDNA band (Fig. 3.3b). If this is the case, repeat the second round PCR. Adjust the DNA concentration to 10 ng/μl and go to next round of SELEX (i.e., the next DNA-binding reaction followed by PCR).

14. For the second and later rounds of SELEX, use 10 ng of the PCR-amplified DNA described above as the starting DNA probe pool, and repeat the same procedure from step 1 of this section.
15. To monitor the increasing amounts of bound DNA during SELEX rounds, use Q-PCR with a SYBR kit and bound fractions of successive SELEX rounds as templates (see Note 6).

3.4. Isolation, Cloning, and Sequencing of Large Numbers of Binding Sequence Units

This method was modified from a SAGE (serial analysis of gene expression) protocol available at http://www.sagenet.org/protocol/index.htm.

1. Amplify a large amount of SELEX-bound DNA for cloning by two round PCR using 5′-biotinylated (BTN) Forward and Reverse primers. Typical condition for the first round PCR is: 80 μl of 10× PCR buffer, 80 μl of 2.5 mM dNTP mix, 1 μl of SELEX PCR DNA (~10 ng/μl), 80 μl of 10 μM BTN-Forward primer, 80 μl of 10 μM BTN-Reverse primer, 8 μl of *Taq* DNA polymerase, and 480 μl of water in a total volume of 800 μl. Put 50 μl of each reaction in 16 PCR tubes. Run 15 PCR cycles for 30 s at 94°C, 1 min at 65°C (62°C for "R" design), 1 min at 72°C, followed by 7 min at 72°C (see Note 5).
2. Add 50 μl each of the following supplement mix: 80 μl of 10× PCR buffer, 80 μl of 2.5 mM dNTP mix, 160 μl of 10 μM BTN-Forward primer, 160 μl of 10 μM BTN-Reverse primer, 16 μl of *Taq* DNA polymerase, and 320 μl of water in a total volume of 800 μl. Run one PCR cycle for 5 min at 94°C, 5 min at 65°C (62°C for "R" design), and 9 min at 72°C.
3. Purify the PCR-amplified DNA (Biotinylated SELEX DNA) using a QiaQuick Gel Extraction kit with 4–8 Spin columns, and pool the elution in one tube.
4. Ensure that the DNA gives a single sharp 72 bp band by loading 1 μl of the sample on a 15% PAGE (Fig. 3c). Measure the sample DNA concentration by NanoDrop. From a 800 μl PCR, approximately 10 μg of Biotin-PCR DNA should be obtained.
5. Digest 10 μg of the Biotin-PCR DNA with the appropriate restriction enzyme (*Bam*HI, *Bgl*II, or *Hin*dIII) overnight in a 500 μl reaction at 37°C.
6. Add 20 μl of Streptavidin Sepharose equilibrated with the same restriction enzyme buffer to the reaction. Rock the suspension at room temperature for 15 min.
7. Transfer the suspension to an Ultrafree centrifugal filter unit, and centrifuge for 5 min to remove the beads. Take the pass-through fraction (Fig. 3c).

8. Extract the pass-through fraction with an equal volume of phenol/chloroform/isoamylalcohol (50:49:1), then ethanol-precipitate with a co-precipitant. This DNA is a mixture of 22 bp DNAs enriched for factor DNA-binding sequences.
9. Dissolve the DNA pellet in 10 μl of LoTE and check its purity by PAGE and amount by NanoDrop.
10. Mix 1 μg of the 22 bp DNAs with 1 μl of 10× ligation buffer and 1 μl T4 DNA Ligase in a total volume of 10 μl, and incubate at 16°C overnight to concatenate the 22 bp DNAs.
11. Extract the ligation reaction with an equal volume of phenol/chloroform/isoamylalcohol (50:49:1), then ethanol-precipitate the DNA with a co-precipitant.
12. Dissolve the concatemer DNA in 50 μl of water and perform an end filling-in reaction as follows: 50 μl of Concatemer DNA, 10 μl of 10× NEB2 buffer, 3 μl of 25 mM dNTPs, 3 μl of T4 DNA polymerase, 3 μl of Klenow fragment, and 30 μl of water in a total volume of 100 μl. Incubate the reaction at room temperature for 1 h.
13. Heat at 65°C for 5 min, then chill on ice.
14. Extract the reaction with an equal volume of phenol/chloroform/isoamylalcohol (50:49:1), then ethanol-precipitate the DNA with a co-precipitant.
15. Dissolve the DNA pellet in 10 μl of TE.
16. Load the sample on a 1.5% agarose gel. A DNA smear should be visible in the 0.1–2 kb size range.
17. Cut out the gel containing 0.5–2 kb DNA fragments.
18. Extract the DNA from the gel using a QiaQuick Gel Extraction kit in 100 μl elution buffer.
19. Extract the elution with an equal volume of phenol/chloroform/isoamylalcohol (50:49:1), then ethanol-precipitate the DNA with a co-precipitant. Rinse the DNA pellet with 80% ethanol and dry.
20. Resuspend the DNA in 10 μl of LoTE.
21. Ligate the resulting concatemer DNA with *Sma*I-digested pUC19 plasmid as follows: 1 μl of 10× ligation buffer, 0.25 μl of 5 ng *Sma*I-digested pUC19 (20 ng/μl), 5 μl of concatemer, 3 μl of water, 1 μl of T4 DNA Ligase in a total volume of 10 μl. Incubate at 16°C for more than 2 h.
22. Ethanol-precipitate the DNA with a co-precipitant. Dissolve the DNA pellet in 2 μl of water.
23. Use 1 μl of the DNA solution to electroporate competent cells and plate on an LB plate containing Ampicillin, X-gal, and IPTG (10).

24. Pick white colonies to a master LB plate containing Ampicillin, and resuspend a small amount of the picked colony cells in a PCR mix as follows: 2 μl of 10× PCR buffer, 2 μl of 2.5 mM dNTP mix, 1 μl of 10 μM M13 Universal primer, 1 μl of 10 μM M13 Reverse primer, 0.25 μl of *Taq* DNA polymerase, and 4 μl of water in a total 10 μl.
25. Run precycle PCR 5 min at 94°C, followed by 30 PCR cycles of 30 s at 94°C, 30 s at 54°C, 1 min at 72°C, followed by 7 min at 72°C.
26. Analyze the PCR products on a 1.5% agarose gel or a 10–15% PAGE.
27. Select the clones carrying inserts >220 bp and sequence them.

3.5. Data Analysis

This protocol results in DNA sequences of many bound oligonucleotides for each round of SELEX. One simple approach to analyze this data is to use MEME or a similar method to derive position weight matrices (PWMs) from each round separately (11, 12). This approach results in different models for factor specificity from each round, however. In addition, PWMs do not capture the non-additive effect that adjacent nucleotides have on the affinity of a protein for a DNA sequence (13). For these reasons, more sophisticated mathematical methods will be needed in future that use data from all rounds to estimate the affinities of each oligonucleotide (Atherton, Boley, Brown, Ogawa, Davidson, Eisen, Biggin and Bickel, unpublished data).

4. Notes

1. TALON immobilized metal affinity chromatography resin contains cobalt instead of Nickel and has a lower affinity to oligonucleotides than typical Ni-resins under low salt (50 mM NaCl) conditions, resulting in lower backgrounds. Magnet beads have a higher affinity to DNA.
2. All three designs have a 16 base random sequence in the center, which is located between a pair of six cutter restriction enzyme sites (either *Bam*HI, *Bgl*II, or *Hin*dIII). Flanking these are 20 base ends that are priming sequences for PCR amplification with Forward and Reverse primers. The flanking sequences of Random72 are from Pollard et al. (14). The sequence of Random72R is the same as the oligonucleotides described in Roulet et al. (8) except that its random sequence length is different.
3. The proper amount of protein on the bead and the concentration of poly(dIdC) in the binding reaction have to be determined

by a test titration before the experimental SELEX run. Make serial dilutions of renatured protein starting at 25 μg/250 μl and bind each to the same volume (250 μl) of TALON beads. If a typical DNA-binding sequence for a factor is available in the literature, use that as a test target sequence. Alternatively, use information for homologous proteins to define the test DNA. Incorporate these sequences into a probe DNA with same sequence as the Random oligoDNA except replace the random region with a test target binding site. Perform DNA-binding assays using this test probe or the Random oligoDNA and the dilution series protein-beads. Measure the amount of bound and eluted probes by Q-PCR or DNA labeling (biotin or radioactivity), comparing the amount of DNA present on the resin versus the input and wash fractions. Ideally, less than 1% of the random probe should be bound to the resin, while more than three times more test binding site probe should be bound than the Random oligoDNA probe.

4. The flanking DNA sequences of the Random oligoDNA probe may contain binding sequences of the transcription factor protein you wish to study. If you know the typical binding sequence of the factor, select one of the three sequences given which contain no similar sequence to the binding sequence. If the flanking sequences contain a binding site for the protein, a high background binding will be observed when the titration test described in Note 3 is performed.
5. The number of PCR amplification cycles should be limited to minimize possible PCR bias. One additional cycle of PCR using fresh materials after the usual PCR cycles helps make the proper double-stranded DNA from a template DNA containing semirandom sequences.
6. As an alternative method to monitor enrichment, amplify eluted DNA by a limited number of PCR cycles (5–15 cycles). Compare the amount of amplified DNA with amplified DNA from 10, 1, and 0.1% of the SELEX input DNA using PAGE. Then, estimate the percent of input DNA trapped and eluted from the protein resin based on DNA band intensities.

Acknowledgments

The authors would like to thank Michael B. Eisen, David Nix, Stuart Davidson, Juli Atherton, Nathan Boley, Ben Brown, and Peter Bickel for SELEX data analysis. Dr. Mark Stapleton for cloning of recombinant transcription factor genes, and Lucy Zeng for performing many SELEX experiments. This work was supported by the US National Institutes of Health (NIH) under grant

GM704403. Work at Lawrence Berkeley National Laboratory was conducted under Department of Energy contract DE-AC02-05CH11231.

References

1. Carey, M. E., and Smale, S. (1999) *Transcriptional Regulation in Eukaryotes: Concepts, Strategies, and Techniques.*, Cold Spring Harbor Laboratory Press, Cold Spring Harbor, New York.
2. Bradley, R. K., Li, X. Y., Trapnell, C., Davidson, S., Pachter, L., Chu, H. C., Tonkin, L. A., Biggin, M. D., and Eisen, M. B. (2010) Binding site turnover produces pervasive quantitative changes in transcription factor binding between closely related Drosophila species, *PLoS biology* *8*, e1000343.
3. MacArthur, S., Li, X. Y., Li, J., Brown, J. B., Chu, H. C., Zeng, L., Grondona, B. P., Hechmer, A., Simirenko, L., Keranen, S. V., Knowles, D. W., Stapleton, M., Bickel, P., Biggin, M. D., and Eisen, M. B. (2009) Developmental roles of 21 Drosophila transcription factors are determined by quantitative differences in binding to an overlapping set of thousands of genomic regions, *Genome biology* ***10***, R80.
4. Visel, A., Blow, M. J., Li, Z., Zhang T., Akiyama, J. A., Holt, A., Plajzer-Frick, I., Shoukry, M., Wright, C., Chen, F., Afzal, V., Ren, B., Rubin, E. M., and Pennacchio, L. A. (2009) ChIP-seq accurately predicts tissue-specific activity of enhancers, *Nature 457*, 854–858.
5. Biggin, M. D. (2010) MyoD, a lesson in widespread DNA binding, *Dev Cell. 18*, 505–506.
6. Berman, B. P., Pfeiffer, B. D., Laverty, T. R., Salzberg, S. L., Rubin, G. M., Eisen, M. B., and Celniker, S. E. (2004) Computational identification of developmental enhancers: conservation and function of transcription factor binding-site clusters in *Drosophila melanogaster* and *Drosophila pseudoobscura*, *Genome biology 5*, R61.
7. Gold, L. (1995) Oligonucleotides as research, diagnostic and therapeutic agents, *J. Biol. Chem. 270*, 13581–13584.
8. Roulet, E., Busso, S., Camargo, A. A., Simpson, A. J., Mermod, N., and Bucher, P. (2002) High-throughput SELEX SAGE method for quantitative modeling of transcription-factor binding sites, *Nat Biotechnol* ***20***, 831–835.
9. Terpe, K. (2003) Overview of tag protein fusions: from molecular and biochemical fundamentals to commercial systems, *Applied Microbiol and Biotech* ***60***, 523–533.
10. Sambrook, J., and Russell, D. W. (2001) *Molecular Cloning: A Laboratory Manual*, Cold Spring Harbor Laboratory, Cold Spring Harbor, New York.
11. Bailey, T. L., Boden, M., Buske, F. A., Frith, M., Grant, C. E., Clementi, L., Ren, J., Li, W. W., and Noble, W. S. (2009) MEME SUITE: tools for motif discovery and searching, *Nucleic acids research 37*, W202–208.
12. Bailey, T. L., Williams, N., Misleh, C., and Li, W. W. (2006) MEME: discovering and analyzing DNA and protein sequence motifs, *Nucleic Acids Res. 34*, 369–373.
13. Benos, P. V., Bulyk, M. L., and Stormo, G. D. (2002) Additivity in protein-DNA interactions: how good an approximation is it? *Nucleic Acids Res. 30*, 4442–4451.
14. Pollard, J., Bell, S. D., and Ellington, A. D. (2000) Design, Synthesis, and Amplification of DNA Pools for Construction of Combinatorial Pools and Libraries, In *Curr. Protoc. Mol. Biol*, pp 24.22.21–24.22.24., John Wiley & Sons, inc.

Reference URLs

BDTNP: http://bdtnp.lbl.gov/Fly-Net/

SAGE protocol: http://www.sagenet.org/protocol/index.htm

Chapter 4

Convenient Determination of Protein-Binding DNA Sequences Using Quadruple 9-Mer-Based Microarray and DsRed-Monomer Fusion Protein

Min-Jeong Kim, Pil Joong Chung, Tae-Ho Lee, Tae-Hoon Kim, Baek Hie Nahm, and Yeon-Ki Kim

Abstract

Protein-binding DNA microarray (PBM) is one of the high-throughput methods to define DNA sequences which potentially bind to a given DNA-binding protein. Quadruple 9-mer-based protein-binding DNA microarray, named Q9-PBM, is designed in such a way that target probes are synthesized as quadruples of all possible 9-mer combinations. Also, recombinant proteins fused with DsRed-monomer fluorescent protein are conveniently constructed. Q9-PBM confirms the well-known DNA-binding sequences of Cbf1 and CBF1/DREB1B transcription factors, and also identifies the adjacent sequences. Moreover, Q9-PBM is applied to elucidate the unidentified *cis*-acting element of the OsNAC6 rice transcription factor. This technology will facilitate greater understanding of genome-wide interactions between proteins and DNA.

Key words: Transcription factor, Protein-binding DNA microarray, DsRed-monomer, *cis*-acting element, Q9-PBM

1. Introduction

Determination of DNA sequences which interact with specific proteins is a critical aspect to understand the related biological process. Especially, transcription factors bind to specific DNA sequences (*cis*-acting elements) of target genes and thereby can modulate their transcriptional rate (1). Protein-DNA binding properties have been investigated by several procedures (2, 3) and, more recently, by high-throughput methods (4, 5). One of these high-throughput methods, the universal protein-binding microarray (PBM) can be used to determine the DNA-binding motif of a protein in a genome-wide manner (Fig. 1) (6, 7).

Bart Deplancke and Nele Gheldof (eds.), *Gene Regulatory Networks: Methods and Protocols*,
Methods in Molecular Biology, vol. 786, DOI 10.1007/978-1-61779-292-2_4, © Springer Science+Business Media, LLC 2012

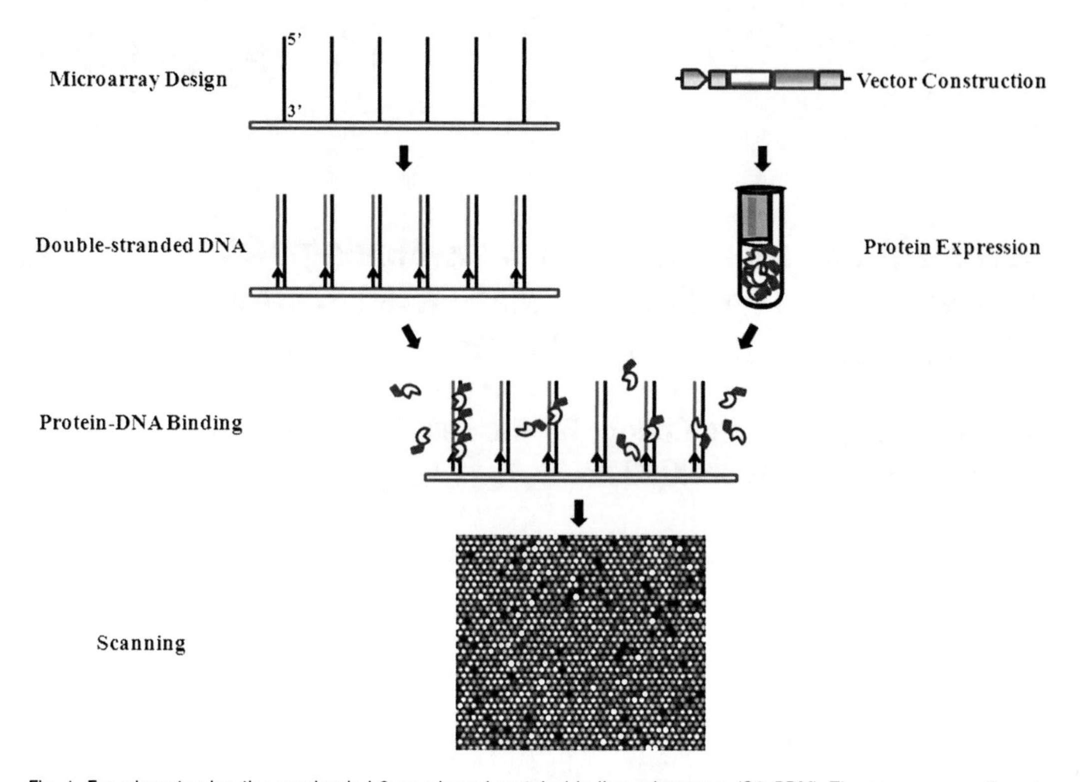

Fig. 1. Experiment using the quadrupled 9-mer-based protein-binding microarray (Q9-PBM). The reverse complimentary strand is synthesized on a slide by thermo-stable DNA polymerase. Separately, protein is prepared with a fusion to the DsRed-monomer fluorescent protein. Purified protein is incubated with the double-stranded microarray. The consensus sequence can be determined from the fluorescence intensity of the spot. The presented sequence logo is the identified sequence of the Cbf1 transcription factor.

Here, we describe a method that uses PBM technology facilitated by using a DsRed fluorescent protein and a concatenated sequence of oligonucleotides (8, 9). Target probes are synthesized as quadruples of all possible 9-mer combinations, creating the so-called Q9-PBM and permitting unequivocal interpretation of the *cis*-acting elements. Q9-PBM can be conveniently performed and it may increase the binding interaction between protein and DNA by adopting quadruple 9-mers compared to single sequences. Also, DsRed fusion makes it possible to shorten the wash and hybridization procedures by eliminating incubation with an antibody against the bound proteins.

2 Materials

2.1. Expression of DsRed-Monomer Fusion Protein

1. pDsRed-monomer vector (Clontech).
2. pET-32a(+) protein expression vector (Novagen).
3. Gene-specific oligonucleotides (Bioneer; see Table 1).

Table 1
Oligonucleotide sequences used for cloning

Name	Sequence
DsRed_F-*Eco*RI	5′-**GAC AAA** GAA TTC ATG GAC AAC ACC GAG GAC GT-3′
DsRed_R-*Sal*I	5′-**AGC TGA** GTC GAC CTA CTG GGA GCC GGA GTG GC-3′
Cbfl_F-*Sac*I	5′-**GAC AAA** GAC CTC ATG AAC TCT CTG GCA AAT AA-3′
Cbfl_R-*Xho*I	5′-**AGC TGA** CTC GAG TCA AGC CTC ATG TGG ATT AT-3′
CBF1/DREB1B_F-*Sac*I	5′-**GAC AAA** GAG CTC ATG AAC TCA TTT TCA GCT-3′
CBF1/DREB1B_R-*Xho*I	5′-**AGC TGA** CTC GAG TTA GTA ACT CCA AAG CGA CA-3′
OsNAC6_F_*Eco*RI	5′-**GAC AAA** GAA TTC ATG AGC GGC GGT GAG GAC CT-3′
OsNAC6_R_*Eco*RI	5′-**AGC TGA** GAA TTC CTA GAA TGG CTT GCC CCA GT-3′

The restriction enzyme sites added are underlined. The additional nucleotides to facilitate enzyme digestion are bolded

4. QIAquick PCR purification kit (Qiagen).
5. DH5α (Invitrogen) and BL21-CodonPlus(DE3)-RIPL (Stratagene) *Escherichia coli* competent cells.
6. Luria-Bertani (LB) culture medium supplemented with ampicillin (50 μg/ml), chloramphenicol (34 μg/ml), streptomycin (75 μg/ml), and 1 mM IPTG.
7. Solid medium was made with 1.5% agar.
8. *Eco*RI, *Sac*I, *Xho*I, and *Sal*I restriction enzymes (Takara).
9. T4 DNA ligase and Taq DNA polymerase (Takara).
10. Complete, EDTA-free protease inhibitor cocktail tablets (Roche Applied Science): dissolve one tablet in 50 ml of phosphate-buffered saline (PBS) buffer just before use and store on ice.
11. TALON His-Tag purification resin adapted with IMAC (Clontech).
12. Equilibration/wash buffer: 50 mM sodium phosphate, 300 mM NaCl, pH 7.0.
13. Elution buffer: 50 mM sodium phosphate, 300 mM NaCl, 100 mM imidazole, pH 7.0.
14. Poly-prep chromatography columns (Bio-Rad): 9 cm high, empty polypropylene and 2 ml bed volume.
15. Slide-A-Lyzer Dialysis Cassettes, 20K MWCO, 3 ml (Thermo Scientific).
16. VCX130 sonicator (Sonics & Materials Inc.).

2.2. Primer Extension of Protein-Binding DNA Microarray

1. Quadruple 9-mer-based protein-binding DNA microarray (Q9-PBM) containing quadrupled all possible combination of 9-mer oligonucleotides and priming sequence (5′-CGG AGT CAC CTA GTG CAG-3′) (Agilent Technology).
2. 2.5 mM dNTPs (Takara).
3. Cy5-dUTP (GE Healthcare): prepare 0.1 mM stock and store at −20°C.
4. 10 μM Extension primer (5′-CTG CAC TAG GTG ACT CCG-3′, Bioneer).
5. Thermo Sequenase cycle sequencing kit (USB Corp.).
6. SureHyb hybridization chamber and SureHyb gasket slide (Agilent Technology).
7. Microarray wash buffer I: 1× PBS, 0.01% Triton X-100.
8. Microarray wash buffer II: 1× PBS, 0.1% Tween 20.

2.3. Protein-Binding Microarraying

1. 200 nM DsRed fusion protein mixture supplemented with incubation buffer (1× PBS, 2% bovine serum albumin (BSA, Amresco)) and 50 ng/μl salmon-testes DNA (Sigma).
2. Microarray wash buffer I.
3. Microarray wash buffer II.
4. Microarray wash buffer III: 1× PBS, 0.5% Tween 20.

2.4. Electrophoretic Mobility Shift Assay

1. 5′-Biotin-oligonucleotides (see Note 1).
2. LightShift chemiluminescent EMSA kit (Thermo Scientific). This kit includes all buffers for electrophoretic mobility shift assay (EMSA).
3. 40% ACRYL/BIS 29:1 solution (Amresco).
4. Tris-borate-EDTA (TBE) buffer.
5. *N*,*N*,*N*′,*N*′-Tetramethylethylenediamine (TEMED, Sigma).
6. Ammonium persulfate (APS, Sigma).
7. Hybond-N + membrane (GE Healthcare).
8. BLX-254 UV crossliker (Vilber Lourmat).
9. TE 70 semidry gel transfer unit (GE Healthcare Bio-science).
10. Las3000 CCD (charge-coupled device) camera (Fujifilm).

3. Methods (see Note 2)

Proteins are expressed with an N-terminal or C-terminal fusion to the polyhistidine-tag and DsRed-monomer fluorescent protein. The DsRed protein is cloned from *Discosoma* sp. based on homology with the green fluorescent protein (GFP). DsRed-monomer possesses a similar spectrum (excitation: 556 nm; emission: 586 nm) to Cy3 that is compatible with the microarray scanner. The DsRed-monomer coding sequence is transferred from pDsRed-monomer vector to pET-32a(+) expression vector and named pET-DsRed expression vector. The full-length cDNA sequences of the transcription factors are cloned into the pET-DsRed expression vector. Separately, the reverse complimentary strand is synthesized on a slide by thermo-stable DNA polymerase. A small quantity of Cy5 fluorescent dUTP is incorporated to confirm successful elongation and also to identify spot positions. Purified DsRed fusion protein is incubated with the double-stranded microarray. The consensus sequence is determined from the fluorescence intensity of the spot without any further step like antibody labeling. The identified consensus is finally confirmed by an EMSA.

Sometimes, the position of DsRed-monomer against transcription factors affects the binding affinity of the protein. This position effect might be verified by tagging DsRed to the other side of the target protein. Also, the activities of some proteins are temperature-sensitive. Therefore, the purified proteins have to be stored on ice and used for binding assay as soon as possible. In addition, some proteins require specific ions such as Zn^{2+} and Ni^{2+} to stabilize their functional structures. In this case, it is needed to supply ions in a medium during the protein expression step.

3.1. Design of the Q9-Protein-Binding Microarray

1. We design a PBM, which we refer to as Q9-PBM, in such a way that target probes are synthesized as quadruples of all possible 9-mer combinations. A total of 131,072 features are selected after consideration of the reverse complimentary sequences of all 9-mer combinations, and 101,073 features are replicated to confirm the binding consistency. Each 9-mer is quadrupled and linked to an extension primer binding site following five thymidine linkers to the slide. Q9-PBM is designed by GreenGene Biotech.

3.2. Preparation of DsRed-Monomer Fusion Protein

1. The coding sequence of the DsRed-monomer fluorescent protein is amplified from the pDsRed-monomer vector by PCR using DsRed_F-*Eco*RI and DsRed_R-*Sal*I gene-specific primers followed by digesting with *Eco*RI and *Sal*I restriction enzymes, and, separately, pET-32a(+) expression vector is digested with the same restriction enzymes.
2. Digested fragments of PCR product and pET-32a(+) are purified using QIAquick PCR purification kit, respectively, and ligated together. This recombinant vector is transformed to DH5α *E. coli* strain competent cells and named pET32-DsRed (see Note 3).

3. Full-length *Cbf1* (GenBank accession number NM_001181718) is amplified by PCR from *Saccharomyces cerevisiae* genomic DNA using Cbf1_F-*Sac*I and Cbf1_R-*Xho*I gene-specific primers. Full-length *CBF1/DREB1B* (NM_118681) is amplified from *Arabidopsis thaliana* genomic DNA using CBF1/DREB1B_F-*Sac*I and CBF1/DREB1B_R-*Xho*I gene-specific primers. Also, *OsNAC6* (NM_001051551) is amplified from cDNA of *Oryza sativa* using OsNAC6_F_*Eco*RI and OsNAC6_R_*Eco*RI gene-specific primers.
4. The amplified DNA fragments and pET32-DsRed vectors are digested with the specified restriction enzymes in the primer names, respectively. Each PCR fragments are ligated with the corresponding pET32-DsRed recombinant vector and transformed to DH5α *E. coli.* strain competent cell. All clones are sequenced to verify the absence of mutations in the DNA-binding domains.
5. The proteins are expressed in *E. coli* strain BL21-CodonPlus. 2 ml of the overnight cultured cells are inoculated in 100 ml of fresh liquid LB medium (see Note 3), grown at 37°C to an OD_{600} of 0.6 and induced with 1 mM isopropyl β-D-1-thiogalactopyranoside (IPTG) at 25°C for 5 h.
6. Cell pellets are obtained by centrifugation at 4°C for 5 min at 5,000 × *g* and after removal of the supernatant.
7. The pellets are resuspended in 10 ml of cold PBS buffer including a protease inhibitor cocktail.
8. Cell pellets are collected by centrifugation, resuspended in 5 ml of cold PBS buffer containing a protease inhibitor cocktail and sonicated until lysis for 5 min at 45 s intervals with 60% power on ice using a sonicator. The supernatant is retained after centrifugation at 4°C for 30 min at 9,000 × *g* (see Note 4).
9. Recombinant proteins are enriched using TALON resins adapted with immobilized metal affinity chromatography (IMAC) (see Note 5). 2 ml of the TALON resin is transferred to a culture tube and centrifuged at room temperature for 2 min at 700 × *g*.
10. After removing the supernatant, 10 ml of Equilibration/Wash buffer is added to the resin and mixed well by inverting. The tube is centrifuged at 700 × *g* for 2 min to pellet the resin and the supernatant is removed. Repeat this washing step once again.
11. Add 4.5 ml of the protein sample to the resin and gently agitate at room temperature for 20 min at 60 rpm on a rotary shaker (see Note 6). To remove the unbound soluble fraction, the suspension is centrifuged at 700 × *g* for 5 min at room temperature and the supernatant is discarded.

12. The resin is washed by adding 10 ml of Equilibration/Wash buffer and agitated at room temperature for 20 min at 60 rpm on a rotary shaker. The tube is centrifuged at 700 × *g* for 5 min to pellet the resin and the supernatant is removed. Repeat this washing step once again.
13. 1 ml of Equilibration/Wash buffer is added to the resin and mixed well by pipetting. The suspension is transferred to a chromatography column and allowed to settle down. Then, the buffer is allowed to drain until it reaches the top of the resin bed. 5 ml of Equilibration/Wash buffer is added to wash the resin.
14. The protein is eluted by adding 5 ml of Elution buffer. The protein fractions are collected in a volume of 500 μl in 1.5 ml tubes and the protein concentrations are then determined using the Bradford protein assay method.
15. The final protein is verified by SDS-polyacrylamide gel electrophoresis (SDS-PAGE) and dialyzed for 12–16 h to equilibrate with PBS using Slide-A-Lyzer dialysis cassette (see Note 7).
16. Remove the protein from the cassette and store at 4°C until binding step.

3.3. Primer Extension of Protein-Binding DNA Microarray

1. Reaction solution containing 40 μM dNTPs, 1 μM Cy5-dUTP, 1 μM extension primer, 1× Thermo Sequenase buffer, and 40 U Thermo Sequenase is prepared in a volume of 800 μl (see Note 8).
2. A custom-designed protein-binding DNA microarray (Q9-PBM) (as described in Subheading 3.1) is combined with the reaction solution in a SureHyb hybridization chamber. A SureHyb gasket slide is placed face up on the base of the chamber, 780 μl of reaction mixture is applied onto the center of the cover slide, the microarray is lowered face down onto the gasket slide and the hybridization chamber is fastened to seal in the liquid.
3. The assembled hybridization chamber is incubated at 85°C for 10 min to denature oligonucleotides and then at 60°C for 90 min for extension reaction.
4. The microarray is washed in microarray wash buffer I at 37°C for 1 min, microarray wash buffer I once again at 37°C for 10 min, PBS at room temperature for 3 min, and the slide was dried by centrifugation at 500 × *g* for 2 min (see Note 9).
5. The double-stranded microarray is scanned to verify successful synthesis.
6. The double-stranded microarray is rehydrated with microarray wash buffer I for 5 min at room temperature and blocked with 10 ml of incubation buffer for 1 h at room temperature.

After blocking, the microarray is washed once with microarray wash buffer II for 5 min and once with microarray wash buffer I for 2 min.

3.4. Protein-Binding Microarraying

1. During the double-stranded microarray blocking step, prepare a protein-binding mixture containing 200 nM DsRed-monomer fusion protein and 50 ng/μl salmon-testes DNA in 10 ml of incubation buffer.
2. The prepared protein mixture is incubated at 25°C for 1 h for stabilization and combined with the double-stranded microarray at 25°C for 1 h (see Note 10).
3. After the protein-binding reaction, the microarray is washed once with microarray wash buffer III for 10 min, microarray wash buffer II for 2 min, and then PBS for 2 min.
4. Fluorescence image is obtained with a 4000B microarray scanner. Each microarray is scanned three to five times at full laser power intensity and pixel resolution 5 μm. In order to get 0.01–0.05% (20–100 spots) of Cy3 saturated spots, different photomultiplier tube (PMT) gain settings are applied ranging from 550 to 780 for Cy3 intensity. The microarray has to be rescanned whenever the number of saturated spots is not in this range. However, use maximum Cy5 PMT gain setting to identify the spot positions.
5. The fluorescence is quantified and bad spots are excluded automatically using GenePix Pro version 5.1 software.
6. Using GenePix Pro version 5.1 software, the background-subtracted probe signal intensities are computed for all scanned image files. This requires a GenePix Array List file containing information regarding the coordinates and median pixel identities of all spots within each subgrid which is supplied by Agilent with each microarray. Each block of spots are aligned over the corresponding subgrid and automatically flag problematic spots as "bad" (i.e., spots with obvious scratches, dust flecks, and so on). The intensities for each subgrid are saved as a separate GenePix results (GPR) file.

3.5. Determination of Binding DNA Motifs

The 29,999 single and the 101,073 replicated features are subjected to the motif extraction. For the replicated ones, those of which intensity differences are less than 40,000 are chosen for subsequent analysis. The rank-ordered signal distribution demonstrates a deep leftward slope followed by a heavy right tail as shown in Fig. 2 and as the one observed by Berger et al. (10). The 9-mers in the deep leftward region are used to extract motifs. Binding motif can be formatted at http://www.ggbio.com/onlogo/.

1. Input files are the GPR file described above. The file must be obtained by setting the channels as the two wave lengths of

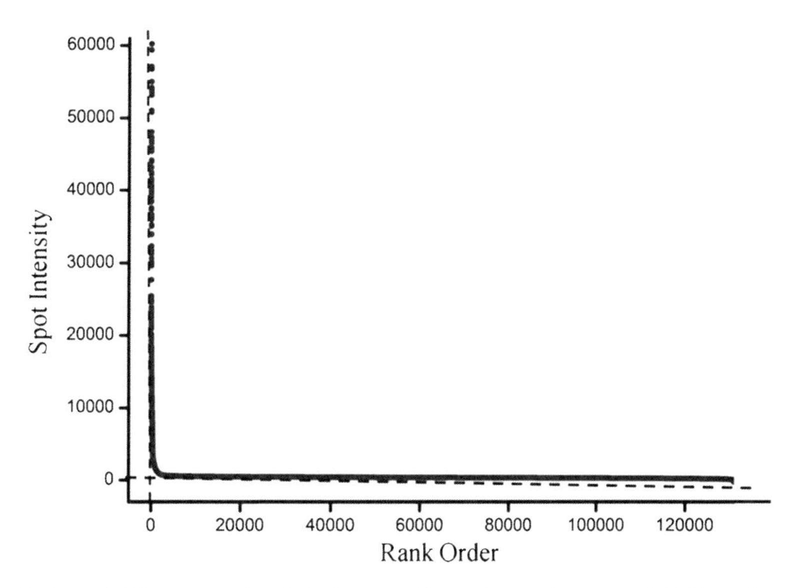

Fig. 2. DsRed fluorescent signal distribution of the spot intensity from Cbf1 PBM results. The rank-ordered signal distribution of the Cbf1 PBM shows a deep leftward slope followed by a heavy right tail because the signal distribution is due to a specific interaction between the protein and features on the microarray.

635 and 532 nm. The GPR file contains 56 column files. The 47th column, background-subtracted intensities (F532 Median − B532) is used in subsequent analysis.

2. A design format file also has to be uploaded. It has three columns of SEQ_ID, SEQ_NAME, and SEQUENCES.

3. Finally, a match score to align with the seed (a spot with the highest intensity among a cluster) and the other 9-mers sequences is provided. By setting the match score as 7 (and recommended), it takes several minutes.

4. As the probes in the deep slope region differed by only one base, we assume that the signal distribution is due to a specific interaction between the protein and features on the microarray. Two independent linear models, $y = ax + b$, are applied in the deep left and the heavy right tail region using the R statistical language (Fig. 2). In one of the examples with OsNAC6, the slope and y-axis intercept of the deep slope are −68.3 and 53,074.8, respectively; those of the heavy tail are 0.0207 and 2,283, respectively. These values provide an extrapolated rank of 745 for OsNAC6. The preferred elements are extracted using the following two steps. First, feature sequences from ranks 1 to 745 are grouped using Perl script language. Any sequence that possessed at least six bases is matched; among these, at least three bases are contiguously matched to the highest one. These groups are ranked according to the highest sequence. At this step, G-rich sequences are omitted.

CBF1 (*S. cerevisiae;* CACGTG)

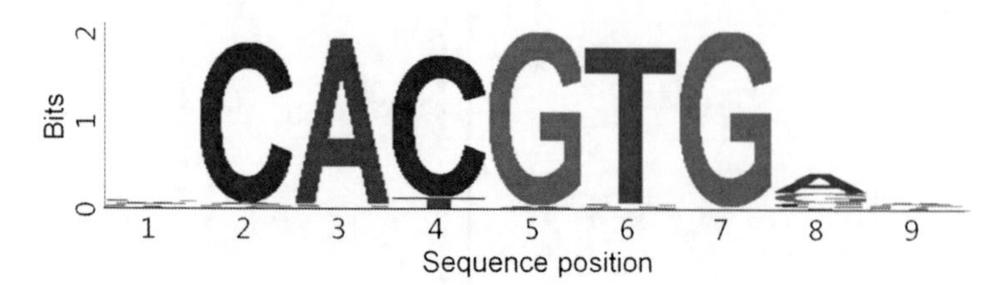

CBF1/DREB1B (*A. thaliana;* CCGAC)

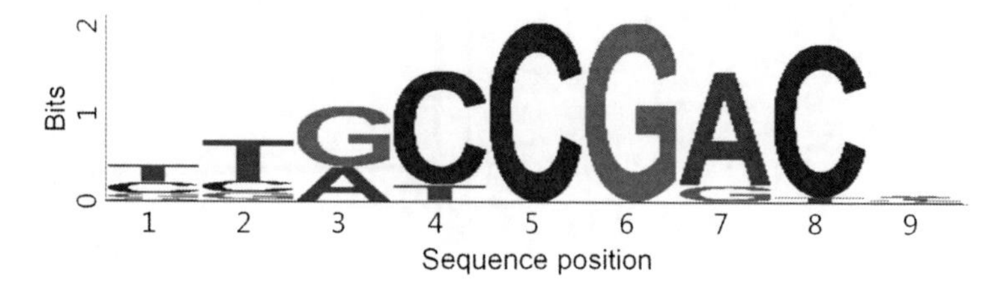

OsNAC6 (*O. sativa;* unkown)

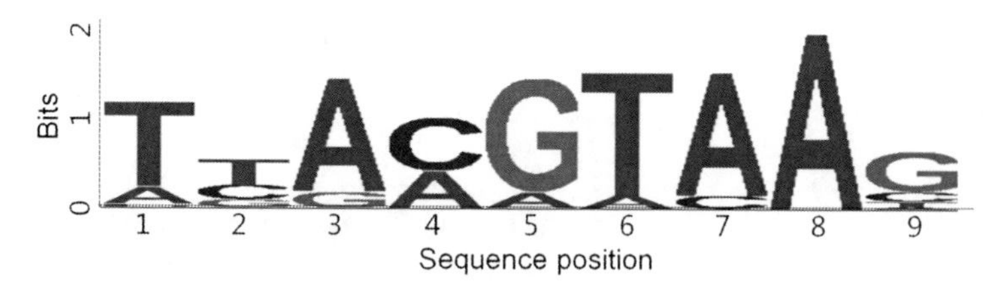

Fig. 3. The determined consensus binding sequences according to the PBM results that bind robustly to each transcription factor. To determine binding motifs, two independent linear models are applied in the deep and the heavy right tail region using the R statistical language as described in Subheading 3.5. Organisms and previously identified consensus sequences are denoted in *parentheses*.

5. The Web site gives an intensity profiling figure, a sequence logo, and its related statistics. *P*-values from Wilcoxon–Mann–Whitney test and position weigh matrix are given. Several sequence logos are shown in Fig. 3.

3.6. Confirmation of Binding Sequence using EMSA

1. Prepare a nondenaturing 6% polyacrylamide gel: 1,500 μl of 40% ACRYL/BIS 29:1 solution, 500 μl of 10× TBE buffer, 10 μl of TEMED, 120 μl of 10% APS, and 7,870 μl of distilled water.
2. Incubate 2 μg of protein with 20 fmol of 5′-biotin-labeled double-stranded oligonucleotides, 1 μg of poly dI-dC, 1× binding buffer, 2.5% glycerol, and 0.05% NP40 in a 20 μl reaction volume for 30 min at room temperature (see Note 1).
3. Prerun the gel with 0.5× TBE buffer at 50 V for 30 min. The reaction mixture is then analyzed by electrophoresis at 50 V for 1 h 30 min.

4. The DNA–protein complexes in the gel are then transferred to Hybond-N+membrane by electrophoretic transfer in 0.5× TBE at 380 mA for 30 min and cross-linked at 120 mJ/cm^2 using a UV-light cross-linker.
5. Biotin is detected using the Lightshift™ chemiluminescent EMSA kit. The membrane is soaked in 20 ml of blocking buffer and incubated for 15 min with gentle shaking. The membrane is transferred to conjugate/blocking buffer containing stabilized streptavidin-horseradish peroxidase conjugate in 20 ml of blocking buffer and incubated for 15 min with gentle shaking. The membrane is washed four times in 20 ml of wash solution for 5 min and then submerged in 30 ml substrate equilibration buffer for 5 min.
6. After removing from the equilibration buffer, the working solution containing 50% luminal/enhancer solution and 50% stable peroxide solution is applied to the membrane for 5 min. Finally, the membrane is exposed using LAS-3000 to acquire CCD images (Fig. 4).

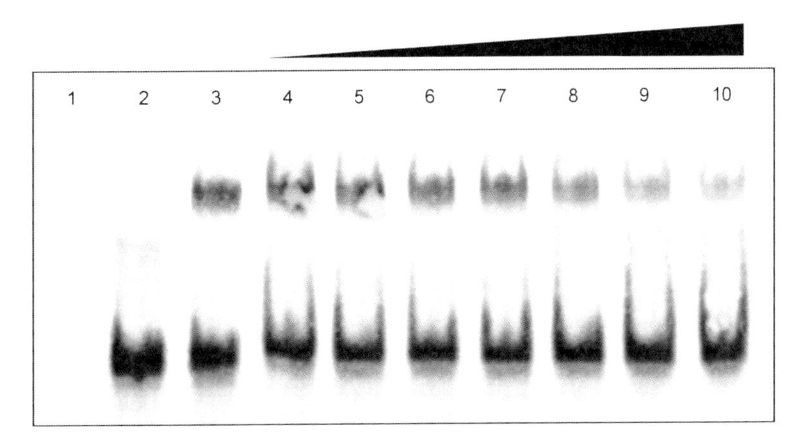

Fig. 4. EMSA-based competition analysis of OsNAC6 with "A[A/C]GTAA" motif. Candidate duplexes (5′-ACG TAA GTT ACG TAA GTT ACG TAA GTT ACG TAA GTT-3′; 5′-AAC TTA CGT AAC TTA CGT AAC TTA CGT AAC TTA CGT-3′) containing "A[A/C]GTAA" is used as core binding sequence for the interaction. The mobility shifts are performed using 2 μg of purified OsNAC6 protein and 20 fmol of biotinylated oligonucleotides. The shift is assessed by competition using unlabeled oligonucleotides. The molar excess of competitor used is over 100-fold. *Lane 1*, OsNac6; *Lane 2*, biotin-labeled DNA; *Lane 3*, OsNAC6 + biotin-labeled DNA; *Lane 4*, OsNAC6 + biotin-labeled DNA + unlabeled DNA (2 pmol); *Lane 5*, OsNAC6 + biotin-labeled DNA + unlabeled DNA (3 pmol); *Lane 6*, OsNAC6 + biotin-labeled DNA + unlabeled DNA (4 pmol); *Lane 7*, OsNAC6 + biotin-labeled DNA + unlabeled DNA (5 pmol); *Lane 8*, OsNAC6 + biotin-labeled DNA + unlabeled DNA (20 pmol); *Lane 9*, OsNAC6 + biotin-labeled DNA + unlabeled DNA (40 pmol); *Lane 10*, OsNAC6 + biotin-labeled DNA + unlabeled DNA (60 pmol).

4. Notes

1. Oligonucleotides have to be dissolved in TE (pH 8.0) buffer to be stabilized. For a competition assay, the excess amount of unlabeled double-stranded oligonucleotides can be used depending on the binding affinity of a protein. We use different amount of unlabeled double-stranded oligonucleotides as described in the legend of Fig. 4.
2. CBF1, CBF1/DREB1B, and OsNAC6 transcription factors used in this study are expressed with an N-terminal fusion to a polyhistidine-tag and DsRed-monomer fluorescent protein (11–13). Gene-specific primers should be carefully designed according to reading frame.
3. Culture medium has to be supplemented with the appropriate concentration of ampicillin, chloramphenicol, and streptomycin antibiotics as decribed in Subheading 2.1: pET-32a(+) carries an Amp^R gene, and BL21-CodonPlus(DE3)-RIPL cells are streptomycin, tetracycline, and chloramphenicol resistant.
4. DsRed protein is light sensitive. So, store DsRed protein in a dark place.
5. IMAC is one of the standard methods to purify histidine-tagged protein. Other methods can be applied in this step, if applicable.
6. Save and store 0.5 ml of the crude protein sample on ice to confirm purification efficiency by SDS-PAGE.
7. To increase dialysis efficiency, remove air from the dialysis cassette using a syringe after protein sample injection.
8. Cy5-dUTP is light sensitive. So, keep the solution in a dark place if it contains Cy5-dUTP.
9. 50 ml Conical tubes can be used for microarray washing steps. To avoid the microarray being dried, washing solutions have to be prepared prior to opening the hybridization chamber and the slide has to be transferred to the solutions as quickly as possible.
10. Make sure not to dry the microarrays after the washing steps as this would result in high background intensity. Transfer the microarray to the prepared protein-binding mixture from microarray wash buffer I. This binding step can be done in any type of container to soak a microarray slide in binding solution. We have used a plastic container (26 mm × 8 mm × 15 mm) with a lid purchased from a local vendor.

Acknowledgments

This work was supported by the Crop Functional Genomics Center of the Frontier Research Program, funded by the Ministry of Science and Technology (B.H.N., grant no, CG1122) and by the BioGreen21 Program (Y.-K.K., grant no. 20070401034008; B.H.N., grant no. 20090101060028), funded by RDA of the Republic of Korea. Also, B.H.N. was supported by BK21.

References

1. Ptashne, M. (1988) How eukaryotic transcriptional activators work, *Nature* **335**, 683–689.
2. Garner, M. M., and Revzin, A. (1981) A gel electrophoresis method for quantifying the binding of proteins to specific DNA regions: application to components of the *Escherichia coli* lactose operon regulatory system, *Nucleic Acids Res* **9**, 3047–3060.
3. Soderman, K., and Reichard, P. (1986) A nitrocellulose filter binding assay for ribonucleotide reductase, *Anal Biochem* **152**, 89–93.
4. Lieb, J. D., Liu, X., Botstein, D., and Brown, P. O. (2001) Promoter-specific binding of Rap1 revealed by genome-wide maps of protein-DNA association, *Nat Genet* **28**, 327–334.
5. van Steensel, B., Delrow, J., and Henikoff, S. (2001) Chromatin profiling using targeted DNA adenine methyltransferase, *Nat Genet* **27**, 304–308.
6. Bulyk, M. L., Gentalen, E., Lockhart, D. J., and Church, G. M. (1999) Quantifying DNA-protein interactions by double-stranded DNA arrays, *Nat Biotechnol* **17**, 573–577.
7. Zhu, C., Byers, K. J., McCord, R. P., Shi, Z., Berger, M. F., Newburger, D. E., Saulrieta, K., Smith, Z., Shah, M. V., Radhakrishnan, M., Philippakis, A. A., Hu, Y., De Masi, F., Pacek, M., Rolfs, A., Murthy, T., Labaer, J., and Bulyk, M. L. (2009) High-resolution DNA-binding specificity analysis of yeast transcription factors, *Genome Res* **19**, 556–566.
8. Kim, M. J., Lee, T. H., Pahk, Y. M., Kim, Y. H., Park, H. M., Choi, Y. D., Nahm, B. H., and Kim, Y. K. (2009) Quadruple 9-mer-based protein binding microarray with DsRed fusion protein, *BMC Mol Biol* **10**, 91.
9. Matz, M. V., Fradkov, A. F., Labas, Y. A., Savitsky, A. P., Zaraisky, A. G., Markelov, M. L., and Lukyanov, S. A. (1999) Fluorescent proteins from nonbioluminescent Anthozoa species, *Nat Biotechnol* **17**, 969–973.
10. Berger, M. F., Philippakis, A. A., Qureshi, A. M., He, F. S., Estep, P. W., 3 rd, and Bulyk, M. L. (2006) Compact, universal DNA microarrays to comprehensively determine transcription-factor binding site specificities, *Nat Biotechnol* **24**, 1429–1435.
11. Kent, N. A., Eibert, S. M., and Mellor, J. (2004) Cbf1p is required for chromatin remodeling at promoter-proximal CACGTG motifs in yeast, *J Biol Chem* **279**, 27116–27123.
12. Novillo, F., Alonso, J. M., Ecker, J. R., and Salinas, J. (2004) CBF2/DREB1C is a negative regulator of CBF1/DREB1B and CBF3/DREB1A expression and plays a central role in stress tolerance in Arabidopsis, *Proc Natl Acad Sci USA* **101**, 3985–3990.
13. Nakashima, K., Tran, L. S., Van Nguyen, D., Fujita, M., Maruyama, K., Todaka, D., Ito, Y., Hayashi, N., Shinozaki, K., and Yamaguchi-Shinozaki, K. (2007) Functional analysis of a NAC-type transcription factor OsNAC6 involved in abiotic and biotic stress-responsive gene expression in rice, *Plant J* **51**, 617–630.

Chapter 5

Analysis of Specific Protein–DNA Interactions by Bacterial One-Hybrid Assay

Marcus B. Noyes

Abstract

The DNA-binding specificity of transcription factors allows the prediction of regulatory targets in a genome. However, very few factor specificities have been characterized and still too little is known about how these proteins interact with their targets to make predictions a priori. To provide a greater understanding of how proteins and DNA interact, we have developed a bacterial one-hybrid system that allows the sensitive, high-throughput, and cost-effective assay of the interaction at the protein–DNA interface. This system makes survival of the bacteria dependent on activation of the reporter gene and therefore dependent on the protein–DNA interaction that recruits the polymerase. We have used this system to characterize DNA-binding specificities for representative members of the most common DNA-binding domain (DBD) families. We have also been able to engineer DBDs with novel specificity to be used as artificial transcription factors and zinc finger nucleases. The B1H assay provides a simple and inexpensive method to investigate protein–DNA interactions that is accessible to essentially any laboratory.

Key words: Transcription factors, DNA-binding domains, DNA-binding specificity, Hybrid systems, Selection, Gene regulation, Zinc finger nuclease

1. Introduction

Understanding the set of DNA sequences to which a transcription factor is able to bind allows the prediction of regulatory targets in a genome (1–3). However, the great majority of transcription factors in higher eukaryotes remain undefined with regard to their DNA-binding specificities. Toward this end, we have developed a bacterial one-hybrid (B1H) system based on the omega subunit of bacterial RNA polymerase (4). This system is similar to other bacterial hybrid systems with the exception that we express the TF of interest as a direct fusion to omega, the only nonessential subunit

Bart Deplancke and Nele Gheldof (eds.), *Gene Regulatory Networks: Methods and Protocols*,
Methods in Molecular Biology, vol. 786, DOI 10.1007/978-1-61779-292-2_5, © Springer Science+Business Media, LLC 2012

of the core polymerase (5–7). This allows us to perform selections in a bacterial strain with the omega subunit knocked out, therefore providing a uniform loading of the TF-omega fusion into the polymerase in the absence of competition from the endogenous omega. This lack of competition provides increased sensitivity over previously described bacterial hybrid methods (8). As a result, the omega-based B1H system provides a simple, high-throughput, and cost-effective assay of the interaction at the protein–DNA interface.

The B1H system has been successfully used to assay protein–DNA interactions in multiple ways. The assay allows for the characterization of transcription factor specificity (4, 9). In fact, we have used this system to characterize representative members of the most common DNA-binding domain (DBD) families utilized by the vast majority of eukaryotic transcription factors. Moreover, we have shown that an exhaustive characterization of a single DBD family allows the prediction of specificities for like factors in other genomes (9). The system also enables the engineering of DBDs with novel specificities to be used as artificial transcription factors or zinc finger nucleases (10–12). Finally, the relative activity of a small set of complementary DBDs and/or potential target sequences can easily be assayed by challenging with different concentrations of inhibitor. The B1H assay provides a simple, inexpensive, and robust method to investigate protein–DNA interactions that is accessible to essentially any laboratory.

2. Materials

2.1. Selection Cells and Base Plasmids

Selection cells and the required base plasmids are available through Addgene under the Laboratory of Scot Wolfe.

1. USO omega selection cells.

   Addgene name: USO hisB- pyrF- rpoZ-.

   Genotype: SB3930 *lac-, DhisB463, DpyrF, DrpoZ*:zeo [F' *proAB lacIZΔM15* Tn*10* (Tet)].

2. Plasmids

   Reporter: pH3U3-mcs.

   Omega fusion expression vectors: pB1H2w2-Prd, pB1H2w5-Prd, pB1H2wL-Prd.

2.2. Medium for Selection Procedure

1. SOB medium

   Purchased from Becton, Dickinson and Company, cat. no. 244310. Add 28.086 g SOB powder to 1 l purified water. Heat while stirring. Allow to boil for 1 min. Autoclave at 121°C for

15 min. After sterilization, store at room temperature. The SOB medium contains the following components per liter: 20 g typtone, yeast extract 5 g, sodium chloride 0.5 g, magnesium sulfate, anhydrous 2.4 g, potassium chloride 0.186 g.

2. SOC medium
   Contains a final concentration of 0.5% filter sterilized glucose in autoclaved SOB.
3. 2× YT medium
   Purchased from Becton, Dickinson and Company, cat. no. 244020. Add 31.0 g 2xYT powder to 1 l purified water. Heat while stirring. Allow to boil for 1 min. Autoclave at 121°C for 15 min. After sterilization, store at room temperature. The 2xYT medium contains the following components per liter: 16 g Pancreatic Digest of Casein, yeast extract 10 g, sodium chloride 5 g.

2.3. YM Medium

Make fresh as required. Stocks of the individual components can be made ahead of time. Prepare the following solution: 1xM9 Salts, 4 mg/ml glucose, 200 mM uracil, 0.1% histidine, 0.01% yeast extract, 1 mM $MgSO_4$, 10 mg/ml thiamine, 10 mM $ZnSO_4$, 100 mM $CaCl_2$, 25 μg/ml kanamycin, and 10 μM IPTG. Filter sterilize through a 0.22 mm filter.

2.4. NM Medium

Make fresh as required. Stocks of the individual components can be made ahead of time. Prepare the following solution: 1xM9 salts, 4 mg/ml glucose, 200 mM adenine–HCl, 1× amino acid mixture (below), 1 mM $MgSO_4$, 10 mg/ml thiamine, 10 mM $ZnSO_4$, 100 mM $CaCl_2$, 25 μg/ml kanamycin, 100 μg/ml carbenicillin, and 10 μM IPTG. Filter sterilize through a 0.22 mm filter.

2.5. YM Counter-Selective Plates

Make fresh as required. Stocks of the individual components can be made ahead of time. Counter-selective plates contain: 1.5% autoclaved agar, 1xM9 salts, 4 mg/ml glucose, 200 mM uracil, 0.1% histidine, 0.01% yeast extract, 1 mM $MgSO_4$, 10 mg/ml thiamine, 10 mM $ZnSO_4$, 100 mM $CaCl_2$, 25 μg/ml kanamycin, and 10 μM IPTG and the desired concentration of 5-fluoroorotic acid (5-FOA).

2.6. NM Selective Plates

Make fresh as required. Stocks of the individual components can be made ahead of time. Selective plates contain: 1.5% autoclaved agar, 1xM9 salts, 4 mg/ml glucose, 200 mM adenine–HCl, 1× amino acid mixture (below), 1 mM $MgSO_4$, 10 mg/ml thiamine, 10 mM $ZnSO_4$, 100 mM $CaCl_2$, 25 μg/ml kanamycin, 100 μg/ml carbenicillin, 10 μM IPTG and the desired concentration of 3-amino-1,2,4-triazole (3-AT).

2.7. Amino Acid Mixture (33.3×)

Can be made ahead of time but store at 4°C. 100 ml aliquots help reduce contamination.

Contains 17 of the 20 amino acids, omitting His, Met, and Cys. Prepare the following six solutions (all percentages are wt/vol):

Solution I (200×): dissolve 0.99 g Phe (0.99%), 1.1 g Lys (1.1%), and 2.5 g Arg (2.5%) in water to a final volume of 100 ml.
Solution II (200×): dissolve 0.2 g Gly (0.2%), 0.7 g Val (0.7%), 0.84 g Ala (0.84%), and 0.41 g Trp (0.41%) in water to a final volume of 100 ml.
Solution III (200×): dissolve 0.71 g Thr (0.71%), 8.4 g Ser (8.4%), 4.6 g Pro (4.6%), and 0.96 g Asn (0.96%) in water to a final volume of 100 ml.
Solution IV (200×): add 1.04 g Asp (1.04%) and 18.7 g potassium-Glu (18.7%) to water, bring to a final volume of 100 ml.
Solution V (200×): add 14.6 g Gln (14.6%) and 0.36 g Tyr (0.36%) to roughly 90 ml of water. Add solution V to solution IV. Add NaOH pellets slowly until all amino acids go into solution. Bring final volume to 200 ml.
Solution VI (200×): dissolve 0.79 g Ile (0.79%) and 0.77 g Leu (0.77%) in water to a final volume of 100 ml.
Mix solutions I –VI together and filter sterilize through a 0.22 mm filter and store at 4°C. This results in a 33.3× amino acid mixture.

2.8. Plates

Titration plates consist of 25 ml of media (rich 2xYT or selective NM agar) in 100 × 15 mm polystyrene Petri dishes.

Selection plates consist of 50 ml of media (NM or YM agar) in 150 × 15 mm polystyrene Petri dishes.

2.9. Colony PCR Primers

HU100 (5′ binding site primer): 5′-GAAATATGTATCCGCT CATGAC-3′.

OK181 (3′ binding site primer): 5′-CAGAGCATGTATCATAT GGTCCAGAAACCC-3′.

OMG5 (5′ DBD primer): 5′-CAAGAGCAGGAAGCCGCTG-3′.

OK60 (3′ DBD primer): 5′-CGCCAGGATCTTCAGCGGAG-3′.

3. Methods

Whether selecting binding sites to characterize TF specificity or screening a library of DBDs that offer novel specificity, the goal of the experiment determines the method. What controls to run or which plasmid contains the library may change but the general technique is the same. A DBD-omega fusion is required to interact with a sequence upstream of the HIS3 gene in order for the bacteria

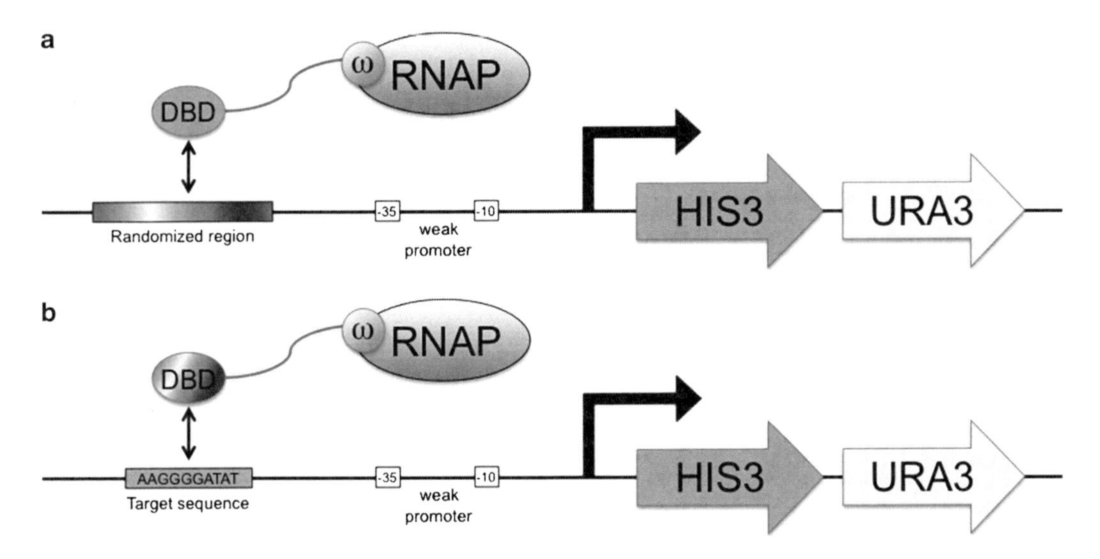

Fig. 1. Sampling complementary interactions at the protein–DNA interface. A DBD-omega fusion is expressed in the USO omega selection strain of *Escherichia coli*. If the DBD-omega fusion interacts with a unique sequence upstream of the HIS3 gene, it will recruit RNA polymerase to the weak promoter that drives HIS3 expression. (**a**) TF specificity is characterized by installing a library of randomized DNA sequences upstream of the reporter and selecting those sequences that are able to activate the reporter in the presence of the TF. (**b**) DBDs with novel specificity are engineered by installing a fixed sequence of interest upstream of the reporter. A library of DBD-omega fusions are then expressed in the USO selection strain. Only those library members able to interact with the desired sequence will be able to recruit RNA polymerase to the promoter and survive the selection.

to survive (Fig. 1). The strength of this interaction can be tuned by challenging with varied concentrations of 3-AT, a competitive inhibitor of the HIS3 product. The bacteria that survive represent a positive protein–DNA interaction and can be analyzed further (Fig. 2).

There are two critical components to a successful selection: extremely competent cells and a counter-selected library that contains enough members to reasonably cover the desired interaction. How to perform a selection is detailed below followed by a step-by-step instruction of how to make the individual components and retrieve results; but first, a note about the cells.

The Omega-based Bacterial one-hybrid system provides a very sensitive assay of the interaction at the protein–DNA interface. One of the keys to this sensitivity is the lack of competition that is provided by expressing the DBD-omega fusion in a strain of bacteria with omega knocked out. Therefore, the cell line is absolutely critical to the success of the selection. Second, characterizing the specificity of TFs with extended target sequences requires binding site libraries with enough randomized bases to accommodate this extended specificity. Furthermore, even for TFs that recognize relatively small sequences (6–10 bp), predicting the position the factor will bind relative to the weak promoter that drives the

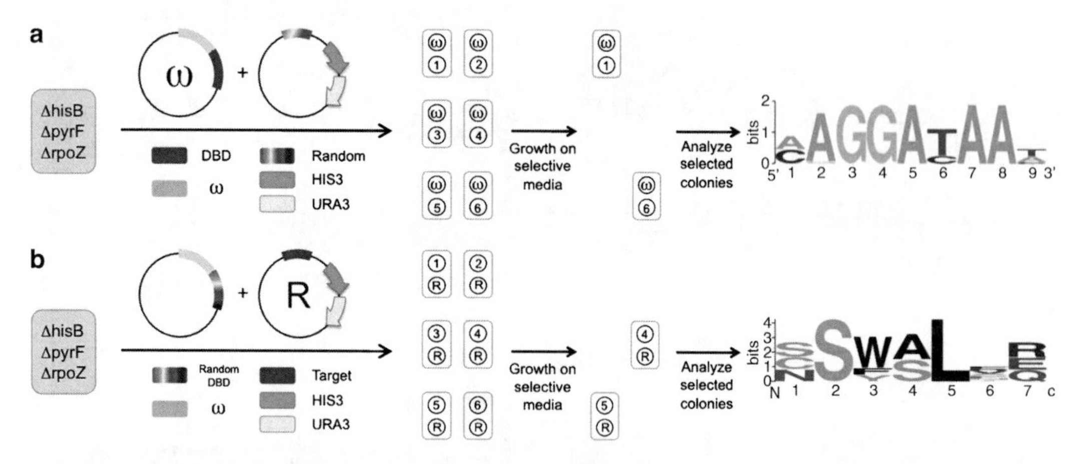

Fig. 2. Bacterial one-hybrid selection procedure. The DBD-omega expression vector (ω) and the reporter plasmid (*R*) are transformed into the USO omega selection strain so that a single library member is transformed into each cell. Each cell represents a compartmentalization of the DBD of interest (or target sequence of interest) and a unique member of the library. When these transformants are grown on media that lack histidine, survival is dependent on activation of the reporter and therefore, a complementary protein–DNA interaction. Unique clones are recovered for further analysis. (**a**) TF specificity is characterized by transforming a library of random DNA sequences in the reporter plasmids along with the TF-omega fusion expression vector. Surviving clones are analyzed by sequencing the region of the reporter plasmid that had been randomized and screening the recovered sequences for an over represented motif. (**b**) DBDs with novel specificity are engineered by transforming a library of randomized DBD-omega fusion expression vectors along with a reporter plasmid that has fixed the sequence of interest upstream of the HIS3 gene. Surviving clones are analyzed by sequencing the region of the randomized DBD-omega vector and assessing amino acid enrichment.

HIS3 reporter gene can be extremely difficult. Increasing the library window over which the factor can bind improves the chances of success. However, this naturally increases the binding site library size. Therefore, *Escherichia coli* with extremely high transformation efficiency is not only required to build the libraries, but also to ensure that enough of the library is being sampled in a given selection. In summary, to take full advantage of the system is to take advantage of the high transformation efficiency of bacteria. The electrocompetency of the USO omega cells is one of the most important components of a successful selection.

3.1. Selection Procedure

Day 1

1. Determine how many new selections to do, including a positive and negative control (see Note 1).
2. Thaw 80 μl of USO omega cells, with tested dual transformation efficiencies near 1×10^8 or greater, for each sample to test (see Subheading 3.7 for cell preparation).
3. Add 10 ml of SOC to one 15 ml polypropylene conical tube for each selection to process, warm at 37°C.
4. Put one cuvette (1 mm, sterile) and one clean Eppendorf on ice for each sample to process (see Note 2).

5. Combine all of the cells in one tube as a master mix and add in enough library to cover all of the selections (see Notes 3–5).
6. Vortex for 10 s and put back on ice for 5 min.
7. Add 1 μl of complementary plasmid at a greater concentration than the library to a labeled, sterile, Eppendorf that was put on ice in step 5.
8. Add 81 μl of cell/library mix to each selection tube that now has the complementary plasmid in it.
9. Vortex for 10 s and let stand on ice for 5 min.
10. Add cell/DNA mix to labeled cuvette on ice and retrieve prewarmed SOC.
11. Electroporate (resistance: 200 Ω, voltage: 1.8 kV/1 mm width cuvette, capacitance: 25 μF).
12. Immediately add 500 ml, warmed SOC to shocked cells and transfer to 15 ml conical tube.
13. Repeat, washing the remaining cells out of cuvette, transfer to same 15 ml of SOC.
14. Recover transformations in SOC for 1 h at 37°C while rotating.
15. Centrifuge to pellet for 15 min at 3,000 × *g* and pour off supernatant.
16. Re-suspend in 5 ml NM media + 0.1% histidine, antibiotics, and IPTG.
17. Expand for 1 h at 37°C while rotating.
18. Centrifuge to pellet for 15 min at 3,000 × *g* and pour off supernatant.
19. Wash the pellet by re-suspending in 750 ml NM media without histidine but does include appropriate antibiotics, IPTG, etc. (see Note 6).
20. Transfer to clean Eppendorf.
21. Spin down for 2 min at 9,400 × *g* on tabletop centrifuge.
22. Remove supernatant.
23. Repeat steps 20 through 23 three more times to complete the wash, all in the same Eppendorf.
24. Re-suspend pellet in 1 ml NM media without histidine, remove 20 μl for step 26, and store remainder at 4°C overnight.
25. Titrate 20 μl of re-suspension in tenfold dilutions (Fig. 3).
26. Plate on rich media with kanamycin and carbenicillin.
27. Count transformants on rich media titration plates (Fig. 3).

Day 2

28. Determine number of transformants and stringencies to plate (see Note 7).

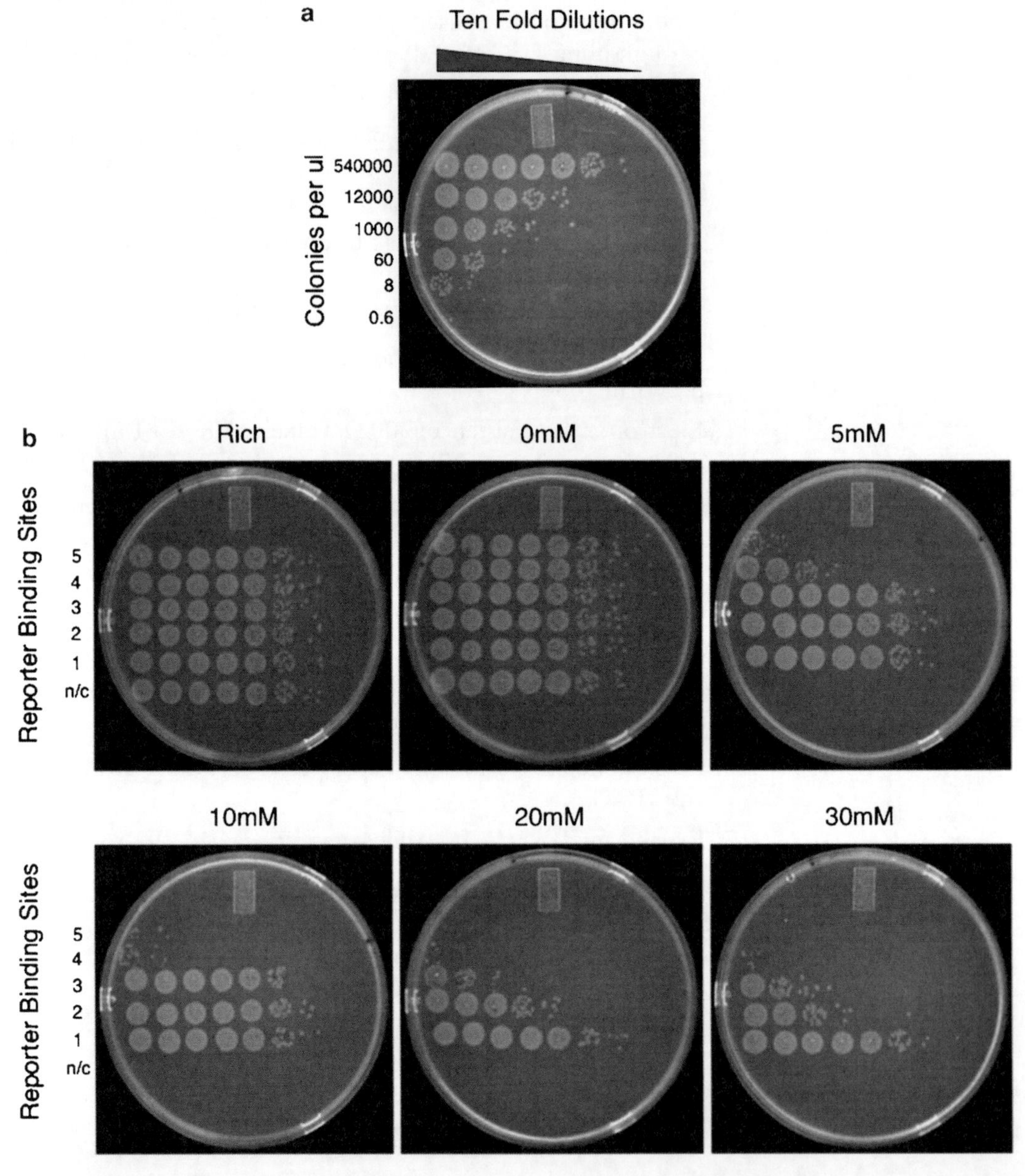

Fig. 3. Titration and activity assay. (**a**) The number of transformants from each electroporation can be calculated by plating a series of tenfold dilutions on rich media supplemented with kanamycin and carbenicillin and by counting the number of colonies at the lowest dilution. For example, in the second lane from the top in the figure above, there are six colonies at the lowest dilution. If the first lane contains neat cells, these six colonies are in the fourth tenfold dilution. Five microliters of transformation was plated on each spot so $(6 \times 10^4)/5 = 12,000$ transformants per microliter of recovered transformation. Shown are a series of transformations with differing amounts of starting material. (**b**) The relative activity of a TF on a series of binding sites can be assayed by challenging activation of the reporter with inhibitor (mM 3-AT) and comparing the fraction of survivors to the unchallenged titration. From the figure, we can see that binding sites 1–3 have similar activity at low stringencies (5 and 10 mM) but only the combination of TF and binding site 1 demonstrates 100% survival at high stringency (30 mM).

29. Determine appropriate volume of overnight NM re-suspension from Day 1, step 25, in order to include the number of desired transformants per plate as calculated in step 29.
30. Add determined volume to selection plate and bring volume up to 1.5 ml NM (no His) on the plate.
31. Spread cells slowly and evenly over the plate with sterilized glass beads (see Note 8).
32. Allow to dry under a flame.
33. Spread the beads one last time and remove beads from the plate (see Note 9).
34. Wrap the plates with parafilm and incubate for approximately 48 h at 37°C (see Note 10).
35. Count surviving colonies for all sample, positive and negative control plates. For the same number of transformants plated, a tenfold increase in survival in comparison to the negative control is a strong indication of a successful selection (Fig. 4).

3.2. Colony PCR and Sequencing

Individual clones can be recovered from the selection by colony PCR (see Note 11).

1. Select one colony with a sterile toothpick or pipette tip and inoculate 25 μl of the following PCR mix: 1 μl of 25 μM 5′ primer, 1 μl of 25 μM 3′ primer, 1 μl of 300 μM dNTP Mix, 2.5 μl of 5× NEB ThermoPol Buffer, 0.5 μl NEB Taq Polymerase, and 19 μl sterile water. PCR primers used for binding site recovery or DBD recovery are listed in Subheading 2.9.
2. Denature for 2 min at 94°C, followed by 30 cycles of 30 s at 94°C, 30 s at 52°C, and 1 min at 68°C, and final extension at 68°C for 7 min. Hold at 4°C.
3. To confirm successful PCRs, run 5 μl of each 25 μl PCR on a 1.5% agarose gel and compare to a DNA ladder standard (see Note 12).
4. Successful reactions can be sequenced with HU100 (binding site selections) or OMG5 (DBD selections).

3.3. Sequence Alignment and Motif Discovery

1. Since position of a binding site within the library window is unknown, all unique library sequences recovered from a binding site selection are analyzed using MEME motif discovery tool (http://meme.sdsc.edu/meme/intro.html).
2. We use an expectation value of $<e^{-5}$ as the standard for a high quality, overrepresented motifs. However, the vast majority of successful selections will have expectation values of $<e^{-10}$ and smaller.
3. The aligned motif sequences recovered by MEME can be used to generate a Sequence Logo using the WebLogo tool (http://weblogo.berkeley.edu/logo.cgi).

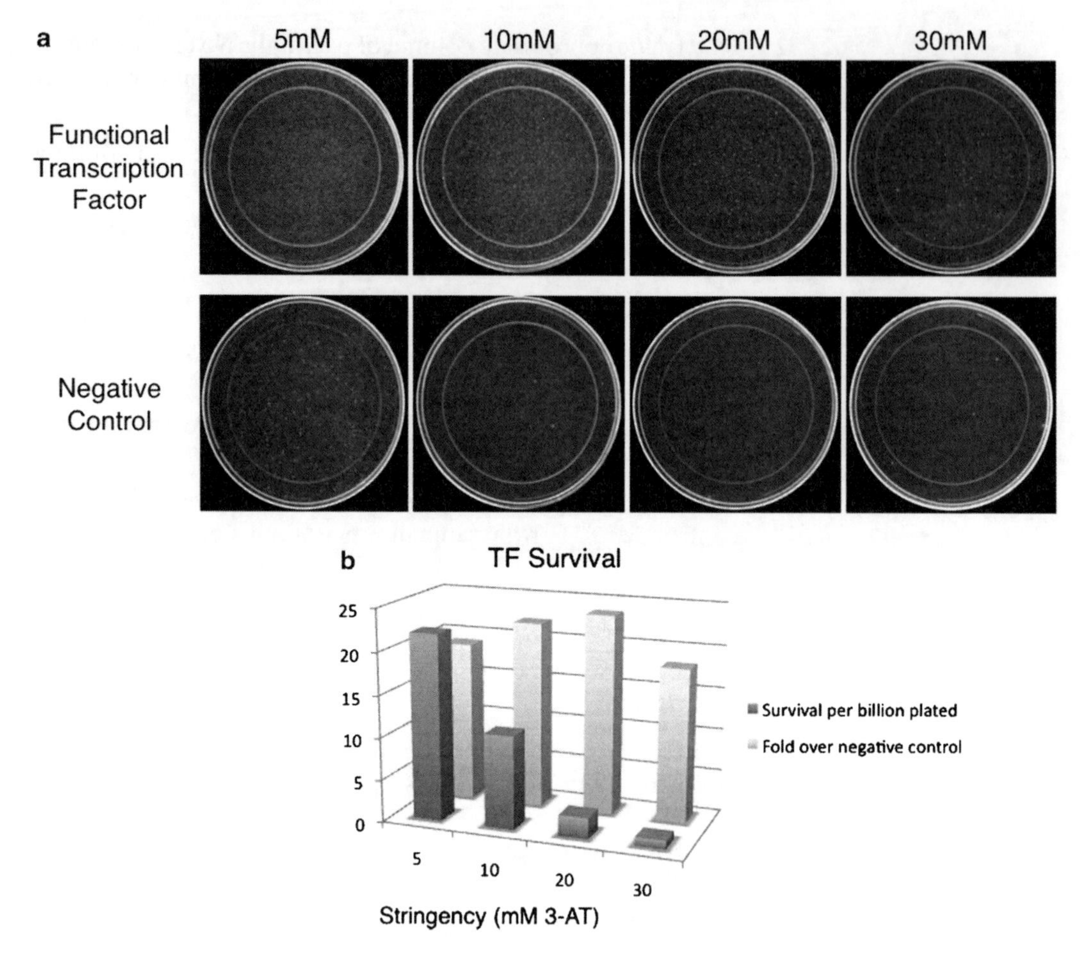

Fig. 4. Example selections. Selections are tested at multiple stringencies when the relative strength and/or specificity of the TF is unknown. The absolute number of surviving colonies can vary from one selection to the next; however, the true indicator of a successful selection is the increase of surviving colonies in comparison to the negative control at each stringency. (**a**) Selections of binding sites from a library of reporter plasmid with a functional transcription factor and a negative control are shown at multiple stringencies. The number of colonies decreases with increased stringency but there are clearly more colonies when a functional TF is present. (**b**) The number of surviving colonies per billion plated for the functional TF decreases with increased stringency. However, the fold over the number of negative control survivors remains consistently high.

4. When selecting a DBD from a library, the placement of the library is known. Therefore, sequences can be recovered and aligned without MEME, and a Sequence Logo can be generated by WebLogo directly.

3.4. Activity Assay

The relative activity demonstrated by a small set of DBDs and/or binding sites can be quickly assayed by testing their ability to sufficiently activate the reporter (Fig. 3). The strength of the protein–DNA interaction should positively correlate with the amount of reporter transcribed and therefore the ability to survive various levels of inhibitor.

Day 1

1. To test the activity of a DBD on a set of target sequences (or a set of DBDs with specific target), transform the DBD and each of the target sequence reporter plasmids into the USO omega cells, independently.
2. Plate on 2xYT + kanamycin and carbenicillin plates. Grow overnight at 37×.

Day 2

3. Select an individual colony for each DBD-target sequence pair and start a 5 ml 2xYT kan/carb overnight culture, rotate at 37°C.

Day 3

4. Centrifuge each culture to pellet (see Note 13).
5. Pour off supernatant.
6. Treat as a selection starting at step 17 (see steps 17–25 in Subheading 3.1).
7. Titrate in tenfold dilution using NM media without histidine in a 96-well plate (see Note 14).
8. Spot 5 μl of each dilution for each DBD-target sequence pair on plates of interest (see Notes 15 and 16).
9. Allow to dry. Grow plates for 24–48 h at 37°C.

3.5. Installation of Libraries and Target Sequences in the Reporter Plasmid

1. How to build libraries of random DNA sequences is outside the scope of this chapter. However, libraries of random DNA sequences have been built into the reporter plasmid by primer extension (4, 5, 13) and annealed oligonucleotides (9) followed by ligation.
2. Library reporters are counter-selected with 5-FOA in YM media to remove false positives as described (4, 5).
3. Target sequences and libraries of randomized sequences are cloned into the pH3U3-mcs reporter plasmid by standard ligation between the *Not*I and *Eco*RI restriction sites (see Notes 17 and 18).
4. The ligations are transformed into standard cloning bacteria such as XL1-Blue (Stratagene) and grown on rich media plates supplemented with kanamycin.
5. Installation of correct targets sequences is confirmed by sequencing with HU100.

3.6. TF-Omega Fusion

1. How one determines what portion of a TF to test for DNA-binding specificity is beyond the scope of this chapter. However, we have successfully characterized full-length proteins as well as truncated DBDs with this system (see Note 19).

2. DBD-omega fusions are created by cloning the coding sequence of the DBD C-terminal to the omega coding sequence between the *Kpn*I and *Xba*I sites of any of the pB1H2w plasmids by standard ligation (see Note 20).
3. A stop codon should be placed immediately internal to the *Xba*I site (see Note 21).
4. Ligations should be transformed into standard cloning bacteria such as XL1-Blue (Stratagene) and grown on rich media plates supplemented with carbenicillin.
5. Installation of correct DBDs is confirmed by sequencing with OMG5 and/or OK60.
6. The resulting expression construction will have the following format: Omega-*Not*I-Flag tag linker-*Kpn*I-DBD-TAA-*Xba*I.
7. Choosing the expression vector: all of the expression vectors utilize some form of lac promoter and expression is therefore induced with IPTG. However, large changes in expression levels appear more readily generated by increasing the promoter strength. Therefore, we have created three expression vectors that differ only in the strength of the promoter that drives expression of the DBD-omega fusion. These vectors, listed in order of promoter strength from low to high, are named pB1H2w2, pB1H2w5, and pB1H2wL (see Notes 22 and 23).

3.7. Electrocompetent Cell Preparation

One of the keys to a successful selection is having cells that are competent enough to efficiently transform two plasmids simultaneously and still cover a large fraction of the library to be assayed. We routinely produce cells that will transform two plasmids into 10^8 or greater cells using the following protocol (see Note 24).

Day 1

1. Streak out USO omega cells from stock on tetracycline/zeocin plate rich media plates (10 μg/ml Tet, 50 μg/ml Zeo) and grow at 37°C overnight.
2. Sterilize 2 l of water and 2 l of 10% glycerol.

Day 2

3. Put water, 10% glycerol, 1 ml filter tips, and Eppendorf tubes at 4°C to chill.
4. Pick a single colony with a sterile tip and inoculate 5 ml of 2xYT + tetracycline. Rotate the culture at 37°C overnight.

Day 3

5. Fill a 2 l baffle flask with 1 l of 2xYT and tetracycline. Inoculate with 1 ml of the overnight culture. Incubate with 190 rpm of agitation at 37°C.
6. Grow the cells until the OD600 reads between 0.5 and 0.6 (see Note 25).

7. When the cells are close to the desired OD, prepare a large ice water bath.
8. Immerse the bottom of the flask in the water bath and agitate in a circular motion. This is done to cool the cells to 4°C as quickly as possible. Stop agitation when the top of the flask feels cool to the touch. Allow the flask to sit in the ice water bath for about 30 min in a cold room.
9. While the flask is chilling, place the bottles to be used to centrifuge and four 50 ml pipettes in the cold room. Typically, we use four 250 ml centrifuge bottles.
10. In a cold room, split the culture between the four centrifuge bottles (see Note 26).
11. Spin at 3,000 × *g* for 12 min at 4°C.
12. Pour off the supernatant and place the bottles on ice. Re-suspend the pellets with 100 ml chilled, autoclaved H_2O. Combine two bottles in one for a final volume of 200 ml in each of the remaining two bottles.
13. Spin at 3,000 × *g* for 10 min.
14. Pour off the supernatant. Re-suspend each pellet in 100 ml of H_2O.
15. Spin at 3,000 × *g* for 10 min.
16. Pour off the supernatant. Re-suspend each pellet in 20 ml of 10% glycerol.
17. Spin at 3,000 × *g* for 10 min.
18. Pour off the supernatant. Re-suspend one pellet in 15 ml of 10% glycerol. Transfer to the other bottle and re-suspend the second pellet. Balance with another bottle. All the cells should now be in the one 15 ml re-suspension.
19. Spin at 3,000 × *g* for 10 min.
20. Aspirate all of the supernatant and place the cell pellet back on ice.
21. Gently re-suspend the pellet in 2 ml of 10% glycerol.
22. Aliquot 100 μl into each prechilled Eppendorfs.
23. Place Eppendorfs in crushed dry ice for 10 min before storing the cells at −80°C.
24. Test the transformation efficiency of the cells by electroporating the DBD and Reporter plasmids you wish to test in 80 μl of cells (to conserve material, it is also possible to substitute comparable plasmids at similar concentrations). Follow the selection procedure outlined in Subheading 3.1 through step 15. Titrate tenfold dilutions of the transformation on rich media that contain kanamycin and carbenicillin to determine the total number of dual transformants produced from a single electroporation.

4. Notes

1. Multiples of 5 are convenient since we often do selections at two stringencies and get ten plates out of one bottle (500 ml) of selective NM agar. So five samples = 1 bottle of agar.
2. Any trace plasmid DNA can produce misleading results. Therefore, we never use a cuvette twice. Even the slightest possibility that a different TF plasmid or reporter plasmid might be transformed in with the rest of the selection confounds the results. To be 100% confident in the results, always use a new cuvette.
3. No more than 1–2 μl of library material should be added per transformation. We have found that a volume greater than 3 μl of Library + complementary plasmid per transformation can reduce the transformation efficiency. Therefore, concentrations need to be high to accommodate the small volumes that lead to high efficiency.
4. The B1H system can be used to select DBDs or binding sites, therefore the Library may be on either of the plasmids employed. For a binding site selection, the library is on the reporter plasmid and the "complementary" plasmid expresses the DBD-omega fusion. For a DBD selection, the library is on the omega plasmid (pB1H2w) and the "complementary" plasmid is the reporter plasmid with the fixed sequence of interest.
5. Library material should be tested beforehand to ensure that it provides the desired number of transformants. We often test the library material, along with a complementary plasmid, to test the transformation efficiency of the cells made. Typically, library material in the 25–100 ng range will produce 10^8 dual transformants if: (a) the cells have been properly made and (b) the complementary plasmid is in excess.
6. Steps 20–25 are used to wash out any remaining histidine before plating on selective media. Any trace of histidine can lead to survival that is independent of reporter activation.
7. The theoretical library size is a key determinant of how many transformants to plate. If the library is relatively small, plating too many cells could lead to a lawn on the selection plates. However, in most cases, we are looking at large libraries where it may be nearly impossible to get enough transformants to cover the library on one or multiple stringencies. In these cases, we put at least 5×10^7 on each selection plate. This number often leads to good results. Plating in the 10^6 range for a large library often leads to failure.
8. Like the electroporation cuvettes, any trace contamination can produce misleading results. We never use glass beads twice.

9. After the plates have dried, there is often moisture trapped below the beads. We spread one last time to dissipate the moisture and then remove the beads.
10. The length of incubation will often depend on the strength of the DBD, the promoter used, and temperature. Forty-eight hours is a standard time point, but we have seen examples of factors that have grown up in 24–36 h or have taken up to 72 h. Moisture in the plates is critical for the selection stringency so we re-wrap the plates with parafilm every 24 h to maintain moisture.
11. Twelve to 24 sequences from unique clones are usually enough to provide a high-quality profile of selected binding sites or selected DBDs. However, if more sequences are desirable, it is also possible to recover the selected clones in pools for analysis by next generation sequencing (unpublished data).
12. Binding sites amplified by HU100 and OK181 result in a fragment of roughly 400 bp. The size DBDs amplified by OMG5 and OK60 depends on the DBD assayed.
13. To continue at this point, we are assuming that each culture has roughly the same number of cells. If growth in rich media is varied between cultures, possibly due to toxicity of one factor but not another, another approach is to take an aliquot of each overnight culture and start a day culture in the morning. Then, grow each culture to a common OD, storing on ice as they finish, before moving on to step 5.
14. Often it is helpful to have a frame of reference when determining what fold dilution to use. Do you anticipate the pairings to be drastically different from one another or somewhat similar? Typically tenfold differences are a good starting point. Fivefold differences are also reasonable. However, differences in activation smaller than fivefold are not reliable measured by this method.
15. Spotting several replicates of each pair on each inhibitor stringency controls for pipetting error.
16. By this method, activation of the reporter is measured by the fraction of transformants that survive at different inhibitor stringencies and comparing this fraction between binding sites or DBDs. A titration on nonselective rich media is critical for comparison. Including a nonfunctional pair of DBD-target sequence, negative control is critical to confirm selection for reporter activation is functioning properly.
17. *Not*I and *Eco*RI are start restriction enzymes used for cloning binding sites and binding site libraries. These sites allow for target sequences to be as close as five bases from the −35 box of the weak promoter that drives the reporter. However, there

are other options in the multiple cloning site if these enzymes are problematic.

18. To screen for a DBDs ability to interact with a specific sequence, the sequence must be placed in a function position relative to the −35 box in the reporter plasmid. For example, to establish the functional positions for Cys_2His_2 Zinc fingers, we tested the activity of the zinc finger Zif268 fused to omega when its known target sequence was spaced from 5 to 21 bp of the −35 box (4) (see Subheading 3.4). We found two peaks of activity centered at 10 and 21 bp, not surprisingly one and two turns of the DNA from the −35 box. Furthermore, in binding site selections performed with zif268, we find the large majority of the sequences recovered fall in these regions with a strong bias toward the proximal position, 10 bp from the −35 box. Bind sites placed 10 bp upstream of the −35 box have routinely allowed the selection of zinc fingers from a random library that are able to recognize a desired sequence. However, to select functional DBDs from different DBD families that recognize a single specific sequence, the same type of positional analysis would be required to determine the optimal position for that domain.
19. Full-length proteins from other organisms can present folding challenges in bacteria, which may decrease the success rate in comparison to simple DBDs. We have tested a handful of factors as full-length proteins and were able to recover DNA-binding specificities, implying that at least this subset of factors are able to fold properly. In addition, we tested the specificity of this same set of factors truncated to just the DBDs and found no difference in their specificity. As a result, we have primarily focused on the DBDs alone to avoid potential folding issues.
20. Though NEB suggests Buffer 1 for a dual digest of *Kpn*I and *Xba*I, we have found using NEB Buffer 2 is far more successful for cloning.
21. Often we will insert a single base between the stop codon TAA and the *Xba*I site when designing the 3′ primer to amplify a DBD. This results in two tandem stop codons: TAAnTCTAGA.
22. pB1H2w2 has two mutations in the −10 box of the lacUV5 promoter that reduce activity. pB1H2w5 is the lacUV5 promoter. pB1H2wL has a dual promoter, back-to-back lacUV5 promoters.
23. Choosing the promoter to use is largely dependent on the assumed strength of the DBD to be tested. High-affinity DBDs can be toxic if overexpressed, and therefore the pB1H2w2 vector may be appropriate. Low-affinity DBDs may require high expression levels for activity making the pB1H2wL vector

necessary. Either way, the first vector is a starting point. If the selection fails, it may be due to toxicity or low activity and a modified expression level is a simple subclone away.

24. If at first you do not succeed, try, try again. Most people have found that it takes several attempts before they are able to consistently produce highly efficient cells.
25. Try not to sample the flask too often to avoid contamination. 1 l of USO omega cells is normally ready after roughly 5 h of growth.
26. The cells should remain at 4°C from this point forward. Re-suspensions should all be done in a cold room.

References

1. Harbison, C.T., Gordon, D.B., Lee, T.I., Rinaldi, N.J., Macisaac, K.D., Danford, T.W., et al. (2004) Transcriptional regulatory code of a eukaryotic genome. *Nature* **431**, 99–104.
2. Schroeder, M. D., Pearce, M., Fak, J., Fan, H., Unnerstall, U., Emberly, E., et al. (2004) Transcriptional control in the segmentation gene network of *Drosophila*. *PLoS Biology* **2**, 1396–1410.
3. Zhu, C., Byers, K.J.R.P., McCord, R.P., Shi, Z., Berger, M.F., Newburger, D.E., et al. (2009) High-resolution DNA-binding specificity analysis of yeast transcription factors. *Genome Research* **19**, 556–66.
4. Noyes, M.B., Meng, X., Wakabayashi, A., Brodsky, M.H., and Wolfe, S.A. (2008) A systematic characterization of factors that regulate *Drosophila* segmentation via a bacterial one-hybrid system. *Nucleic Acids Research* **38**, 2547–60.
5. Meng, X., Brodsky, M.H., and Wolfe, S.A. (2005) A bacterial one-hybrid system for determining the DNA-binding specificity of transcription factors. *Nature Biotechnology* **23**, 988–94.
6. Joung, J.K., Ramm, E.I., and Pabo, C.O. (2000) A bacterial two-hybrid selection system for the studying protein-DNA and protein-protein interactions. *Proceedings of the National Academy of Sciences of the United States of America* **97**, 7382–87.
7. Dove, S.L., Joung, J.K., and Hochschild, A. (1997). Activation of prokaryotic transcription through arbitrary protein-protein contacts. *Nature* **386**, 627–30.
8. Meng, X., and Wolfe, S.A. (2006). Identifying DNA sequences recognized by a transcription factor using a bacterial one-hybrid system. *Nature protocols* **1**, 30–45.
9. Noyes, M.B., Christensen, R.G., Wakabayashi, A., Stormo G.D., Brodsky, M.H., and Wolfe, S.A. (2008). Analysis of homeodomain specificities allows the family-wide prediction of preferred recognition sites. *Cell* **133**, 1277–89.
10. Meng, X., Noyes, M.B., Zhu, L., Lawson, N.D., and Wolfe, S.A. (2008) Targeted gene inactivation in zebrafish using engineered zinc finger nucleases. *Nature Biotechnology* **26**, 695–701.
11. Siekmann A.F., Standley, C., Fogarty, K.E., Wolfe, S.A., and Lawson, N.D. (2009) Chemokine signaling guides regional patterning of the first embryonic artery. *Genes and Development* **23**, 2272–77.
12. Cifuentes, D., Xue, H., Taylor, D.W., Patnode, H., Mishima, Y., Cheloufi, S., et al. (2010) A novel miRNA processing pathway independent of dicer requires Argonaute2 catalytic activity. *Science* **328**, 1694–98.
13. Meng, X., Thibodeau-Beganny, S., Jiang, T., Joung, J.K., and Wolfe, S.A. (2007) Profiling the DNA-binding specificities of engineered Cys2His2 zinc finger domains using a rapid cell-based method. *Nucleic Acids Research* **35**, e81.

Chapter 6

MITOMI: A Microfluidic Platform for In Vitro Characterization of Transcription Factor–DNA Interaction

Sylvie Rockel, Marcel Geertz, and Sebastian J. Maerkl

Abstract

Gene regulatory networks (GRNs) consist of transcription factors (TFs) that determine the level of gene expression by binding to specific DNA sequences. Mapping all TF–DNA interactions and elucidating their dynamics is a major goal to generate comprehensive models of GRNs. Measuring quantitative binding affinities of large sets of TF–DNA interactions requires the application of novel tools and methods. These tools need to cope with the difficulties related to the facts that TFs tend to be expressed at low levels in vivo, and often form only transient interactions with both DNA and their protein partners. Our approach describes a high-throughput microfluidic platform with a novel detection principle based on the mechanically induced trapping of molecular interactions (MITOMI). MITOMI allows the detection of transient and low-affinity TF–DNA interactions in high-throughput.

Key words: Microarrays, Transcription factor binding sites, High-throughput, Binding affinities, DNA array, Protein array, Surface chemistry, Two-step PCR, Microfluidics

1. Introduction

Transcription factors (TFs) are proteins that bind to specific DNA sequences and regulate the level of gene expression by either promoting or blocking the transcription of specific genes. These specific TF–DNA interactions are part of a dynamic gene regulatory network (GRN) which is beginning to be understood by generating integrated models from data of both experimental and computational methods (1, 2).

The comprehensive characterization of GRNs requires large-scale quantitative measurements of TF–DNA interactions in a high-throughput format. However, conventional experimental methods to study molecular interactions are limited by being either nonquantitative (3, 4) or relatively low-throughput (5). Generating microarrays of immobilized double-stranded DNA sequences and

Bart Deplancke and Nele Gheldof (eds.), *Gene Regulatory Networks: Methods and Protocols*,
Methods in Molecular Biology, vol. 786, DOI 10.1007/978-1-61779-292-2_6, © Springer Science+Business Media, LLC 2012

probing them with off-chip purified proteins has been widely used for the detection of TF–DNA interactions.

The more recent introduction of microfluidics in the field of protein and DNA microarrays is a promising approach to scale-down and parallelize biological assays and study individual molecular interactions in a miniaturized format. The use of microfluidics platforms has many advantages: samples can be detected with high-precision and sensitivity while at the same time decreasing the amount of consumables, and time needed, as compared to more conventional methods (6). With the development of multilayer soft lithography for the rapid prototyping of microfluidic systems (7, 8) and microfluidic large-scale integration (MLSI) (9) microfluidic technology has become appealing to the field of biology. These MLSI devices are generally fabricated from elastomeric materials, such as polydimethylsiloxane (PDMS) and harbor micron-sized channels with thousands of integrated micromechanical valves. Among the increasing number of applications in biology, microfluidic platforms emerged as powerful screening tools to study molecular interactions, which show the potential of realizing high-throughput and high-precision measurements (10, 11).

With the highly integrated microfluidic device described in this protocol, a novel detection method has been established based on the mechanically induced trapping of molecular interactions (MITOMI). MITOMI allows the capture of transient and low-affinity interactions between DNA sequences and TFs at equilibrium, and thus the measurement of absolute binding affinities. In short, picoliter-sized reaction chambers within the array of the microfluidic chip are aligned to spots of double-stranded DNA (dsDNA) sequences printed onto an epoxy-coated glass slide using standard DNA microarray instruments. After loading an in vitro transcription/translation (ITT) mixture together with genomic DNA coding for the TF, the TF of interest is synthesized on-chip and can bind to freely diffusing target DNA. These binding events are separated by micromechanical valves and thus will be detected independently by MITOMI.

2. Materials

2.1. Mask and Wafer Fabrication

2.1.1. Instruments for Mask and Wafer Fabrication

1. Mask writing on Heidelberg DWL200 laser lithography system (Heidelberg Instruments Mikrotechnik GmbH).
2. Mask and wafer development using DV10 (Süss MicroTec AG).
3. Wafer cleaning with oxygen plasma before processing using Tepla300 (PVA Tepla AG).
4. MA6 Mask Aligner (Süss MicroTec AG) for exposure of wafers.

5. Sawatec LMS200, programmable coater for negative resist and Sawatec HP401Z, programmable hot plate for soft bake (Sawatec AG).
6. Süss RC-8 THP, manual coater and hotplate for positive resist (Süss MicroTec AG).

2.1.2. Materials for Mask and Wafer Fabrication

Chemicals used in mask and wafer fabrication are from Rockwood Electronic Materials, Gréasque, France, and of Metal-Oxyde-Semiconductor (MOS) quality, unless otherwise stated.

1. Masks: square blank 5″ Nanofilm SLM 5 (Nanofilm).
2. Silicon wafers (diameter: 100 ± 0.5 mm, thickness: 525 ± 25 μm, conductivity type: P, dopant: Boron, resistivity range: 0.1–100 Ω cm; Okmetic).
3. Photoresists: AZ9260 positive photoresist (MicroChemicals GmbH); SU-8 negative photoresist GM1060 and GM1040 (Gersteltec).
4. Chrome etching of masks: CR7 consisting of $(NH_4)_2 Ce(NO_3)_6$; $HClO_4$.
5. Developers: MP 351 for mask and AZ 400K for AZ9260coated wafers (AZ Electronic Materials); PGMEA (1-methoxy-2-propyl-acetate) for manual development of SU-8 wafers.
6. Solvents: Remover 1165 composed of 93% NMP (*N*-methyl-2-pyrrolidone; Sotrachem Technic) and 7% PGMEA for masks; isopropyl alcohol (IPA); acetone for wafers.

2.2. MITOMI Device Fabrication by Multilayer Soft Lithography

1. PDMS resin: Heat curable silicone elastomer (Dow Corning Sylgard 184).
2. Trimethylchlorosilane (TMCS) (Sigma-Aldrich).
3. Mixing and degassing of PDMS: Thinky Mixer ARE-250 equipped with adaptor for 100 ml disposable PP beakers (C3 Prozess- und Analysentechnik GmbH).
4. Degassing of PDMS control layer: Vacuum desiccator (Fisher Scientific AG).
5. Spin coating of PDMS flow layer: Programmable spin coater SCS P6700 (Specialty Coating Systems Inc.).
6. Stereomicroscope, SMZ1500 (Nikon AG).
7. Manual hole punching machine and pin vises, 21 gauge (0.04″ OD) (Technical Innovations, Inc.).

2.3. Epoxy Slide Preparation

Chemicals for epoxy-coating of glass slides are from Sigma-Aldrich.

1. Standard (76 × 26 × 1 mm) microscope glass slides (VWR).
2. Milli-Q water.
3. Ammonium hydroxide (NH_4OH; 30%).
4. Hydrogen peroxide (H_2O_2; 30%).

5. Solvents: Acetone, toluene, IPA.
6. 3-glycidoxypropyl-trimethoxymethysilane (3-GPS; 97%).
7. Nitrogen gas supply.

2.4. DNA Synthesis

All chemicals used for synthesis of DNA are from Sigma-Aldrich. For DNA primer sequences, see Table 1.

2.4.1. Synthesis of Linear Template DNA

1. DNA primers (Integrated DNA Technologies, IDT).
2. dNTPs (Roche Diagnostics AG).
3. Yeast genomic DNA (Merck Chemicals Ltd.).
4. Polymerase enzyme, Expand High Fidelity PCR system (Roche Diagnostics AG).
5. Elution buffer: 10 mM Tris–HCl, pH 8.5.

Table 1
Primer sequences used to generate ITT linear templates and target DNA library

Name (Comment)	Sequence
5′CompCy5 (Extension primer for target DNA synthesis)	5′-[Cy5]GTCATACCGCCGGA-3′
Target DNA (Design of target DNA library. Variable binding sites of interest are bracketed by constant linker sequences. 3′ ends consist of reverse complementary sequence of CompCy5 extension primer)	5′-[5′LINKER]-[BINDING SITE OF INTEREST]-[3′LINKER]-TCCGGCGGTATGAC-3′
Forward-ORF-HIS (Design gene-specific sequence to Tm of 60°C. Start codon is underlined. First five codons code for 5×His-tag. Alternatively, 5×His-tag can be added to Reverse_ORF primer)	5′-CTCGAGAATTCGCCACCATGCACCAC CACCACCAC-[GENE SPECIFIC]-3′
Reverse-ORF (Design gene-specific sequence to Tm of 60°C. Stop codon is underlined)	5′-GTAGCAGCCTGAGTCGTTATTA-[GENE SPECIFIC]-3′
Forward extension (T7 promoter sequence and transcription start site is indicated in italic and bold, respectively)	5′-GATCTTAAGGCTAGAGTA*CTAATACGA CTCACTATA*G GGAATACAAGCTACTT GTTCTTTTTGCACTCGAGAATTC GCCACC-3′
Reverse extension (Poly(A) track is underlined)	5′-CAAAAAACCCCTCAAGACCCGTTTAGAGG CCCCAAG GGGTTATGCTAGTTTTTT TTTTTTTTTTTTTTTTTTTTTTTTGTA GCAGCCTGAGTCG-3′
Forward final	5′-GATCTTAAGGCTAGAGTAC-3′
Reverse final	5′-CAAAAAACCCCTCAAGAC-3′

2.4.2. Synthesis of Target DNA

1. Primers, 5′CompCy5, (Integrated DNA Technologies, IDT).
2. dNTP (Roche Diagnostics AG).
3. Klenow fragment (3′→5′ exo-) (Bioconcepts).
4. Dithiothreitol (DTT).
5. Magnesium chloride ($MgCl_2$).
6. Buffer: Tris–HCl, pH 7.9.
7. 0.5% bovine serum albumin (BSA) in dH_2O (Table 1).

2.5. Microarraying/Spotting

1. QArray2 microarrayer (Genetix GmbH) equipped with a NanoPrint™ microarray printer printhead and a 946MP3 microspotting pin (Arrayit Corporation).
2. BSA (Sigma-Aldrich) resuspended in DI water to 2 mg/ml.
3. Synthesized target DNA templates (see Subheading 2.4.2).
4. Epoxy-coated microscope glass slides (see Subheading 2.3).

2.6. Microfluidic Control Elements

2.6.1. Pressure Regulation and Control

1. Precision pressure regulator, BelloFram Type 10, 2–25 psi; 1/8″ port size (part. no. 960-001-000; Bachofen SA).
2. Bourdon tube pressure gauges, 0–30 psi (0–2.5 bar), G ¼ male connection (part. no. NG 63-RD23-B4; Kobold Instruments AG).
3. Custom-designed manual manifolds (rectangular metal casing: 14.5 × 1 × 1″) for control line regulation with 16 detented toggles and barbs for 1/16″ tubing ID; 1/4 NPT connection (Pneumadyne Inc.).
4. Fittings to connect regulators to gauges and to luer manifolds (Serto AG): tee union (SO 03021-8), male adaptor union (SO 01121-8-1/8).
5. Tubing: Tygon ¼″ OD × 1/8″ ID (Fisher Scientific AG).
6. Polycarbonate luer fittings (Fisher Scientific AG): multiport luer manifolds for flow inlet regulation (e.g., part. no. Cole Parmer 06464-87); male luer to luer connector (part. no. Cole Parmer 06464-90).

2.6.2. Fluidic Connections

Disposable stainless steel dispensing needles to connect to syringe 23 gauge, 1/2″ long, ID 0.33 mm (part. no 560014; I and Peter Gonano).

1. Tubing (Fisher Scientific AG): For liquid/gas; flexible plastic tubing for fluidic connections, Tygon S54HL, ID 0.51 mm;
2. Steel pins for chip-to-tube interface: Tube AISI 304 OD/ID × L Ø0.65/0.30 × 8 mm, cut, deburred, passivated (USA).

2.7. Mechanically Induced Trapping of Molecular Interactions (MITOMI)

2.7.1. Surface Chemistry

1. Biotinylated BSA (Sigma-Aldrich), reconstituted to 2 mg/ml in DI water (referred to as: BSA-biotin).
2. Neutravidin (Thermo Scientific Pierce), reconstituted to 0.5 mg/ml in PBS (referred to as: NA/PBS).
3. 1:1 solution of biotinylated Penta-histidine antibody (Qiagen AG) in 2% BSA.

2.7.2. Transcription Factor Synthesis

1. TNT® T7 coupled wheat germ extract mixture (Promega AG).
2. FluoroTect™ GreenLys tRNA (Promega AG).
3. Linear template DNA coding for transcription factor (see Subheading 2.4.1).

2.8. Data Acquisition

1. Modified ArrayWorx scanner (Applied Precision) for detection and softWoRx® software.
2. Axon GenePix (Molecular Devices) for analysis.

3. Methods

The entire process involves seven distinct steps that are described in details (see Fig. 1). Molds for the two-layer microfluidic device are fabricated on silicon wafers patterned from laser-written chrome masks in a class 100 clean-room environment. A set of control and flow molds is used to produce a double-layer device by multilayer soft lithography (10).

Libraries of Cy5-labeled target DNA sequences are synthesized and spotted onto epoxy-coated microscope slides. The DNA arrays are aligned to the microfluidic device containing 768 unit cells and bonded overnight. Assembled chips are mounted on a microscope stage and connected to a pneumatic setup. Each unit cell of the device can be controlled by three individually addressable micromechanical valves, which allow compartmentalization of each unit cell, control of DNA chamber access and the detection area. A circular button membrane is used to mask a precisely defined area during surface derivatization and for the MITOMI.

Upon surface derivatization the device is loaded with a mixture of wheat germ ITT extract and the linear DNA template coding for the TF of interest. Spotted target DNA and the synthesized TF localize to an antibody deposited underneath the circular button area. MITOMI is performed by actuating the button membrane and trapping bound complexes in equilibrium. These complexes can subsequently be visualized by scanning the device with a DNA array scanner. Binding affinities are determined by quantifying the detected signals.

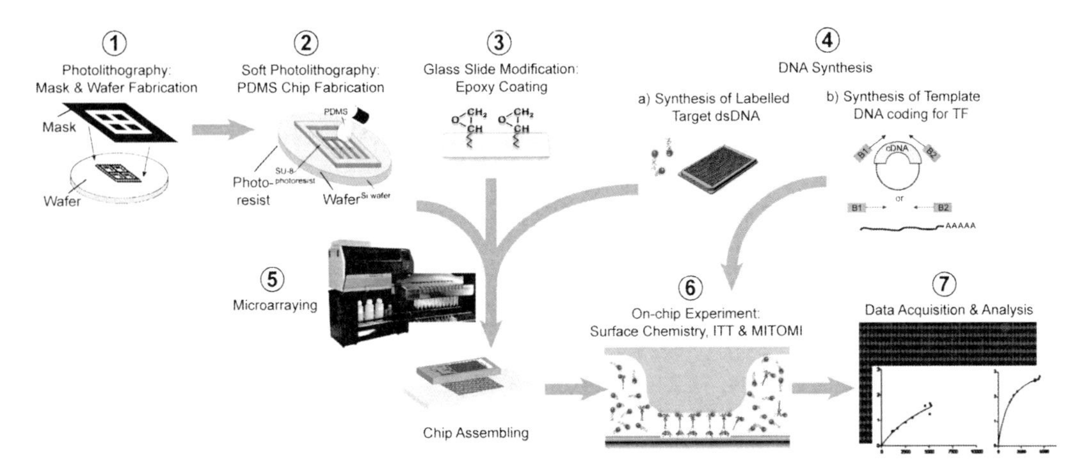

Fig. 1. Workflow of an MITOMI experiment. (1) Molds for the two-layer microfluidic device are produced on silicon wafers reproduced from chrome masks. (2) A double-layer device is fabricated of polydimethylsiloxane (PDMS) by multilayer soft lithography using a control and flow mold as template. (3) Microscope glass slides are surface modified with an epoxysilane coating. (4a) Short, fluorescently labeled target DNA sequences are synthesized and (5) spotted onto the epoxy-coated glass slides using a microarrayer before aligning and bonding them to microfluidic chip generated in step (2). (6) After surface derivatization of the glass slide, a mixture is loaded containing wheat germ in vitro transcription/translation (ITT) extract and the synthesized linear DNA template (4b) coding for the TF of interest. On-chip synthesized TFs are pulled down by immobilized antibody. TF–DNA interactions are captured by a mechanism based on mechanically induced trapping of molecular interactions (MITOMI) and (7) binding affinities quantified from detected interactions after scanning.

Note for researchers without clean-room and/or PDMS fabrication facilities: The MITOMI devices may also be obtained directly for a nominal fee from the California Institute of Technology (http://kni.caltech.edu/foundry/) and Stanford Microfluidic Foundries (http://www.stanford.edu/group/foundry/index.html).

Up-to-date protocols and microfluidic design files can be found on our laboratory homepage (http://microfluidics.epfl.ch).

3.1. Mask and Wafer Fabrication

All processes in this section are performed in a class 100 clean room.

3.1.1. Mask Fabrication

1. The two-layer device is designed in CleWin4 (WieWeb software).
2. Each layer is reproduced as a chrome mask using a Heidelberg DWL200 laser lithography system with a 10 mm writing head and a solid state wavelength stabilized laser diode (max. 110 mW at 405 nm).
3. For the development of masks, first the dispenser arm within the DV10 development chamber is purged for 10 s, then a developer mixture (MP 351:DI water 1:5) is applied twice to the mask (15 s), agitated for 45 s and drained, before rinsing and drying (50 s).
4. Developed masks are chrome etched for 110 s, rinsed, cleaned twice 15 min in 1165 remover bath, quick rinsed and air dried.

3.1.2. Flow Mold Fabrication

1. A 3″ silicon wafer is cleaned in a plasma stripper (Tepla 300) with 400 ml/min oxygen gas at 500 W and a frequency of 2.45 GHz for a period of 7 min.
2. A 1–2 μm thin layer of GM1040 negative resist is deposited on the oxygen plasma cleaned silicon wafer by spin coating first for 10 s at 500 rpm (ramp 100 rpm/s), then for 46 s at 1,100 rpm (ramp 100 rpm/s), followed by a short quick spin for 1 s at 2,100 rpm, and finally for 6 s at 1,100 rpm.
3. The precoated wafer is baked for 15 min at 65°C and 15 min at 105°C with a ramp of 4°C/min, and allowed to cool down to room temperature.
4. The wafer is exposed for 2 s in flood exposure mode with an alignment gap of 15 μm, using a lamp with a light intensity of 10 mW/cm^2 (further settings: WEC type: cont, N2 purge: no, WEC-offset: off).
5. The exposed wafer is baked on a hotplate for 35 min at 100°C.
6. Positive resist AZ9260 is spin coated on the precoated wafer for 10 s at 800 rpm, followed by 40 s at 1,800 rpm (ramp 1,000 rpm/s) to yield a substrate height of around 14 μm.
7. The coated wafer is then baked on a hotplate for 6 min at 115°C.
8. The soft-baked positive resist is allowed to rehydrate for 1 h.
9. The wafer is exposed on an MA6 mask aligner for two intervals of 18 s with 15 s waiting time in photolithography (soft contact) mode at 360 mJ/cm^2 with a light intensity of 10 mW/cm^2 (broad band spectrum lamp). The alignment gap is set to 15 μm (further settings: WEC type: cont, N2 purge: no, WEC - offset: off).
10. Following 1 h relaxation time, the wafer is developed in a development chamber (DV10) for 8–12 min based on visual observation after each cycle of the following routine with a total time of 4 min: A development mixture (AZ 400 K:DI water 1: 4) is dispensed and agitated on the wafer in three cycles, drained, rinsed (total time: 3:15 min), and finally dried.
11. In a final step, channels of the flow mold are annealed and rounded at 160°C for 20 min to create a geometry that allows full valve closure.

3.1.3. Control Mold Fabrication

1. Negative photoresist SU-8 GM1060 is spin coated on an oxygen plasma cleaned silicon wafer (see step 1 in Subheading 3.1.2) first for 10 s at 500 rpm (ramp 100 rpm/s), then for 50 s at 1,500 rpm (ramp 100 rpm/s), followed by a short quick spin for 1 s at 2,500 rpm, and finally for 6 s at 1,500 rpm to yield a height of ~14 μm.
2. The coated wafer is baked for 30 min at 130°C and 25 min at 30°C on a hotplate.

3. The wafer is exposed on an MA6 mask aligner for three intervals of 20 s with 15 s waiting time (all other settings see step 9 in Subheading 3.1.2).
4. The exposed wafer is developed manually for 2×5 min in PGMEA, rinsed with IPA and dried with an air gun.

3.2. MITOMI Device Fabrication by Multilayer Soft Lithography

1. The control layer mold is placed in a glass Petri dish lined with aluminum foil to facilitate easy removal. Care must be taken that the aluminum foil lining does not contain any holes.
2. To generate a hydrophobic surface, both flow and control mold are exposed to vapor deposits of TMCS for 30 min by placing them into a sealable plastic container with 1 ml TMCS filled into a plastic cap. TMCS treatment is repeated for 10 min each time prior to PDMS chip fabrication.
3. For the control layer, 60 g of a 5:1 Sylgard mixture (50 g Part A:10 g Part B) is prepared, mixed for 1 min at 2,000 rpm ($\sim 400 \times g$) and degassed for 2 min at 2,200 rpm ($\sim 440 \times g$) in a centrifugal mixer.
4. The mixture is poured onto the control layer mold and degassed in a vacuum desiccator for 10 min.
5. For the flow layer, 10.5 g of a 20:1 Sylgard mixture (10 g Part A:0.5 g Part B) is prepared, mixed for 1 min at 2,000 rpm ($\sim 400 \times g$) and degassed for 2 min at 2,200 rpm ($\sim 440 \times g$) in a centrifugal mixer.
6. The mixture is spin coated onto the flow layer with a 15 s ramp and a 35 s spin at 2,200 rpm.
7. After removing the control layer mold from the vacuum chamber any residual surface bubbles are destroyed by blowing on top of the PDMS layer. Any visible particles on top of the control channel grid are carefully removed using a toothpick.
8. Both layers are cured in an oven for 30 min at 80°C.
9. Following polymerization, both molds are taken from the oven and allowed to cool for 5 min.
10. The control layer is then diced with a scalpel and holes (1–8 and B, S, C, O in Fig. 2a) are punched at the control input side using a hole puncher or a 21 gauge luer stub.
11. The channel side of the control layer is thoroughly cleaned with Scotch Magic Tape.
12. The cleaned control layer is then aligned to the flow layer on the stereomicroscope.
13. The device is bonded for 90 min at 80°C in an oven.
14. Bonded devices are removed from the oven and allowed to cool for 5 min.
15. Following the outline of the control layer each individual device is cut with a scalpel and peeled off the flow layer.

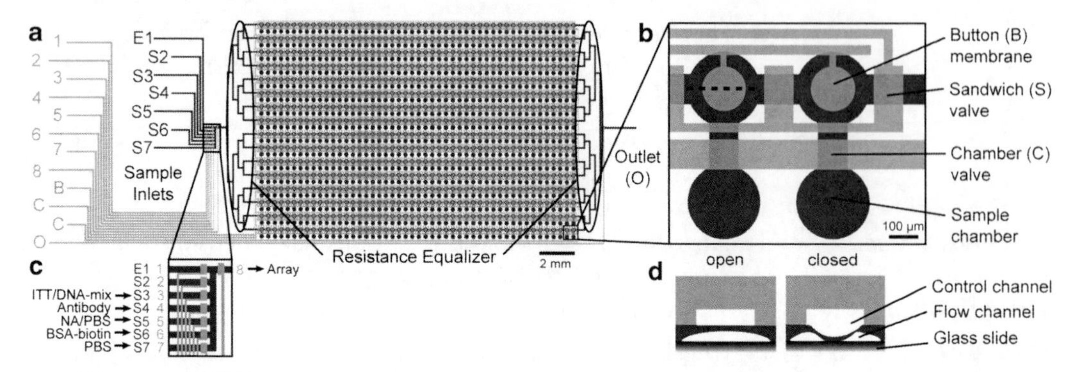

Fig. 2. (**a**) Drawing of an MITOMI device with 768 unit cells in the flow layer (*dark grey*) which are controlled (*light grey*) by 2,388 valves. Resistance equalizers toward solution inlets and outlet ensure equal flow velocities and even derivatization in each row of the channels. (**b**) The magnified view shows two individual unit cells each controlled by three separate micromechanical valves: the chamber in each dumbbell-shaped unit cell hosts a different target DNA sequence and is isolated with the chamber valve during surface modification steps for subsequent pull-down to an immobilized antibody in the detection area underneath the button membrane. For diffusion of samples to the immobilized antibody chamber valves are opened while sandwich valves between individual unit cells are closed in order to prevent cross-contamination between different samples. (**c**) Tygon tubing is loaded with different sample solutions and connected to a metal pin that is plugged into a hole at the end of each channel within the sample inlet tree (also see magnified insert of Fig. 3b). Loading of the device with samples via the inlets (S2–S7) is controlled by opening and closing the corresponding control valves. (**d**) A cross-section of one of the unit cells is shown to illustrate the detection mechanism based on mechanically induced trapping of molecular interactions (MITOMI). A deflectable membrane can be pushed down by actuating the water-filled control channel and consequently physically trap surface bound target DNA.

16. Holes are punched for the sample inlet and outlet (S1–S7 and O in Fig. 2a) using a hole puncher.
17. The flow channel side is cleaned thoroughly with tape before aligning the device to a spotted glass slide (see Subheading 3.5).
18. The flow mold is cleaned of any residual polymerized PDMS either by peeling off the thin layer of PDMS using a pair of tweezers or by an additional PDMS layer. For the latter, 11 g of a 10:1 Sylgard mixture (10 g Part A:1 g Part B) is mixed for 1 min at 2,000 rpm (~400 × *g*), degassed for 2 min at 2,200 rpm (~440 × *g*), poured on the flow mold cured in the oven for 30 min at 80°C, and peeled off after cooling down to room temperature. The control mold is cleaned with a nitrogen air gun of any PDMS debris.

3.3. Glass Slide Preparation

3.3.1. Cleaning Procedure

1. All glassware is prepared by rinsing with Milli-Q water.
2. 750 ml Milli-Q water and 150 ml ammonium hydroxide are heated to 80°C in a staining bath.
3. 150 ml hydrogen peroxide is carefully poured to the ammonium solution.
4. Glass slides are added into the staining bath and incubated for 30 min.

5. After removal from the staining bath, the glass slides are allowed to cool for 5 min.
6. Glass slides are then rinsed with Milli-Q water in the staining bath.
7. Clean glass slides are dried with nitrogen and stored in a dust free box.

3.3.2. Epoxysilane Deposition

1. Before epoxysilane deposition, all glassware is rinsed with acetone and dried at 80°C.
2. Cleaned glass slides are incubated for 20 min in 891 ml toluene with 9 ml 3-GPS.
3. After rinsing with fresh toluene to remove unbound 3-GPS, the glass slides are dried with nitrogen.
4. Glass slides are baked at 120°C for 30 min.
5. Following sonication in toluene for 15 min, glass slides are rinsed with fresh IPA.
6. Coated glass slides are dried with nitrogen and stored in a dust-free box under oxygen free conditions until usage.
7. In case of systematic PDMS chip delamination: Prior to DNA spotting, glass slides are rinsed with toluene and dried with nitrogen.

3.4. DNA Synthesis

3.4.1. Synthesis of Linear Template DNA

Generation of linear templates from genomic DNA or cDNA (see Note 1) of the TF of interest by a two-step polymerase chain reaction (PCR) method in which the first step amplifies the target sequence and the second step adds required 5′UTR and 3′UTR for efficient ITT.

1. For the first PCR step, a mixture of the following components is prepared in a final volume of 50 μl:

1 μM	Forward-ORF primer
1 μM	Reverse-ORF primer
100 ng	Yeast genomic DNA
200 μM	of each dNTP of a nucleotide mix
2.5 units	HiFi Polymerase enzyme mixture

2. After initial denaturation for 4 min at 94°C, the first PCR amplification is performed with 30 cycles as follows:

(a) Denaturation	94°C for 4 min
(b) Annealing	53°C for 60 s
(c) Elongation	72°C for 90 s and finished with a final extension at 72°C for 7 min

The correct product of this step should be ascertained on a 1% agarose gel.

3. For the second PCR step, a mixture of the following components is prepared to yield a final volume of 100 μl:

   5 nM Forward extension

   5 nM Reverse extension

   1 μl PCR product (from previous step)

   200 μM of each dNTP of a nucleotide mix

   2.5 units HiFi Polymerase enzyme mixture

4. After initial denaturation for 4 min at 94°C, the second PCR amplification is performed with ten cycles as follows:

(a) Denaturation	94°C for 4 min
(b) Annealing	53°C for 60 s
(c) Elongation	72°C for 90 s and finished with a final extension at 72°C for 7 min

5. To each reaction, 1 μl of final primer mix is added (Forward final + Reverse final; each at 1 μM final concentration) and cycling continued for 30 cycles after 4 min at 94°C as follows:

(a) Denaturation	94°C for 4 min
(b) Annealing	50°C for 60 s
(c) Elongation	72°C for 90 sand finished with a final extension at 72°C for 7 min

6. The final product can be used directly in ITT reactions or purified on spin columns and eluted in 100 μl 10 mM Tris–Cl (pH 8.5).

3.4.2. Synthesis of Target DNA

1. Small Cy5 labeled, dsDNA oligos are synthesized by isothermal primer extension in a reaction of a total volume of 30 μl containing:

6.7 μM	5′CompCy5
10 μM	Library primer
5 units	Klenow fragment (3′→5′ exo-)
1 mM	of each dNTP from a nucleotide mix
1 mM	Dithiothreitol (DTT)
50 mM	NaCl
10 mM	$MgCl_2$
10 mM	Tris–HCl, pH 7.9

2. All reactions are incubated at 37°C for 1 h followed by 20 min at 72°C and a final annealing gradient down to 30°C at a rate of 0.1°C/s.
3. After the synthesis, 70 μl of a 0.5% BSA dH_2O solution are added to each reaction.
4. The entire volume is then transferred to a 384 well plate and a sixfold dilution series prepared with final concentrations of 5′CompCy5 of 2 μM, 600 nM, 180 nM, 54 nM, 16 nM, and 5 nM (see Note 2).

3.5. Microarraying/Spotting

Spotting target DNA onto epoxy-coated microscope slides is performed by a QArray2 DNA microarrayer.

1. Before each spotting routine, an Eppendorf or Falcon tube filled with DI water and the spotting pins is submerged in a sonicator water bath to clean the pins. During the spotting routine, a sterilizing loop (1 s DI water, followed by 1 s air drying) between different DNA samples keeps pins clean throughout the spotting procedure.
2. Sample plate(s) of target DNA are placed in an external source plate stacker before starting the spotting routine (see Note 3).
3. The dilution series for each target DNA sequence is spotted as a microarray with a column and row pitch of 373 μm and 746 μm, respectively.
4. Spotted arrays are aligned manually to a tape-cleaned PDMS device (see Subheading 3.2) on a Nikon SMZ1500 stereoscope and bonded overnight in an incubator at 40°C.
5. DNA arrays can be stored in a sealed box protected from light and dust at room temperature for several weeks.

3.6. On-Chip Experiment (Surface Chemistry and MITOMI)

3.6.1. Mounting MITOMI Device to Microfluidic Control Elements

1. The assembled device is mounted on a light microscope and connected to tubing (for details see Fig. 3).
2. Control channels are filled with DI water by actuation with ~5 psi of pneumatic pressure which forces the air from the dead-end channels into the bulk porous silicone. This procedure eliminates subsequent gas transfer into the flow layer upon valve actuation, as well as prevents evaporation of the liquid contained in the flow layer.
3. Devices are actuated with 15–20 psi in the control lines and between 5 and 8 psi for the flow line.
4. Upon actuation button membrane and sandwich valves are opened again; chamber valves remain closed during the following initial surface derivatization steps to prevent liquid from entering the sample containing chambers.

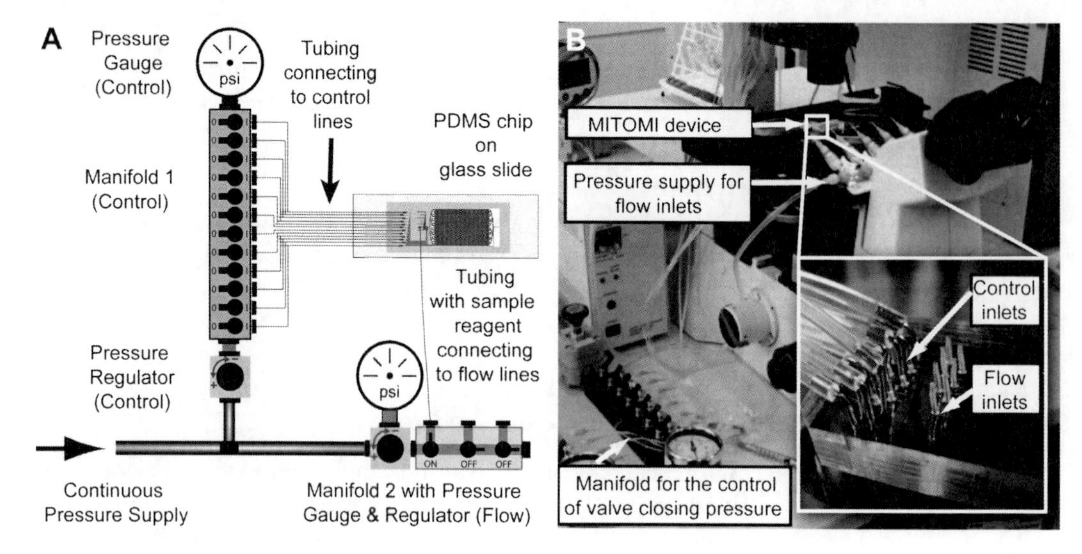

Fig. 3. Setup of the experiment. (**a**) Schematic of fluidic control set-up using regulated pressure and manually controlled manifolds. (**b**) Photograph of an assembled MITOMI device placed on the stage of a microscope, where flexible tubing (Tygon) is connected via metal pins to the inlets that actuate the control lines on the device (*magnified inset*). The tubing is filled with DI water displacing the air in the channels when actuated with pneumatic pressure that is controlled with manifolds. Each reagent for the on-chip experiment is filled into a pin-end flexible tube (Tygon) and connected to the flow inlets. The pressure of the flow can be controlled with a gauge.

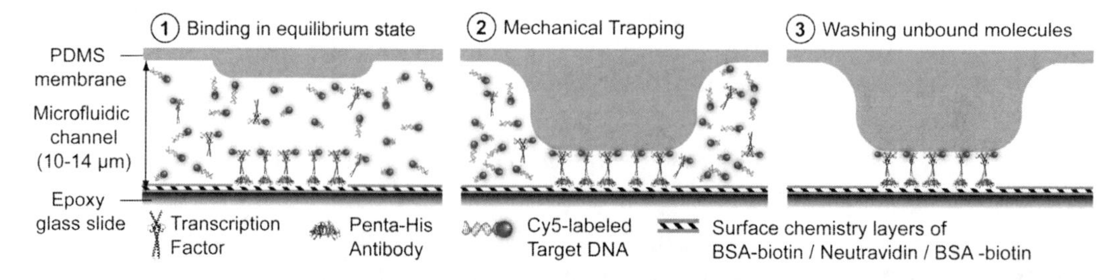

Fig. 4. Schematic of the MITOMI process. The *gray structure* at the top of each panel represents the deflectable button membrane that can be brought into contact with the glass surface (also see Fig. 2d). (1) His_5-tagged TFs are localized to immobilized penta-His antibody at the epoxy-coated glass slide. Specific binding between Cy5-labeled target DNA and TFs are at steady state when (2) the button membrane is actuated and brought into contact with the surface. Any molecules in solution are expelled while surface-bound material is mechanically trapped. (3) Unbound material that was not physically protected is washed away, and the remaining molecules are quantified.

3.6.2. Surface Chemistry

The surface area within the flow channels of the device is modified by depositing layers of BSA-biotin, NA/PBS, and biotinylated antibody onto the epoxy coated glass slide (see Fig. 4). Using a syringe, the different sample solutions are loaded into short pieces (30–40 cm) of clean Tygon tubing which are hooked to a metal pin that is then pushed into the corresponding flow sample inlet holes (see Fig. 2c).

1. Tubing with 30 μl of BSA-biotin is inserted into the sample inlet hole S6. The port on manifold 2 for flow inlet regulation (see Fig. 3) is actuated and valves are opened by switching the

corresponding toggles on manifold 1 in the following order: valve 6 controlling sample flow inlet S6, equalizer (1), array inlet (8).

2. Once the air in the channels of the array is displaced with liquid, the outlet valve (O) is opened and the equalizer valve (1) opened.
3. After flowing BSA-biotin for ~20 min the array inlet valve (8) and the sample valve (6) are closed again and the port on manifold 2 turned off.
4. The tubing is disconnected from the port of manifold 2.
5. The process of valve and port opening/closing is performed for the following samples.
6. After BSA-biotin derivatization, the surface area is washed for 2–3 min with 5 μl PBS (S7).
7. A 30 μl solution of NA/PBS (S5) is flowed for ~20 min and washed again for 2–3 min with 5 μl PBS.
8. The button membrane (B) is closed and PBS washing continued for 1 min (~2 μl) making sure button is closed.
9. The remaining surface area is passivated with 30 μl BSA-biotin (~20 min) and washed with 5 μl PBS (2–3 min) while button is actuated.
10. 30 μl of a 1:1 solution of biotinylated penta-His antibody in 2% BSA is loaded (S4). To ensure that all channels are saturated with antibody solution, the button membrane is opened only after flowing 5 μl.
11. After finishing the antibody deposition (total ~20 min), the surface is washed again with 5 μl PBS for 2–3 min.
12. The surface derivatization procedure is finished with a final 5 μl PBS washing step (see Note 4).

3.6.3. DNA Pull-Down and Detection of Interactions On-Chip

1. 25 μl TNT T7 coupled wheat germ extract is prepared and spiked with 1 μl $tRNA_{Lys}$-Bodipy-Fl and 2 μl of linear expression ready template coding for the appropriate transcription factor.
2. The mixture is immediately loaded onto the device (S3) and flushed for 10 μl (around 5–7 min) while the button membrane is closed.
3. Chamber valves (C) are opened and the outlet valve (O) is closed to allow for dead end filling of the chambers with wheat germ extract.
4. Chamber valves are closed and the outlet valve opened again and flushing is continued for an additional 10 μl (5–7 min).
5. Sandwich valves (S) that separate each unit cell are closed.
6. After ensuring that all sandwich valves are closed the chamber valves and button membranes are opened.

7. The device is incubated for 90 min at room temperature to allow for protein synthesis and diffusion of the samples to the immobilized antibody under the button membrane.
8. After the incubation period, the device is scanned on a modified ArrayWoRx microarray scanner.
9. Button membranes are closed to trap bound samples.
10. Chamber valves are closed, sandwich valves opened, and the channels washed with 10 μl PBS (5–7 min).
11. The washed device is scanned once more with closed button membranes to detect the trapped molecules.

3.7. Data Acquisition and Analysis

1. For each experiment, two images (see Fig. 5a) are analyzed with GenePix3.0 (Molecular Devices): The first image, taken directly after the 60–90 min incubation period before washing, is used to determine the concentration of solution phase or total target DNA concentration (Cy5 channel). The second image, taken after MITOMI and the final PBS wash, is used to determine the concentration of surface bound protein (FITC channel) as well as surface bound target DNA (Cy5 channel).
2. Dissociation equilibrium constants can be calculated for each experiment using a curve fitting software (e.g., Graphpad Prism4 or Mathematica) by performing global nonlinear regression fits using a one site binding model to the data plotted as surface bound target DNA (RFU) divided by surface protein concentration (RFU) (or effectively fractional occupancy) as a function of total target DNA concentration in RFU (see Fig. 5b).

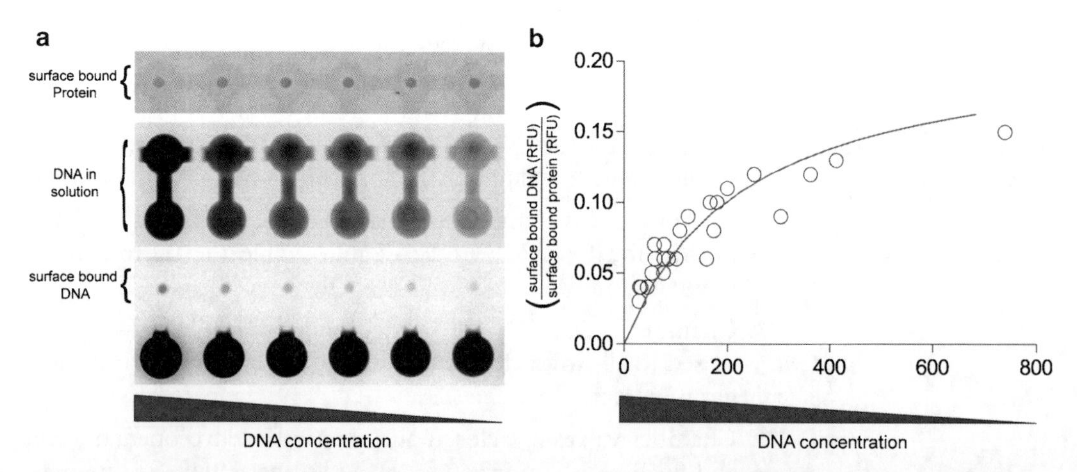

Fig. 5. Steps to analysis of MITOMI experiments. (**a**) Fluorescence scans of subsequent MITOMI steps. Scan of immobilized, fluorescent labeled protein (*Top*), solubilized target DNA (*Middle*), and surface bound target DNA after mechanical trapping and washing step (*Bottom*). (**b**) The fraction of surface bound target DNA is plotted against concentration of target DNA in solution. Dissociation constants are determined by performing a nonlinear regression fit using a one site binding model.

3. Relative K_ds (RFU^{-1}) must be transformed into absolute K_ds (M^{-1}) using a calibration curve previously established by measuring known quantities of 5′CompCy5 primer.
4. The change of free Gibbs energy ($\Delta\Delta$Gs) with $\Delta\Delta G = RT \times \ln(K_d / K_{d,ref})$ is calculated at a temperature of 298 K, where the highest affinity sequence serves as the reference.

4. Notes

1. Linear expression templates can also be synthesized from bacterial cDNA clones after lysing them in 2.5 μl Lyse n'Go buffer (Pierce) at 95°C for 7 min. The lysate serves as template in an Expand High Fidelity PCR reaction (Roche). The first PCR product is then purified using the Qiaquick 96 PCR purification kit (Qiagen) and eluted in 80 μl of 10 mM Tris–HCl, pH 8.5.
2. The on-chip DNA concentration can be increased by raising the numbers of repetitive stamps of sample DNA per spot during the spotting routine (multiple returns of spotting pin to the same spot).
3. The humidity inside the spotter is set to 65–80% to prevent the samples in the source plate from evaporating during long spotting routines (>2 h).
4. An additional passivation step can be included by flowing 5 μl BSA-biotin for 2–3 min after the antibody immobilization, then closing the button, followed by 2–3 min of 5 μl BSA-biotin. This additional BSA-biotin passivation step was found to reduce background signal.

References

1. Kim, H. D., Shay, T., O'Shea, E. K., and Regev, A. (2009) Transcriptional Regulatory Circuits: Predicting Numbers from Alphabets, *Science* **325**, 429–432.
2. Beer, M. A., and Tavazoie, S. (2004) Predicting gene expression from sequence, *Cell* **117**, 185–198.
3. Berger, M. F., Philippakis, A. A., Qureshi, A. M., He, F. S., Estep, P. W., 3 rd, and Bulyk, M. L. (2006) Compact, universal DNA microarrays to comprehensively determine transcription-factor binding site specificities, *Nat Biotechnol* **24**, 1429–1435.
4. Bulyk, M. L., Huang, X., Choo, Y., and Church, G. M. (2001) Exploring the DNA-binding specificities of zinc fingers with DNA microarrays, *Proc Natl Acad Sci USA* **98**, 7158–7163.
5. Majka, J., and Speck, C. (2007) Analysis of protein-DNA interactions using surface plasmon resonance, *Adv Biochem Eng Biotechnol* **104**, 13–36.
6. Whitesides, G. M. (2006) The origins and the future of microfluidics, *Nature* **442**, 368–373.
7. Duffy, D. C., McDonald, J. C., Schueller, O. J. A., and Whitesides, G. M. (1998) Rapid Prototyping of Microfluidic Systems in Poly(dimethylsiloxane), *Analytical Chemistry* **70**, 4974–4984.
8. Xia, Y., and Whitesides, G. M. (1998) Soft Lithography, *Angewandte Chemie International Edition* **37**, 550–575.
9. Thorsen, T., Maerkl, S. J., and Quake, S. R. (2002) Microfluidic large-scale integration, *Science* **298**, 580–584.

10. Maerkl, S. J., and Quake, S. R. (2007) A systems approach to measuring the binding energy landscapes of transcription factors, *Science* **315**, 233–237.

11. Maerkl, S. J., and Quake, S. R. (2009) Experimental determination of the evolvability of a transcription factor, *Proc Natl Acad Sci USA* **106**, 18650–18655.

Chapter 7

Detecting Protein–Protein Interactions with the Split-Ubiquitin Sensor

Alexander Dünkler, Judith Müller, and Nils Johnsson

Abstract

A detailed understanding of a cellular process requires the knowledge about the interactions between its protein constituents. The Split-Ubiquitin technique allows to monitor and detect interactions of very diverse proteins, including transcription factors and membrane-associated proteins. The technique is based on unique features of ubiquitin, the enzymes of the ubiquitin pathway, and the reconstitution of a native-like ubiquitin from its N- and C-terminal fragments. Using Ura3p as a reporter for the reconstitution of the ubiquitin fragments, methods are presented that enable to screen in yeast for interaction partners of a given protein with either a randomly generated expression library or a defined but more limited array of protein fusions.

Key words: Protein–protein interaction, Split-protein sensor, Split-ubiquitin, Protein interaction screens, Systems biology, Protein complementation assay, Yeast, Binary interactions

1. Introduction

Ubiquitin, a compact protein of 76 amino acids, can be artificially separated into an N-terminal peptide of 35 (N_{ub}) and a C-terminal peptide of 41 residues (C_{ub}). Both ubiquitin halves reassociate into a native-like ubiquitin once coexpressed in the same cell. Proteins that are attached to the C-terminus of ubiquitin are rapidly cleaved off in vivo by the ubiquitin-specific proteases (UBP). Because the UBPs identify features of the folded ubiquitin, C_{ub}-fusions are only recognized as substrates when they are reassembled with N_{ub} to form a native-like ubiquitin. The spontaneous reassociation of N_{ub} and C_{ub} is slow and inefficient. Connecting N_{ub} and C_{ub} to interacting proteins X (N_{ub}-X) and Y (Y-C_{ub}) forces them into close proximity, and thereby accelerates the formation of the native-like ubiquitin and the cleavage of the reporter activity from the C-terminus of C_{ub} (1, 2).

Bart Deplancke and Nele Gheldof (eds.), *Gene Regulatory Networks: Methods and Protocols*,
Methods in Molecular Biology, vol. 786, DOI 10.1007/978-1-61779-292-2_7, © Springer Science+Business Media, LLC 2012

Ura3p (Orotidine-5′-phosphate decarboxylase) is instrumental in the synthesis of uracil in yeast and has become a universal reporter for detecting the N_{ub}/C_{ub} reassociation. It was used in yeast cells for measuring interactions of membrane-bound and cytosolic proteins as well as transcription factors (TFs) and DNA modifying enzymes from different organisms (3–12). The junction between C_{ub} and Ura3p in the C_{ub}-Ura3p fusion was configured in a way that after cleavage from C_{ub} an arginine becomes the N-terminal amino acid of Ura3p (RUra3p). As a destabilizing residue according to the N-end-rule pathway of protein degradation, the new N-terminus induces the rapid destruction of Ura3p. Thus, in cells expressing an interacting pair of N_{ub}/C_{ub}-fusions, the RUra3p moiety is rapidly degraded resulting in uracil auxotrophy. A positive selection for interacting proteins is achieved by introducing 5-fluoro-orotic acid (5-FOA) into the media. 5-FOA is converted to the toxic compound 5-fluoro-uracil (5-FU) by Ura3p. As the interaction of the N_{ub}/C_{ub}-coupled proteins leads to the rapid degradation of the RUra3p, growth of the transformed yeast on 5-FOA is indicative of cells harboring interacting proteins (3).

Proteins that constitutively localize to membranes provide a challenge for the application of the split-protein sensor technology. The restriction of their free movement to a two dimensional plane and a predefined orientation of the membrane-coupled N_{ub}- and C_{ub}-moieties relative to each other will favor their spontaneous reassociation and thus interfere with the robust detection of "real" interaction-driven N_{ub}/C_{ub} reassembly. The use of N_{ub}-mutants, where the isoleucine at position 13 was replaced by alanine (N_{ua}) or glycine (N_{ug}), reduces the affinity of the ubiquitin halves to each other (1). Consequently, the cleavage of the reporter from C_{ub} becomes more strictly dependent on the interaction rather than a mere proximity between the two membrane proteins that are linked to them (2, 3, 13).

Using the Split-Ub technique in monitoring the interactions between transcription factors allows analyzing the proteins when bound to their target DNA within complex assemblies. This is an advantage over the conventional yeast two hybrid system (Y2H), where self-activation biases against interactions between transcription factors or other DNA bound proteins. This increased flexibility is traded against a less straight interpretation of the Split-Ub derived interaction data. Whereas the Y2H system is considered to predominantly report on direct protein interactions, the split protein sensors might equally well detect the assembly of two proteins within a larger protein complex without requiring a direct contact between them. Both considerations are strictly relevant only when screening proteins in their native hosts. To analyze, for example, a TF from *Drosophila* in yeast under quasi-native conditions would require the integration of its DNA target sites into the yeast genome to achieve an approximation to the situation encountered in *Drosophila* nuclei.

2. Materials

2.1. Saccharomyces cerevisiae and Escherichia coli Strains

1. Yeast strains: JD47: *MATa*, *leu2-3*, *112 lys2-801*, *his3-Δ200*, *trp1-Δ63*, *ura3-52*; JD53: *MATα*, *leu2-3*, *112 lys2-801*, *his3-Δ200*, *trp1-Δ63*, *ura3-52*.
2. *E. coli* strain: DH5α (luxS supE44 lacU169 (φ80dlacZ M15) hsdR17 recA1 endA1 gyrA96 thi-1 relA1) (Invitrogen, Carlsbad, CA).

2.2. Media

1. Yeast SD minimal media: 6.7 g/L yeast nitrogen base without amino acids (e.g., DIFCO, Detroit, MI), 60 mg/L L-isoleucine, 20 mg/L L-arginine, 40 mg/L L-lysine, 60 mg/L L-phenylalanine, 10 mg/L L-threonine, 10 mg/L L-methionine, 50 mg/L adenine sulfate, 20 mg/L L-histidine, 60 mg/L L-leucine, 40 mg/L L-tryptophan, 50 mg/L uracil, and 20 g/L glucose. Leave out amino acids or uracil for the selection of the corresponding auxotrophic markers.
2. YPD medium: 10 g/L yeast extract, 20 g/L peptone, 20 g/L glucose.
3. Solid media are obtained from liquid media by addition of 20 g agar per liter medium.
4. Media for selection of the *kanMX* marker: 200 μg/mL Geneticin (Gibco BRL Life Technologies, Island, NY). Selection of diploid zygotes after mating of N_{ub} and C_{ub} expressing strains is achieved by 400 μg/mL geneticin.
5. 5-FOA agar plates: 5-FOA (Fermentas, St. Leon-Rot, Germany) is added to SD agar in a concentration of 0.1%. To obtain 500 mL FOA medium, dissolve 0.5 g 5-FOA in 250 mL water, filter-sterilize and heat up to 50°C. Simultaneously, twofold concentrated SD minimal medium (e.g., SD-His^-) is prepared, autoclaved, cooled down, and mixed with the preheated 5-FOA solution.
6. LB medium: 5 g/L yeast extract, 10 g/L tryptone, 10 g/L NaCl. Add 100 μg/mL ampicillin for plasmid selection.

2.3. Reagents for Yeast Transformation

1. 1 M lithium acetate.
2. 50% PEG3000 solution.
3. 10× TE-buffer (100 mM Tris–HCl pH 8.0, 10 mM EDTA pH 8.0).
4. Single-stranded carrier DNA 2 mg/mL: Dissolve 200 mg of salmon sperm DNA (sodium salt) in 100 mL sterile 1×TE buffer and heat to approximately 50°C while vigorously stirring. Aliquots are boiled 10 min at 95°C, cooled in an ice-water bath and stored at –20°C. Boiling and cooling may be repeated before transformation to increase transformation efficiency.

2.4. Reagents for DNA Preparation from Yeast Cells

1. STE-buffer (1 M Sorbitol, 50 mM Tris–HCl pH 8.0, 100 mM EDTA pH 8.0).
2. 10% SDS solution.
3. 25 mg/mL Zymolyase-100 T solution (Sigma-Aldrich, St. Louis, MO).
4. 3 M potassium acetate.
5. Isopropanol.
6. 70% ethanol.

2.5. C_{ub}-Gene Fusion Expression Plasmids

Several yeast shuttle plasmids were adapted to the split-Ub system (3, 5). The C_{ub}-*RURA3* (CRU) fusion genes (baits) were constructed either directly in yeast by homologous recombination or by inserting the full length open-reading frame (ORF) between the P_{MET17} promoter and the C_{ub}-*RURA3* cassette, followed by transformation into the strain JD47.

1. pC_{ub}-RURA3-303, an integrative plasmid for the construction of full-length CRU fusions integrated into the yeast genome. The 3′ends of yeast genes are amplified and inserted in frame 5′ to the C_{ub}-*RURA3* cassette. A single restriction site in the amplified gene is used for linearization. After transformation, the homologous recombination between the chromosomal copy of the gene and the introduced C_{ub}-*RURA3* gene fusion fragment leads to the formation of the full-length CRU gene fusion at the chromosomal locus (3) (Fig. 1).
2. pP_{MET17} C_{ub}-RURA3-313, a centromere-based vector which drives the ORF-C_{ub}-*RURA3* fusion of interest by the regulatable P_{MET17} promoter. Stable, low copy number maintenance of the plasmid is provided by a CEN6 and ARSH4 element. Introduction of the ORF is achieved by a multiple cloning site between the sequence of the P_{MET17} promoter and the C_{ub}-*RURA3* cassette. Both plasmids contain an *HIS3* marker gene to allow selection in yeast and an antibiotic resistance gene for the selection against ampicillin in *E. coli* (3, 10). The option of using the Gateway system for N_{ub}- and C_{ub}-gene fusion construction was realized by Deslandes et al. (5).
3. pYM-N35 is used for genomic integration of the P_{MET17} promoter in front of C_{ub}-*RURA3* fused genes by PCR-based gene targeting. (*natNT2* and *amp* marker) (14) (see Note 1).

2.6. N_{ub}-Fusion Expression Plasmids

The N_{ub}-fusion genes are expressed under control of the copper-inducible P_{CUP1} promoter. These are obtained either by homologous recombination of PCR-amplified P_{CUP1}-N_{ub} cassettes or by cloning of the full length ORF behind the P_{CUP1} promoter. N_{ub}-fusion expressing strains are generated in the *MAT*α strain JD53 to allow mating with the C_{ub}-*RURA3* fusion expressing strains.

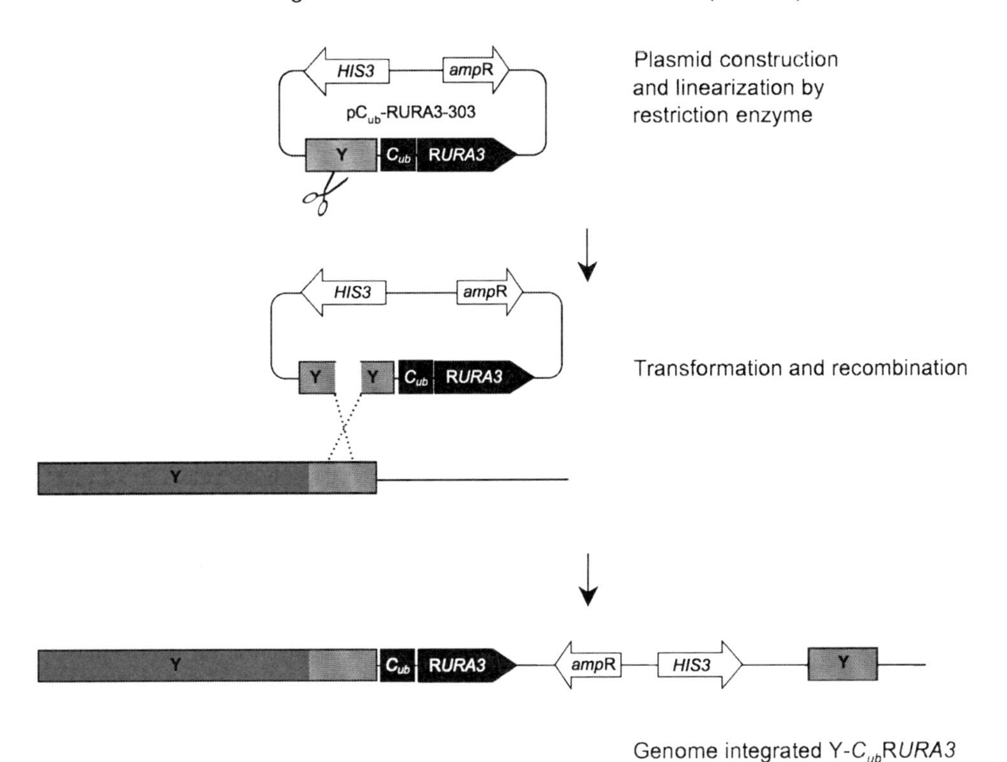

Fig. 1. Construction of the yeast genome-integrated C_{ub}-fusion genes (*Y-C_{ub}-RURA3*). A PCR-derived fragment of the 3′ end of the ORF of interest without its stop codon is ligated in front and in frame of the *C_{ub}-RURA3* sequence on a pRS303 vector. The fragment harbors a unique restriction site situated a minimal 100 bp from the 5′ and 3′ ends of the fragment. After linearization at the unique restriction site, the vector is transformed into yeast cells where homologous recombination leads to the insertion of the *C_{ub}-RURA3* module into the yeast genome.

1. pP_{CUP1}-N_{ub}-HA-kanMX4 is used as a template for the PCR-amplification and site-specific integration of *P_{CUP1}-N_{ub}-HA* sequences, where the sequence *HA* codes for the HA epitope. Using primers composed of sequences up- and downstream of the start codon of the respective yeast gene and sequences annealing with the *kanMX6-P_{CUP1}-N_{ub}-HA* cassette (N_{ub}-FW, N_{ub}-RV; see primers in Subheading 2.8) a PCR product is obtained that will integrate in front or within the gene of interest via homologous recombination (Fig. 2). Successful recombination is verified in geneticin-resistant strains by a diagnostic PCR using primers annealing with the *N_{ub}* sequence and sequences 300–600 bp downstream of the integration site within the respective gene (15).
2. pP_{CUP1}-N_{ub}-HA-kanMX4 CEN is a centromere-based vector (CEN6 and ARSH4 element). Common restriction sites are used to insert the ORF of interest behind the P_{CUP1}-*N_{ub}-HA* cassette. Expression of the N_{ub} fusion in yeast is obtained after transformation of the plasmid and selecting for its presence by Geneticin (15).

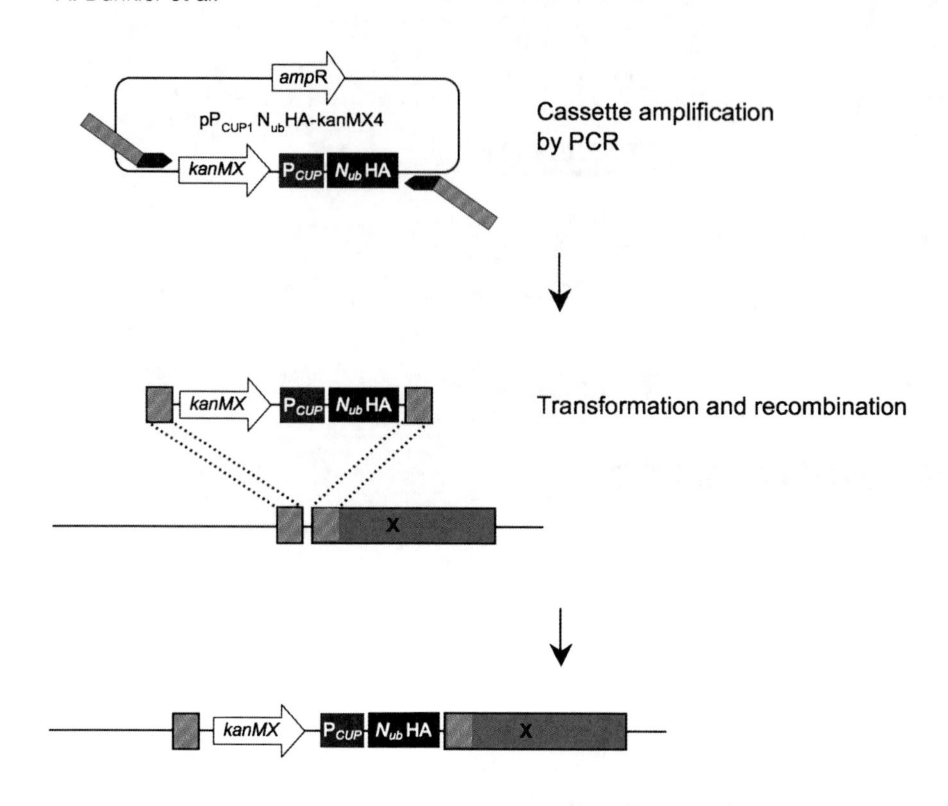

Fig. 2. Construction of the yeast genome integrated N_{ub} fusion genes (P_{CUP1}-N_{ub}-*HA-X*). A PCR with two primers annealing at the 5′ and 3′ end of the P_{CUP1}-N_{ub}-*HA-kanMX* cassette yields a fragment harboring beside the N_{ub}-module and the geneticin resistance gene a stretch of approximately 45 bp at each of its ends identical to the sequences within the yeast genome, where the integration of the cassette will occur after transformation and selection on Gen+ medium.

2.7. N_{ub}-Libraries

N_{ub}-libraries are available as translational fusions between N_{ub}, N_{ua}, N_{ug}, and various sources of genomic and cDNA from various tissues and organisms (8, 16–19). N_{ug}-libraries from different sources are commercially available from Dualsystems Biotech (Schlieren, Switzerland). The N_{ub}-library exemplarily employed in this protocol is a genomic yeast library on a high copy plasmid bearing an *LEU2* auxotrophy gene as selectable marker. Random genomic fragments were generated by partial Tsp509I digestion and inserted behind a P_{ADH1} N_{ub}-*HA* cassette in the three possible reading frames (18, 19). The vector chosen for library construction should carry a different selection marker for bacterial transformation from the C_{ub}-vector (i.e., chloramphenicol and ampicillin).

2.8. Primers

1. N_{ub}-SEQ: CATTGGAAGTTGAATCTTCCG
2. N_{ub}-FW: 45 bp homology upstream of the start codon + GCATAGGCCACTAGTGGATC
3. N_{ub}-RV: 45 bp homology downstream of the start codon + GGTCGACCCCGCATAGTCAGG

2.9. Technical Equipment

1. For the systematic interaction screen, a Singer RoToR HAD robot system (Singer Instruments, Somerset, UK) was used. This system requires the Singer Plus Plates with standard SBS footprint dimensions and RePads (disposable plastic pads of pins). The robot system allows replicating and mating of one C_{ub}-*RURA3* expressing yeast strain against 384 N_{ub}-fusion strains in quadruplicate with finally 1,536 positions on the array.
2. Alternatively, a 96-pin steel replicating tool can be used to manually transfer colonies from one plate to another. This less expensive alternative is however time consuming and less reproducible.

3. Methods

3.1. Protein Interaction Screen with Random N_{ub}-Libraries

For a successful interaction screen, the following steps are recommended (Fig. 3):

1. Adjusting the expression level of the Y-C_{ub}-RUra3p fusion protein and selecting the appropriate N_{ub} mutant with Y being the ORF coding for the protein bait of interest.
2. Introducing the N_{ub}-library into the Y-C_{ub}-RUra3p expressing strain and selecting for interacting N_{ub}-fusions on plates containing 5-FOA.

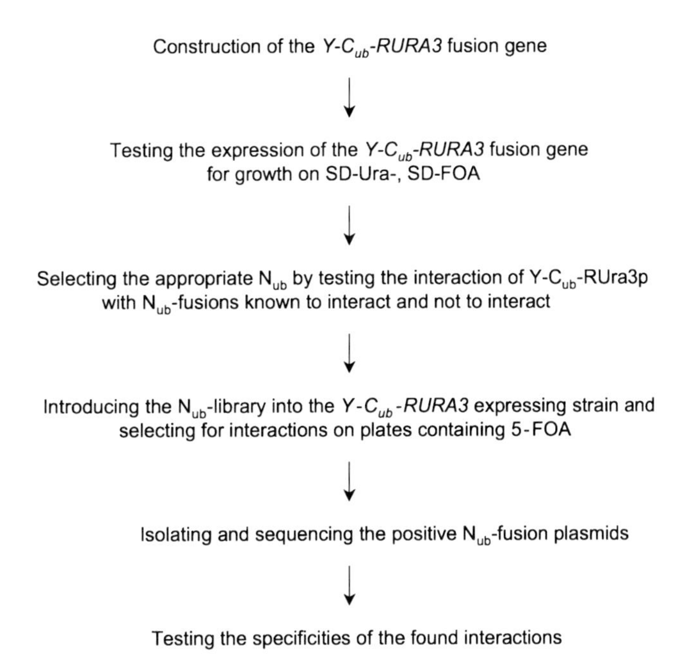

Fig. 3. Flowchart for a Split-Ub-based protein interaction screen with random N_{ub}-libraries. See the text for details.

3. Isolating and sequencing the N_{ub}-fusion genes.
4. Testing the specificity of the found interactions.

3.1.1. Adjusting the Expression Level of the Y-C_{ub}-RUra3p Fusion Protein and Selecting the Appropriate N_{ub}-Mutant

1. Grow yeast cells expressing the Y-C_{ub}-RUra3p fusion (*HIS3*) under the native promoter or under the control of the P_{MET17} promoter in 2 mL of liquid SD-His⁻ medium selecting for the presence of the *Y-C_{ub}-RURA3* fusion to an OD_{600} of 1.
2. Spot 4 μl of the culture and of three successive one-tenth dilutions onto SD-Ura⁻ and SD-FOA medium selecting for the presence of *Y-C_{ub}-RURA3*. The media for the cells expressing *Y-C_{ub}-RURA3* from the P_{MET17} promoter contain 70 μM, 10 μM, or no methionine.
3. After 3 days, solid growth on SD-Ura⁻ and no growth on SD-FOA indicate sufficient amounts of the Y-C_{ub}-RUra3p fusion and thus appropriate conditions for interaction analysis. *Y-C_{ub}-RURA3* constructs yielding weak or absent Ura3p activity are not suitable for interaction assays (see Note 1).
4. Transform into the yeast JD53 the *Y-C_{ub}-RURA3* construct in combination with N_{ub}-, N_{ua}-, and N_{ug}-fusions of suitable marker proteins. Soluble proteins should be tested against other soluble proteins while membrane-bound proteins have to be evaluated in combination with membrane proteins of different subcellular compartments (see Note 2). Provided that known interaction partners of Y are available, they should be included as positive controls as N_{ub}-, N_{ua}-, and N_{ug}-fusions. The promoter for the expression of the N_{ub}-fusions should differ from the promoter controlling the expression of the *Y-C_{ub}-RURA3*. P_{CUP1} or P_{ADH1} were shown to be appropriate choices.
5. Pick and restreak colonies on selective media.
6. Inoculate in 2 mL of selective media. Let grow at 30°C to an OD_{600} of 1.
7. Spot 4 μl of each culture and dilutions of 1/10, 1/100, and 1/1,000 on selective 5-FOA and SD-Ura⁻ media under the optimal conditions determined for the *Y-C_{ub}-RURA3* expression (see step 3).
8. Incubate for up to 5 days at 30°C.
9. The N_{ub} variant with the highest affinity to C_{ub} that does not provide growth on 5-FOA should be selected for library construction (compare Note 2).

3.1.2. Introducing the N_{ub}-Library into the Y-C_{ub}-RUra3p Expressing Strain and Selecting for Interacting N_{ub}-Fusions on Plates Containing 5-FOA

1. Inoculate a fresh culture from a plate with yeast containing the Y-C_{ub}-RURA3 (*HIS3*) construct into the appropriate SD-His⁻ medium.
2. To obtain a mid-log phase culture on the day of transformation, inoculate fresh medium with the overnight culture to an

OD_{600} of 0.2–0.25 and incubate at 30°C with shaking to a final OD_{600} of 0.6–0.8.

This last step may be performed in YPD medium, but should be kept to 2–3 h to reduce the risk of Ura3p activity loss.

3. Determine exact cell density before transformation. One OD_{600} corresponds roughly to 2×10^7 cells/mL. Plate $0.5–1 \times 10^9$ cells per large plate (22.5 cm × 22.5 cm).
4. Prepare sample DNA for library transformation: Per large plate, pipette 4.5 μg of library DNA (here: represented on an *LEU2* plasmid), 100 μl carrier DNA (2 mg/mL), 0.5 mL cells ($0.5–1 \times 10^9$), and 3 mL PEG/LiAc/TE solution into 15 mL sterile tubes; mix vigorously.
5. Incubate for 30 min at 30°C with gentle agitation.
6. Heat-shock samples for 20 min in 42°C water bath; invert tubes several times to keep cells in suspension.
7. Let samples adjust to room temperature.
8. Adjust volume of each transformation to 5 mL with TE and plate transformation mixtures onto large SD-FOA-His^-Leu^- plates. Spreading of yeast cells is conveniently performed using 10–20 sterile glass beads with 5 mm diameter.
9. Alternatively, spin down transformation mix and resuspend cells carefully in 5 mL of SD-His^-Leu^- containing the tested concentrations of methionine and copper for the proper expression of the fusion proteins. Incubate at 30°C for 1 h, spin down, and resuspend in 5 mL of TE. Spread cells as described on SD-FOA-His^-Leu^-.
10. Determine the transformation efficiency by plating appropriate dilutions of the transformation mixtures onto small plates, selecting for the presence of the two plasmids (i.e., SD-His^-Leu^-; 1:100 and 1:2,000 dilutions work well). Expect $0.5–1 \times 10^6$ transformants per large plate in an experiment using this protocol.
11. Incubate plates at 30°C (it will take up to 3 days for the first colonies to be visible, but more colonies may appear during the course of the next 2 weeks).

3.1.3. Isolating the N_{ub}-Fusion Plasmids and Testing Their Interactions Against the Original Y-C_{ub}-RUra3p

1. Pick colonies from the 5-FOA transformation plate and restreak on SD-FOA-His^-Leu^- to separate the positive colonies from the background cell layer (see Note 3).
2. Grow at 30°C and pick freshly from plate to inoculate 5 mL SD-Leu^- to select for the presence of the N_{ub}-fusion plasmid only.
3. Grow in SD-Leu^- medium for 2–3 days to high cell density.
4. Spin down complete culture at 10,000 × *g* for 5 min. Yeast cells are resuspended in 500 μl STE buffer, supplemented with 8 μl zymolyase to a final concentration of 0.4 mg/mL, and incubated for 30 min at 37°C to digest cell walls.

5. To lyse the cells, add 50 μl SDS (10%), mix, and incubate 30 min at 65°C.
6. Continue plasmid isolation by addition of 200 μl 3 M potassium acetate, and ethanol precipitation following the standard alkaline lysis protocol for the isolation of plasmid DNA from *E. coli*. When high throughput is required, vacuum-based plasmid preparation kits from various suppliers may be used instead. Follow the manufacturer's instructions continuing with alkaline lysis buffer.
7. Transform a 10% aliquot of the plasmid preparation into *E. coli* by standard high efficiency procedures (e.g., electroporation) and plate on selective media. If yields of plasmid preparations are low, which may be the case when using CEN plasmids, the DNA may have to be precipitated and resuspended in a smaller volume.
8. Pick four colonies from each transformation plate and proceed with DNA preparation by standard methods.
9. Prepare a restriction digest of an aliquot of the DNA with enzymes that cut out the insert. Continue only with those samples that show a uniform restriction pattern in all four preparations.
10. Co-transform a small aliquot of the DNA with the original *Y-C_{ub}-RURA3* expressing plasmid into JD53.
11. Test the cells for FOA resistance and Ura^- phenotype as detailed under Subheading 3.1. Continue only with those N_{ub}-X plasmids (with X being the ORF coding for the detected prey protein) that confer FOA resistance and at the same time significantly reduce the growth of the cells on SD-Ura^-. This will eliminate several classes of false positives:
    (a) Clones that are picked from background colonies,
    (b) Contaminating *HIS3*-containing plasmids,
    (c) Plasmids that were rescued from yeast containing more than one N_{ub}-X plasmid, and
    (d) N_{ub}-X fusions that eliminate the toxic action of 5-FOA by other means (transporters, pumps).
12. Use a further aliquot of the prepared DNA and the primer "N_{ub} seq" to determine the nature of the insert by sequencing.

3.1.4. Testing the Specificity of the Found Interactions

A class of false positives might arise from N_{ub}-X fusions whose high local concentrations in proximity to the Y-C_{ub}-RUra3p are not caused by direct interactions with Y. Such high local concentrations might, for example, be the result of an exceptionally strong expression level of X in combination with a relatively unspecific colocalization with Y.

1. Construct a *Z*-C_{ub}-*RURA3* expressing plasmid, where Z displays roughly the same expression level and cellular localization as Y but should be involved in a different activity than Y (for example: Y is a receptor and Z is an ion transporter, both localized at the plasma membrane).
2. Co-transform N_{ub}-*X* with the original *Y*-C_{ub}-*RURA3* expressing plasmid or with *Z*-C_{ub}-*RURA3* into JD53.
3. Pick single colonies from both transformation plates and subject them to an interaction assay as described under Subheading 3.1.
4. Continue the analysis of those N_{ub}-X fusions that show a preferential interaction with Y over Z (see Note 4).

3.2. Systematic Interaction Screen Using an Ordered N_{ub}-Array

Methods for systematic interaction screening using a large yet limited set of N_{ub}-fusion genes and the mating of yeast strains of different mating types expressing either the C_{ub}- or the N_{ub}-fusion gene, respectively, were already described for a transcriptional readout (20–22).

A very similar procedure using *RURA3* as the reporter for the N_{ub}/C_{ub} assembly was recently established (15). Here, 383 yeast strains each expressing a different N_{ub}-fusion gene were arrayed as quadruplicates onto a single media plate. The array was mated with a single yeast strain expressing the *Y*-C_{ub}-*RURA3* gene fusion. The resultant 1,532 strains were transferred onto selective 5-FOA medium and scored for growth. Arraying, mating, and transfer of the yeast strains onto different media is achieved by automatic colony picking. The procedure is described for 383 different N_{ub} expressing strains but can easily be scaled up or down (Fig. 4).

1. Construct the *Y*-C_{ub}-*RURA3* fusion gene in vector pP_{MET17} C_{ub}-RURA3-313 or pC_{ub}-RURA3-303 and transform into strain JD47.
2. Construct as many different N_{ub}-X fusions as desired by either inserting the ORFs of interest into the plasmid pP_{CUP1}-N_{ub}-HA-kanMX4 CEN followed by transformation into yeast JD53 or by direct homologous integrations of the P_{CUP1}-N_{ub}-*HA*-*kanMX4* cassette into JD53 using the plasmid pP_{CUP1}-N_{ub}-HA-kanMX4 as template for the PCR.
3. Once the desired collection of N_{ub}-fusion genes is represented as different JD53 strains (here 383 strains) each strain is grown in liquid YPD-Gen^{+} to an OD_{600} between 1 and 3.
4. Transfer 100 μl of each culture into a well of a 96-well plate.
5. Using four 96 long-pin pads (Singer robot) transfer the 383 N_{ub} expressing yeast strains from four 96-well plates onto 383 positions (+ one empty control position) of one YPD-Gen^{+} master PlusPlate.

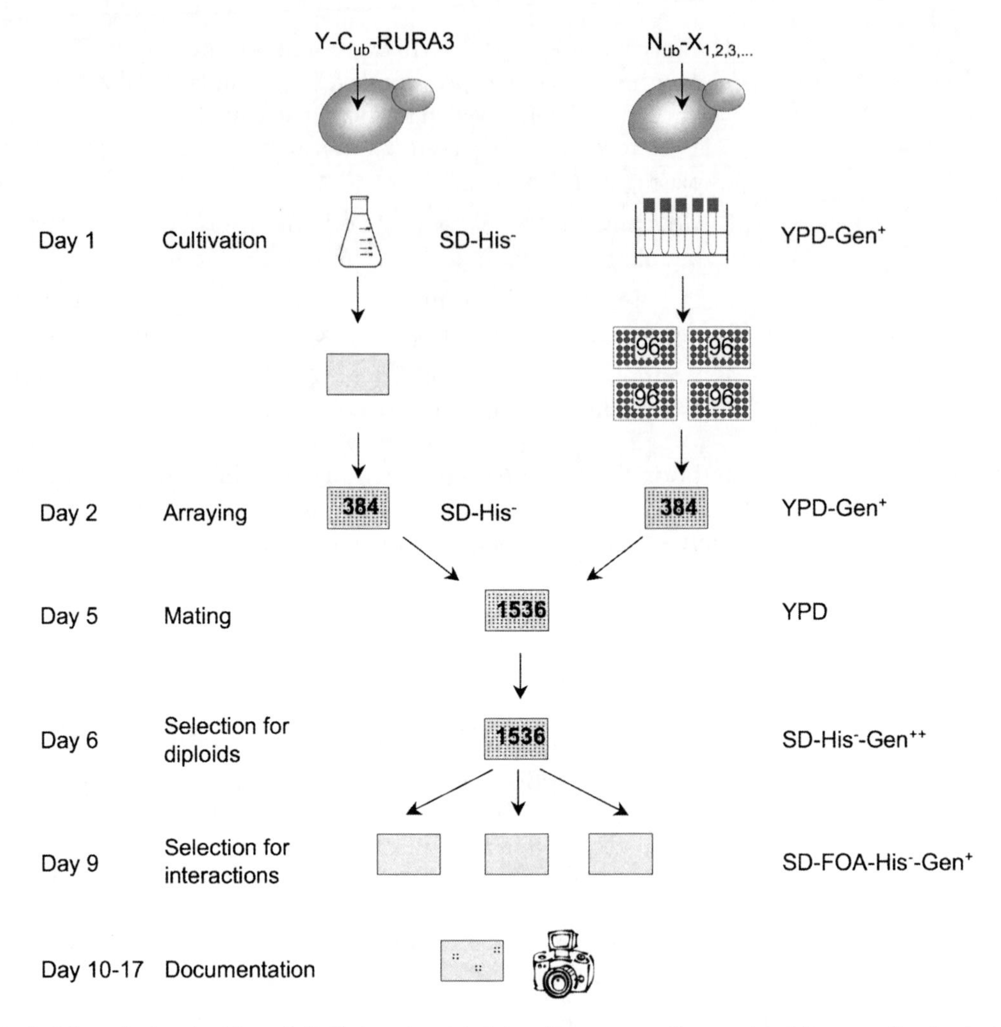

Fig. 4. Schematic flowchart for a Split-Ub-based protein interaction screen with an ordered N_{ub}-array. See the text for details.

6. After addition of glycerol to a final concentration of 15%, the N_{ub}-array displayed on the four 96-well plates can be stored at −80°C (see Note 5).
7. Inoculate 50 mL SD-His⁻ with the *Y-C_{ub}-RURA3* gene fusion strain and incubate the culture for 2–3 days at 30°C to high cell density.
8. Transfer the liquid yeast culture into an empty PlusPlate and pin the liquid using a 96 long-pin pad onto an SD-His⁻ plate at 384-colony density.
9. Incubate the C_{ub}- and N_{ub}-plate (see step 5) for 2 days at 30°C.

10. First, replicate the N_{ub}-array by quadruplication from 383 to 1,532 colonies on YPD-Gen^{+} using pads with 384 short pins ("384 short-pin pads"). This agar plate serves as target plate for the next step.
11. Transfer the cells of the C_{ub}-array onto the target plate (see step 10) and mix them with the used 384 short-pin pads on the agar surface. Let N_{ub}- and C_{ub}-strains mate for 18–24 h at 30°C.
12. Transfer the diploid cells by a 1536 short-pin pad on SD-His^{-}-Gen^{++} (400 μg/mL geneticin) and additionally incubate for 2 or 3 days at 30°C.
13. For the identification of protein–protein interactions, transfer the mated cells with a 1536 short-pin pad onto SD-FOA-His^{-}-Gen^{+} agar plates containing 70 μM and no methionine (in case the *Y-C_{ub}-RURA3* is under the control of P_{MET17} promoter) in combination with no or 50 μM copper sulfate. One plate without 5-FOA serves as control for equal cell transfer and growth of the cells.
14. Document the growth of the cells for the next 7 days (see Note 6 and Fig. 5).

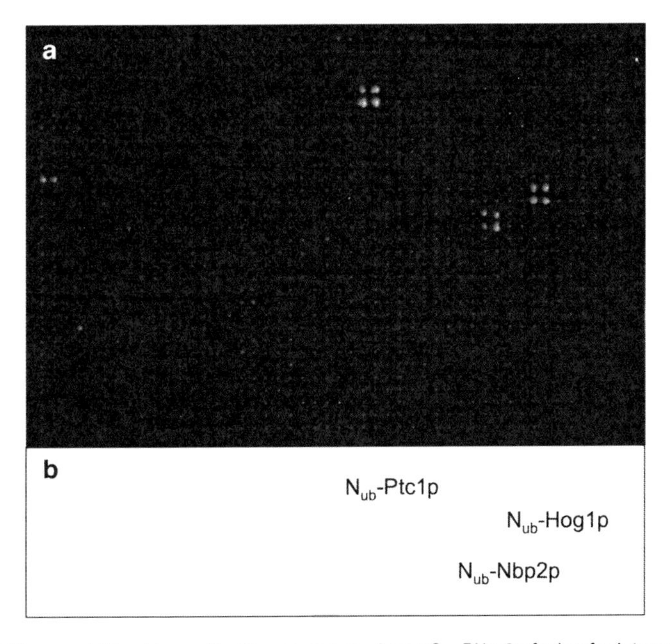

Fig. 5. The yeast Map kinase Pbs2p was screened as a C_{ub}-RUra3p fusion for interactions against 383 yeast N_{ub}-fusions. (**a**) Shown is the growth of the 1,532 mated yeast strains on SD-FOA-His^{-}Gen^{+} after 3 days. (**b**) The growth of four colonies per positive interaction indicates Ptc1p, Nbp2p, and Hog1p as binding partners of Pbs2p (15). We consider three colonies out of four still an indication of interaction. Two out of four is not considered as evidence of interaction and will require an independent repetition of the experiment.

4. Notes

1. To increase Ura3p activity, recloning of the *Y-C_{ub}-RURA3* by changing fusion junctions, expressing subdomains, or switching to high copy-number plasmids (2 μ) are three alternative strategies. Insertion of the P_{MET17} promoter in front of a genome integrated *Y-C_{ub}-RURA3* fusion that does not produce sufficient Ura3p activity from the native promoter often alleviates the problem.
2. The comprehensive interaction screening of membrane-localized proteins with the RUra3p reporter system might require more than one screening run. A first screen with an N_{ub}-library (high affinity for C_{ub}) might reveal its interactions with cytosolic proteins. Here, the membrane-localized N_{ub}-fusions have to be automatically excluded as false positives. In contrast, an N_{ua}- or N_{ug}- (lower affinity for C_{ub}) based library screen might detect only the membrane-bound fraction of interaction partners. Here, cytosolic partners that only temporarily interact are very probably missed (8).
3. The use of more complex N_{ub}-libraries than those derived from genomic yeast DNA might necessitate additional steps to reduce the isolation of false positive clones. A detailed protocol is given by Dirnberger et al. (18).
4. Novel interactions discovered with the Split-Ub systems should be validated by independent means. Expressing one of the interaction partners as a GST fusion in bacteria and precipitating the partner protein from yeast- or other cell culture extracts with the immobilized protein proved to be a reliable strategy (12, 15).
5. Test the N_{ub}-array in regular intervals with the same panel of C_{ub}-fusions that interact with different N_{ub}-fusions on the array. Constant interaction patterns testify the integrity of the array. Once the patterns change over time, one has to go back to the original master plate.
6. During the testing of a large amount of different N_{ub}-fusion proteins, we observed that a fraction interacts with multiple C_{ub}s in an unspecific manner. Before the use of any newly created N_{ub}-array, it is therefore recommended to first define these sticky or promiscuous N_{ub}-fusions. One option is to test several C_{ub}-fusions with a very low likelihood to interact with any of the N_{ub}-fusions of the array, for example C_{ub}-fusions derived from a very different organism than the N_{ub}-fusions.

References

1. Johnsson, N., and Varshavsky, A. (1994). Split ubiquitin as a sensor of protein interactions in vivo. *Proc Natl Acad Sci USA* **91**, 10340–10344.
2. Müller, J., and Johnsson, N. (2008). Split-ubiquitin and the split-protein sensors: chessman for the endgame. *Chembiochem* **9**, 2029–2038.
3. Wittke, S., Lewke, N., Müller, S., and Johnsson, N. (1999). Probing the molecular environment of membrane proteins in vivo. *Mol Biol Cell* **10**, 2519–2530.
4. Laser, H., Bongards, C., Schüller, J., Heck, S., Johnsson, N., and Lehming, N. (2000). A new screen for protein interactions reveals that the Saccharomyces cerevisiae high mobility group proteins Nhp6A/B are involved in the regulation of the GAL1 promoter. *Proc Natl Acad Sci USA* **97**, 13732–13737.
5. Deslandes, L., Olivier, J., Peeters, N., Feng, D.X., Khounlotham, M., Boucher, C., Somssich, I., Genin, S., and Marco, Y. (2003). Physical interaction between RRS1-R, a protein conferring resistance to bacterial wilt, and PopP2, a type III effector targeted to the plant nucleus. *Proc Natl Acad Sci USA* **100**, 8024–8029.
6. Xue, X., and Lehming, N. (2008). Nhp6p and Med3p regulate gene expression by controlling the local subunit composition of RNA polymerase II. *J Mol Biol* **379**, 212–230.
7. Krick, R., Bremer, S., Welter, E., Schlotterhose, P., Muehe, Y., Eskelinen, E.L., and Thumm, M. Cdc48/p97 and Shp1/p47 regulate autophagosome biogenesis in concert with ubiquitin-like Atg8. *J Cell Biol* **190**, 965–973.
8. Cailleteau, L., Estrach, S., Thyss, R., Boyer, L., Doye, A., Domange, B., Johnsson, N., Rubinstein, E., Boucheix, C., Ebrahimian, T., Silvestre, J.S., Lemichez, E., Meneguzzi, G., and Mettouchi, A. alpha2beta1 integrin controls association of Rac with the membrane and triggers quiescence of endothelial cells. *J Cell Sci* **123**, 2491–2501.
9. Fusco, D., Vargiolu, M., Vidone, M., Mariani, E., Pennisi, L.F., Bonora, E., Capellari, S., Dirnberger, D., Baumeister, R., Martinelli, P., and Romeo, G. The RET51/FKBP52 complex and its involvement in Parkinson disease. *Hum Mol Genet* **19**, 2804–2816.
10. Eckert, J.H., and Johnsson, N. (2003). Pex10p links the ubiquitin conjugating enzyme Pex4p to the protein import machinery of the peroxisome. *J Cell Sci* **116**, 3623–3634.
11. Vargiolu, M., Fusco, D., Kurelac, I., Dirnberger, D., Baumeister, R., Morra, I., Melcarne, A., Rimondini, R., Romeo, G., and Bonora, E. (2009). The tyrosine kinase receptor RET interacts in vivo with aryl hydrocarbon receptor-interacting protein to alter survivin availability. *J Clin Endocrinol Metab* **94**, 2571–2578.
12. Tiedje, C., Holland, D.G., Just, U., and Hofken, T. (2007) Proteins involved in sterol synthesis interact with Ste20 and regulate cell polarity. *J Cell Sci* **120**, 3613–3624.
13. Stagljar, I., Korostensky, C., Johnsson, N., and te Heesen, S. (1998). A genetic system based on split-ubiquitin for the analysis of interactions between membrane proteins in vivo. *Proc Natl Acad Sci USA* **95**, 5187–5192.
14. Janke, C., Magiera, M.M., Rathfelder, N., Taxis, C., Reber, S., Maekawa, H., Moreno-Borchart, A., Doenges, G., Schwob, E., Schiebel, E., and Knop, M. (2004). A versatile toolbox for PCR-based tagging of yeast genes: new fluorescent proteins, more markers and promoter substitution cassettes. *Yeast* **21**, 947–962.
15. Hruby, A., Zapatka, M., Heucke, S., Rieger, L., Wu, Y., Nussbaumer, U., Timmermann, S., Dünkler, A., and Johnsson, N. (2010) A constraint network of interactions: Protein-protein interaction analysis of the yeast type II phosphatase Ptc1p and its adaptor protein Nbp2p. *J Cell Sci.* **123**, 2491–2501.
16. Thaminy, S., Auerbach, D., Arnoldo, A., and Stagljar, I. (2003). Identification of novel ErbB3-interacting factors using the split-ubiquitin membrane yeast two-hybrid system. *Genome Res* **13**, 1744–1753.
17. Wang, B., Pelletier, J., Massaad, M.J., Herscovics, A., and Shore, G.C. (2004). The yeast split-ubiquitin membrane protein two-hybrid screen identifies BAP31 as a regulator of the turnover of endoplasmic reticulum-associated protein tyrosine phosphatase-like B. *Mol Cell Biol* **24**, 2767–2778.
18. Dirnberger, D., Messerschmid, M., and Baumeister, R. (2008). An optimized split-ubiquitin cDNA-library screening system to identify novel interactors of the human Frizzled 1 receptor. *Nucleic Acids Res* **36**, e37.
19. Reichel, C., and Johnsson, N. (2005). The split-ubiquitin sensor: measuring interactions and conformational alterations of proteins in vivo. *Methods Enzymol* **399**, 757–776.
20. Grefen, C., Obrdlik, P., and Harter, K. (2009). The determination of protein-protein interactions by the mating-based split-ubiquitin system (mbSUS). *Methods Mol Biol* **479**, 217–233.

21. Obrdlik, P., El-Bakkoury, M., Hamacher, T., Cappellaro, C., Vilarino, C., Fleischer, C., Ellerbrok, H., Kamuzinzi, R., Ledent, V., Blaudez, D., Sanders, D., Revuelta, J.L., Boles, E., Andre, B., and Frommer, W.B. (2004). K+ channel interactions detected by a genetic system optimized for systematic studies of membrane protein interactions. *Proc Natl Acad Sci USA* **101**, 12242–12247.
22. Miller, J.P., Lo, R.S., Ben-Hur, A., Desmarais, C., Stagljar, I., Noble, W.S., and Fields, S. (2005). Large-scale identification of yeast integral membrane protein interactions. *Proc Natl Acad Sci USA* **102**, 12123–12128.

Chapter 8

Genome-Wide Dissection of Posttranscriptional and Posttranslational Interactions

Mukesh Bansal and Andrea Califano

Abstract

Transcriptional interactions in the cell are modulated by a variety of posttranscriptional and posttranslational mechanisms that make them highly dependent on the molecular context of the specific cell. These include, among others, microRNA-mediated control of transcription factor (TF) mRNA translation and degradation, transcription factor activation by phosphorylation and acetylation, formation of active complexes with one or more cofactors, and mRNA/protein degradation and stabilization processes. Thus, the ability of a transcription factor to regulate its targets depends on a variety of genetic and epigenetic mechanisms, resulting in highly context-dependent regulatory networks. In this chapter, we introduce a step-by-step guide on how to use the MINDy systems biology algorithm (Modulator Inference by Network Dynamics) that we recently developed, for the genome-wide, context-specific identification of posttranscriptional and posttranslational modulators of transcription factor activity.

Key words: Posttranslational interaction, Posttranscriptional interaction, Gene expression profile, Conditional mutual information, Systems biology, Reverse engineering

1. Introduction

Reverse engineering of cellular networks in prokaryotes and lower eukaryotes (1–3), as well as in mammals (4–9), has shown the remarkable complexity of transcriptional programs. These programs, however, are not static but rather dependent on the availability of specific factors that affect the synthesis of transcription factors (TFs), their posttranslational modification, or their participation in transcriptional complexes. These factors may include microRNAs (miRNAs) as well as phosphorylation, acetylation, and ubiquitination enzymes (10). The result is that not only the presence

Bart Deplancke and Nele Gheldof (eds.), *Gene Regulatory Networks: Methods and Protocols*,
Methods in Molecular Biology, vol. 786, DOI 10.1007/978-1-61779-292-2_8, © Springer Science+Business Media, LLC 2012

of an interaction but even its strength and activity (activator or repressor) can be completely dependent on the specific molecular context. For instance, the ability of the MYC transcription factor to repress specific targets is dependent on the availability of the MIZ1 cofactor (11).

Although the large-scale reprogramming of the cell's transcriptional logic was studied qualitatively in yeast (12, 13), the identification of the specific gene products that may rewire the cell's regulatory logic remains elusive. Indeed, at the time we introduced the Modulator Inference by Network Dynamics (MINDy) algorithm (14), there was only one experimentally validated algorithm which, by allowing substrate interference of 73 kinases, enabled the dissection of signaling networks in a mammalian context (15). Here, we discuss the MINDy algorithm in greater detail, providing specific examples for its application and offering further insights into the way it can be used as an effective tool for the dissection of posttranscriptional and posttranslational interactions. MINDy is a gene expression profile-based method, for the systematic identification of genes that modulate a TF's transcriptional program at the posttranslational level, i.e., those genes encoding proteins that affect the activity of a TF without changing its mRNA abundance. These proteins may posttranslationally modify the TF (e.g., kinases), affect its cellular localization or turnover, be its cognate partners in transcriptional complexes, or compete for its DNA binding sites. They may also include proteins that do not physically interact with the TF, such as those in its upstream signaling pathways. However, provided for instance that miRNA and gene expression profiles are available for the same samples, MINDy could also be used to identify proteins that modulate the interaction between miRNAs and their targets. Thus, we also present some additional applications of the MINDy algorithm, that may rely on additional data types, including the inference of proteins that modulate miRNA target interactions, better prediction of TF target pairs, and the identification of genomic aberrations with a functional effect.

2. Methods

The probability distribution of the expression state of an interaction network can be written as the product of functions of the individual genes, their pairs, and higher order combinations (16, 17). Most reverse engineering techniques are based on pairwise statistics (16–18), thus failing to reveal third and higher order interactions. Alternatively, some methods attempt to address the full dependency model (19) based on Bayesian networks, where they try to identify the causal relationship between genes using the probabilistic dependencies between them. This makes the problem computationally intractable to be applied on genome-wide scale and also under-sampled as lots of samples are required to evaluate the probabilistic dependencies between the genes. Given these limitations, MINDy addresses the intermediate task of identifying a specific type of third-order multivariate statistical dependencies that are biologically relevant yet still computationally tractable, even in a mammalian context. Specifically, MINDy is restricted to the inference of third-order interactions, where the ability of a TF to transcriptionally activate or repress its target(s) is a function of a third gene-product.

3. The MINDy Algorithm

MINDy (Modulator Inference by Network Dynamics) interrogates a large gene expression profile dataset to identify *candidate modulator* genes whose expression strongly correlates with changes in a TF's transcriptional activity. MINDy is based on a multivariate statistical dependence model designed to capture a particular type of three-way interaction, where the ability of a transcription factor to control its target gene, t, is influenced by a (possibly large) number of other proteins, which are referred to as modulators (M). These modulators do not need to directly interact with the TF, but may regulate another gene or protein that subsequently interacts with this TF downstream. This can be efficiently accomplished by computing an information theoretic measure known as the *Conditional Mutual Information* (CMI), $I[\mathrm{TF};t \mid M]$ (20), between the expression profile of a transcription factor and one of its putative targets (t), given the expression of a modulator gene (M). Accurate estimation of the CMI requires exceedingly large datasets. Thus, MINDy infers candidate modulators using a related, yet simpler estimator, which is denoted as ΔI. Given a triplet (TF, M, t), with $t \neq \mathrm{TF}$ and $t \neq M$, MINDy assesses whether the CMI, $I\left[\mathrm{TF};t \mid M\right]$, is constant as a function of M. This can be efficiently tested by measuring

$$\Delta I = I\left[\mathrm{TF};t \mid M \in L_m^+\right] - I\left[\mathrm{TF};t \mid M \in L_m^-\right] \neq 0$$

where L_m^+ and L_m^- represent two subsets of the samples where M is respectively most and least expressed (Fig. 1).

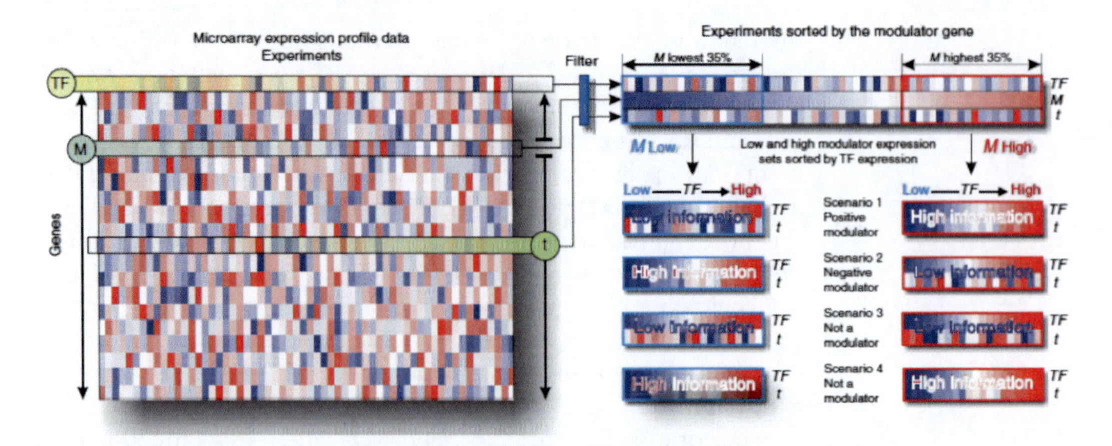

Fig. 1. A gene expression profile dataset is represented as a matrix, where *columns* indicate different samples and *rows* indicate different genes. Given a transcription factor of interest, TF, a modulator and a target to be tested by MINDy (*M* and *t* respectively) are chosen among the remaining genes. Some modulator-target combinations may be eliminated a priori based on functional or statistical constraints (*blue rectangle*). For instance, one may want to consider only ubiquitin ligases as candidate modulators. All the samples are then sorted according to the expression of the selected modulator *M*. The set of samples (e.g., 35%) with the lowest and highest expression of the modulator are then selected. These sample sets are labeled *M*-low and *M*-high. In each of the two sample sets, samples are finally sorted according to the *TF* expression. Three cases are possible when comparing the TF-target correlation in M-high vs. *M*-low. *Scenario 1* (*Positive Inferred Modulator*): Significant increase in mutual information (i.e., more correlation in the *M*-high set than in the *M*-low set); *Scenario 2* (*Negative Inferred Modulator*): Significant decrease in mutual information (i.e., less correlation in the *M*-high set than in the *M*-low set); *Scenario 3 and 4* (*Not a Modulator*): No significant mutual information change is observed.

4. Procedure

4.1. Step 1: Selection of Candidate Modulator Genes

Any modulator is considered as a candidate modulator, *M*, according to two criteria:

1. *M* must have a sufficient expression range to separate its two expression tails compared to the experimental noise level. This criterion requires an assessment of the measurement error in microarray data. For each candidate modulator, *M*, in our dataset, the *x*th and *y*th quantiles of its expression values are used to define its two expression tails, $L_m^-(M < m^{(x)})$ and $L_m^+(M > m^{(y)})$, where $m^{(x),(y)}$ represents, respectively, the *x*th and *y*th quantile expression level of the modulator (see Note 1). The gene expression measurement error model in (21) is used, where the authors quantified the measurement noise in the set of microarray experiments used in (14, 17) (254 B-cell microarrays, GEO accession number GSE2350) to evaluate if *M* has significant expression range or not. At each expression level, the standard deviation of the measurement error can be quantified as:

$$\sigma^2(\theta) = \beta e^{-\Upsilon e}$$

where θ is the logarithm transformed gene expression measurement, and $\beta = 4.6 \pm 0.2$ ⊲ and $\gamma = 1.1 \pm 0.1$ based on dataset used in (14, 17) (254 B-cell microarrays).

Given this model, a statistical test is devised by constructing the following statistic

$$z = \frac{\theta^{(y)} - \theta^{(x)}}{\sqrt{\sigma^2(\theta^{(y)}) + \sigma^2(\theta^{(y)})}} \sim N(0,1)$$

where $\theta^{(y)} = \log(m^{(y)})$ and $\theta^{(x)} = \log(m^{(x)})$. If the one-sided p-value associated with this statistic (computed using the normal distribution with mean 0 and standard deviation 1) is smaller than 0.05, then the two tails are considered as statistically separated in the presence of measurement noise. This test is very conservative, since statistical separation is tested between the two close boundaries of the expression tails, which guarantees that all data points in the two expression tails can be statistically distinguished.

2. Any statistical dependence of M on TF can cause difficulty in estimating the significance of ΔI, as the value of M conditions the range of TF. So any M that is not statistically independent of TF is excluded. To evaluate if TF and M are statistically independent, we compute mutual information between TF and M and if mutual information has a p-value$\geq 10^{-5}$ (see Note 2), then they are considered statistically independent (16). This criterion removes genes that may transcriptionally interact with TF, which can be trivially detected by pair-wise coexpression analysis, e.g., using mutual information. A substantial increase of false positives is not expected in this condition. In fact, it is reasonable to expect that the expression of a posttranscriptional modulator of a TF should be statistically independent of the TF's expression.

4.2. Step 2: Evaluation of Conditional Mutual Information Statistics

Each candidate modulator M is then used to partition the complete set of expression profiles into two equal-sized, nonoverlapping subsets L_m^- and L_m^+. For each subset, mutual information between TF and t, I^- and I^+ is computed using a computationally efficient Gaussian Kernel estimator (22). Given a set of two-dimensional measurements, $\vec{z}_i \equiv \{x_i, y_i\}, i = 1 \ldots M$, the mutual information is computed as:

$$I(\{x_i\},\{y_i\}) = \frac{1}{M}\sum_i \log\frac{(f(x_i,y_i))}{f(x_i)f(y_1)}$$

where $f(x,y)$ is the joint probability distribution, $f(x)$ and $f(y)$ are the marginals of $f(x,y)$ and they are defined as:

$$f(x_i) = \frac{1}{\sqrt{2\pi}Mh_1}\sum_j e^{-\frac{|x_i - x_j|^2}{2h_1^2}}$$

$$f(y_i) = \frac{1}{\sqrt{2\pi} M h_2} \sum_j e^{-\frac{|y_i - y_j|^2}{2h_2^2}}$$

$$f(x_i, y_i) = \frac{1}{2\pi M h_1 h_2} \sum_j e^{-\left(\frac{|x_i - x_j|^2}{2h_1^2} + \frac{|y_i - y_j|^2}{2h_2^2}\right)}$$

Here, h_1 and h_2 are called the smoothing parameters or kernel widths. Mutual information has the property that it is reparameterization invariant (for proof see (23)), so before computing mutual information we reparameterize the data using a rank transformation. This rank transformation projects the M measurements for each gene into equally spaced real numbers in the interval [0, 1], preserving their original order. This transformation is called copula-transformation (24). Copula-transformation decreases the influence of arbitrary transformations involved in microarray data preprocessing and removes the need to consider position-dependent kernel widths, h_1 and h_2, and we can consider a unique kernel width h for all genes. To obtain the optimal value of h, we used Monte Carlo simulations, using a wide range of bivariate normal probability densities. Details of how to compute h can be found in the supplementary material of (25). A MATLAB script to compute h as well as to compute mutual information using Gaussian kernel method can be downloaded from http://wiki.c2b2.columbia.edu/califanolab/index.php/Software.

Once I^- and I^+ are computed, the statistical significance of I^- and I^+ is calculated. ΔI is computed if either I^- or I^+ is significant (p-value ≤ 0.05, Bonferroni corrected for the total number of targets, see Note 2). Significance of ΔI is then evaluated as a function of I, using the extended exponentials fit (see Note 2). Gene pairs with a statistical p-value ≤ 0.05, Bonferroni corrected for the total number of targets times the number of modulators for each TF, are retained for further analysis.

4.3. Step 3: Modulator Pruning Using Data Processing Inequality

Significant modulators identified by CMI testing are further pruned if the interaction between a TF and a target gene is inferred as an indirect one in both conditions of the modulator, based on the Data Processing Inequality (DPI). DPI is a well-known property of mutual information which states that the interaction between two genes is likely to be indirect (i.e., mediated through a third gene), if it holds true that $I[\text{TF};t] < (1-\varepsilon)\min(I[\text{TF};\text{TF}_2], I[\text{TF}_2;t])$ (16, 17), where ε is DPI tolerance threshold and TF_2 is another transcriptional factor which is predicted to be regulating the same target t. This step eliminates some specific cases, illustrated in Fig. 2, where a modulator can produce a significant CMI even though it does not directly affect the $\text{TF} \rightarrow t$ interaction. To apply DPI,

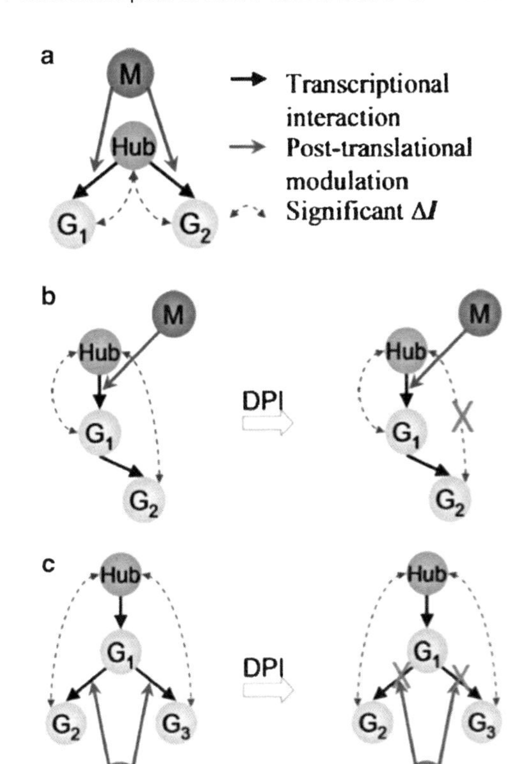

Fig. 2. Schematic diagram of the effect of DPI on eliminating indirect regulatory relationships. (**a**) Correct modulation model where the modulator (*M*) significantly changes the regulatory relationship between the TF (Hub) and its direct targets (G_1 and G_2). (**b**) Removal of indirect connections to the hub eliminates the detection of a significant ΔI on indirect targets. (**c**) Modulators that affect the downstream targets of the TF hub, thus causing significant ΔI between the TF and its indirect neighbors, will be removed by applying DPI.

examine each gene triplet (in both conditions of the modulator) for which all three mutual information are statistically significant and remove the edge if it satisfies the above inequality. Each triplet is analyzed irrespective of whether its edges are marked for removal by prior DPI applications to triplets. Thus, the method is independent of the order in which the triplets are examined. As discussed in (16, 17), the DPI was applied with a 10% tolerance to minimize the impact of potential mutual information estimation errors.

4.4. Step 4: Modulator Selection Based on the Number of TF Target Support

Once MINDy inference is made on individual (TF, M, t) sets, modulators are selected based on their support, i.e., the number of TF targets they modulate. The minimum support is determined using a permutation test procedure, where an identical MINDy run is performed on the same set of candidate modulators and candidate targets except that L_m^+ and L_m^- are chosen at random by randomly selecting the samples for two tails from the entire sample set.

This produces a permutated set of MINDy inferences, based on which the modulator support under the null hypothesis is computed. The minimum support is then selected as the support that gives the smallest ratio between the numbers of selected modulators in the permutated run versus the real run (FDR).

4.5. Step 5: Inference of Modulator Mode of Action and Biological Activity

For each significant triplet (TF, t, M) identified by MINDy, the modulator M's *mode of action* with respect to the TF$\rightarrow t$ interaction is defined as:

$$\begin{cases} \text{positive modulator, if } \Delta I\left[\text{TF}; t \mid M\right] > 0 \\ \text{negative modulator, if } \Delta I\left[\text{TF}; t \mid M\right] < 0 \end{cases}$$

This indicates only whether an increase in M's abundance is associated with a corresponding increase or decrease in the mutual information between TF and t. However, it does not necessarily reflect the biological nature of this modulatory interaction. To this end, one must consider instead the modulator's *biological activity*: a modulator is considered an activator if it enhances the TF's ability to transcriptionally control its target gene(s), and an antagonist otherwise. See Table 1 for the summary of relationship between the MINDy mode of action, the correlation between TF and its target gene, and the biological activity of the modulator. The biological activity of M with respect to TF$\rightarrow t$ interaction is determined as:

$$\begin{cases} \text{activator} & \text{if } \rho\left(\mu_t^+ - \mu_t^-\right) > 0 \\ \text{antagonist} & \text{if } \rho\left(\mu_t^+ - \mu_t^-\right) < 0 \\ \text{undetermined} & \text{if } \rho\left(\mu_t^+ - \mu_t^-\right) \approx 0 \end{cases}$$

where ρ is the Pearson correlation between the TF and the candidate target t, and $\mu_t^{\pm}$ is the mean expression of t in $L_m^{\pm}$. These differences are assessed using a two-sample (two sided) Student t-test with 10% type-I error rate. For modulators affecting more than one target, the biological mode is labeled as undetermined if the undetermined triplets make up the majority (>50%) or if neither mode dominates the other by more than 30%. Otherwise, it is assigned the dominant mode.

5. Applications

MINDy has already been applied successfully to the genome-wide identification of modulators of the MYC proto-oncogene, using a previously assembled collection of 254 gene expression profiles (17, 26), representing 17 distinct cellular phenotypes derived from normal and neoplastic human B lymphocytes (14). MYC is a

Table 1
Inferring the biological activity of a MINDy modulator

MoA	ρ	$\mu_t^+ - \mu_t^-$	Plot	BA	Sign $(\rho(\mu_t^+ - \mu_t^-))$
+	+	+	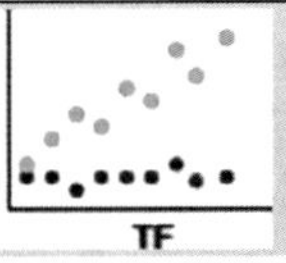	Activator	+
+	+	−	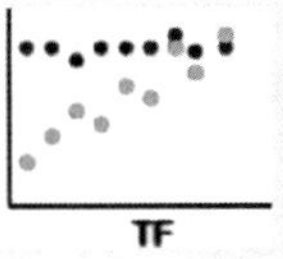	Antagonist	−
+	−	−	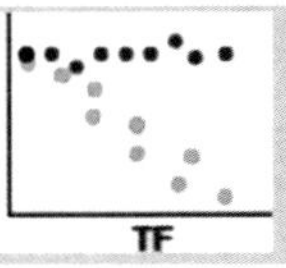	Activator	+
+	−	+	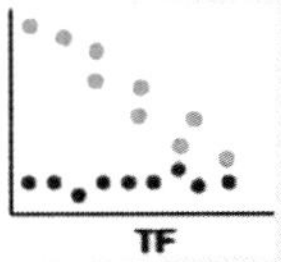	Antagonist	−
−	+	−	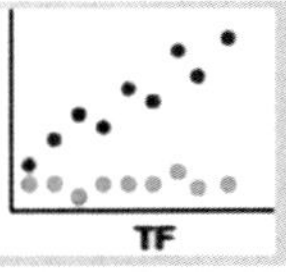	Antagonist	−
−	+	+	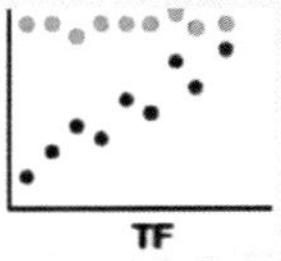	Activator	+
−	−	+	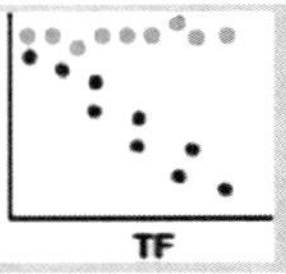	Antagonist	−
−	−	−	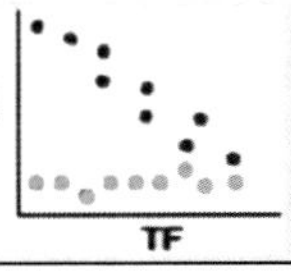	Activator	+

MoA: MINDy mode of action, ρ: Pearson correlation between the TF and the target gene *t*, $\mu_t^{\pm}$: mean expression of *t* in the most and least expressed condition of the modulator; *BA*: Biological activity. The schematic scatter plots shown in the table demonstrate the relationship between TF and *t* when the modulator is most (*gray dots*) and least (*black dots*) expressed. BA is determined by considering the sign of product of ρ (*column 2*) and $\mu_t^+ - \mu_t^-$ (*column 3*) as shown in *column 6*

proto-oncogene encoding a basic helix-loop-helix/leucine zipper (bHLH/Zip) TF, which controls many cellular processes, including cell growth, differentiation, apoptosis, and DNA replication (27, 28), through the activation and repression of a large number of targets, depending on cellular context (29). It has been implicated in the pathogenesis of several human cancers (30).

First, we performed an unbiased analysis, where gene expression profile of all genes satisfying the range and independence constraints were included in the candidate modulator/target lists. From a list of 3,146 genes meeting the candidate modulator selection criteria, MINDy inferred a repertoire of 296 modulators (at a 5% statistical significance level, Bonferroni corrected). Gene Ontology (GO) molecular function enrichment analysis of the inferred modulators revealed only three enriched categories, at a false discovery rate (FDR) of less than 5%, including transcription factors ($FDR < 1.4 \times 10^{-3}$), acetyltransferases ($FDR < 1.5 \times 10^{-2}$), and protein kinases ($FDR < 3.4 \times 10^{-2}$). Other significant categories were either parents or children of these in the GO hierarchy. Thus, inferred modulators were enriched in functional categories known to include effectors of MYC activity (31–33). Among the initial set of 3,146 candidate genes, only 25 were reported to interact with MYC in the Human Protein Reference Database, HPRD (34). Eight of the 25 (32%) were inferred by MINDy as MYC modulators ($p < 2.7 \times 10^{-4}$). These include Cyclin T1 (CCNT1) (35), the kinases MAP2K1 (MEK1) (36) and CSNK2A1 (37), the DNA mismatch repair protein MLH1 (38), the transcription factors SMAD2/SMAD3 (39), RELA (40), and the regulator of MYC-related apoptosis TIAM1 (41). This shows that MINDy is effective at identifying TF modulators even in the absence of any prior knowledge about posttranslational interaction mechanisms. 83 upstream modulators of MYC, reported in the Ingenuity pathway tool™ (42) were studied, which were expressed in the gene expression profile data and viable for MINDy analysis. Of these, 29 (35%) could be inferred as MYC modulators by MINDy ($p \leq 0.0059$ by Fisher Exact Test, FET). Targeted analysis of 542 signaling genes (protein kinases, phosphatases, acetyltransferases, and deacetylases) and 598 TFs identified 116 and 123 MYC modulators, respectively. Among these, 10 of the top 30 signaling genes (33.3%) appeared as validated MYC modulators in the literature. Similarly, 15 of the top 35 TFs (42.9%) were either validated in the literature or had highly enriched DNA-binding sites in the associated MYC targets. Four of the newly inferred MYC modulators were selected for biochemical validation, and were demonstrated their ability to affect MYC activity without affecting its expression (i.e., the definition of a posttranslational modulator). For instance, Fig. 3a shows that silencing STK38 by lentiviral vectormediated shRNA expression dramatically reduces MYC protein levels while MYC mRNA is unaffected, see Fig. 3b. Finally, coimmunoprecipitation (co-IP) of endogenous MYC and STK38 using specific antibodies in the Ramos cell line showed that the two proteins can interact in vivo (Fig. 3c).

Following this study, MINDy has been applied to the discovery of all regulatory interactions between signaling molecules

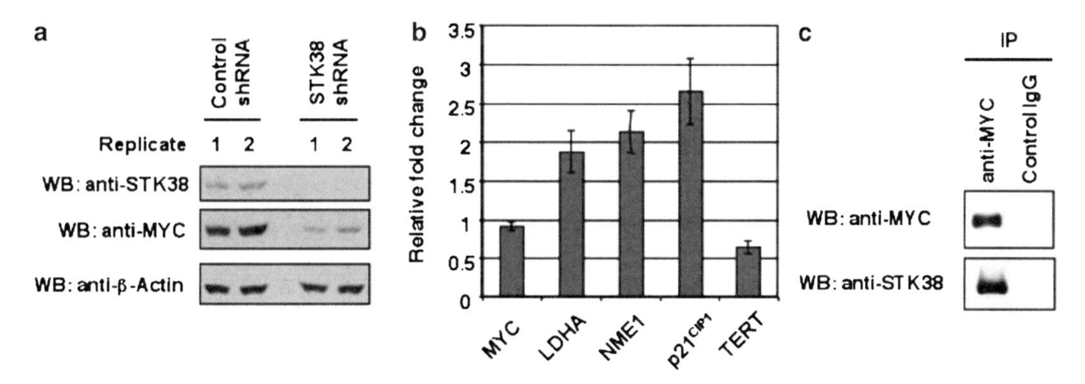

Fig. 3. (**a**) shRNA mediated silencing of STK38 decreases MYC protein levels but not (**b**) its mRNA levels while the mRNA levels of MINDy predicted MYC targets are affected as predicted. (**c**) MYC and STK38 could be co-immunoprecipitated endogenously by specific antibodies.

and transcription factors (43), as well as to the inference of all protein-protein interactions in the cell (26, 44). More specifically, MINDy has been successful in the identification of key posttranslational regulatory relationships that are involved in brain morphogenesis and tumorigenesis (7).

6. Other Applications

Clearly, third-order multivariate interactions are not restricted to those between posttranslational acting proteins and TFs. Indeed, MINDy can be equally effective in dissecting a variety of additional modulatory molecular interactions depending on the availability of additional data modalities. These include, among others, the study of the impact of epigenetic and genetic variants/mutations, as well as the role of miRNAs in transcriptional regulation. In the following paragraph, we provide ideas for the specific application of MINDy in these areas.

miRNA data: As in the case of protein-coding genes, miRNA transcription regulation is a key component of miRNA biogenesis. Validated miRNA transcriptional regulators in tumor cells include p53 which regulates miR-34 (45), and MYC which regulates the expression of members of the miR-17-92 miR cluster (46). Recently, miR-26a has been implicated in glioblastoma multiforme (GBM) cell growth (47), and amplification at its host loci was associated with increased tumor proliferation. It has already been shown that STAT3, which is implicated in regulating the mesenchymal-specific gene expression data signature (48), is a transcriptional regulator of miR-26a. MINDy could be applied to the inference of additional

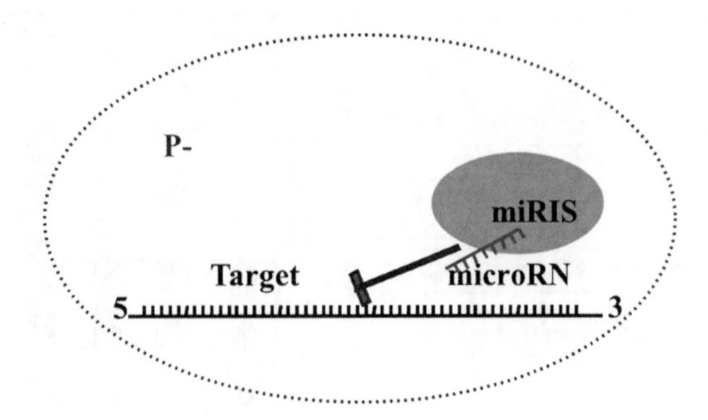

Fig. 4. miR-mediated degradation depends on the availability of other degradation factros (miR-modulators).

proteins that modulate the interaction between miRNAs and their targets. For instance, the potential for miRNA regulation is dependent on the availability of machinery to synthesize mature miRNAs from their precursors and to bind and regulate their targets; see Fig. 4 for an illustration. The expression of a battery of genes implicated in miRNA biosynthesis and miRNA-target processing can be used as candidate modulators of miRNA regulation (miRNA-modulators). MINDy could be applied on each validated miRNA target interactions to search for their modulators using all genes as candidate modulators.

Other data types: MINDy could also be applied on Copy Number Variation (CNV) and SNP data to uncover the functional role of genetic variants or mutations. For instance, a SNP in the coding region of a TF may not affect its expression level. However, it may result in a change in the TF's activity. MINDy can help in detecting the activity change of a TF as a function of the presence or absence of the specific SNP. To apply MINDy, unlike gene expression profile data where the most and least expressed conditions for candidate modulator are considered as two tails, samples with homozygous major and heterozygous in a SNP dataset and amplified and deleted samples in a CNV dataset could be considered as two tails and ΔI could be evaluated to infer the functional impact of these aberrations.

Inference of pathways: Predictions of MINDy could be effectively used to map physical and pathway-mediated posttranslational interactions between signaling proteins and transcription factors (43). Specifically, signaling proteins that are in the same pathway should affect a significant number of the same TFs. Additionally, a signaling protein should affect a superset of the TFs affected by its substrate. This may be helpful in resolving the topology of signaling programs, which is only marginally understood.

7. Notes

1. Optimal selection of $L_m^{\pm}$

   The size of $L_m^{\pm}$ can be chosen between 0 and 50% of all samples such that two subsets contain the same number of samples. The trade-off is intuitive: if it is too small, there are not enough samples in each subset to accurately estimate I^+, I^-, and ΔI; on the other hand, if $|L_m^{\pm}|$ is close to 50%, then the modulator's expression levels in the two subsets are not sufficiently separated, mitigating the modulation effect, resulting in the fact that too many modulators will fail to satisfy the *Range Constraint*. The optimal choice should lie somewhere in between and should be determined empirically.

   To estimate $L_m^{\pm}$, we computed (14) the number of known MYC target genes being modulated by genes in the B-cell receptor (BCR) pathway, which are known to harbor MYC modulators because upstream of GSK3, JNK, and ERK kinases directly phosphorylate MYC. For each $|L_m^{\pm}| = 10\%, 15\%, \ldots, 45\%$, we ran MINDy to search for MYC modulators among the BCR pathway genes on the set of known MYC target genes. The 35% threshold corresponded to the optimal number of MYC-targets modulated by at least one BCR pathway gene (Fig. 5). The function has a broad maximum, with values between 25 and 40% producing in excess of 700 modulated targets. As shown in (14), this threshold is not MYC specific and performs well for a variety of TFs. We tested the performance of using a threshold

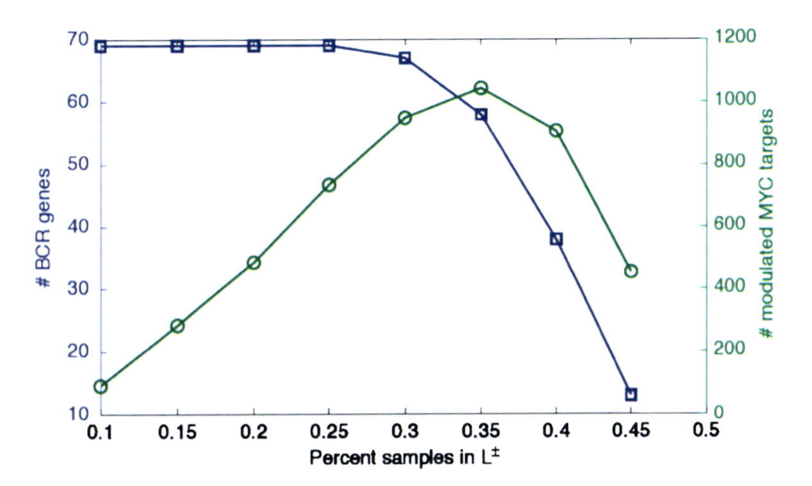

Fig. 5. Optimal size of the expression tail for modulator identification. The *blue curve* using the left *y*-axis plots the number of BCR genes that satisfy the MINDy range constraint. The *green curve* using the right *y*-axis shows the total number of known MYC target genes modulated by at least one BCR pathway gene. 35% is selected as the optimal tail size for this dataset.

of 35% on various other datasets and found that it gives an optimal performance, so we recommend it to be fixed as a default threshold.

2. Statistical threshold for mutual information and ΔI

   Mutual information is always non-negative, i.e., its evaluation from random samples gives a positive value even for variables that are, in fact, mutually independent. Therefore, we eliminate all edges for which the null hypothesis of mutually independent genes cannot be ruled out. To this extent, we randomly shuffle the expression of genes across the various microarray profiles similar to (18), and evaluate the mutual information for such manifestly independent genes and assign a p-value, p, to the mutual information threshold, I_o, by empirically estimating the fraction of the estimates below I_o. This is done for different sample sizes M and for 10^5 gene pairs so that reliable estimates of $I_o(p)$ are produced up to $p = 10^{-4}$. Extrapolation to smaller p-values is done using $p(I \geq I_0 \mid \bar{I} = 0) \propto e^{-\alpha M I_o}$, where the parameter α is fitted from the data. This formula is based on the intuition of the large deviation theory (20), which for discrete data and unbiased estimators suggests $p(I \geq I_0 \mid \bar{I} = 0) \propto e^{-M I_o}$. The log of $p(I \geq I_0 \mid \bar{I} = 0) \propto e^{-\alpha M I_o}$ is $\log p = \beta + \gamma I_o$ which is a linear equation. Therefore, $\log p$ can be fitted as a linear function of I_o and the slope, γ, should be proportional to M. Based on the above equation, we analyzed the empirical distribution of the mutual information estimated using 10^5 statistically independent variables, by randomizing the samples for each gene independently, for samples of variable sizes (between 30 and 100% of M). For each sample size, we select the largest 10% of mutual information values and fit it to the above linear equation. We repeat this step three times by randomly selecting samples and the resulting fits are averaged to avoid biased samples. We compute the constant β as the average β for all sample sizes and fit γ as a linear function of the number of samples ($(\gamma = a + Nb)$, where N is the sample size. The resulting fitting values are used in $\log p = \beta + \gamma I_o$ to obtain the threshold for mutual information for the desired p-value.

   To assess the statistical significance of a particular ΔI value, a series of null hypotheses are generated, and the ΔI distribution is measured across 10^4 distinct (TF, t) pairs with random condition. For each gene pair, the expression profiles in the non-overlapping subsets, $L_m^{\pm}$ used to measure $I^{\pm}$ and ΔI, are chosen at random from the entire dataset and 1,000 ΔI from random subsamples are generated for each gene pair. Since the statistics of ΔI depends on I, bin I into 100 equiprobable bins, resulting in 100 gene pairs and 10^5 ΔI measurements per bin. Within each bin, model the distribution of ΔI

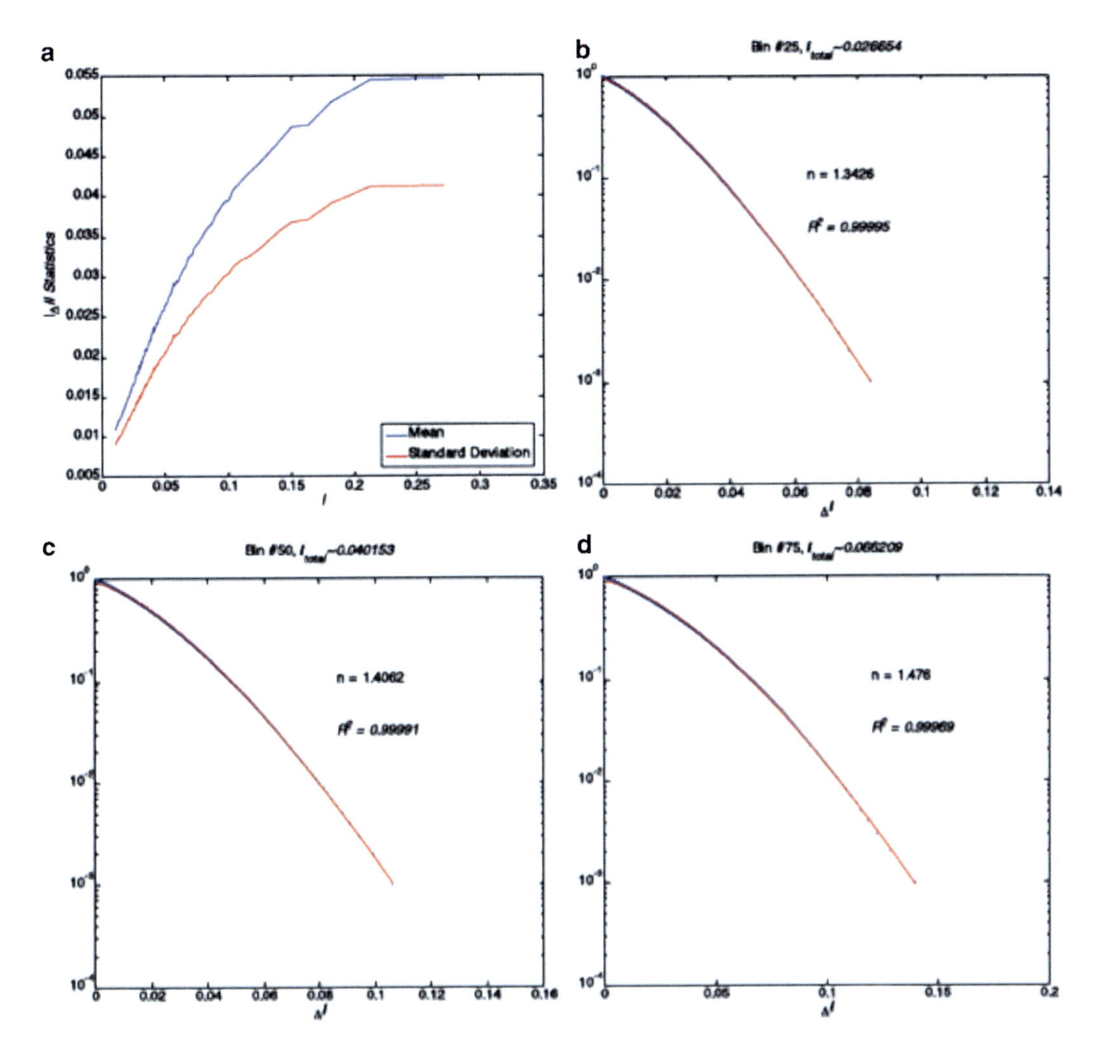

Fig. 6. Null distributions for the ΔI statistics. (**a**) Mean and standard deviation of the ΔI statistics in each bin as a function of I. (**b**–**d**) Examples of distributions of the ΔI statistics (*blue curves*) and the extended exponential function (*red curves*), $(I \geq I_o \mid \overline{I} = 0) = e^{(-\alpha MI_o + \beta)}$, obtained by the least squares fitting in bins 25, 50, and 75; a goodness-of-fit measure, R^2, and the value of n are also shown for each bin.

as an extended exponential, $p(\Delta I) = \exp(-\alpha \Delta I^n + \beta)$, which allows to extrapolate the probability of a given ΔI under the null hypothesis model. Both mean and standard deviation of ΔI increase monotonically with I (Fig. 6), and the extended exponentials produce an excellent fit for all bins. Specifically, for small I, the exponent of the fitted function is $n \approx 1$ because both I^+ and I^- are close to zero and ΔI is dominated by the estimation error, which falls off exponentially according to large deviation theory (16, 49). For a large I, the error estimation becomes smaller than the true mutual information difference between the two random subsamples, hence $n \approx 2$ from the central limit theorem.

8. Limitations of MINDy

1. MINDy cannot be applied on a modulator if its expression is statistically correlated with the expression of the TF. This constraint leads to a decrease of the false positive rate but at the expense of an increase in the false negative rate as the expression of some TFs may be correlated with the expression of the respective modulators.
2. MINDy cannot be applied on a modulator if the expression of the modulator is not varying across samples and if the samples cannot be separated into the most and least expressed condition of the modulator. This can, however, be mitigated by using larger datasets.
3. Modulators of TFs identified by MINDy are always going to be functional, but they may not necessarily be direct. For example, if two modulators are in the same pathway, then both of them could be predicted to be modulating the downstream TF.
4. One of the major limitations of MINDy is the required number of samples (at least 200). If there are fewer samples, then MINDy cannot be applied on that dataset.
5. Application of MINDy on a genome-wide scale requires extensive computational resources.
6. MINDy may detect false positives if the expression of a gene is highly correlated with the expression of a bona fide modulator.

Additional Information

MINDy executables and MATLAB scripts to compute mutual information, kernel width and the statistical threshold for mutual information and ΔI can be downloaded from http://wiki.c2b2.columbia.edu/califanolab/index.php/Software.

Acknowledgments

We would like to thank Pavel Sumazin for providing the insight into the application of MINDy on miRNAs and providing the Figure for it and Paolo Guarnieri for proof reading the document.

References

1. Faith, J. J., Hayete, B., Thaden, J. T., Mogno, I., Wierzbowski, J., Cottarel, G., Kasif, S., Collins, J. J., and Gardner, T. S. (2007) Large-Scale Mapping and Validation of Escherichia coli Transcriptional Regulation from a Compendium of Expression Profiles, *PLoS Biol* 5, e8.
2. Friedman, N. (2004) Inferring cellular networks using probabilistic graphical models, *Science* 303, 799–805.
3. Gardner, T. S., di Bernardo, D., Lorenz, D., and Collins, J. J. (2003) Inferring genetic networks and identifying compound mode of action via expression profiling, *Science* 301, 102–105.
4. Basso, K., Margolin, A. A., Stolovitzky, G., Klein, U., Dalla-Favera, R., and Califano, A. (2005) Reverse engineering of regulatory networks in human B cells, *Nat Genet* 37, 382–390.
5. Elkon, R., Linhart, C., Sharan, R., Shamir, R., and Shiloh, Y. (2003) Genome-Wide In Silico Identification of Transcriptional Regulators Controlling the Cell Cycle in Human Cells, *Genome Res.* 13, 773–780.
6. Stuart, J. M., Segal, E., Koller, D., and Kim, S. K. (2003) A Gene-Coexpression Network for Global Discovery of Conserved Genetic Modules, *Science* 302, 249–255.
7. Zhao, X., D, D. A., Lim, W. K., Brahmachary, M., Carro, M. S., Ludwig, T., Cardo, C. C., Guillemot, F., Aldape, K., Califano, A., Iavarone, A., and Lasorella, A. (2009) The N-Myc-DLL3 Cascade Is Suppressed by the Ubiquitin Ligase Huwe1 to Inhibit Proliferation and Promote Neurogenesis in the Developing Brain, *Dev Cell* 17, 210–221.
8. Carro, M. S., Lim, W. K., Alvarez, M. J., Bollo, R. J., Zhao, X., Snyder, E. Y., Sulman, E. P., Anne, S. L., Doetsch, F., Colman, H., Lasorella, A., Aldape, K., Califano, A., and Iavarone, A. (2009) The transcriptional network for mesenchymal transformation of brain tumours, *Nature*.
9. Della Gatta, G., Bansal, M., Ambesi-Impiombato, A., Antonini, D., Missero, C., and di Bernardo, D. (2008) Direct targets of the TRP63 transcription factor revealed by a combination of gene expression profiling and reverse engineering, *Genome Res* 18, 939–948.
10. Zeitlinger, J., Simon, I., Harbison, C. T., Hannett, N. M., Volkert, T. L., Fink, G. R., and Young, R. A. (2003) Program-Specific Distribution of a Transcription Factor Dependent on Partner Transcription Factor and MAPK Signaling, *Cell* 113, 395.
11. Si, J., Yu, X., Zhang, Y., and DeWille, J. W. (2010) Myc interacts with Max and Miz1 to repress C/EBPdelta promoter activity and gene expression, *Mol Cancer* 9, 92.
12. Luscombe, N. M., Babu, M. M., Yu, H., Snyder, M., Teichmann, S. A., and Gerstein, M. (2004) Genomic analysis of regulatory network dynamics reveals large topological changes, *Nature* 431, 308–312.
13. Segal, E., Shapira, M., Regev, A., Pe'er, D., Botstein, D., Koller, D., and Friedman, N. (2003) Module networks: identifying regulatory modules and their condition-specific regulators from gene expression data, *Nat Genet* 34, 166–176.
14. Wang, K., Saito, M., Bisikirska, B. C., Alvarez, M. J., Lim, W. K., Rajbhandari, P., Shen, Q., Nemenman, I., Basso, K., Margolin, A. A., Klein, U., Dalla-Favera, R., and Califano, A. (2009) Genome-wide identification of post-translational modulators of transcription factor activity in human B cells, *Nat Biotechnol* 27, 829–837.
15. Linding, R., Jensen, L. J., Ostheimer, G. J., van Vugt, M. A., Jorgensen, C., Miron, I. M., Diella, F., Colwill, K., Taylor, L., Elder, K., Metalnikov, P., Nguyen, V., Pasculescu, A., Jin, J., Park, J. G., Samson, L. D., Woodgett, J. R., Russell, R. B., Bork, P., Yaffe, M. B., and Pawson, T. (2007) Systematic discovery of in vivo phosphorylation networks, *Cell* 129, 1415–1426.
16. Margolin, A. A., Nemenman, I., Basso, K., Wiggins, C., Stolovitzky, G., Favera, D., and Califano, A. (2006) ARACNE: An Algorithm for the Reconstruction of Gene Regulatory Networks in a Mammalian Cellular Context, *BMC Bioinformatics* 7 Suppl 1, S1–7.
17. Basso, K., Margolin, A. A., Stolovitzky, G., Klein, U., Dalla-Favera, R., and Califano, A. (2005) Reverse engineering of regulatory networks in human B cells, *Nat Genet* 37, 382–390.
18. Butte, A. J., and Kohane, I. S. (2000) Mutual information relevance networks: functional genomic clustering using pairwise entropy measurements, *Pac Symp Biocomput*, 418–429.
19. Friedman, N., Linial, M., Nachman, I., and Pe'er, D. (2000) Using Bayesian networks to analyze expression data, *J Comput Biol* 7, 601–620.
20. Cover, T. M., and Thomas, J. (2006) *Elements of Information Theory, 2nd edition*, Wiley Interscience.
21. Tu, Y., Stolovitzky, G., and Klein, U. (2002) Quantitative noise analysis for gene expression microarray experiments, *Proc Natl Acad Sci USA* 99, 14031–14036.
22. Nemenman, I. (2004) Information theory, multivariate dependence, and genetic network

inference, In *Technical Report NSF-KITP-04-54, KITP*, http://arxiv.org/abs/0904.1587, UCSB.

23. Kraskov, A., Stogbauer, H., and Grassberger, P. (2004) Estimating mutual information, *Phys Rev E Stat Nonlin Soft Matter Phys* 69, 066138.
24. Joe, H. (1997) *Multivariate models and dependence concepts.*, Chapman & Hall, Boca Raton, FL.
25. Margolin, A. A., Wang, K., Lim, W. K., Kustagi, M., Nemenman, I., and Califano, A. (2006) Reverse engineering cellular networks, *Nat Protocols* 1, 663–672.
26. Mani, K. M., Lefebvre, C., Wang, K., Lim, W. K., Basso, K., Dalla-Favera, R., and Califano, A. (2008) A systems biology approach to prediction of oncogenes and molecular perturbation targets in B-cell lymphomas, *Mol Syst Biol* 4, 169.
27. Dang, C. V., O'Donnell, K. A., Zeller, K. I., Nguyen, T., Osthus, R. C., and Li, F. (2006) The c-Myc target gene network, *Seminars in Cancer Biology* 16, 253–264.
28. Dominguez-Sola, D., Ying, C. Y., Grandori, C., Ruggiero, L., Chen, B., Li, M., Galloway, D. A., Gu, W., Gautier, J., and Dalla-Favera, R. (2007) Non-transcriptional control of DNA replication by c-Myc, *Nature* 448, 445–451.
29. Zeller, K. I., Jegga, A. G., Aronow, B. J., O'Donnell, K. A., and Dang, C. V. (2003) An integrated database of genes responsive to the Myc oncogenic transcription factor: identification of direct genomic targets, *Genome Biol* 4, R69.
30. Pelengaris, S., Khan, M., and Evan, G. (2002) c-MYC: more than just a matter of life and death, *Nat Rev Cancer* 2, 764–776.
31. Levens, D. L. (2003) Reconstructing MYC, *Genes Dev* 17, 1071–1077.
32. Patel, J. H., Du, Y., Ard, P. G., Phillips, C., Carella, B., Chen, C. J., Rakowski, C., Chatterjee, C., Lieberman, P. M., Lane, W. S., Blobel, G. A., and McMahon, S. B. (2004) The c-MYC oncoprotein is a substrate of the acetyltransferases hGCN5/PCAF and TIP60, *Mol Cell Biol* 24, 10826–10834.
33. Sears, R., Nuckolls, F., Haura, E., Taya, Y., Tamai, K., and Nevins, J. R. (2000) Multiple Ras-dependent phosphorylation pathways regulate Myc protein stability, *Genes Dev* 14, 2501–2514.
34. Peri, S., Navarro, J. D., Amanchy, R., Kristiansen, T. Z., Jonnalagadda, C. K., Surendranath, V., Niranjan, V., Muthusamy, B., Gandhi, T. K., Gronborg, M., Ibarrola, N., Deshpande, N., Shanker, K., Shivashankar, H. N., Rashmi, B. P., Ramya, M. A., Zhao, Z., Chandrika, K. N., Padma, N., Harsha, H. C., Yatish, A. J., Kavitha, M. P., Menezes, M., Choudhury, D. R., Suresh, S., Ghosh, N., Saravana, R., Chandran, S., Krishna, S., Joy, M., Anand, S. K., Madavan, V., Joseph, A., Wong, G. W., Schiemann, W. P., Constantinescu, S. N., Huang, L., Khosravi-Far, R., Steen, H., Tewari, M., Ghaffari, S., Blobe, G. C., Dang, C. V., Garcia, J. G., Pevsner, J., Jensen, O. N., Roepstorff, P., Deshpande, K. S., Chinnaiyan, A. M., Hamosh, A., Chakravarti, A., and Pandey, A. (2003) Development of human protein reference database as an initial platform for approaching systems biology in humans, *Genome Res* 13, 2363–2371.
35. Kanazawa, S., Soucek, L., Evan, G., Okamoto, T., and Peterlin, B. M. (2003) c-Myc recruits P-TEFb for transcription, cellular proliferation and apoptosis, *Oncogene* 22, 5707–5711.
36. Tsuneoka, M., and Mekada, E. (2000) Ras/MEK signaling suppresses Myc-dependent apoptosis in cells transformed by c-myc and activated ras, *Oncogene* 19, 115–123.
37. Luscher, B., Kuenzel, E. A., Krebs, E. G., and Eisenman, R. N. (1989) Myc oncoproteins are phosphorylated by casein kinase II, *EMBO J* 8, 1111–1119.
38. Mac Partlin, M., Homer, E., Robinson, H., McCormick, C. J., Crouch, D. H., Durant, S. T., Matheson, E. C., Hall, A. G., Gillespie, D. A. F., and Brown, R. (2003) Interactions of the DNA mismatch repair proteins MLH1 and MSH2 with c-MYC and MAX, *Oncogene* 22, 819–825.
39. Feng, X. H., Liang, Y. Y., Liang, M., Zhai, W., and Lin, X. (2002) Direct interaction of c-Myc with Smad2 and Smad3 to inhibit TGF-beta-mediated induction of the CDK inhibitor p15(Ink4B), *Mol Cell* 9, 133–143.
40. Chapman, N. R., Webster, G. A., Gillespie, P. J., Wilson, B. J., Crouch, D. H., and Perkins, N. D. (2002) A novel form of the RelA nuclear factor kappaB subunit is induced by and forms a complex with the proto-oncogene c-Myc, *Biochem J* 366, 459–469.
41. Otsuki, Y., Tanaka, M., Kamo, T., Kitanaka, C., Kuchino, Y., and Sugimura, H. (2003) Guanine nucleotide exchange factor, Tiam1, directly binds to c-Myc and interferes with c-Myc-mediated apoptosis in rat-1 fibroblasts, *J Biol Chem* 278, 5132–5140.
42. Ingenuity Systems, I. www.ingenuity.com.
43. Wang, K., Alvarez, M. J., Bisikirska, B. C., Linding, R., Basso, K., Dalla Favera, R., and Califano, A. (2009) Dissecting the interface between signaling and transcriptional regulation in human B cells, *Pac Symp Biocomput*, 264–275.
44. Lefebvre, C., Rajbhandari, P., Alvarez, M. J., Bandaru, P., Lim, W. K., Sato, M., Wang, K.,

Sumazin, P., Kustagi, M., Bisikirska, B. C., Basso, K., Beltrao, P., Krogan, N., Gautier, J., Dalla-Favera, R., and Califano, A. (2010) A human B-cell interactome identifies MYB and FOXM1 as master regulators of proliferation in germinal centers, *Mol Syst Biol* 6, 377.

45. He, L., He, X., Lowe, S., and Hannon, G. (2007) microRNAs join the p53 network – another piece in the tumour-suppression puzzle., *Nat Rev Cancer* 7, 819–822.

46. Mestdagh, P., Fredlund, E., Pattyn, F., Schulte, J., Muth, D., Vermeulen, J., Kumps, C., Schlierf, S., De Preter, K., Van Roy, N., Noguera, R., Laureys, G., Schramm, A., Eggert, A., Westermann, F., Speleman, F., and Vandesompele, J. (2009) MYCN/c-MYC-induced microRNAs repress coding gene networks associated with poor outcome in MYCN/c-MYC-activated tumors., *Oncogene*.

47. Kim, H., Huang, W., Jiang, X., Pennicooke, B., Park, P., and Johnson, M. (2010) Integrative genome analysis reveals an oncomir/oncogene cluster regulating glioblastoma survivorship., *Proc Natl Acad Sci USA*.

48. Carro, M., Lim, W., Alvarez, M., Bollo, R., Zhao, X., Snyder, E., Sulman, E., Anne, S., Doetsch, F., Colman, H., Lasorella, A., Aldape, K., Califano, A., and Iavarone, A. (2009) The transcriptional network for mesenchymal transformation of brain tumours., *Nature*.

49. Cover, T. M., and Thomas, J. A. (1991), John Wiley & Sons, New York.

Chapter 9

Linking Cellular Signalling to Gene Expression Using EXT-Encoded Reporter Libraries

Anna Botvinik and Moritz J. Rossner

Abstract

Intracellular signalling initiated by extracellular ligands that activate cell surface receptors is a complicated process that involves multiple interconnected biochemical steps. Protein–protein interactions are often regulated by activated kinases via phosphorylation of specific residues. Such transient regulated interactions are central to many signalling cascades. Downstream signalling converges at the level of transcription factors to finally regulate adaptive transcriptional responses. There are powerful methods available to study transcriptional changes even at a global level; however, measuring upstream regulatory mechanisms is still challenging. We designed an experimental approach termed *ex*pressed oligonucleotide *t*ag (EXT)assay that enables the parallel analysis of signalling events upstream of gene expression. We make use of different types of reporter gene assays that are invariably linked to unique EXTs serving as quantitative decoders of respective assays. EXT reporters can be introduced into living cells and analyzed in pools by microarray hybridization or sequencing.

Key words: Cellular signalling, Regulated protein–protein interactions, Phosphorylation, Split TEV assays, *cis*-Regulatory assays, EXT, Microarray

1. Introduction

Cellular signalling cascades are considered as complex networks of interacting proteins that transduce extracellular signals via activated membrane receptors and adapter proteins located in the cytoplasm to the nucleus causing finally gene expression changes (1). A prominent canonical example is the Neuregulin–ErbB signalling pathway that has been extensively studied (2). Mainly biochemical methods but more recently also novel techniques, such as protein arrays and mass spectrometry-coupled pull-down experiments,

Bart Deplancke and Nele Gheldof (eds.), *Gene Regulatory Networks: Methods and Protocols*,
Methods in Molecular Biology, vol. 786, DOI 10.1007/978-1-61779-292-2_9, © Springer Science+Business Media, LLC 2012

have been applied to study Neuregulin–ErbB and other signalling cascades(3, 4). These approaches enabled a deeper understanding of receptor activation mechanisms and signalling by adapter protein recruitment of phosphorylated receptors. These methods can, however, provide only a partial view onto the molecularly very different signalling events that are associated with an extracellular stimulus.

In an attempt to overcome the limitation of existing techniques that mostly only cover a particular biochemical event (e.g., by studying receptor activation or protein–protein interaction or transcription factor activation each separately), we have developed a novel integrated and highly scalable reporter system, which we termed *ex*pressed oligonucleotide *t*ag (EXT)assay. The system relies on different cellular reporter gene assays that can be combined with RNA-based *ex*pressed *t*ag reporters in a flexible way. The rational design of highly complex EXT libraries by combinatorial synthesis of defined building blocks (5) ensures that each EXT reporter has identical GC contents, highly similar melting temperatures, and is virtually free of secondary structures (6). As a consequence, EXT reporters have been shown to function with high performance and predictability using DNA microarrays or next-generation sequencing as readout techniques (6). As a biological proof-of-principle showing the applicability of EXTassays, split tobacco etch virus (TEV) (7, 8) and *cis*-regulatory reporter gene assays were combined to monitor ERBB receptor activation and downstream signalling simultaneously (6). The split TEV technique relies on the functional complementation of inactive fragments of the TEV fused to interacting proteins activating TEV cleavage-dependendent reporter proteins (7). Split TEV assays can be applied to study constitutive as well as regulated protein–protein interactions at the membrane and in the cytosol of mammalian cells (see Fig. 1a for details) (7, 8). Here, we describe a detailed protocol that allows for integrated EXTassay measurements with a novel multiplexed reporter system based on EXT reporters. EXT-coupled split TEV assays can be used to monitor for different types of regulated protein–protein interactions while EXT-coupled *cis*-regulatory assays can be applied to study transcription factor activation simultaneously within one experiment using, e.g., microarrays as readout system (Fig. 1b). EXT abundances measured with complementary probes on custom arrays reflect the intensities of particular EXT-encoded assays. Standard bioinformatic tools can be used to analyze and visualize the data (see Fig. 1c for an example). We report on the synthesis and subcloning strategy of EXT libraries, a simple transfection strategy for multiplexed assays and guidelines for microarray analyses.

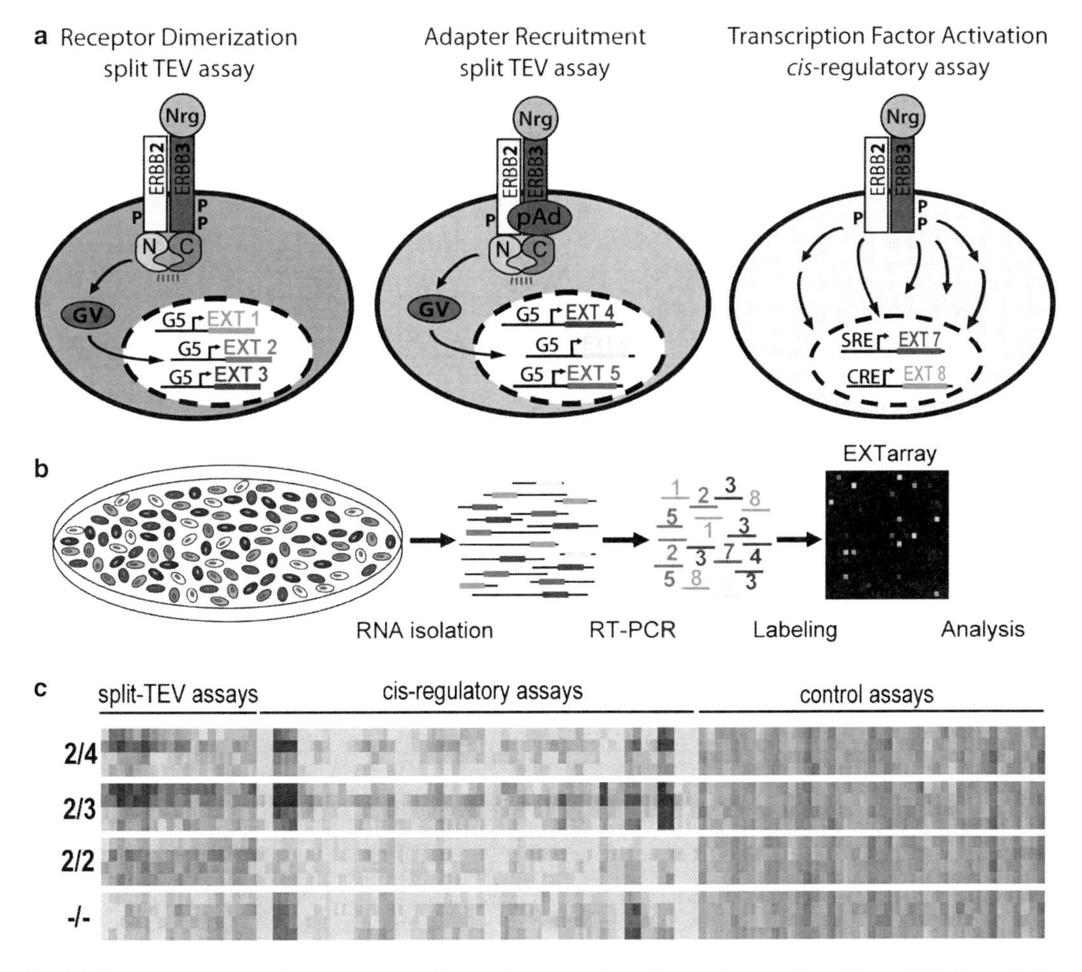

Fig. 1. (**a**) Schematic drawing of representative cells carrying three different types of assays. To monitor receptor dimerization, the left "cell" expresses ERBB–TEV protease N- and C-terminal fragment fusion proteins which upon dimerization cause the proteolytic release of transcription factor GV that binds the corresponding *cis*-element (G5) to activate EXT reporter expression. The middle "cell" expresses a C-TEV-fused phosphotyrosine adapter protein (pAd) that upon recruitment to an N-TEV-fused ERBB receptor leads to activation of different EXT reporters. The right "cell" carries *cis*-element reporter constructs exemplified by serum response element (SRE) and cAMP response element (CRE) to measure the activity of endogenous transcription factors with the help of corresponding unique EXTs. (**b**) Assay components are coexpressed in different pools of cells and coplated in the same dish (indicated by differently *grey-shaded cell populations*). RNA is isolated, EXTs are amplified by RT-PCR, labelled, and hybridized to a "decoding" microarray (EXTarray). (**c**) Microarray data are normalized and analyzed following standard procedures and can be visualized, e.g., by heat maps. Here, a *grey-scaled heat map* shows the time-dependent activation of different ERBB receptor dimers by Neuregulin-1 monitored by split TEV assays (*left*), the activation of transcription factors monitored with *cis*-regulatory assays (*middle*), and unregulated control assays monitored with EXT reporter constructs driven by a constitutively active promoter (*right*).

2. Materials

2.1. Cloning of EXT Libraries

1. EXT oligodeoxynucleotide synthesis is performed according to the phosphoramidite method (9) on a solid support using phosphoramidite DNA synthesis columns and, e.g., a 394 DNA Synthesizer (Applied Biosystems).
2. PAGE gels and electrophoresis setup (e.g., from Invitrogen).

3. Primers:

EXTlib-F	AGCTAGTTGCTAAGTCTGCCGAGTAGAATTAACCCTCACTAAAGGGTAGGTGACACTAT
EXTlib-R	TCGTACATGCATTGACTCGCGTCTACTAATACGACTCACTATAGG
EXT_B3	GGGGCAACTTTGTATAATAAAGTTGAGCTAGTTGCTAAGTCTGCCGAGTAG
EXT_B2	GGGGCCACTTTGTACAAGAAAGCTGTCGTACATGCATTGACTCGCGTCTAC
pENTR_s	CGCGTTAACGCTAGCATGGATCTC
Dec1	AGCTAGTTGCTAAGTCTGCCGAGTAG
Dec2	TCGTACATGCATTGACTCGCGTCTAC
Luc_as	GGCGTCTTCCATGGTGGCTTTACC

4. High-fidelity or proofreading PCR reagents (e.g., HotStarTaq from Qiagen).
5. Multisite Gateway Pro 3.0 Kit, Gateway BP clonase II, and LR clonase II Plus enzyme mix.
6. ElectroMax DH10b or One Shot Mach1 competent cells (all reagents from Invitrogen).
7. LB medium and LB agar plates (50 μg/ml kanamycin, 100 μg/ml ampicillin, or 100 μg/ml carbenicillin).
8. Donor and Destination vectors for entry and expression constructs can be purchased from Invitrogen (pDONR-P1P4, pDONR-P4rP3r, pDONR-P3P2) or generated by cloning using a Gateway recombination cassette (Invitrogen) into any expression vectors of choice (e.g., pGL3 from Promega yielding pDEST_GL3_Luci).
9. Plasmid-DNA purification kit at the Mini and Midiprep scale (Qiagen).

2.2. Transfection and RNA/DNA Isolation

1. PC12 rat adrenal pheochromocytoma cells or any other mammalian cell line or primary cells amenable to transient transfection.
2. Dulbecco's Modified Eagle's Medium (DMEM) (Gibco) supplemented with 10% fetal bovine serum (FBS) (Gibco) and 5% horse serum (HS) (Gibco). Alternatively, use appropriate growth medium for the cell line of choice.
3. Transfection reagents, e.g., Lipofectamine 2000 (Invitrogen) and OptiMEM reduced serum medium (Invitrogen).
4. RNA preparation reagents, including the Trizol reagent (Invitrogen) and the RNeasy kit (Qiagen).
5. 3 M sodium acetate, pH 5.2.

6. 7.5 M ammonium acetate, molecular biology grade.
7. Reverse transcription reagents: Random nanomer primer, Superscript III reverse transcriptase (RT) (Invitrogen), deoxyribonucleotides (Roche).
8. PCR reagents, e.g., HotStarTaq (Qiagen).
9. Primers:

Dec1	AGCTAGTTGCTAAGTCTGCCGAGTAG
Dec2	TCGTACATGCATTGACTCGCGTCTAC

2.3. EXT Labelling and Hybridization to EXTarrays

1. T7-RNA polymerase MEGAscript kit (Ambion).
2. 5-(3-aminoallyl)-UTP (aaUTP).
3. Total RNA isolation kit.
4. Cy3 and Cy5 monoreactive dyes (GE Healthcare). One tube of the lyophilized dye is dissolved in 55 μl DMSO and used for five labelling reactions. Remaining dye solution can be stored at −20°C protected from humidity for over 6 months.
5. Coupling buffer (Ambion): 0.5 M $NaHCO_3$, pH 9, could also serve as 10x coupling buffer.
6. 4 M hydroxylamine.
7. Custom microarrays from Agilent (http://www.agilent.com) or Roche/Nimblegen (www.nimblegen.com).
8. Hybridization reagents (Agilent): HI-RPM microarray hybridization buffer and blocking agent.
9. Formamide.
10. 20× SSPE buffer.
11. Hybridization reagents (Nimblegen): Hybridization Kit, Sample Tracking Control Kit, Wash Buffer Kit.
12. Acetonitrile.
13. GenePix 4200A microarray scanner (Axon Instruments) and Analysis Software GenePix Pro 6.1 (Axon).
14. Feature extraction software (Agilent).
15. NimbleScan (Roche NimbleGen).
16. R statistical computing environment (open source at http://www.bioconductor.org).

3. Methods

3.1. Cloning of EXT Libraries and Expression Constructs

The synthesis follows a mix and divide strategy combining defined building blocks (depicted in Fig. 2). The combinatorial synthesis can be performed following standard procedures, e.g., on a solid

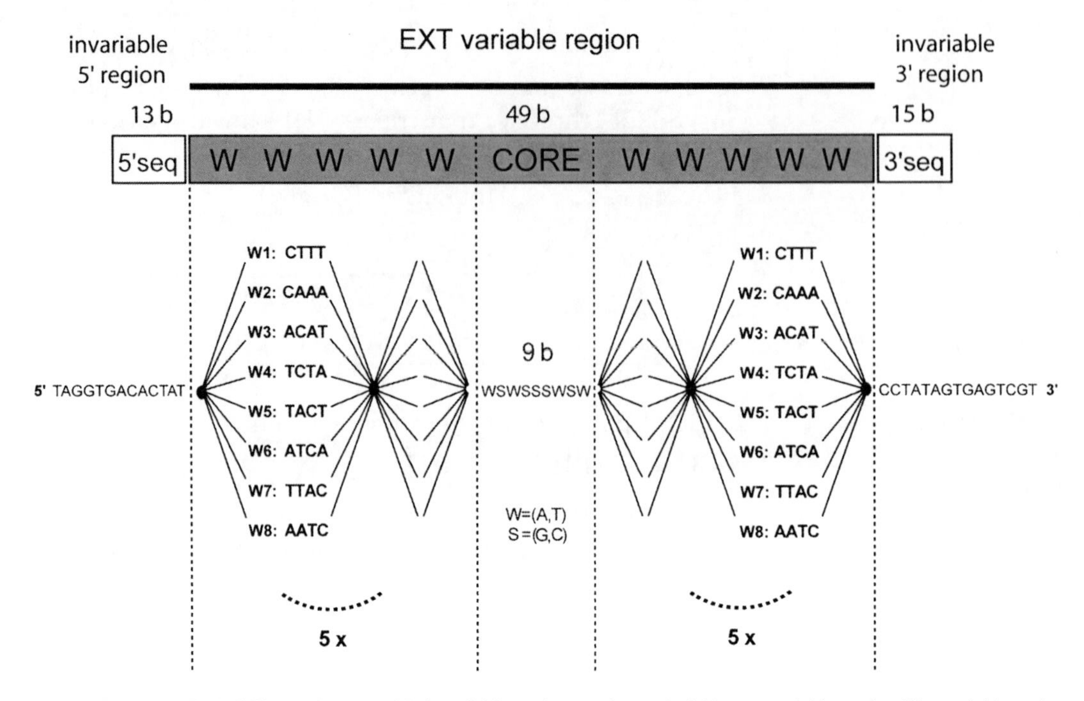

Fig. 2. EXT design. Each EXT contains 5′ and 3′ invariable regions and a central 49-mer variable region. The variable region has a symmetric organization and consists of ten structural modules called "words" (W) that flank a central core region. Each word consists of four nucleotides. Out of all possible four-nucleotide combinations, only eight are used that are built of three adenosine (A) or thymidine (T) residues, only one cytosine (C), and no guanine (G) residues at all. Each of the eight possible words (CTTT, CAAA, ACAT, TCTA, TACT, ATCA, TTAC, and AATC) occurs randomly at each of the ten possible positions. The core region comprises nine bases of alternating A,T (W) or G,C (S) residues with three centrals G,C (S) residues. The total theoretical complexity of an EXT library following this design can be calculated as: $8^{10} \times 2^{9} \approx 5.5 \times 10^{11}$.

support using phosphoramidite DNA synthesis columns and any applicable DNA synthesizer.

1. A 13-mer 5′ invariable region (TAGGTGACACTAT) is synthesized in eight cartridges in parallel followed by the first set of the eight different words. Subsequently, by mixing and dividing the samples (see Note 1), another set of the eight different words are synthesized. The cycle is repeated to introduce three additional word blocks (five in total) followed by a symmetrical 9-mer core sequence that are generated by splitting the samples into two equal portions after each next nucleotide (WSWSSSWSW, where W=A/T and S=C/G). The synthesis is completed by five additional word structures and adding a 3′ invariable region (CCTATAGTGAGTCGT) (summarized in Fig. 1).
2. The resulting library of 77-mer oligonucleotides is PAGE gel purified and diluted to concentrations not below 50 μM for storage.

3. The EXT oligonucleotide library is diluted to 1 nM, and 1 fmol (1 μl) is PCR amplified for 10–15 cycles with the primers EXTlib_F and EXTlib_R complementary to the 5′ and 3′ invariable regions extended by T3 (optional) and T7 promoter sequences for later in vitro transcription purposes (see Note 2).
4. The PCR product is reamplified for 10–15 cycles with a second set of primers EXT_B3 and EXT_B2 to introduce attB3 and attB2 recombination sites.
5. PCR products are agarose gel purified and recombined with the pDONR-P3P2 vector using BP Clonase enzyme mix to generate entry shuttle clone libraries (see Fig. 2). The corresponding "BP" reaction is set up in 20 μl volume with 4 μl pDONR-P3P2 vector (50 ng/μl), 12 μl PCR product (5 ng/μl), and 4 μl BP clonase II enzyme mix.
6. After overnight incubation at 25°C, the recombination product is purified by phenol chloroform extraction and ethanol precipitation, resuspended in 5 μl molecular biology grade H_2O, and electroporated into DH10b cells.
7. Electroporation is performed with 1 mm gap cuvettes using, e.g., a BioRad-Gene Pulser II and standard settings. The bacteria are allowed to recover in 2 ml SOC medium for 1 h. 10 and 100 μl of the bacterial suspension is plated to estimate the entry library complexity.
8. Adjust the complexity to exceed your future needs by at least tenfold to avoid unwanted duplicates in further subclonings (see Note 3). Remaining bacteria are transferred to 200 ml LB medium, 50 μg/ml kanamycin, and are allowed to grow at 37°C for 1 h with mild shaking.
9. After subcloning, the EXT library has to be screened for full-length unique EXTs to exclude repeating sequences and products of incomplete oligonucleotide synthesis.
10. *Option 1: Analysis of individual entry clones.* A portion of the bacteria are kept at 4°C overnight, and appropriate volumes are plated according to the determination of recombinants adjusted to 100–500 clones per 10 cm LB agar dish.
11. From these plates, e.g., 96 colonies are screened by "colony PCR" by dipping a sterile tooth pick into wells of a 96-well PCR plate containing all reagents, including the primers pENTR_s and Dec2. Subsequently, the tooth pick is transferred to the corresponding positions in 96-deep well dishes (with ~1 ml LB/well) and grown at 37°C for at least 16 h. The PCR products are separated on 4% agarose gels, and clones are discarded that show reduced insert length (<280 bp). The PCR products of proper size can be directly sequenced using the pENTR_s primer. Bacterial colonies corresponding to per-

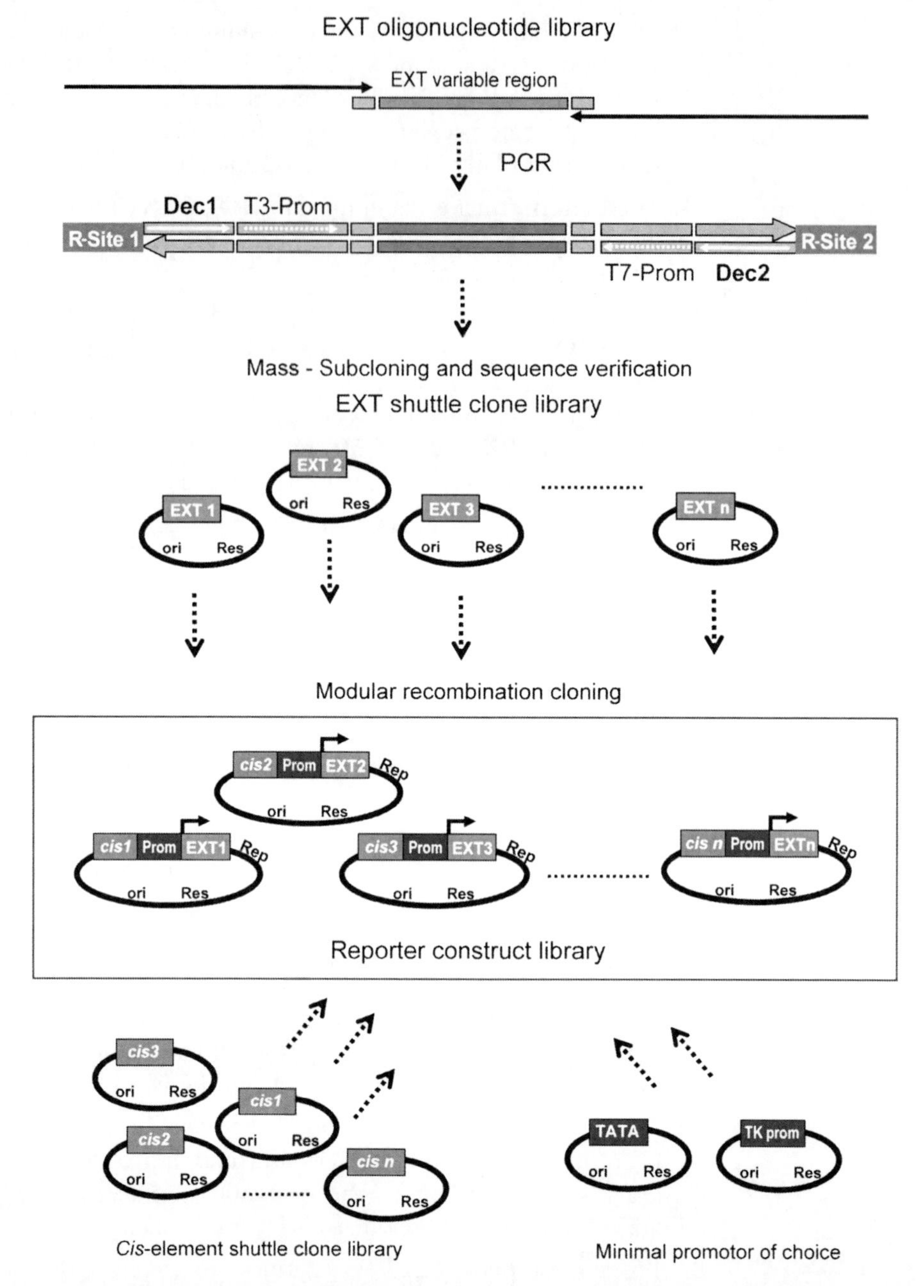

Fig. 3. Subcloning of the EXT library. The EXT oligonucleotide library is PCR amplified and subcloned to generate a pool of shuttle clones. Several functional elements are introduced flanking the EXT sequence: recombination sites (R-Site1/2), "decoder" primer-binding sites (Dec1/2), and viral promoters (T3/T7). Three shuttle clones harboring different functional elements are used in one recombination reaction to place a *cis*-regulatory element, a minimal promoter, and an EXT 5′ to a firefly luciferase reporter gene (Rep). Depending on the shuttle clone combination, defined *cis*-regulatory elements are linked to unique EXTs.

fect clones (expected success rate of nearly 50%) are subjected to plasmid preparations. These can be used in individual subclonings to generate expression clones with predefined EXTs (illustrated in Fig. 3).

12. *Option 2: Mass subcloning of entry clones.* The culture is centrifuged and the pellet is used for a midi-sized plasmid DNA preparation and results in an entry clone library of a given complexity. Imperfect EXTs are excluded during PCR screening and sequencing of final reporter constructs through a similar procedure as described above.
13. The entry clone EXT library (harboring attL3/L2 sites) is subsequently recombined along with *cis*-element (cloned into the pDONR-P1P4 vector) and minimal promoter (cloned into the pDONR-P4P3) entry clones of choice and a destination plasmid harboring a reporter gene, such as firefly luciferase (e.g., pDEST_GL3_Luci), to generate a library of *cis*-regulatory EXT reporter constructs (see (6) for selected *cis*-element and minimal promoter sequences) as illustrated in Fig. 3.
14. The "LR" recombination reaction is set up in 20 μl with 40 fmol of the destination vector and 20 fmol of each of the entry clone and incubated at room temperature overnight followed by a Proteinase K digestion according to the Multisite Gateway manual.
15. The reaction is purified by PEG precipitation and transformed into DH10b (see Note 4). Cells are transferred to 200 ml of the respective LB selection medium and grown for 4 h at 37°C. 10 and 100 μl of the suspension are plated out to estimate the complexity and the rest is kept at 4°C overnight.
16. The next day, the volume corresponding to 2,000–4,000 clones is plated on square (25 × 25 cm) agar plates with antibiotics (in the case of pDEST_GL3_Luci, use 50 μg/ml carbenicillin) and cultured overnight.
17. The colonies are picked and transferred into 96-well plates containing 100 μl LB per well with antibiotics. Clone picking can be done manually or semiautomated using any Colony Picking Robot.
18. The bacteria are grown overnight and frozen at −80°C. The clones are screened by colony PCR using 0.5 μl of the culture as a template. The inserts are amplified with Dec1 and Luc_as primers (when pDEST_GL3_Luci was used) and the PCR products are directly sequenced with the antisense primer.

3.2. Transfection, RNA/DNA Isolation, and cDNA Synthesis

For some multiplexed assays (e.g., *cis*-element activity profiling under defined conditions), the cells are transfected with the entire *cis*-EXT reporter library using standard protocols. However, for some experimental designs (e.g., integrated measurement of pathway activation, including protein–protein interaction assays), multiple transfections are required. To ensure equal experimental conditions and to enable simultaneous analysis of all samples, the cells should be transfected in suspension and combined immediately after transfection. Alternatively, adherent cells could be trypsinized after transfection which, however, increases the workload and may pose additional stress to the cells.

1. For transfection in suspension, e.g., PC12 cells are trypsinized and triturated through a 24-G needle five times to remove any cell clumps and the cell number is determined. Desired numbers of cells are pelleted by centrifugation for 5 min at 800 × *g* and resuspended in DMEM supplemented with 1% horse serum and antibiotics to reach a cell density of 100,000 per 100 μl.
2. The following transfection components are assembled (adjusted to 10^5 cells): Lipofectamine 2000 0.4 μl, plasmid DNA 100 ng, OptiMEM 10 μl. The DNA and the Lipofectamine 2000 reagent each are diluted in one half of the OptiMEM, mixed, and combined together (see Note 5).
3. The Lipofectamine–DNA complexes are allowed to form within 20 min at room temperature and are immediately applied to the cell suspension. The tubes are incubated at 37°C for 4 h without shaking.
4. Treated cells are subsequently centrifuged to remove the transfection mixture, resuspended in fresh growth medium, mixed, and plated out according to the experimental design.
5. Following this procedure, cells from several transfections can be combined and cultured under identical experimental conditions. As estimation, 10^5 to 10^6 cells should be transfected with 2–5 ng of each EXT reporter constructs depending on the overall complexity of the EXTassay. We successfully used 20 to 100 different plasmids in one transfection although further upscaling may be possible.
6. *Cis*-reporter plasmids at miniprep DNA quality are premixed and used for transfection to minimize pipetting errors.
7. Plasmids required for the protein–protein interaction assays are derived from midi-prep DNA preparations, and 100 ng of DNA is used per 10^5 cells for each multiplexed experiment. Figure 4 illustrates control experiments demonstrating high efficiency of transfection in suspension measured with firefly Luciferase reporter gene assays.
8. RNA and input plasmid DNA can be isolated simultaneously by a combined Trizol-RNA column purification (see Note 6). Care should be taken (and properly controlled, see below) not to carry over extensive amounts of supercoiled plasmid into the RNA samples.

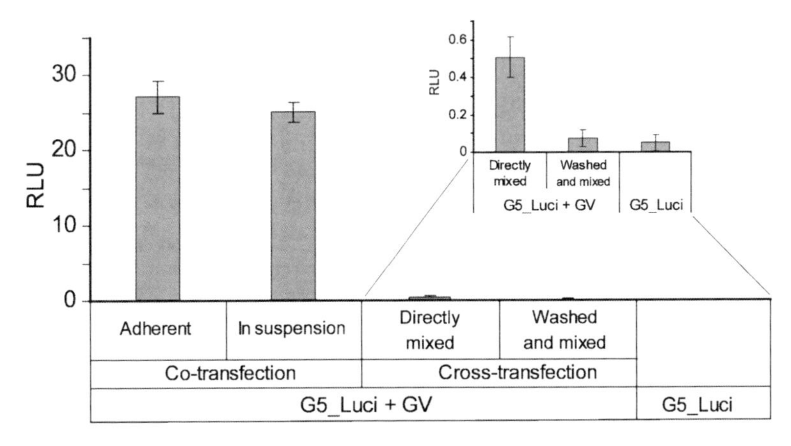

Fig. 4. Transfection of PC12 cells in suspension. PC12 cells were cotransfected with the Gal4-VP16 (GV) synthetic transcription activator and the corresponding GV responsive firefly luciferase reporter construct "G5_Luci". 24 h after transfection, the cells were lysed and Luciferase expression levels were quantified in a chemiluminescence assay. Data are given as relative luminescence units (RLUs). The Luciferase assays show that PC12 cells could be transfected on a plate (adherent) or in suspension with similar efficiencies. Independent transfections with GV and G5_Luci and subsequent pooling of cell suspensions after 4 h yielded only slight cross-transfection effects upon direct mixing of both samples. Slight activation of the G5-Luci reporter resulting from the cross-transfection of the GV was completely abolished by introducing a single "washing" step before mixing the samples (see inset).

9. For the first step of RNA/DNA isolation, follow the standard Trizol protocol according to the manufacturer, avoiding PBS washing steps to reduce the risk of cell loss for 96-well plates.
10. The following amounts of the Trizol reagent and chloroform should be used:

Plate	Area	Cell number	Trizol	Chloroform
96-well	0.3 cm^2	50,000	125 μl	25 μl
24-well	2 cm^2	250,000	500 μl	100 μl
6-well	10 cm^2	1,000,000	1 ml	200 μl

11. Trizol reagent is added to the wells, left for at least 1 min, lysates are transferred into 2-ml test tubes, vortexed to ensure complete lysis, and 1/5th volume of chloroform is added.
12. Each sample is vortexed thoroughly and centrifuged for 15 min at 4°C, 12,000×*g*. The aqueous phase of the Trizol lysate is transferred to a new tube. One-third of the sample is used for DNA and two-third for RNA isolation.
13. The DNA is precipitated with sodium acetate (final concentration 0.3 M) and 1 volume isopropanol.
14. The RNA sample is subjected to an RNeasy column purification, including an on-column DNase treatment, according to the manufacturer's protocol.

15. The RNA is precipitated with ammonium acetate (final concentration 2.5 M), glycogen as carrier, and 2 volumes ethyl alcohol.
16. The RNA is washed with 100 μl of 70% EtOH, briefly dried and resuspended in 4–6 μl of H_2O, and quantified with appropriate instruments (e.g., nanodrop or bioanalyzer).
17. For cDNA synthesis, 0.5–1 μg total RNA is reverse transcribed using Superscript III reverse transcriptase (see Note 7). Equal amounts of RNA for each sample are used. The sample volumes are adjusted to 4.5 μl with RNase free water, and 1 μl (120 pmol) of the random nanomer primer is added. The samples are denatured for 2 min at 70°C and placed on ice. Then, other components are added in order: 5× First Strand buffer, 2 μl; 0.1 M DTT, 1 μl; dNTP mix (10 mM each), 0.5 μl; Superscript III reverse transcriptase (200 U/μl), 1 μl.
18. The samples are incubated for 10 min at 25°C to allow for the annealing of random nanomer primers, then for 1 h at 50°C for cDNA synthesis, and the enzyme is heat inactivated for 5 min at 85°C.
19. The cDNA should be diluted 1:10 with water and 1–2 μl should be used for PCR amplification.
20. EXTs are amplified from both cDNA and plasmid DNA samples with Dec1 and Dec2 primers and standard PCR reagents (to avoid unnecessary overcycling, quantitative PCR, e.g., using SYBR Green, could be performed).
21. In parallel, a PCR with Dec1 and Dec2 primers is run on the RT-negative control samples to determine the success of the cDNA synthesis and potential carry over of plasmid DNA during RNA purification.

3.3. EXT Labelling and Hybridization to EXTarrays

Unpurified PCR products carrying T7-promotor sequences can be used without further purification as a template for in vitro transcription (IVT) incorporating 5-(3-aminoallyl)-modified UTP as illustrated in Fig. 5. The IVT reaction volume can be downscaled relative to the manufacturer's protocol still yielding sufficient material.

1. For the IVT reaction, assemble the following reagents: PCR product 4 μl; 10× buffer 1 μl; ATP (7.5 mM) 1 μl; CTP (7.5 mM) 1 μl; GTP (7.5 mM) 1 μl; UTP (7.5 mM) 0.5 μl; 5-(3-aminoallyl-UTP) (50 mM, Ambion) 0.75 μl; T7 RNA polymerase 1 μl.
2. Incubate the reaction overnight at 37°C using a heated lid at 42°C to prevent evaporation.
3. Degrade the template DNA subsequently by adding 1 μl DNase to each sample and incubating at 37°C for 15 min.

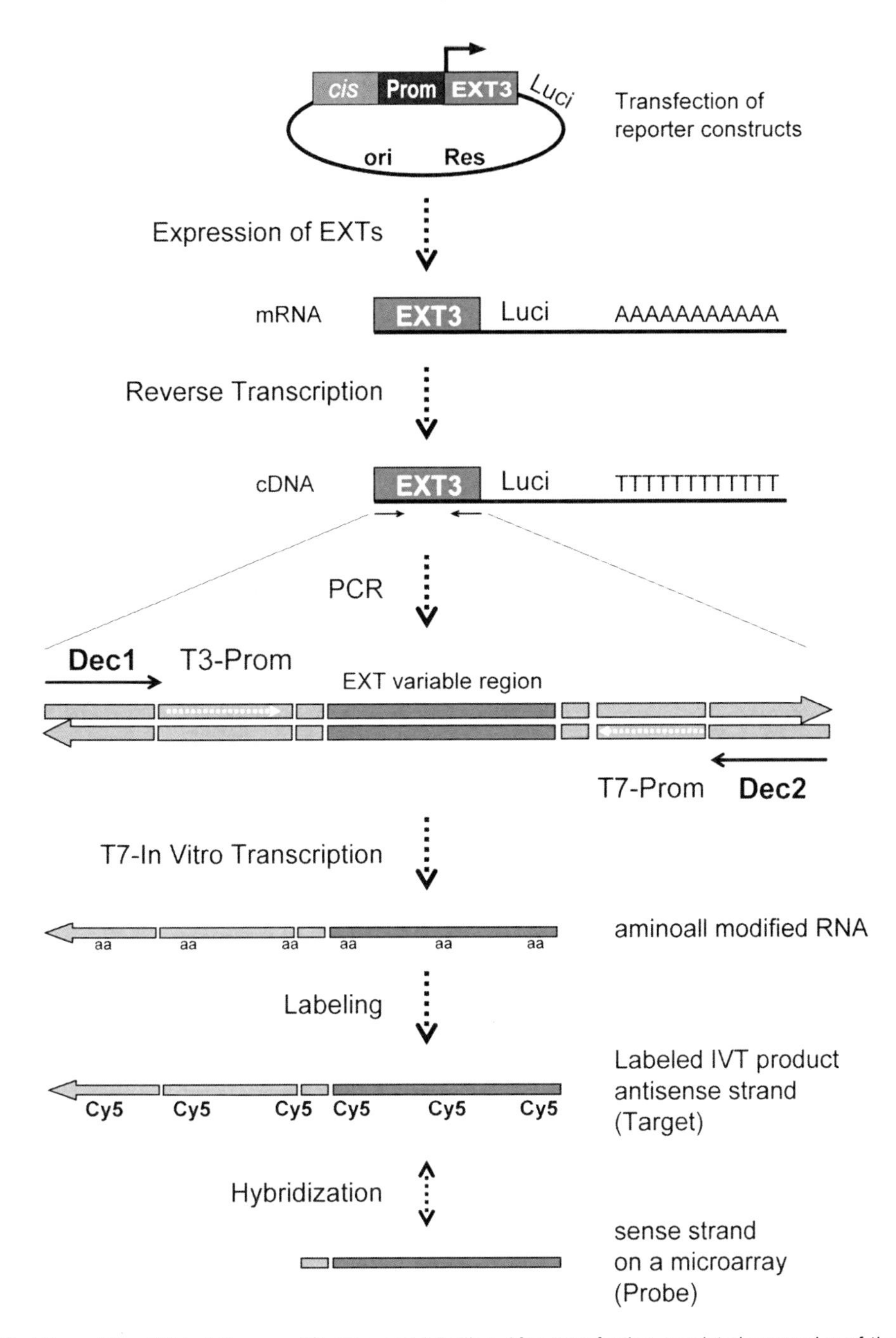

Fig. 5. Workflow of the EXT isolation, amplification, and labelling. After transfection, regulated expression of the *cis*-EXT reporter constructs generates mRNAs carrying *cis*-element-specific EXT and the firefly Luciferase reporter gene. To quantify the expression levels of the individual reporters, total RNA is isolated. After reverse transcription, expressed EXT reporters are amplified by PCR with Dec1/Dec2 primers. To generate single-stranded targets for microarray hybridization, T7 in vitro transcription (IVT) is performed in the presence of 5-(3-aminoallyl)-UTP. The resulting aminoallyl-modified RNA (aaRNA) is labelled by coupling of cyanine dye esters (Cy3 or Cy5) to the aminoallyl moieties. The labelled product (= target, antisense strands) is hybridized to a microarray harboring complementary probes (sense orientation).

4. The typical yield of an EXT-IVT is 5–10 μg of aminoallyl-modified RNA (aaRNA) at 111 bases in length.
5. The aaRNA is purified over an RNeasy column using the following modified protocol (see Note 8). The volume of all samples is adjusted to 100 μl with RNase-free water. Per each sample, 100 μl buffer RLT is added and the samples are mixed. Then, 400 μl of isopropanol is added, the samples are quickly mixed by vortexing, immediately transferred to the RNeasy columns, and centrifuged for 15 s at 10,000 × *g*. The membranes are washed twice with 500 μl RPE buffer and dried by centrifugation for 2 min. The aaRNA is eluted in two steps with 50 μl H_2O each time to maximize the yield.
6. To generate target samples for hybridization to custom EXT-decoding microarrays (EXTarrays), aaRNA is coupled to cyanine dye esters Cy3 and Cy5 (we routinely label cDNA-derived samples with Cy5 and plasmid DNA-derived with Cy3).
7. For each labelling reaction, 5–10 μg of the aaRNA is lyophilized and resuspended in 4.5 μl coupling buffer. 5.5 μl of the Cy3/Cy5 dye is added to each sample. After 30 min incubation at room temperature in the dark, the reaction is neutralized by the addition 2.3 μl 4 M hydroxylamine.
8. The tubes are incubated in the dark for an additional 15 min, and the labelled RNA is purified using the modified RNeasy protocol as described above.
9. The dye incorporation efficiency (E) can be calculated as follows:

   $E(\text{pmol}/\mu\text{g}) = \text{Dye}(\text{pmol}/\mu\text{l}) \,/\, \text{RNA Conc}\ (\mu\text{g}\,/\,\mu\text{l})$. Successful labelling should result in a dye incorporation of 50–100 pmol/μg, and corresponding samples can be used for the microarray hybridizations.
10. The amount of the target should be adjusted to the amount of the dye and not the amount of RNA in order to compensate for the slight differences in labelling efficiencies. The amount of target should correspond to 10–30 fmol of the Cy-dye per each unique EXT.
11. The hybridization cocktail for an 8-plex Agilent custom microarray (see Note 9) contains: 2× hybridization buffer 25 μl; 10× blocking agent 5 μl; formamide (final concentration 15%) 7.5 μl; target 12.5 μl (or water up to 50 μl final volume). The samples are mixed by pipetting, centrifuged for 1 min at 10,000 × *g*, and 40 μl are loaded per subarray. After hybridization at 65–68°C for 16–20 h, the arrays are washed according to the manufacturer's recommendations and dried by dipping for 30 s into 250 ml acetonitrile.

12. NimbleGen 4-plex arrays (see Note 9) are hybridized according to the standard protocol provided by the manufacturer. Each target should comprise 10–30 fmol of the Cy-dye per EXT. The arrays are hybridized at 48–50°C. Washing procedures are carried out using the NimbleGen wash buffer kit following the manufacturer's recommendations.
13. The slides are dried by dipping for 30 s in acetonitrile and scanned, e.g., with the GenePix 4200A microarray scanner or any other scanner with at least 5 μm resolution. Data were extracted according to the manufacturer's specifications.
14. Raw data analysis can be performed, e.g., with "R" statistical computing environment (http://www.r-project.org) or any other suitable analysis platform. Microarray data are normalized using, e.g., the RMA algorithm.
15. The average probe intensities are calculated across all EXT replicate features depending on the design of the array. Average Cy5 signals (RNA/cDNA derived) represent the relative EXT expression levels and are normalized to the Cy3 signals originating from the plasmid DNA input. Plasmid input-normalized EXT expression data are either given as relative differences (fold changes) between samples or as arbitrary normalized expression values.

4. Notes

1. Before staring with the synthesis of the combinatorial EXT oligonucleotide library, ensure that the repeated opening and tight closure of the coupling cartridges are possible. After each synthesis step, the resins from all cartridges are mixed on a clean glass surface and split into eight equal portions, e.g., using razor blades. These portions are reloaded to the cartridges for the next synthesis step.
2. When amplifying the EXT oligonucleotide library by PCR, do not use more than 1 fmol of template to avoid competition of the library oligonucleotides with the amplification primers.
3. The number of colonies to be screened should be determined by the final size of the library. Calculate with a maximum of 25% correct final expression clones.
4. Commercial electrocompetent DH10b from Invitrogen (Electromax) performs by far better compared to our self-made competent cells according to standard procedures. Bacteria from the XL-1 blue strain may not be compatible with the Multisite Gateway system, but chemically competent MACH1 as well as DH10b cells (Invitrogen) can be used.

5. In-suspension transfection conditions are only given for PC12 cells. Principally, the described multiplexed assays should by applicable to any genetically amenable cell type; however, optimal conditions need to be adjusted with appropriate control experiments. In any case, circumvent shaking of cells during transfection to avoid increased cell death.
6. Avoid unnecessary washing step while harvesting cells to reduce losses particularly when using 96- or 384-well plates.
7. To control the cDNA synthesis with respect to plasmid DNA contamination, always analyze for each sample a control lacking reverse transcriptase (containing all reagents, except reverse transcriptase).
8. The amplified antisense RNA has to be purified over an RNeasy column using a modified protocol for small-sized species. Do not change the conditions as you may reduce the yield or even lose the RNA.
9. Custom microarray should contain complementary EXT replicates (optimally $n > 4$ features per EXT) and a set of hybridization specificity controls, including EXT sequence permutations carrying mismatches at different sites along the sequence and varying numbers of mismatches to determine optimal hybridization stringency (see (6)).

References

1. Papin JA, Hunter T, Palsson BO, Subramaniam S (2005) Reconstruction of cellular signalling networks and analysis of their properties. Nat Rev Mol Cell Biol 6, 99–111.
2. Citri A, Yarden Y (2006) EGF-ERBB signalling: towards the systems level. Nat Rev Mol Cell Biol 7, 505–516.
3. Jones RB, Gordus A, Krall JA, MacBeath G (2006) A quantitative protein interaction network for the ErbB receptors using protein microarrays. Nature 439, 168–174.
4. Schulze WX, Deng L, Mann M (2005) Phosphotyrosine interactome of the ErbB-receptor kinase family. Mol Syst Biol 1, 2005 0008.
5. Brenner S, Williams SR, Vermaas EH, Storck T, Moon K, McCollum C, et al. (2000) In vitro cloning of complex mixtures of DNA on microbeads: physical separation of differentially expressed cDNAs. Proc Natl Acad Sci USA 97, 1665–1670.
6. Botvinnik A, Wichert SP, Fischer TM, Rossner MJ (2010) Integrated analysis of receptor activation and downstream signaling with EXTassays. Nat Methods 7, 74–80.
7. Wehr MC, Laage R, Bolz U, Fischer TM, Grunewald S, Scheek S, et al. (2006) Monitoring regulated protein-protein interactions using split TEV. Nat Methods 3, 985–993.
8. Wehr MC, Reinecke L, Botvinnik A, Rossner MJ (2008) Analysis of transient phosphorylation-dependent protein-protein interactions in living mammalian cells using split-TEV. BMC Biotechnol 8, 55.
9. Dorper T, Winnacker EL (1983) Improvements in the phosphoramidite procedure for the synthesis of oligodeoxyribonucleotides. Nucleic Acids Res 11, 2575–2584.

Chapter 10

Sample Preparation for Small RNA Massive Parallel Sequencing

Willemijn M. Gommans and Eugene Berezikov

Abstract

High-throughput sequencing has allowed for a comprehensive small RNA (sRNA) expression analysis of numerous tissues in a diverse set of organisms. The computational analysis of the millions of generated sequencing reads has led to the discovery of novel miRNAs and other sRNA species, and resulted in a better understanding of the roles these sRNAs play in development and disease. This chapter describes the generation of sRNA deep-sequencing libraries for the Illumina massively parallel sequencing platform by using a cloning method that anneals specific RNA sequences to the 5′- and 3′-ends of the sRNA molecules.

Key words: Deep sequencing, Small RNA expression, Solexa, Illumina, miRNA, Library construction, Profiling

1. Introduction

There is steadily increasing evidence that small, noncoding RNAs play regulatory roles within a wide variety of cellular pathways (1). For example, there are small RNAs (sRNAs) with a size around 20 nt that are associated with promoter elements (promoter-associated sRNAs, PASRs) and even tinier RNAs with a size around 18 nt that seem to be specifically located adjacent to transcription start sites (transcription-initiation RNAs, tiRNAs). Other sRNAs can help to protect the genome against transposon incorporations which is a function fulfilled by PIWI-associated RNAs (piRNAs). In addition, small RNA molecules can influence (post)transcriptional gene regulation or are involved in the host viral defense systems (including endo-siRNAs and microRNAs (miRNAs)) (1). Thus, there is a growing list of sRNA molecules that were previously just considered transcriptional noise, but seem to fulfill similar functions that were solely ascribed to proteins.

Bart Deplancke and Nele Gheldof (eds.), *Gene Regulatory Networks: Methods and Protocols*,
Methods in Molecular Biology, vol. 786, DOI 10.1007/978-1-61779-292-2_10, © Springer Science+Business Media, LLC 2012

One sRNA species can play a role in different pathways, for example miRNAs can influence endogenous gene expression levels as well as directly function in the host viral defense system. miRNAs constitute a class of short (20–22 nt), noncoding RNA molecules that predominantly effect protein synthesis through binding to complementary sites within the 3′-UTR of mRNAs and subsequent inhibition of translation or mRNA degradation (2). There are currently over 900 miRNAs annotated in the miRNA database miRBase (3). To determine the expression profiles of all these miRNAs in a tissue or a cell line, the most straightforward approach is deep sequencing. Because of the high number of miRNAs, deep sequencing is less laborious than other methods that determine miRNA expression levels, such as quantitative PCR or Northern blotting. Moreover, deep sequencing allows for the unbiased discovery of novel miRNAs, miRNA modifications, or other sRNA species.

It is nowadays possible to obtain several gigabases of DNA sequence in one sequencing run, allowing for in-depth genome (4) or transcriptome (5) analyses. There are three commonly used deep-sequencing platforms, the Roche 454, AB SOLiD, and Illumina/Solexa sequencing technology. Each platform has its own strength, such as the number of obtained reads or read length. The platform should, therefore, be chosen dependent on the research question.

The Solexa platform can generate more than 20 million raw, short sequencing reads per sequencing lane, and is therefore extremely suitable for the sequencing of sRNA species (6). There are several methods for generating the libraries for deep sequencing (7). Importantly, it was recently shown that different library preparation methods result in distinct miRNA expression profiles (8). Therefore, the individual samples should be compared to one another using the same library cloning method.

The method described in this chapter deals with the generation of sRNA expression libraries for the Illumina deep-sequencing platform and is based on the ligation of specific RNA adaptor sequences. To this end, an RNA sequence is annealed on each site of the RNA molecule (see Fig. 1). These so-called 5′- and 3′-adaptors contain the sequences needed to anneal the library to the surface of the sequencing flow cell and subsequent solid-phase amplification. The adaptors are ligated to the sRNA molecules prior to the synthesis of the cDNA. After reverse transcription, the sequences in the adaptors are also used to amplify the library to obtain a sufficient DNA amount for subsequent deep sequencing (see Fig. 1). The final generated deep-sequencing data can provide information on the sRNA expression levels and modifications, but can also reveal the presence of novel sRNA species. The generation of this enormous amount of information in one sequencing run is, therefore, a major strength of the research method described below.

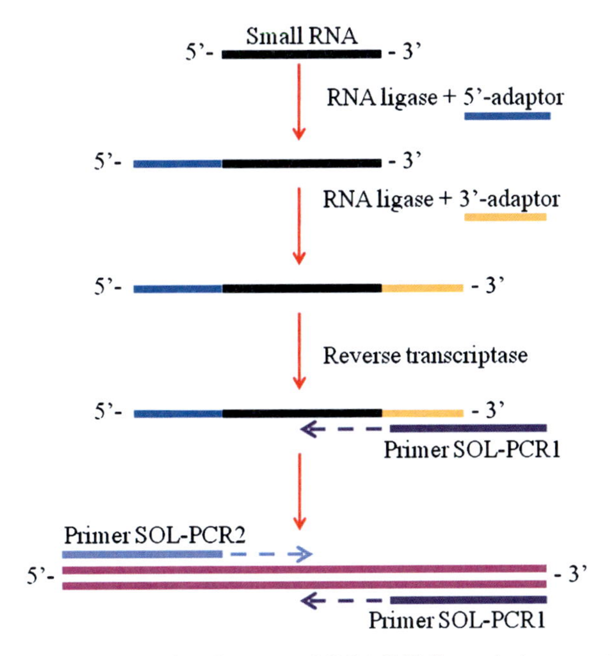

Fig. 1. Schematic overview of the Solexa small RNA (sRNA) library cloning procedure. After isolating the sRNA fraction from the total RNA, a synthetic RNA molecule is ligated on each site of the sRNA (the 5′- and 3′-adaptor). These two sequences are subsequently used for the reverse transcription and library amplification.

2. Materials

2.1. General

1. Spin-X Cellulose Acetate tube filters (Sigma-Aldrich).
2. Isopropanol.
3. Glycogen: We use glycoBlue (Applied Biosystems) or Pellet Paint (Merck), as this results in clearly visible blue pellets.
4. 70% EtOH: Mix 100% EtOH with diethylpyrocarbonate (depc)-treated H_2O to a final percentage of 70%.

2.2. Denaturing Polyacrylamide Gel

1. Depc-treated H_2O: Note that depc is carcinogenic and work in the flow hood with gloves. Add 1 ml depc (Sigma-Aldrich) per liter Millipore-Q water. Firmly shake the bottle and let stand overnight at room temperature. Autoclave the next day to inactivate the depc.
2. Urea.
3. 40% Acrylamide/bis solution 19:1 (Bio-rad). Always use gloves when handling acrylamide, as this is a neurotoxin when unpolymerized.
4. Mini protean tetra electrophoresis system (Bio-Rad).

5. 1× Tris/borate/EDTA (TBE) buffer: Dilute a concentrated solution of TBE buffer to 1× with depc-treated H_2O.
6. 10% ammonium persulfate (APS): Dissolve 1 g of APS to a final volume of 10 ml with depc-treated H_2O. Store at 4°C.
7. *N,N,N'*,N'-Tetramethylethylenediamine (Temed): Wear gloves and work in the hood as this is a toxic compound. Store at 4°C.
8. RNA loading buffer: 0.1% bromophenol blue, 0.1% xylene cyanol, 5 mM EDTA, and 95% formamide. Formamide is irritating and teratogenic; wear gloves when handling.

   DNA loading buffer: 6× orange DNA loading dye (Fermentas).
9. Markers: Use two RNA oligonucleotides for the isolation of the sRNA fraction. One sRNA oligo indicates the upper bound of where to excise the gel (e.g., 26 nt for miRNA isolation), and one indicates the lower bound (e.g., 19 nt for miRNA isolation). For the remaining RNA isolation steps, use the Low Range ssRNA ladder (NEB). For the final isolation of the library, use the O'Range Ruler 20 bp DNA ladder (Fermentas).
10. Sybr gold staining solution (Invitrogen).
11. Elution buffer for RNA: 0.3 M NaCl. Dissolve the NaCl in depc-treated H_2O. Elution buffer for DNA: 1× NEB restriction enzyme buffer 2 (50 mM NaCl, 10 mM Tris–HCl, 10 mM $MgCl_2$, 1 mM dithiothreitol, pH 7.9); store at –20°C.

2.3. Ligation of the Small RNA Adaptor Oligonucleotides

1. Adaptor RNA oligonucleotides (see Note 1).

   5′-adaptor sequence

   5′-5AmMC6 GUU CAG AGU UCU ACA GUC CGA CGA UC-3′

   (5AmMC6 = 5′ Amino Modifier C6)

   3′-adaptor sequence

   5′-5Phos UCG UAU GCC GUC UUC UGC UU 3AmMO-3′

   (5Phos = 5′phosphorylation, 3AmMO = 3′ Amino Modifier)
2. rRNAsin (Promega).
3. T4 RNA ligase and 10× ligase buffer (Promega).
4. Nuclease-free H_2O (Invitrogen).

2.4. Reverse Transcription and Library Amplification

1. DNA oligonucleotides for the PCR:

   SOL-PCR1: 5′-CAA GCA GAA GAC GGC ATA CGA-3′

   SOL-PCR2: 5′-TAA TGA TAC GGC GAC CAC CGA CAG GTT CAG AGT TCT ACA GTC CG-3′
2. Reverse transcriptase and 5× reaction buffer (Promega).
3. rRNAsin (Promega).

4. 25 mM dNTP mix: Mix 100 mM dATP, dCTP, dGTP, and dTTP (Promega) in a 1:1 ratio.
5. AmpliTaq gold DNA polymerase enzyme and 10× PCR buffer (Applied Biosystems).
6. FlashGel System, 2.2% Agarose (16 + 1 double-tier) (Lonza).
7. 3 M NaOAc: Dissolve 20.4 g NaOAc-3H_2O in nuclease-free H_2O and adjust with glacial acetic acid to pH 5.2.
8. Resuspension buffer for the final library: 10 mM Tris–HCl, pH 8.5 (see Note 2).

3. Methods

3.1. Isolation of the sRNA Fraction

1. For each individual RNA sample (see Notes 3 and 4), prepare a denaturing 15% acrylamide/bis gel in a mini protean tetra electrophoresis system. Clean the 0.75-mm spacers, glass plates, and comb with soap, water, an RNAse decontamination solution, and Millipore-Q water. Assemble the gel system according to the manufacturer's instructions. For one gel, dissolve 2.1 g urea in depc-treated H_2O to a total volume of 2.6 ml (see Note 5), add 500 µl 10× TBE, 1.87 ml 40% acrylamide/bis, 50 µl 10% APS, and 2 µl Temed. Let polymerize for about 30 min.
2. Pre-run the gel in 1×TBE buffer for 30 min at 100 V.
3. Mix 10 µg of total RNA in a 1:1 volume ratio with the RNA loading buffer. Heat at 65°C for 5 min, and snap cool on ice. Briefly spin down prior to loading to collect the sample at the bottom of the tube, and load each sample in two consecutive lanes. Make sure to wash the lanes with a syringe and 1× TBE before loading the sample.
4. For the marker, load a mixture of two sRNAs with a random sequence (5 µM each). The sizes of these two oligonucleotides are described in item 9 in Subheading 2.2. The marker should contain the same volume and RNA loading dye as one lane of the sample. Make sure to keep two lanes empty between the marker and the sample to avoid contamination.
5. Run the gel at 100 V until the bromophenol blue reaches the bottom of the gel (approximately 2 h).
6. Stain each gel separately using Sybr gold staining solution (see Note 6). Add 2.5 µl Sybr gold to 25 ml 1×TBE and stain the gel for 10 min in the dark.
7. Visualize your RNA by using a blue-light transilluminator (see Note 7) and use a razor blade to cut out your sample from the gel. Use a clean razor blade for each sample. The two marker

RNAs serve as a guide where to excise the RNA from the gel. In human RNA samples, we often observe a strong RNA band around 30 nt and one around 19 nt. These are tRNAs and rRNAs, so make sure not to take these along in the isolated sRNA sample.

8. Puncture the bottom of a 500-μl Eppendorf tube with a needle. Cut the isolated gel fraction into small pieces such that it fits into the 500-μl Eppendorf tube. Place the 500-μl tube into a 1.5-ml RNAse-free nonstick Eppendorf tube (see Note 8) and spin down in a tabletop centrifuge for 2 min at maximum speed. The gel piece is now shredded and in the 1.5-ml tube. If there are still some pieces of gel left in the 500-μl Eppendorf tube, puncture the tube again and spin down a second time. However, if it is only a small piece, just use a pipet tip to transfer it to the 1.5-ml Eppendorf tube.
9. Add 500 μl sterile 0.3 M NaCl to the shredded gel piece and rotate overnight at 4°C.
10. The next day, transfer the gel pieces onto a Spin-X Cellulose acetate filter by using a 1-ml pipet tip, from which the tip is cut off. Spin the elution through the filter using a tabletop centrifuge for 2 min at maximum speed.
11. Add another 100 μl 0.3 M NaCl on top of the gel pieces to wash. Spin down an additional 2 min at maximum speed.
12. Add to the flow-through an equal volume (600 μl) of isopropanol and 3 μl of glycoBlue.
13. Store the RNA at –20°C for at least 1 h.
14. Spin down at 4°C in a tabletop centrifuge for 30 min at maximum speed.
15. A clear blue pellet should be visible. Remove the supernatant, and wash the pellet once with 750 μl 70% EtOH.
16. Dissolve the pellet in 5.7 μl depc-treated H_2O.

3.2. Ligation of the 5′-Adaptor Sequence

1. Prepare a 15% denaturing acrylamide/bis gel as described in step 1 in Subheading 3.1. Two samples can now be loaded onto one gel.
2. Place the RNA from step 16 in Subheading 3.1 for 30 s at 90°C and snap cool on ice. This is to make sure that the RNA is completely denatured.
3. Set up the following ligation reaction in a nonstick, RNAse-free 1.5-ml Eppendorf tube: 1 μl 10× ligation buffer, 1.3 μl 5′-adaptor oligonucleotide (5 μM), 5.7 μl RNA from step 16 in Subheading 3.1, 1 μl RNAsin, and 1 μl RNA ligase. Flick the tube, briefly spin down, and incubate at 37°C for 1 h. As a ligation control, make an extra reaction without the RNA, but instead add 4.7 μl depc-treated H_2O and 1 μl of the 3′-adaptor oligonucleotide (10 μM).

4. After 1 h, add 10 µl RNA loading buffer to the reaction mix. Incubate at 65°C for 5 min and snap cool on ice.
5. For the marker, mix 0.5 µl low-range ssRNA ladder with 9.5 µl depc-treated H_2O and 10 µl RNA loading buffer. Incubate at 65°C for 5 min and snap cool on ice. Pre-run the 15% acrylamide/bis gel made in step 1 in Subheading 3.2 for 30 min at 100 V in 1×TBE and wash the lanes with a syringe and 1×TBE running buffer. If there are several samples, make sure to keep two lanes empty between each sample and between a sample and the ladder. This is to avoid contamination.
6. Run the gel at 100 V until the bromophenol blue reaches the bottom of the gel, and stain the gel with Sybr gold staining solution. Stain each gel separately in 45 ml 1× TBE and 4.5 µl of Sybr gold. Stain the gel for 10 min in the dark.
7. Visualize the RNA on the gel using a blue-light transillumineator. The 5′-adaptor is 26 nt long, and is now annealed to the sRNA fraction. For miRNAs, the RNA length therefore shifts from 20 to 24 nt to approximately 50 nt. This band is sometimes visible on the gel. Cut the gel with a clean razor blade between 40 and 60 nt. Make sure not to cut below 40 nt to avoid isolation of the 5′-adaptor oligonucleotide as well.
8. Place the gel pieces in a punctured 500-µl tube, spin down, and add 500 µl 0.3 M NaCl. Elute overnight and precipitate the RNA as described in steps 8–15 in Subheading 3.1.
9. Dissolve the RNA pellet in 6.4 µl depc-treated H_2O.

3.3. Annealing of the 3′-Adaptor Sequence

1. Prepare a denaturing 10% acrylamide/bis gel. Again, two samples can now be loaded onto one gel. Clean the glass plates, comb, and spacers as described in step 1 in Subheading 3.1, and assemble the gel system according to the manufacturer's instructions. For one gel, dissolve 2.1 g urea in depc-treated H_2O to a total volume of 3.2 ml (see Note 5), add 500 µl 10× TBE, 1.25 ml 40% acrylamide/bis, 50 µl 10% APS, and 2 µl Temed. Let polymerize for about 30 min.
2. Put the RNA from step 9 in Subheading 3.2 at 90°C for 30 s and snap cool on ice.
3. Set up the following ligation reaction in a nonstick, RNAse-free 1.5-ml Eppendorf tube: 1 µl 10× ligation buffer, 0.6 µl 3′-adaptor oligonucleotide (10 µM), 6.4 µl RNA from step 9 in Subheading 3.2, 1 µl RNAsin, and 1 µl RNA ligase. Flick the tube, briefly spin down, and incubate at 37°C for 1 h. As a ligation control, make an extra reaction without the RNA, but instead add 5.4 µl depc-treated H_2O and 1 µl of the 5′-adaptor oligonucleotide (5 µM).
4. After 1 h, add 10 µl RNA loading dye to the reaction and heat for 5 min at 65°C. Snap cool on ice.

5. Pre-run the gel made in step 1 in Subheading 3.3 for 30 min at 100 V in 1×TBE. For the marker, use the low-range ssRNA ladder as described in step 5 in Subheading 3.2. Wash the lanes with 1×TBE and make sure to keep two empty lanes between the samples and between the sample and the marker.
6. Run the gel at 100 V until the xylene cyanol has just reached the bottom of the gel. Stain the gel with Sybr gold staining solution as described in step 6 in Subheading 3.2.
7. Visualize the marker with a blue-light transilluminator. The sample is not visible anymore at this step. The 3′-adaptor sequence is 21 nt long, and is annealed to both your sRNA sample containing the 5′-adaptor sequence, as well as any free 5′-adaptor oligonucleotide still present in the reaction. The 5′- to 3′-adaptor dimer has a length of 47 nt, so make sure to isolate your sample above that size. Your sample containing the 5′- and 3′-adaptors should be around 70 nt.
8. Isolate the gel fraction, shred it, add 500 μl 0.3 M NaCl, elute overnight, and precipitate the RNA as described in steps 8–15 in Subheading 3.1.
9. Dissolve the RNA in 3 μl nuclease-free H_2O (see Note 9).

3.4. Reverse Transcription

1. Put together in a PCR tube 3 μl RNA from step 9 in Subheading 3.3 and 0.5 μl primer SOL-PCR1. Using a thermocycler, heat at 70°C for 5 min and place on ice for 5 min.
2. Add into the same PCR tube 10.1 μl nuclease-free H_2O, 4.0 μl 5× reaction buffer, 0.4 μl dNTP mix (25 mM), 1 μl RNAsin, and 1 μl reverse transcriptase. Incubate at 55°C for 1 h.
3. The cDNA can now be stored at –20°C.

3.5. Library Amplification

1. Set up the following PCR reaction: 86.1 μl nuclease-free H_2O, 10 μl 10× PCR buffer 1, 0.5 μl primer SOL-PCR1 (100 μM), 0.5 μl primer SOL-PCR2 (100 μM), 0.8 μl 25 mM dNTP mix, 0.92 μl cDNA, and 1.2 μl AmpliTaq gold polymerase (5 units/μl).
2. Use the following cycling conditions: 98°C for 30 s, followed by 12 cycles of 10 s at 98°C, 30 s at 58°C, and 30 s at 72°C.
3. After 12 cycles, put the thermocycler on hold and place the PCR tube on ice.
4. Add 1 μl Flash gel loading dye to 4 μl of the PCR product and run on a FlashGel DNA cassette for 6 min at 200 V. For the marker, use the Flash gel DNA (50–1,500 bp) marker.
5. Check if there is a band appearing above the adaptor-dimer band. The size of the adaptor dimer is around 70 bp while the library size is around 92 bp. If you see only one band around this size, it is most likely the adaptor-dimer band (see Fig. 2).

Fig. 2. Example of a 2.2% agarose gel, visualizing the PCR products after the library amplification step (step 5 in Subheading 3.5 in the protocol). *Lanes 1–7* are seven different libraries. The two bands in each sample represent the adaptor dimer (~70 nt) (*star*) and the cloned sRNA with the adaptors (~92 nt) (*arrow*). The template in the positive control reaction is a purified library and the PCR product is, thus, missing the adaptor band (*lane 8*). The negative control does not contain any template (*lane 9*). As soon as the upper band is visible, there is enough product for subsequent deep sequencing. Sample 2, therefore, is overamplified, and sample 6, although hardly visible, already contains enough material.

6. If there is a second (library) band visible, stop the PCR reaction. If there is a very faint higher band visible, add one or two more cycles and put another 4 μl of the PCR product on gel to check. Continue the PCR until there is a second band appearing (see Note 10).
7. Precipitate the PCR product by adding 10 μl 3M NaOAc, 250 μl 100% EtOH, and 1.5 μl glyco-blue.
8. Store at −20°C for at least 30 min.
9. Spin down in a tabletop centrifuge at maximum speed for 30 min at 4°C.
10. Remove the supernatant, and wash the pellet once with 750 μl 70% EtOH.
11. Dissolve the pellet in 20 μl nuclease-free H_2O.

3.6. Final Library Purification

1. Prepare a native 8% acrylamide/bis gel using the mini protean tetra electrophoresis system. Clean the glass plates, spacers, and comb with soap and water, and assemble the gel according to the manufacturer's instructions. Two samples can be loaded onto one gel. For one gel, mix 3.45 ml depc-treated H_2O, 500 μl 10× TBE, 1 ml 40% acrylamide/bis, 50 μl 10% APS, and 2 μl Temed. Let polymerize for about 30 min.
2. Add 4 μl 6× orange DNA loading dye to the sample and load the sample into two consecutive lanes. Leave two lanes empty in between the samples to avoid cross-contamination. Load 2 μl of the O'Range Ruler 20 bp DNA ladder, and add H_2O and extra loading dye such that the final volume is the same as for one lane of the sample (12 μl).
3. Run the gel at 100 V until the xylene cyanol is two-third down the gel.

4. Stain the gel with the Sybr gold staining solution. Add 2 µl Sybr gold to 45 ml 1×TBE and incubate in the dark for 10 min.
5. Visualize the gel using the blue-light transilluminator and excise the band corresponding to 90 bp. Use a clean razor blade for each sample.
6. Shred the gel as described in step 8 in Subheading 3.1.
7. Add 300 µl 1×NEB buffer 2 to the gel debris and rotate overnight at 4°C to elute.
8. The following day, pipet the gel pieces with the elution buffer onto a Spin-X filter using a pipet tip from which the tip is cut off. Spin down for 2 min at maximum speed in a tabletop centrifuge and wash the gel pieces once with 100 µl 1×NEB buffer 2. Spin down again for 2 min at maximum speed.
9. Add to the flow-through 1 µl of pellet-paint (see Note 11), 40 µl of 3 M NaOAc, and 1 ml 100% EtOH.
10. Place at −20°C for at least 30 min and spin down at 4°C for 30 min maximum speed in a tabletop centrifuge.
11. Wash the pellet once with 750 µl 70% EtOH, air dry, and dissolve the pellet in 15 µl resuspension buffer (see Note 12).

4. Notes

1. It is possible to attach a specific "barcode" to the adaptor sequence. This can be a stretch of four or five nucleotides at the 3′-end of the 5′-adaptor. This way, it is feasible to load two or more samples onto one lane of the Solexa flow cell, and later sort the samples according to their barcode.
2. This is the same buffer as the elution buffer in many mini- and midi-prep kits, which can be used instead.
3. General remarks when working with RNA: Always use gloves when handling RNA as the human skin contain RNAses which can degrade the RNA, try to work on ice as much as possible, use non-stick RNAse-free microfuge tubes (Applied Biosystems) and filter tips throughout the procedure, and use RnaseZap (Applied Biosystems) to remove RNAse contamination from the bench, pipets, etc.
4. The quality of the final library is highly dependent on the quality of the starting RNA material. If there is a lot of RNA degradation, these degraded products can end up in the sequencing library which reduces the number of interesting sRNA reads. It is, therefore, recommended to store the RNA at −80°C at all times.

Also, when processing several samples, first isolate the sRNA fraction of all the samples before continuing with the remaining library preparation steps.

5. It is possible to store the gel mixture without the Temed and APS at 4°C for several weeks, thus make a stock solution. Also, briefly microwave to dissolve the urea a bit faster.
6. Sybr gold is a more sensitive and less mutagenic dye than ethidium bromide. It is especially important that it is less mutagenic to reduce the risk of nucleotide modifications during the gel excision.
7. We use a blue-light transilluminator to minimize potential RNA and DNA damage.
8. In case you have a large gel fragment, use a 2-ml Eppendorf tube to collect the shredded gel fragment.
9. Do not use depc-treated water to dissolve the RNA pellet, as this might interfere with the PCR reaction step due to the presence of residual EtOH, which is formed during the autoclaving of depc-treated water.
10. In our hands, a higher PCR cycle number often results in a bad-quality library, with respect to a relatively low number of miRNA reads. We use a cutoff of 22 cycles. If there is still no band visible after 22 cycles, we make the library again with freshly isolated RNA.
11. At this stage, we use Pellet Paint instead of Glyco Blue, as Glyco Blue might interfere with later DNA concentration measurements.
12. To check for the quality of the library, we usually run the sample on an Agilent High Sensitivity DNA chip, together with a known library. If the library has the same size as the control, this is usually a good indication that the library is correct. Alternatively, it is possible to subclone the library and check approximately 100 clones to obtain a snapshot of the composition of the library. In general, we end up with 40–90% of miRNA reads, depending on the quality of the starting RNA material.

References

1. Taft, R. J., Pang, K. C., Mercer, T. R., Dinger, M., and Mattick, J. S. (2009) Non-coding RNAs: regulators of disease, *J Pathol* 220, 126–139.
2. Filipowicz, W., Bhattacharyya, S. N., and Sonenberg, N. (2008) Mechanisms of post-transcriptional regulation by microRNAs: are the answers in sight? *Nat Rev Genet* 9, 102–114.
3. Griffiths-Jones, S., Saini, H. K., van Dongen, S., and Enright, A. J. (2008) miRBase: tools for microRNA genomics, *Nucleic Acids Res* 36, D154–158.
4. Margulies, M., Egholm, M., Altman, W. E., Attiya, S., Bader, J. S., Bemben, L. A., Berka, J., Braverman, M. S., Chen, Y. J., Chen, Z., Dewell, S. B., Du, L., Fierro, J. M., Gomes, X. V., Godwin, B. C., He, W., Helgesen, S., Ho, C. H., Irzyk, G. P., Jando, S. C., Alenquer, M. L., Jarvie, T. P., Jirage, K. B., Kim, J. B., Knight, J. R., Lanza, J. R., Leamon, J. H., Lefkowitz, S. M., Lei, M., Li, J., Lohman, K. L., Lu, H., Makhijani, V. B.,

McDade, K. E., McKenna, M. P., Myers, E. W., Nickerson, E., Nobile, J. R., Plant, R., Puc, B. P., Ronan, M. T., Roth, G. T., Sarkis, G. J., Simons, J. F., Simpson, J. W., Srinivasan, M., Tartaro, K. R., Tomasz, A., Vogt, K. A., Volkmer, G. A., Wang, S. H., Wang, Y., Weiner, M. P., Yu, P., Begley, R. F., and Rothberg, J. M. (2005) Genome sequencing in microfabricated high-density picolitre reactors, *Nature* 437, 376–380.

5. Berezikov, E., Thuemmler, F., van Laake, L. W., Kondova, I., Bontrop, R., Cuppen, E., and Plasterk, R. H. (2006) Diversity of microRNAs in human and chimpanzee brain, *Nat Genet* 38, 1375–1377.
6. Kuchenbauer, F., Morin, R. D., Argiropoulos, B., Petriv, O. I., Griffith, M., Heuser, M., Yung, E., Piper, J., Delaney, A., Prabhu, A. L., Zhao, Y., McDonald, H., Zeng, T., Hirst, M., Hansen, C. L., Marra, M. A., and Humphries, R. K. (2008) In-depth characterization of the microRNA transcriptome in a leukemia progression model, *Genome Res* 18, 1787–1797.
7. Berezikov, E., Cuppen, E., and Plasterk, R. H. (2006) Approaches to microRNA discovery, *Nat Genet* 38 *Suppl*, S2–7.
8. Linsen, S. E., de Wit, E., Janssens, G., Heater, S., Chapman, L., Parkin, R. K., Fritz, B., Wyman, S. K., de Bruijn, E., Voest, E. E., Kuersten, S., Tewari, M., and Cuppen, E. (2009) Limitations and possibilities of small RNA digital gene expression profiling, *Nat Methods* 6, 474–476.

Part II

Genomic Components

Chapter 11

CAGE (Cap Analysis of Gene Expression): A Protocol for the Detection of Promoter and Transcriptional Networks

Hazuki Takahashi, Sachi Kato, Mitsuyoshi Murata, and Piero Carninci

Abstract

We provide here a protocol for the preparation of cap-analysis gene expression (CAGE) libraries, which allows for measuring the expression of eukaryotic capped RNAs and simultaneously map the promoter regions. The presented protocol simplifies the previously published ones and moreover produces tags that are 27 nucleotides long, which facilitates mapping to the genome. The protocol takes less than 5 days to complete and presents a notable improvement compared to previously published versions.

Key words: Cap-analysis gene expression, RNAseq, Transcriptome, Sequencing, RNA

1. Introduction

CAGE (Cap Analysis of Gene Expression) is based on a series of full-length cDNA technologies previously developed at RIKEN. The purpose of the technology is to comprehensively map the vast majority of human transcription starting sites and hence their promoters, and simultaneously decipher the expression of the RNAs produced at each promoter. Thus, CAGE allows for high-throughput gene expression profiling with simultaneous identification of the tissue/cell/condition-specific transcriptional start sites (TSS), including promoter usage analysis. CAGE has various advantages over microarray-based expression analysis. The identification of the promoters used in the analyzed biological phenomena (tissue, cells, treatments, time courses, etc.), together with the determination of the expression level at each promoter, allows for identifying regulatory elements, such as core promoters and the transcription factor binding sites (TFBS) that are responsible for transcription. Bioinformatic analysis allows for analysis of promoters

Bart Deplancke and Nele Gheldof (eds.), *Gene Regulatory Networks: Methods and Protocols*,
Methods in Molecular Biology, vol. 786, DOI 10.1007/978-1-61779-292-2_11, © Springer Science+Business Media, LLC 2012

having similar expression profiles that are analyzed for the presence of common TFBS. Coupled to the determination of the expression of transcription factors, which drive the gene transcription, this analysis allows to reconstruct the networks that drive gene expression (1). By counting the number of CAGE tags for each promoter within a gene, we can determine not only the RNA expression level (this is a digital detection of frequency) but also from which of the various alternative promoters the RNA is transcribed, allowing comprehensive mapping of promoters in mammalian genomes (2). Additionally, sequencing-based methods such as CAGE allow for identifying the transcriptome of expressed retrotransposon elements, a task not previously possible by microarrays hybridization (3). Comprehensive examples of applications have been published elsewhere (4). Here we present an updated protocol for the Illumina GA2X sequencer, in which the CAGE tags are 27 nucleotides long thanks to the use of *Eco*P15I, as compared with the previously used *Mme*I, which allows preparing only 20–21 bp long tags. Longer CAGE tags contribute to high efficiency mapping, if compared to previous versions of CAGE. To prepare a CAGE library, cDNA complementary strands are synthesized from total RNA extracted from cells or tissues, generally by using random primers, or a mixture of random and oligo-dT primers. The 5′ end of cDNA is then selected by using the cap-trapper method (5). Next, a biotinylated linker is attached to the 5′ end of single-strand cDNA. The linker contains the recognition site of the endonuclease *Eco*P15I. After the second cDNA strand is synthesized, the cDNA is cleaved 27 nucleotides away from the *Eco*P15I recognition site, which isolates the DNA derived from the 5′ end of the original RNA, to produce CAGE tags for the molecules present in the reactions. Next, a linker is attached to the 3′ end, to amplify and apply the sample into the Illumina GA2X sequencer to produce up to 20 million tags or more per sequencing lane. By changing the primers, the user may adapt the technology to other sequencing platforms.

2. Materials

2.1. Preparation of Sorbitol-Trehalose (3.3 M/0.66 M) Mix

1. Saturated trehalose solution: dissolve 7.27 g D-trehalose in 10 ml water.
2. 4.9 M sorbitol: dissolve 17.8 g D-Sorbitol in 20 ml water.
3. Mix saturated trehalose and 4.9 M sorbitol in a 50-ml tube. Autoclave at 121°C for 30 min.

2.2. 250 mM $NaIO_4$ for Oxidation of the Diol Groups

Dissolve 530.47 μg $NaIO_4$ in 10 μl water. Store at room temperature, avoiding light. Prepare freshly before use.

2.3. 15 mM Biotin (Long Arm) Hydrazide for Biotinylation

Dissolve 75.06 μg Biotin (Long Arm) Hydrazide (VECTOR Lab) in 13.5 μl water. Prepare freshly before use.

2.4. Preparation of 20 μg/μl E. coli tRNA

1. Dissolve 30 mg *E. coli* tRNA (Ribonucleic acid, transfer from *Escherichia coli* Type XX, Strain W, lyophilized powder (Sigma)) in 400 μl water and add 45 μl of 10× RQ1 DNase buffer and 30 μl of RQ1 RNase-Free DNase. Incubate at 37°C for 2 h.
2. Add 10 μl of 0.5 M EDTA (pH 8.0), 10 μl of 10% SDS, and 10 μl of 10 ng/ml Proteinase K to tRNA solution. Incubate at 45°C for 30 min.
3. Add 500 μl of phenol–chloroform to incubate solution. Centrifuge at 20,000 × *g* for 3 min at room temperature.
4. Collect aqueous phase and add 500 μl of chloroform. Centrifuge at 20,000 × *g* for 3 min at room temperature.
5. Collect supernatant and add 25 μl of 5 M NaCl and 525 μl of Isopropanol. Centrifuge at 20,000 × *g* for 5 min at room temperature.
6. Remove supernatant and add 900 μl of 80% Ethanol to tRNA pellet. Centrifuge at 20,000 × *g* for 5 min at room temperature. Repeat this step.
7. Dissolve the tRNA pellet in 1.5 ml water. Store in aliquots at −20°C.

2.5. Preparation of Wash Buffer for MPG Beads

1. Wash buffer 1: Mix 45 ml of 5 M NaCl and 5 ml of 0.5 M EDTA (pH 8.0). Store at room temperature.
2. Wash buffer 2: Mix 3 ml of 5 M NaCl, 100 μl of 0.5 M EDTA (pH 8.0), and 46.9 ml of water. Store at room temperature.
3. Wash buffer 3: Mix 1 ml of 1 M Tris–HCl (pH 8.5), 100 μl of 0.5 M EDTA (pH 8.0), 25 ml of 1 M NaOAc (pH 6.1), 2 ml of 10% SDS, and 21.9 ml of water. Store at room temperature.
4. Wash buffer 4: Mix 500 μl of 1 M Tris–HCl (pH 8.5), 100 μl of 0.5 M EDTA (pH 8.0), 25 ml of 1 M NaOAc (pH 6.1), and 24.4 ml of water. Store at room temperature.

2.6. Preparation of 3′ Linker Ligation Buffer (5×)

Mix 50 μl of 250 mM Tris–HCl (pH 7.0), 10 μl of 100 mM ATP, and 0.5 μl of 10 mg/ml BSA. Adjust to 200 μl in water. Store at room temperature.

2.7. Other Reagents

1. PrimeScript Reverse Transcriptase (TAKARA, 10,000 U).
2. Agencourt RNAClean XP Kit (BECKMAN COULTER, 40 ml).
3. Agencourt AMPure XP Kit (BECKMAN COULTER, 60 ml).
4. RNase ONE Ribonuclease (Promega, 1,000 U).

5. MPG Streptavidin (TAKARA, 2 ml).
6. OliGreen ssDNA Quantitation Kit (Molecular Probes).
7. Agilent RNA Pico-kit.
8. Agilent DNA1000 kit.
9. DNA Ligation Kit < Mighty Mix > (TAKARA).
10. T4 DNA ligase (NEB, 20,000 U).
11. TaKaRa LA Taq (TAKARA, 125 U).
12. EcoP15I (NEB, 500 U).
13. Sinefungin (Calbiochem-Novabiochem international, 2 mg).
14. Phusion™ High-Fidelity DNA Polymerase (FINNZYMES, 100 U).
15. ExonucleaseI (*E. coli*) (NEB, 3,000 U).
16. MinElute PCR Purification Kit (QIAGEN).
17. Ethanol (70%).
18. 10% SDS.
19. 10 mM dNTPs (Invitrogen).
20. 1 M NaOAc (pH 4.5).
21. 1 M Na citrate (pH 6.0).
22. 0.5 M EDTA (pH 8.0).
23. 40% glycerol.
24. 1 M Tris–HCl (pH 8.5).
25. 1 M Tris–HCl (pH 7.0).
26. 50 mM NaOH.
27. 0.4 M $MgCl_2$.
28. Nuclease-free water (Invitrogen Corp).

2.8. Equipment

1. Micropipettes.
2. Multipipettors.
3. Pipette Tips (Low binding tips).
4. 1.5-ml SnapLock Microtube, Nonsterile, MaxyClear, Maxymum Recovery (AXYGEN).
5. 96-well PCR Plate, 0.2 ml, Nonsterile, Clear (AXYGEN).
6. Agilent 2100 Bioanalyzer.
7. NanoDrop 1000 spectrophotometer.
8. Dynal Magnetic stand (Invitrogen).
9. Centrifugal Concentrator (TOMY Digital Biology Co., Ltd.).
10. Thermal cycler.
11. Genome Analyzer 2-X (Illumina).

2.9. Oligonucleotides

1. Reverse transcription: RT-N15-EcoP primer (EcoP15I site = *Italic*).

   5′-AAGGTCTAT*CAGCAG*NNNNNNNNNNNNNNN-3′
2. 2nd SOL primer.

   5′-Bio CCACCGACAGGTTCAGAGTTCTACAG-3′
3. PCR Forward primer.

   5′-AATGATACGGCGACCACCGACAGGTTCAGAGTTC-3′
4. PCR Reverse primer.

   5′-CAAGCAGAAGACGGCATACGA-3′
5. Sequencing Primer.

   5′-CGGCGACCACCGACAGGTTCAGAGTTCTACAG-3′

2.10. 5′ Linker

1. 5′-N6 upper linker, (bar code = **Bold**, EcoP15I site = *Italic*).

   AGA: 5′-CCACCGACAGGTTCAGAGTTCTACAG**AGA***CAGCAG*NNNNNN Phos-3′

   CTT: 5′-CCACCGACAGGTTCAGAGTTCTACAG**CTT***CAGCAG*NNNNNN Phos-3′

   GAT: 5′-CCACCGACAGGTTCAGAGTTCTACAG**GAT***CAGCAG*NNNNNN Phos-3′

   ACA: 5′-CCACCGACAGGTTCAGAGTTCTACAG**ACA***CAGCAG*NNNNNN Phos-3′

   ACT: 5′-CCACCGACAGGTTCAGAGTTCTACAG**ACT***CAGCAG*NNNNNN Phos-3′

   ACG: 5′-CCACCGACAGGTTCAGAGTTCTACAG**ACG***CAGCAG*NNNNNN Phos-3′

   ATC: 5′-CCACCGACAGGTTCAGAGTTCTACAG**ATC***CAGCAG*NNNNNN Phos-3′

   ATG: 5′-CCACCGACAGGTTCAGAGTTCTACAG**ATG***CAGCAG*NNNNNN Phos-3′

   AGC: 5′-CCACCGACAGGTTCAGAGTTCTACAG**AGC***CAGCAG*NNNNNN Phos-3′

   AGT: 5′-CCACCGACAGGTTCAGAGTTCTACAG**AGT***CAGCAG*NNNNNN Phos-3′

   TAG: 5′-CCACCGACAGGTTCAGAGTTCTACAG**TAG***CAGCAG*NNNNNN Phos-3′

   TGG: 5′-CCACCGACAGGTTCAGAGTTCTACAG**TGG***CAGCAG*NNNNNN Phos-3′

   GTA: 5′-CCACCGACAGGTTCAGAGTTCTACAG**GTA***CAGCAG*NNNNNN Phos-3′

   GAC: 5′-CCACCGACAGGTTCAGAGTTCTACAG**GAC***CAGCAG*NNNNNN Phos-3′

GCC: 5′-CCACCGACAGGTTCAGAGTTCTACAG**GCC***CA GCAG*NNNNNN Phos-3′

2. 5′-GN5 upper linker, (bar code = **Bold**, EcoP15I site = *Italic*).

AGA: 5′-CCACCGACAGGTTCAGAGTTCTACAG**AGA***CA GCAG*GNNNNN Phos-3′

CTT: 5′-CCACCGACAGGTTCAGAGTTCTACAG**CTT***CA GCAG*GNNNNN Phos-3′

GAT: 5′-CCACCGACAGGTTCAGAGTTCTACAG**GAT***CA GCAG*GNNNNN Phos-3′

ACA: 5′-CCACCGACAGGTTCAGAGTTCTACAG**ACA***CA GCAG*GNNNNN Phos-3′

ACT: 5′-CCACCGACAGGTTCAGAGTTCTACAG**ACT***CA GCAG*GNNNNN Phos-3′

ACG: 5′-CCACCGACAGGTTCAGAGTTCTACAG**ACG***CA GCAG*GNNNNN Phos-3′

ATC: 5′-CCACCGACAGGTTCAGAGTTCTACAG**ATC***CA GCAG*GNNNNN Phos-3′

ATG: 5′-CCACCGACAGGTTCAGAGTTCTACAG**ATG***CA GCAG*GNNNNN Phos-3′

AGC: 5′-CCACCGACAGGTTCAGAGTTCTACAG**AGC***CA GCAG*GNNNNN Phos-3′

AGT: 5′-CCACCGACAGGTTCAGAGTTCTACAG**AGT***CA GCAG*GNNNNN Phos-3′

TAG: 5′-CCACCGACAGGTTCAGAGTTCTACAG**TAG***CA GCAG*GNNNNN Phos-3′

TGG: 5′-CCACCGACAGGTTCAGAGTTCTACAG**TGG***CA GCAG*GNNNNN Phos-3′

GTA: 5′-CCACCGACAGGTTCAGAGTTCTACAG**GTA***CA GCAG*GNNNNN Phos-3′

GAC: 5′-CCACCGACAGGTTCAGAGTTCTACAG**GAC***CA GCAG*GNNNNN Phos-3′

GCC: 5′-CCACCGACAGGTTCAGAGTTCTACAG**GCC***CA GCAG*GNNNNN Phos-3′

3. 5′-lower linker, (bar code = **Bold**, EcoP15I site = *Italic*).

AGA: 5′-Phos *CTGCTG***TCT**CTGTAGAACTCTGAACCTG TCGGTGG NH_2-3′

CTT: 5′-Phos *CTGCTG***AAG**CTGTAGAACTCTGAACCTG TCGGTGG NH_2-3′

GAT: 5′-Phos *CTGCTG***ATC**CTGTAGAACTCTGAACCTG TCGGTGG NH_2 -3′

ACA: 5′-Phos *CTGCTG***TGT**CTGTAGAACTCTGAACCTG TCGGTGG NH_2-3′

ACT: 5′-Phos *CTGCTG***AGT**CTGTAGAACTCTGAACCTG TCGGTGG NH_2-3′

ACG: 5′-Phos *CTGCTG***CGT**CTGTAGAACTCTGAACCTG TCGGTGG NH_2-3′

ATC: 5′-Phos *CTGCTG***GAT**CTGTAGAACTCTGAACCTG TCGGTGG NH_2-3′

ATG: 5′-Phos *CTGCTG***CAT**CTGTAGAACTCTGAACCTG TCGGTGG NH_2-3′

AGC: 5′-Phos *CTGCTG***GCT**CTGTAGAACTCTGAACCTG TCGGTGG NH_2-3′

AGT: 5′-Phos *CTGCTG***ACT**CTGTAGAACTCTGAACCTG TCGGTGG NH_2-3′

TAG: 5′-Phos *CTGCTG***CTA**CTGTAGAACTCTGAACCTG TCGGTGG NH_2-3′

TGG: 5′-Phos *CTGCTG***CCA**CTGTAGAACTCTGAACCTG TCGGTGG NH_2-3′

GTA: 5′-Phos *CTGCTG***TAC**CTGTAGAACTCTGAACCTG TCGGTGG NH_2-3′

GAC: 5′-Phos *CTGCTG***GTC**CTGTAGAACTCTGAACCTG TCGGTGG NH_2-3′

GCC: 5′-Phos *CTGCTG***GGC**CTGTAGAACTCTGAACCTG TCGGTGG NH_2-3′

4. Preparation of 5′ linkers by annealing oligonucleotides (see Note 1).
   (a) Dissolve the purified oligonucleotides to 2 μg/μl in 1 mM Tris–HCl (pH 7.5) and 0.1 mM EDTA (pH 8.0).
   (b) N6 linker reaction solution: Mix 1.5 μl of each specific 5′-N6 upper linker (3.0 μg), 1.5 μl of each specific 5′-lower linker (3.0 μg), 0.75 μl of 1 M NaCl, and 3.25 μl of water.
   (c) GN5 linker reaction solution: Mix 6 μl of each specific 5′-GN5 upper linker (12 μg), 6 μl of each specific 5′-lower linker (12 μg), 3 μl of 1 M NaCl, and 15 μl of water.
   (d) The annealing reaction is carried at the following conditions: 95°C, 5 min gradient 0.1°C/s, 83°C, 5 min, gradient 0.1°C/s, 71°C 5 min, gradient 0.1°C/s, 59°C 5 min, gradient 0.1°C/s, 59°C 5 min, gradient 0.1°C/s, 47°C 5 min, gradient 0.1°C/s, 35°C 5 min, gradient 0.1°C/s, 23°C 5 min, gradient 0.1°C/s and hold at 11°C.
   (e) The final annealed linker solution can be kept on hold at 4°C, but for long-term storage it should be frozen at –20°C.
   (f) The "N6" and "GN5" linkers carrying the same bar code should be mixed at this stage. The total volume is 37.5 μl

(0.8 μg/μl), using a final ratio of N6–GN5 at1:4. This was found effective to maximize the ligation efficiency of the cDNAs with cap-trapped cDNA ends. These linkers are used at the concentration of 200 ng/μl when starting from 5 μg of total RNA.

2.11. 3′ Linkers

1. 3′ upper linker.

   5′-Phos NNTCGTATGCCGTCTTCTGCTTG-3′
2. 3′ lower linker.

   5′-CAAGCAGAAGACGGCATACGA-3′
3. Preparation of 3′ linkers by annealing oligonucleotides.
   (a) Dissolve the purified oligonucleotides to 2 μg/μl in 1 mM Tris–HCl pH 7.5 and 0.1 mM EDTA (pH 8.0).
   (b) Linker reaction solution: Mix 2.5 μl of 3′ upper linker (5.0 μg), 2.5 μl of 3′ lower linker (5.0 μg), 1.25 μl of 1 M NaCl, and 6.25 μl of water.
   (c) The annealing reaction is carried at the following conditions: 95°C, 5 min gradient 0.1°C/s, 83°C, 5 min, gradient 0.1°C/s, 71°C 5 min, gradient 0.1°C/s, 59°C 5 min, gradient 0.1°C/s, 59°C 5 min, gradient 0.1°C/s, 47°C 5 min, gradient 0.1°C/s, 35°C 5 min, gradient 0.1°C/s, 23°C 5 min, gradient 0.1°C/s and hold at 11°C.
   (d) The final annealed linker concentration is 0.8 μg/μl. These linkers are used at the concentration of 100 ng/μl when starting from 5 μg of total RNA. It can be kept or held at 4°C, but for long-term storage it should be frozen at –20°C.

2.12. Samples (5 μg Total RNA, PolyA Plus RNA or PolyA Minus RNA)

It is advisable to use RNAs that are isolated with Trizol LS (Invitrogen) or the RNeasy kit (Qiagen) and that have a RIN value over 7 as measured by the Agilent RNA nano kit. In this protocol, RNA amounts are set at 5 μg total RNA. However, we could prepare the CAGE library with 5 μg polyA minus RNA and 1 μg polyA plus RNA, which are separated by Poly(A)Purist mRNA Purification Kits (Ambion).

3. Methods

3.1. First Strand: Reverse Transcription (see Notes 2 and 3)

1. Preparation of RNA–primer mix: Mix 5 μg total RNA and 2.2 μl of 210 μM Reverse Transcription (RT)-N15-EcoP primer. Adjust volume to 7.5 μl in water. Incubate at 65°C for 5 min and then cool on ice immediately.
2. Preparation of Enzyme mix: Mix 7.5 μl of 5× PrimeScript buffer, 1.87 μl of 10 mM dNTPs, 7.5 μl of Sorbitol–Trehalose

mix solution, 3.75 μl of PrimeScript Reverse Transcriptase, and 9.38 μl of water (see Note 4).

3. Add Enzyme mix to RNA–primer mix tube and carefully mix by pipetting on ice (total volume 37.5 μl).
4. Incubate in a thermal cycler as follows: 25°C, 30 s; 42°C, 30 min; 50°C, 10 min; 56°C, 10 min; 60°C, 10 min; keep on hold on ice (see Note 5).

3.2. cDNA Cleanup with the RNAClean XP Kit (see Note 6)

1. Add 67.5 μl of RNAClean XP to 37.5 μl of RT reaction solution and thoroughly mix by pipetting. Incubate at room temperature for 30 min, mixing every 10 min by pipetting.
2. Set the tube on the magnetic stand for 5 min and remove the supernatant.
3. Keep the sample on the magnetic stand and wash the beads with ethanol by pouring 150 μl of 70% EtOH, washing the beads and the tube walls. After checking that the beads are settled on the tube wall, remove the supernatant. Repeat this washing step once more.
4. To the rinsed beads, add 40 μl of water (preheated at 37°C) and extensively pipette (the manufacturers suggest pipetting at least 20 times) to elute the RNA-cDNA hybrid.
5. Incubate for 10 min at 37°C and then set on the magnetic stand for 5 min to separate the beads. Collect the supernatant (40 μl).
6. Keep the remainder of cDNA on ice.

3.3. Oxidation of the Diol Groups, Including the Cap-Site

The cap is modified in a two-step reaction, the first step constituted by oxidation with $NaIO_4$.

1. Mix 40 μl of cDNA, 2 μl of 1 M NaOAc (pH 4.5), and 2 μl of 250 mM $NaIO_4$ by ten times pipetting on ice. Proceed the reaction in the dark by wrapping immediately in aluminum foil to cover the samples and leave on ice for 45 min.
2. Stop the reaction by adding 2 μl of 40% glycerol and mix thoroughly. Add 14 μl of 1 M Tris–HCl (pH 8.5) to bring the pH above 5.6. Subsequently, purify the sample with the RNAClean XP as in Subheading 3.2. Notice that the volume of the cDNA is different: to keep the ratio of RNAClean XP solution–cDNA at 1.8-fold, add 108 μl of RNAClean XP reagents to the 60 μl of cDNA obtained from the oxidation reaction above. At the end, collect the supernatant in 40 μl of water.

3.4. Biotinylation of the RNA (see Note 7)

1. Mix 40 μl of Oxidate cDNA, 4 μl of 1 M Na–Citrate (pH 6.0), and 13.5 μl of 15 mM biotin hydrazide (Long Arm) (total volume 57.5 μl) by pipetting for ten times and incubate at 23°C for 14–15 h (overnight) (see Note 8).

3.5. RNase I Treatment (see Note 9)

1. Add 6 μl of 1 M Tris–HCl (pH 8.5), 1 μl of 0.5 M EDTA (pH 8.0), and 5 μl of RNase ONE Ribonuclease to 57.5 μl of Biotinylated solution (total volume 69.5 μl). Mix by pipetting and incubate at 37°C for 30 min and at 65°C for 5 min (see Note 10).
2. Cool the cDNA on ice for 2 min and proceed with the RNAClean XP purification using 125 μl of the Agencourt reagent with 69.5 μl of cDNA (1.8-fold ratio) and perform all the steps as in Subheading 3.2. Redissolve the cDNA in 40 μl of water.

3.6. Cap-Trapping Using the MPG Streptavidin Beads

Cap-trapping is achieved by capturing the cDNA that have reached the cap-site with MPG streptavidin beads. The vast majority of the truncated cDNAs are left in the solution and eliminated.

1. Prepare the beads by blocking them with tRNAs, to diminish nonspecific interactions. Add 1.5 μl of 20 μg/μl *E. coli* tRNA mix to 100 μl of MPG beads and incubate at room temperature for 30–60 min, mixing every 10 min by pipetting. Separate the beads on a magnetic stand and remove supernatant. Wash the beads with 50 μl of Wash buffer 1 (two times) and resuspend in 80 μl of Wash buffer 1.
2. Add 40 μl of RNase I treated cDNA to the 80 μl of washed MPG beads.
3. Incubate at room temperature for 30 min (mix by ten times pipetting or moderate vortexing every 5 min.). Separate on the magnetic stand for 3 min. Next, remove the supernatant.
4. Extensively wash the beads by multiple pipetting with 150 μl of the washes below, followed by capture with a magnetic stand: Wash buffer 1 (one time), Wash buffer 2 (one time), Wash buffer 3 (two times), and Wash buffer 4 (two times) (see Note 11).

3.7. Release cDNA from Beads

1. Before proceeding with the next reaction, cDNAs have to be removed from the magnetic beads with alkali, which denatures RNA–cDNA hybrids and simultaneously fragments the RNAs. To do this, add 60 μl of 50 mM NaOH to the RNA–cDNA-washed beads tube and incubate at room temperature for 10 min, with occasional mixing.
2. Separate the beads on the magnetic stand for 3 min and collect the supernatant into a new tube.
3. To this new tube, add 12 μl of 1 M Tris–HCl (pH 7.0) to neutralize the alkali solution. The total collected volume will be 72 μl. Keep the cDNA on ice before the next step.
4. Subsequently purify the sample using the AMPure XP using 130 μl of the reagents to 72 μl of eluted cDNA as in Subheading 3.2 RNAClean XP kit. Resuspend the cDNA in 34 μl of water, and keep 3 μl for quality control (QC). The QC

consists in measuring the concentration of the cDNAs with OliGreen and the size of captured cDNAs with the RNA Pico Kit (see Note 12). The cDNA is subsequently concentrated by a centrifugal concentrator at room temperature in a siliconized tube, and finally redissolved in 4 μl water. It is preferable to avoid complete drying of the pellet by measuring the remaining volume of the water during the concentration operation, although this may be a tedious operation.

3.8. Ligation of a Linker to the Single-Stranded cDNA (see Note 13)

1. In a separate tube, prepare 1 μl of the 5′ linker (200 ng/μl) for each sample and incubate at 37°C for 5 min (see Note 14).
2. At the end, cool the linker on ice for 2 min and add 4 μl of cDNA and 10 μl of DNA ligation Mighty Mix to 5′ linker tubes.
3. After extensive mixing, incubate overnight at 16°C.
4. Add 55 μl of water to the 5′ linker ligated cDNA. In case of mixing cDNA, add 10 μl of water to mixed cDNA (total volume 70 μl) (see Note 15). Purify the cDNA with the AMPure XP kit as in Subheading 3.2. Resuspend the cDNA in 30.5 μl of water.
5. To avoid remaining 5′ linkers, repeat twice this purification step.

3.9. Second-Strand cDNA Synthesis

At this stage, the second-strand cDNA is prepared by priming the sequences added in the previous stage. The enzyme for the synthesis is LA-Taq, a thermostable DNA polymerase that is able to amplify long cDNA fragments. The reaction is set up as follows for each sample:

1. Add 5 μl of 10× LA Taq buffer, 5 μl of 25 mM $MgCl_2$, 8 μl of 2.5 mM dNTPs, 1 μl of 2nd SOL primer (200 ng/μl, 24 μM), and 0.5 μl of LA Taq (5 U/μl) to 30.5 μl of 5′ linker ligated sscDNA (total volume 50 μl) and gently mix by pipetting on ice.
2. Incubate at 94°C for 3 min, 42°C for 5 min to anneal the primer, 68°C for 20 min, 62°C for 2 min and then hold at 4°C.
3. The sample is then purified again using the AMPure XP kit as in Subheading 3.2, adding 90 μl of beads to 50 μl of the second-strand cDNA reaction and redissolving the cDNA in 30 μl of water.

3.10. EcoP15I Digestion

At this stage, cDNAs is cleaved with *Eco*P15I, from the end of second-strand primer 27 nt into the cDNA (see Note 16).

1. Preparation of premix solution: Mix 4 μl of 10× NEBuffer 3, 0.4 μl of 10 mg/ml (100×) BSA, 4 μl of 10 mM (10×) ATP, 0.4 μl of 10 mM Sinefungin, 0.1 μl of *Eco*P15I (10 U/μl), and 1.1 μl of water (total volume 10 μl) by pipetting on ice.
2. Add 30 μl of the double-strand cDNA from Subheading 3.9 to 10 μl of premix solution. Incubate at 37°C for 3 h.

3. Add 1 µl of 0.4 M $MgCl_2$ (to 10 mM final concentration) to stabilize the short tags and prevent their denaturation.
4. Incubate at 65°C, 20 min to inactivate the restriction enzyme. The digested cDNA can be kept on ice until the next step.

3.11. Addition of a 3′ Linker to the Cleaved Tags

This step provides to the 5′ cDNA tags a 3′ end linker, suitable for the subsequent PCR for the final preparation of the CAGE tags suitable for sequencing.

1. Add 16 µl of 5× 3′ linker ligation buffer (2.6), 1 µl of 3′ linker (100 ng/µl), 3 µl of T4 DNA ligase (400 U/µl), and 19 µl of water to 41 µl of *Eco*P15I-digested cDNA (total volume 80 µl) by pipetting on ice and incubate overnight at 16°C.

3.12. Removal of Excess of 3′ Linker (see Note 17)

1. Prepare the beads by mixing 10 µl of MPG beads and 1 µl of 20 µg/µl *E. coli* tRNA, followed by moderate vortexing and incubation at room temperature for 30–60 min to coat the surface to avoid nonspecific binding.
2. Separate the beads on a magnetic stand as in Subheading 3.6 and wash with Wash buffer 1 twice as described above. Finally, redissolve the beads in 25 µl of Wash buffer 1.
3. Add the beads to 80 µl of the cDNA ligated from Subheading 3.11, and incubate at room temperature for 30 min with occasional mixing by pipetting or mild vortexing.
4. Wash the beads as in Subheading 3.6, followed by a final quick wash with 100 µl of water. At this step, avoid heating the sample and perform as quickly as possible, to avoid losing the tags due to denaturation. After separating the beads, redissolve them with water, in a final volume of 20 µl. This will be the template for subsequent PCR reactions. Beads/CAGE tags may be kept at –20°C.

3.13. Pilot PCR Experiments

An aliquot of the beads is tested by PCR to verify the number of cycles necessary to amplify the bulk PCR reaction in the subsequent stage. We test multiple cycles.

At first, set up three reactions to check the number of PCR cycles (e.g., 8, 10, 12 cycles). Depending on the experiment, the amount of cDNAs may be lower, so a different number of PCR cycles are recommended when starting.

1. Preparation of PCR premix reaction solution: Mix 10 µl of 5× High-Fidelity buffer, 4 µl of 2.5 mM dNTPs, 0.5 µl of 100 µM PCR Forward primer, 0.5 µl of 100 µM PCR Reverse primer, 0.5 µl of Phusion polymerase (2 U/µl), 32.5 µl of water, and 2 µl of cDNA from Subheading 3.12 (total volume 50 µl) by pipetting on ice.
2. Perform the PCR at the following conditions: 98°C for 30 s, followed by (98°C for 10 s, 60°C for 10 s) times the number of PCR cycles as required, hold at 4°C.

3. Check the product on the Bioanalyzer DNA1000 to measure the concentration and verify the size of the amplified product. The desired product is 96 bp long. Appearance of other contaminants should be minimal (see Note 18).

3.14. Bulk PCR Amplification of the CAGE Library

After selecting the best PCR cycle number (Subheading 3.13), the bulk PCR (6 PCR tubes) is performed for the large part of the 12 μl remaining samples (see Note 19).

3.15. Purification of Primers by Exonuclease I

Rather than performing tedious purification steps, excess of PCR primers are removed by Exonuclease I treatment, which cleaves only single-stranded RNA. The CAGE tags are thus protected, being double strand.

1. Pool the PCR reaction solutions into one 1.5 ml siliconized tube for the equivalent of three PCR reactions.
2. Add 1 μl of ExonucleaseI (20 U/μl) to the 150 μl of PCR reaction solution and mix by pipetting on ice, and then incubate at 37°C for 0.5–1 h.
3. The CAGE tag sample is purified with the MinElute PCR Purification Kit, using the 151 μl of above product for each column, following the manufacturer's instructions.
4. The CAGE tag is eluted in 10 μl EB per column.
5. An aliquot is used to check the DNA concentration with the Agilent Bioanalyzer DNA1000. An example result is shown in Fig. 2d. The remaining sample is ready for the Illumina GA2-X sequencing, using the 36-nt reads cycle. The standard protocol requires a DNA concentration of 10 nM (0.67 ng/μl) (10 μl for a 96 bp CAGE tags), while the final DNA concentration in the sequencing reaction should be in the order of 5.0–7.0 pM.

4. Notes

1. The 5′ linkers are bar-coded for pooling the CAGE libraries. This helps to (a) pool multiple libraries in the same sequencing lane and (b) run the PCR reaction treating all the samples equally, to avoid differences due to different PCR conditions in different tubes. The latter is particularly important when comparing different samples, like in a time course. There are linkers labeled N6 (where the random fraction that will anneal on the cDNA will contain simply a random hexamer) or a GN5, where one of the bases is a G, which preferentially anneals with the first base of the cDNA that is often a C, which is added frequently by the RT in correspondence of the cap site. "Phos" stands for a phosphate group, NH_2 for an amino-link that prevent ligation of 3′ ends. Upper and lower linkers are combined

together according to their bar-code sequence to form a double strand with partial single-strand random protruding ends, which ligates on the terminal end of the cDNA. Linkers are mixed at GN5 to N6 at the ratio of 4:1, to an equimolar amount of lower linker. Annealing should take place by slowly cooling the linkers as described in step 4(d) in Subheading 2.10.

2. The purpose of this step is to convert the RNA to cDNA. The use of random primers allows reverse transcription of all RNAs including poly-A minus RNAs. In fact, a large amount of non-polyadenylated RNAs are constituted by capped molecules, including long noncoding RNAs (6). Additionally, random priming minimizes the risk of underrepresentation of long polyadenylated RNAs in the libraries, which may be due to differences in reverse transcription efficiency of small differences in quality of RNA. Altogether, random priming minimizes the chances to introduce biases in the library that are due to the mRNA size or potential mRNA truncation.
3. It is very important to work under RNase-free conditions from Subheadings 3.1–3.4. Damaging the RNA even after the synthesis of the cDNA may interfere with cap-trapping method.
4. The amount is given for a single tube, but the reaction can be scaled up depending on the number of samples. Add the enzyme to the mixture at the last moment before mixing.
5. The incubation at 25°C is essential to anneal the random primer and the RNA, which is extended at 42°C at first, followed by further extension at higher temperature. In the presence of trehalose and sorbitol (7), the RT preserves its activity at higher temperatures and further extends the cDNA by reverse-transcribing structured RNA regions, which are often present at the 5′ UTRs (untranslated regions) of the mRNAs.
6. It is important to clean up the cDNA reaction and change the buffer before proceeding with the cap-oxidation and biotinylation reactions. Traces of Tris buffer, as well as saccharides or glycerol, interfere with the oxidation and biotinylation, as diol groups are reactive (Subheading 3.3). Additionally, fragments shorter than 100 nt (such as the first-strand cDNA primers) are removed by this purification step.
7. This step adds a biotin group to the cap (and the 3′ end of the RNAs).
8. The reaction might alternatively be kept at 37°C for about 3 h, but some biotin hydrazide batches have shown to degrade nucleic acids at 37°C due to some impurities; at room temperature, we have not observed issues in many years of experience with this reaction.
9. At this stage, the caps (and the 3′ ends of the RNAs) are biotinylated. To perform the cap-trap and eliminate the cDNAs

that did not reach the cap-site, it is mandatory to cleave with RNAse digesting the single strand RNAs at the 3′ ends of the cDNAs and at the 5′ ends, when the cDNAs do not reach the cap site. The RNAseI is an RNAse that cleaves at every base and is relatively easy to be inactivated by high temperature or SDS, thus serving as an ideal reagent for a RNA-dedicated laboratory.

10. It is important to remove, by denaturation, partially digested/nicked RNA from cDNA molecules. This happens at the random priming sites. Multiple random primers may produce multiple cDNA on the same RNA, only one of which would reach the cap site. Since random primers are long, they prime with several mismatches, cleaved by RNAse I. Heat treatment denatures these double-strand nucleic acids where the RNA is nicked, preventing the capture of multiple cDNA hybridized to the same cDNA. Using total RNAs, this step has been important to reduce the ribosomal contamination below 1%.

11. Although multiple washes may seem tedious, we found that this helps to prevent contamination of noncapped molecules in the final library.

12. The cDNA concentration obtained with OliGreen is 3 ng to 30 ng which measures the actual concentration of double-strand cDNA. We expect to see a relatively broad cDNA size range by Agilent RNA pico kit. An example result is shown in Fig. 1.

13. This step is essential to ligate a linker at the 5′ ends, which will later be used to prime the second-strand cDNA. This is

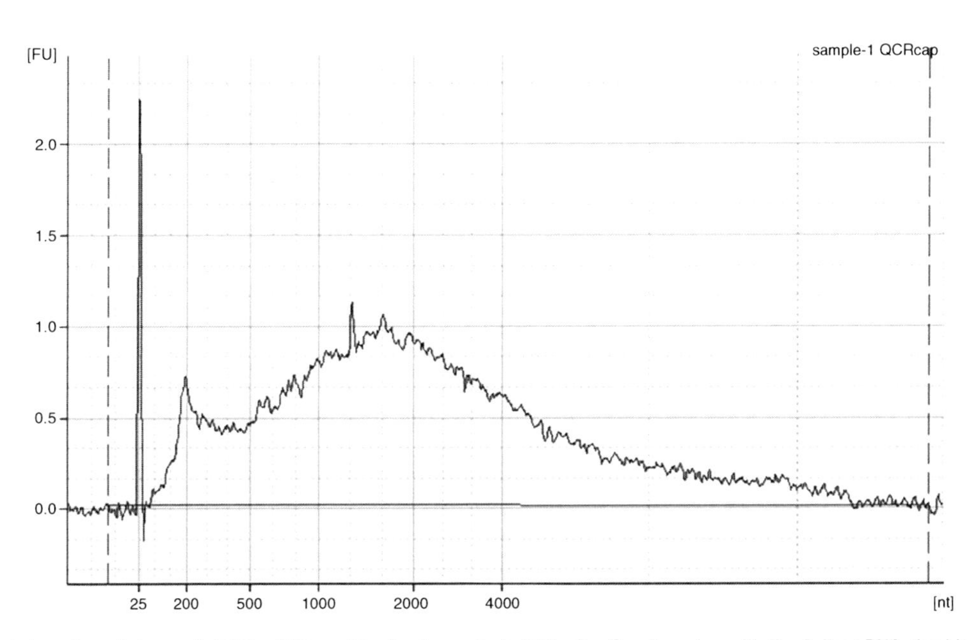

Fig. 1. Length and shape of cDNA. cDNA quality check result of cDNA after Cap-trapping with the Agilent RNA pico kit. One microliter of the reaction was applied. The expected size ranges from a few 100 nts to above 1–2 kb. cDNA concentration is also measured by OliGreen (see Note 12 for details).

obtained with the SSLLM (8), which exploits the ability of a double-strand linker with a protruding single strand random fraction to single strand cDNAs with DNA ligase. 200 ng of linkers are added to the 4 μl of cDNA solution, incubated at 65°C for 5 min to denature the cDNA secondary structures, followed by cooling on ice for 2 min. Sequences of various linkers with bar-code sequencer are listed in the methods section and they should be annealed to each other separately before the step of ligation with the cDNA.

14. Since the linker contains regions of random sequences, they may have annealed during storage of the linker. This step thus helps to melt any dimer structure formed by linkers and make them fully available for the reaction.

15. It is possible at this stage to pool different cDNA mixtures if the 5′ ends linkers, used for each different cDNA, are bar-coded. Sequencing through the bar code will allow distinguishing, after sequencing, the origin of the sample (9). Since the concentration of the linker dimer is high, repeat twice the purification of the cDNAs to avoid linker dimers in the final library. During the second purification step, we can pool another pooled cDNA after binding (just before 70% Ethanol wash) step.

16. Also, the first-strand primer contains a *Eco*P15I sequence: having two sequences in opposite orientation has been found important to increase cleavage by *Eco*P15I, as well as the introduction of sinefungin to the reaction. We calculate carefully the amount of the enzyme, avoiding overdigestion of the cDNA. Some class IIS restriction enzymes have been reported to inhibit the reaction when too large amount of enzymes were used.

17. The excess of 3′ linkers must be removed before performing the final PCR, otherwise the dimers produced by their ligation would heavily contaminate the CAGE library. To do this, we take advantage of the biotinylated primer that was used to prime the second strand (see reagents). The cDNA is retained on the beads, while the 3′ linkers are washed away.

18. When the protocol works properly, the CAGE tags can be applied on the Illumina sequencer without any size fractionation. To minimize contamination, we usually select the lowest number of PCR cycles that produce an acceptable amount of PCR product for the Illumina sequencer. Depending on the number of sequencing runs, or the desire to repeat the run multiple times or at a later time, the number of PCR cycles can be moderately increased. An example result is shown in Fig. 2a–c.

19. Do not amplify in a single tube a large amount of beads, as they may be inhibitory to the PCR reaction (in general, do not amplify more than 2 μl of beads for a 50-μl PCR reaction).

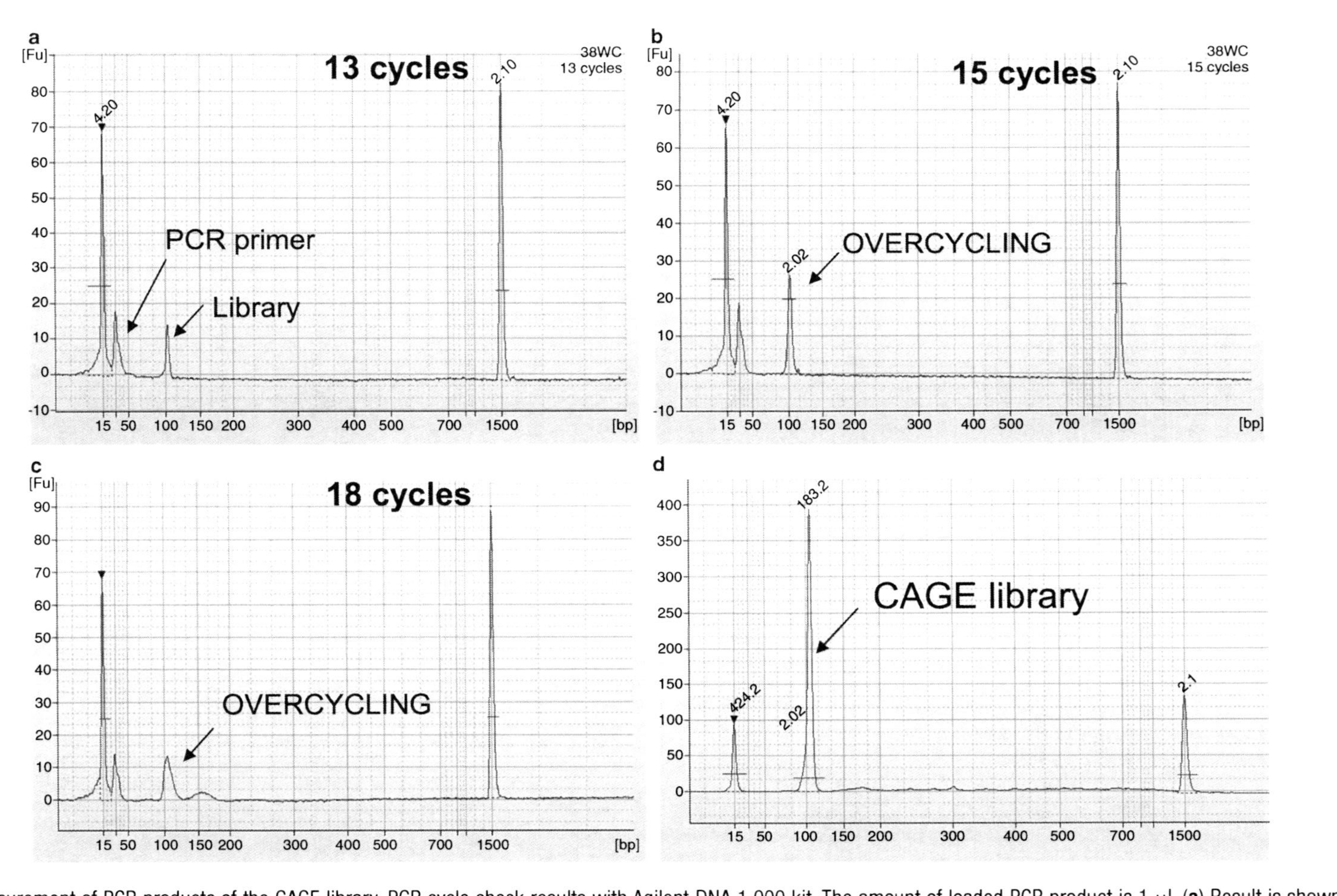

Fig. 2. Measurement of PCR products of the CAGE library. PCR cycle check results with Agilent DNA 1,000 kit. The amount of loaded PCR product is 1 μl. (**a**) Result is shown for 13 cycles, (**b**) 15 cycles, and (**c**) 18 cycles. In case of 13 cycles, the single peak (103 bp) has a FU value between 5 and 10 (molarity: ~10 nmol/l), which is usable for bulk PCR. In case of 15 cycles, the FU exceeds 20 (molarity: ~30 nmol/l) and with 18 cycles the reactions shows an additional broad, larger peaks, due to overcycling. (**d**) Final product is measured with the Agilent DNA 1,000 kit. The molarity of the single peak at 103 bp was estimated to be ~183.2 nmol/l. PCR primers are removed before the assay during Subheading 3.15, consisting of an ExoI treatment and MinElute column purification. The single peak products are ready for sequencing.

5. Conclusions

Using this protocol, we routinely sequence 10–20 million tags per library we sequence with the current Illumina GA2-X sequencer. Depending on the user needs and future developments, it will be possible to sequence even more deeply to detect very rare transcriptional events, which take place only in few cells or cell compartments. Sequencing technology progress is relentless. We foresee that by changing primer sequences, this method will be suitable for other platforms or other versions of the Illumina sequencers. High throughput will allow pooling multiple, bar-coded CAGE libraries for each lane of the Illumina sequencer, allowing to profile RNA by sequencing their 5′ end at a fraction of the cost of a microarray experiment, notably enhancing our capacity to interpret the genome and the significance of expression analysis.

Bioinformatics analysis still poses big challenges, starting from the storage of progressively larger amount of data, to the development of all the interpretation tools. This part goes beyond the scope of this chapter. There are other publications on the CAGE (1, 4). At RIKEN we have prepared a Web site (http://www.osc.riken.jp/english/activity/cage/) which contains the outline of the various versions of the CAGE technology, an updated list of publications and software that can be used to analyze the CAGE data.

We are convinced that this type of analysis, taking into account only the 5′ end of the cDNA, will allow for maximizing the cost/performance of sequencing RNAs to study biology, with the further strength to analyze TSSs and thus the promoters that are responsible for gene expression.

Acknowledgments

This work was founded by a Research Grant for RIKEN Omics Science Center from MEXT and by the National Human Genome Research Institute grants U54 HG004557. We thank all the colleagues at the OSC for the precious feedbacks during the development of the methodology.

References

1. Suzuki, H., Forrest, A. R. R., van Nimwegen, E., Daub, C. O., Balwierz, P. J., Irvine, K. M., Lassmann, T., Ravasi, T., Hasegawa, Y., de Hoon, M. J. L., Katayama, S., Schroder, K., Carninci, P., Tomaru, Y., Kanamori-Katayama, M., Kubosaki, A., Akalin, A., Ando, Y., Arner, E., Asada, M., Asahara, H., Bailey, T., Bajic, V. B., Bauer, D., Beckhouse, A. G., Bertin, N., Bjorkegren, J., Brombacher, F., Bulger, E., Chalk, A. M., Chiba, J., Cloonan, N., Dawe, A., Dostie, J., Engstrom, P. G., Essack, M., Faulkner, G. J., Fink, J. L., Fredman, D., Fujimori, K., Furuno, M., Gojobori, T., Gough, J., Grimmond, S. M., Gustafsson, M., Hashimoto, M.,

Hashimoto, T., Hatakeyama, M., Heinzel, S., Hide, W., Hofmann, O., Hornquist, M., Huminiecki, L., Ikeo, K., Imamoto, N., Inoue, S., Inoue, Y., Ishihara, R., Iwayanagi, T., Jacobsen, A., Kaur, M., Kawaji, H., Kerr, M. C., Kimura, R., Kimura, S., Kimura, Y., Kitano, H., Koga, H., Kojima, T., Kondo, S., Konno, T., Krogh, A., Kruger, A., Kumar, A., Lenhard, B., Lennartsson, A., Lindow, M., Lizio, M., MacPherson, C., Maeda, N., Maher, C. A., Maqungo, M., Mar, J., Matigian, N. A., Matsuda, H., Mattick, J. S., Meier, S., Miyamoto, S., Miyamoto-Sato, E., Nakabayashi, K., Nakachi, Y., Nakano, M., Nygaard, S., Okayama, T., Okazaki, Y., Okuda-Yabukami, H., Orlando, V., Otomo, J., Pachkov, M., Petrovsky, N., Plessy, C., Quackenbush, J., Radovanovic, A., Rehli, M., Saito, R., Sandelin, A., Schmeier, S., Schonbach, C., Schwartz, A. S., Semple, C. A., Sera, M., Severin, J., Shirahige, K., Simons, C., Laurent, G. S., Suzuki, M., Suzuki, T., Sweet, M. J., Taft, R. J., Takeda, S., Takenaka, Y., Tan, K., Taylor, M. S., Teasdale, R. D., Tegner, J., Teichmann, S., Valen, E., Wahlestedt, C., Waki, K., Waterhouse, A., AWells, C., Winther, O., Wu, L., Yamaguchi, K., Yanagawa, H., Yasuda, J., Zavolan, M., Hume, D. A., Arakawa, T., Fukuda, S., Imamura, K., Kai, C., Kaiho, A., Kawashima, T., Kawazu, C., Kitazume, Y., Kojima, M., Miura, H., Murakami, K., Murata, M., Ninomiya, N., Nishiyori, H., Noma, S., Ogawa, C., Sano, T., Simon, C., Tagami, M., Takahashi, Y., Kawai, J., Hayashizaki, Y., Consortium, F., and Ctr, R. O. S. (2009) The transcriptional network that controls growth arrest and differentiation in a human myeloid leukemia cell line. *Nature Genetics* **41**, 553–62.

2. Carninci, P., Sandelin, A., Lenhard, B., Katayama, S., Shimokawa, K., Ponjavic, J., Semple, C. A., Taylor, M. S., Engstrom, P. G., Frith, M. C., Forrest, A. R., Alkema, W. B., Tan, S. L., Plessy, C., Kodzius, R., Ravasi, T., Kasukawa, T., Fukuda, S., Kanamori-Katayama, M., Kitazume, Y., Kawaji, H., Kai, C., Nakamura, M., Konno, H., Nakano, K., Mottagui-Tabar, S., Arner, P., Chesi, A., Gustincich, S., Persichetti, F., Suzuki, H., Grimmond, S. M., Wells, C. A., Orlando, V., Wahlestedt, C., Liu, E. T., Harbers, M., Kawai, J., Bajic, V. B., Hume, D. A., and Hayashizaki, Y. (2006) Genome-wide analysis of mammalian promoter architecture and evolution. *Nature Genetics* **38**, 626–35.
3. Faulkner, G. J., Kimura, Y., Daub, C. O., Wani, S., Plessy, C., Irvine, K. M., Schroder, K., Cloonan, N., Steptoe, A. L., Lassmann, T., Waki, K., Hornig, N., Arakawa, T., Takahashi, H., Kawai, J., Forrest, A. R. R., Suzuki, H., Hayashizaki, Y., Hume, D. A., Orlando, V., Grimmond, S. M., and Carninci, P. (2009) The regulated retrotransposon transcriptome of mammalian cells. *Nature Genetics* **41**, 563–71.
4. Carninci, P. (2010) Cap-analysis gene expression (CAGE): the science of decoding gene transcription, Pan Stanford, Singapore.
5. Carninci, P., Kvam, C., Kitamura, A., Ohsumi, T., Okazaki, Y., Itoh, M., Kamiya, M., Shibata, K., Sasaki, N., Izawa, M., Muramatsu, M., Hayashizaki, Y., and Schneider, C. (1996) High-efficiency full-length cDNA cloning by biotinylated CAP trapper. *Genomics* **37**, 327–36.
6. Carninci, P., Kasukawa, T., Katayama, S., Gough, J., Frith, M. C., Maeda, N., Oyama, R., Ravasi, T., Lenhard, B., Wells, C., Kodzius, R., Shimokawa, K., Bajic, V. B., Brenner, S. E., Batalov, S., Forrest, A. R. R., Zavolan, M., Davis, M. J., Wilming, L. G., Aidinis, V., Allen, J. E., Ambesi-Impiombato, X., Apweiler, R., Aturaliya, R. N., Bailey, T. L., Bansal, M., Baxter, L., Beisel, K. W., Bersano, T., Bono, H., Chalk, A. M., Chiu, K. P., Choudhary, V., Christoffels, A., Clutterbuck, D. R., Crowe, M. L., Dalla, E., Dalrymple, B. P., de Bono, B., Della Gatta, G., di Bernardo, D., Down, T., Engstrom, P., Fagiolini, M., Faulkner, G., Fletcher, C. F., Fukushima, T., Furuno, M., Futaki, S., Gariboldi, M., Georgii-Hemming, P., Gingeras, T. R., Gojobori, T., Green, R. E., Gustincich, S., Harbers, M., Hayashi, Y., Hensch, T. K., Hirokawa, N., Hill, D., Huminiecki, L., Iacono, M., Ikeo, K., Iwama, A., Ishikawa, T., Jakt, M., Kanapin, A., Katoh, M., Kawasawa, Y., Kelso, J., Kitamura, H., Kitano, H., Kollias, G., Krishnan, S. P. T., Kruger, A., Kummerfeld, S. K., Kurochkin, I. V., Lareau, L. F., Lazarevic, D., Lipovich, L., Liu, J., Liuni, S., McWilliam, S., Babu, M. M., Madera, M., Marchionni, L., Matsuda, H., Matsuzawa, S., Miki, H., Mignone, F., Miyake, S., Morris, K., Mottagui-Tabar, S., Mulder, N., Nakano, N., Nakauchi, H., Ng, P., Nilsson, R., Nishiguchi, S., Nishikawa, S., Nori, F., Ohara, O., Okazaki, Y., Orlando, V., Pang, K. C., Pavan, W. J., Pavesi, G., Pesole, G., Petrovsky, N., Piazza, S., Reed, J., Reid, J. F., Ring, B. Z., Ringwald, M., Rost, B., Ruan, Y., Salzberg, S. L., Sandelin, A., Schneider, C., Schonbach, C., Sekiguchi, K., Semple, C. A. M., Seno, S., Sessa, L., Sheng, Y., Shibata, Y., Shimada, H., Shimada, K., Silva, D., Sinclair, B., Sperling, S., Stupka, E., Sugiura, K., Sultana, R., Takenaka, Y., Taki, K., Tammoja, K., Tan, S. L., Tang, S., Taylor, M. S., Tegner, J., Teichmann, S. A., Ueda, H. R., van Nimwegen, E., Verardo, R., Wei, C. L., Yagi, K., Yamanishi, H., Zabarovsky, E., Zhu, S., Zimmer, A., Hide, W., Bult, C., Grimmond, S. M., Teasdale, R. D., Liu, E. T., Brusic, V., Quackenbush, J., Wahlestedt, C., Mattick, J. S., Hume, D. A., Kai, C., Sasaki, D., Tomaru, Y., Fukuda, S.,

Kanamori-Katayama, M., Suzuki, M., Aoki, J., Arakawa, T., Iida, J., Imamura, K., Itoh, M., Kato, T., Kawaji, H., Kawagashira, N., Kawashima, T., Kojima, M., Kondo, S., Konno, H., Nakano, K., Ninomiya, N., Nishio, T., Okada, M., Plessy, C., Shibata, K., Shiraki, T., Suzuki, S., Tagami, M., Waki, K., Watahiki, A., Okamura-Oho, Y., Suzuki, H., Kawai, J., Hayashizaki, Y., Consortium, F., and S, R. G. E. R. G. (2005) The transcriptional landscape of the mammalian genome. *Science* **309**, 1559–63.

7. Carninci, P., Shiraki, T., Mizuno, Y., Muramatsu, M., and Hayashizaki, Y. (2002) Extra-long first-strand cDNA synthesis. *Biotechniques* **32**, 984–5.

8. Shibata, Y., Carninci, P., Watahiki, A., Shiraki, T., Konno, H., Muramatsu, M., and Hayashizaki, Y. (2001) Cloning full-length, cap-trapper-selected cDNAs by using the single-strand linker ligation method. *Biotechniques* **30**, 1250–54.

9. Maeda, N., Nishiyori, H., Nakamura, M., Kawazu, C., Murata, M., Sano, H., Hayashida, K., Fukuda, S., Tagami, M., Hasegawa, A., Murakami, K., Schroder, K., Irvine, K., Hume, D., Hayashizaki, Y., Carninci, P., and Suzuki, H. (2008) Development of a DNA barcode tagging method for monitoring dynamic changes in gene expression by using an ultra high-throughput sequencer. *Biotechniques* **45**, 95–7.

Chapter 12

Detecting DNaseI-Hypersensitivity Sites with MLPA

Thomas Ohnesorg, Stefanie Eggers, and Stefan J. White

Abstract

DNaseI-hypersensitive sites within chromatin are indicative of genomic loci with regulatory function. Several techniques have been described for analyzing these regions, but are either laborious, offer low-throughput possibilities, or are expensive. We have developed a new approach based on a modified version of multiplex ligation-dependent probe amplification (MLPA). Using this method, it is possible to analyse up to 50 defined genomic regions for DNaseI-hypersensitivity in a single PCR-based reaction. This chapter outlines the approach and discusses the critical features of each step of the procedure.

Key words: MLPA, DNaseI hypersensitivity, Gene regulation, Chromatin structure

1. Introduction

During most cellular stages, the genetic material of a cell is packaged into chromatin, which consists of DNA wrapped around histone proteins. The functions of chromatin are to compact DNA to fit in the nucleus, to strengthen the DNA to allow mitosis and meiosis, and to serve as a mechanism to control gene expression and DNA replication. As gene expression requires the binding of DNA-binding proteins such as transcription factors and RNA-polymerases, chromatin undergoes conformational changes in regions of actively transcribed genes. These rearrangements open the chromatin structure, providing access for factors essential for gene expression.

DNaseI-hypersensitive sites are defined as short, open regions of chromatin that show an increased sensitivity (compared to bulk chromatin) to cleavage by DNaseI (1). The identification of DNaseI-hypersensitive sites has been frequently used to identify and analyse regulatory regions such as promoters, enhancers, and silencers (2, 3).

Early methods for the detection of DNaseI-hypersensitive sites included Southern blotting, which is time consuming, usually

Bart Deplancke and Nele Gheldof (eds.), *Gene Regulatory Networks: Methods and Protocols*,
Methods in Molecular Biology, vol. 786, DOI 10.1007/978-1-61779-292-2_12,

requires radioactivity, and is limited to short stretches of DNA. Although PCR-based methods have been published within the last years, they do not allow multiplexing, which makes them unsuitable for large-scale or genome-wide analysis of DNaseI-hypersensitive sites (4, 5). Approaches used for large-scale analysis of DNaseI-hypersensitive sites have primarily been microarrays (6, 7) or deep sequencing (8), which is valuable for genome-wide analysis, but the costs are usually the limiting factor for many applications, such as comparing different developmental stages or tissues. Another disadvantage of these large-scale methods is that they usually require millions of cells, which makes these methods impractical for use in ex vivo studies unless large numbers of tissues are pooled.

We recently described a modified multiplex ligation-dependent probe amplification (MLPA) approach for the identification and analysis of genomic regulatory regions (9). MLPA is a PCR-based approach originally described in 2002 to identify deletions and duplications in genomic DNA (10). The method is based on the ligation of two half-probes after they have annealed to a specific genomic region of interest. Up to 50 different probes can be combined in a mix, with each having a different length. Adding the same forward and reverse priming sequences to each pair of half-probes within a probe mix allows the subsequent amplification of the ligated products with PCR using a single primer pair (with the forward primer having a fluorescent label at the 5′ end).

PCR products are then typically separated by capillary electrophoresis and the different probes can be recognized by their unique length. Advantages of this method are its simplicity, sensitivity, and the multiplex potential of the reaction, as well as the low amount of DNA required. MLPA can be used to analyse up to 50 different loci in just one reaction with as little as 20 ng of genomic DNA as starting material. Furthermore, the only equipment which is required to perform MLPA experiments is a thermocycler and DNA sequencer, which are accessible to most researchers.

DNaseI-MLPA is a quick and straightforward method for the analysis of DNaseI-hypersensitive sites, which overcomes many of the limitations associated with previously described approaches. We have shown that it is possible to get reproducible results with as few as 5×10^4 cells per DNaseI reaction, which allows the analysis of regulatory elements of embryonic tissues.

2. Materials

2.1. DNaseI Treatment

1. RQ1 DNaseI (1 μ/μl; Promega) including DNaseI reaction buffer.
2. NE-PER Nuclear and Cytoplasmic Extraction kit (Thermo Scientific) containing CER I and CER II.

3. Nuclei lysis buffer: 200 mM NaCl, 150 mM Tris–HCl, pH 8, 10 mM EDTA, pH 8, 0.2% SDS.
4. Proteinase K (10 mg/ml stock solution).
5. Glycerol.
6. RNaseA (Sigma; 10 mg/ml).
7. HighPure PCR Purification Kit (Roche) containing Binding Buffer, Wash and Elution Buffer or DNeasy Blood & Tissue Kit (Qiagen).

2.2. Extension MLPA Reaction

All reagents for the basic MLPA protocol can be purchased from MRC-Holland, the Netherlands (http://www.mlpa.com). Each item can be recognized by a cap of a specific color.

1. Synthetic probe mix. Although standard MLPA probes can be used for this method, it is recommended to use extension MLPA probes (with a gap of up to 120 bp between the half-probes) for genomic regions of interest, as they can cover larger regions and give more flexibility in probe design (see Note 1).
2. MLPA buffer (yellow cap).
3. Ligase-65 (green cap: note that this used to be a brown cap in previous versions).
4. Ligase-65 buffer A (transparent cap).
5. Ligase-65 buffer B (white cap).
6. Stoffel fragment 10 U/μl (Applied Biosystems).
7. Nucleotides (10 mM each of dATP, dCTP, dGTP, dTTP).
8. SALSA enzyme dilution buffer (blue cap).
9. SALSA PCR buffer.
10. SALSA PCR primers + dNTPs (brown cap: note that this used to be a purple cap in previous versions). The sequences of the primers are

    forward 5′-GGGTTCCCTAAGGGTTGGA-3′ (fluorescently labeled at the 5′ end)

    reverse 5′-GTGCCAGCAAGATCCAATCTAGA-3′
11. SALSA DNA Polymerase (orange cap).
12. HiDi formamide (Applied Biosystems).
13. ROX-400 HD size standard (Applied Biosystems).

3. Methods

3.1. Isolation of Nuclei

1. Harvest cells (~2×10^5 to 2×10^6) and isolate nuclei by using the NE-PER Nuclear and Cytoplasmic Extraction kit. All required solutions are provided with the kit (see Note 2).

2. Spin down cells at 500 × *g* for 2 min and discard the supernatant, leaving cell pellet as dry as possible.
3. Resuspend cell pellet in ice-cold CER I (100 μl for 10^6 cells) by flicking the tube or gently pipetting up and down, incubate on ice for 10 min (see Note 3).
4. Add 5.5 μl (for 10^6 cells) of ice-cold CER II to the tube, mix by flicking the tube, and incubate on ice for 1 min.
5. Flick tube again, centrifuge in a swing out rotor at 200 × *g* at 4°C for 2 min and discard the supernatant.
6. Wash nuclei with 200 μl of ice-cold DNaseI reaction buffer (provided with the enzyme, containing 2% glycerol) (see Note 4), centrifuge at 200 × *g* for 2 min at 4°C and gently resuspend in 75 μl DNaseI reaction buffer, containing 2% glycerol (at room temperature, see Note 5).
7. Add 25 μl nuclei aliquots to 50 μl of DNaseI solution (reaction buffer, containing 2% glycerol and required amount of DNaseI) in a 2 ml Eppendorf tube and incubate for 20 min at room temperature.
8. Stop the reaction and release digested genomic DNA by adding 250 μl nuclei lysis buffer (containing 50 μg Proteinase K) and incubate at 55°C for 45 min.

   Transfer tubes to 37°C, add 20 μg RNaseA and incubate at 37°C for 30 min.
9. Use the HighPure PCR purification kit to clean up and recover digested DNA (see Note 6).
10. Add 1.6 ml Binding Buffer and mix.
11. Transfer up to 700 μl to a High Pure filter tube and spin at maximum speed for 30 s. Discard flowthrough and repeat this step with remaining solution.
12. Add 500 μl Wash Buffer and centrifuge at maximum speed for 1 min, discard flowthrough.
13. Add 200 μl Wash Buffer and centrifuge at maximum speed for 1 min, transfer filter tube to a clean 1.5 ml Eppendorf tube.
14. Add 50 μl Elution Buffer and centrifuge at maximum speed for 1 min.
15. Keep the eluate and continue with step 3.2.2. (see Note 7).

3.2. Extension MLPA Protocol

1. The MLPA protocol below is based on that described in the original publication (10), and can also be found at http://www.mlpa.com.
2. Add 100–200 ng DNA in a final volume of 5 μl water to a PCR tube (see Note 8). The DNA is denatured at 98°C for 5 min (see Note 9), and then allowed to cool to room temperature for at least 5 min.

3. Add the following to the denatured DNA: 1.5 μl MLPA Probe mix (each oligonucleotide at 4 fmol) + 1.5 μl SALSA MLPA buffer. Mix with care (see Note 10). Incubate 1 min at 95°C, then 16 h at 60°C (see Note 11).
4. Make the Ligase-65 mix at room temperature by mixing: 3 μl Ligase-65 buffer A (transparent cap) + 3 μl Ligase-65 buffer B (white cap) + 25 μl H_2O + 0.2 μl dNTPs (10 mM) + 0.1 μl Stoffel fragment. Add 1 μl Ligase-65 (green cap) and mix again.

   Reduce the temperature of the thermal cycler to 54°C. *Keeping the PCR tubes in the thermal cycler*, add 32 μl ligase-65 mix to each tube and mix. The reactions should be incubated for 15 min at 54°C, then heated for 5 min at 98°C to inactivate the ligase (see Note 12).
5. There have been different procedures described for preparing the PCR amplification, with the most commonly used described below (see Note 13).

   For each reaction, combine:

SALSA PCR buffer (red)	5 μl
SALSA enzyme buffer (blue)	2 μl
SALSA PCR primers (brown)	2 μl
SALSA Polymerase (orange)	0.5 μl
H_2O	30.5 μl

   Aliquot 40 μl into PCR tubes on ice, then add 10 μl ligation mix.
6. The PCR should be carried out with the following settings (see Note 14):

   1 cycle: 1 min at 95°C

   35 cycles: 30 s at 95°C; 30 s at 58°C; 30 s at 72°C

   1 cycle: 20 min at 72°C.
7. Prepare the samples for fragment analysis (see Note 15):

   To 1 ml Hi Di Formamide, add 5 μl size standard, and mix well.

   Into each well of a 96-well, add 10 μl of the Formamide/size standard mix.

   Add 1 μl of PCR product (see Note 16).
8. MLPA reactions are usually processed by fragment analysis on a capillary sequencer, although other possibilities that have been described include agarose gel electrophoresis (11) and microarray hybridization (12). In a fragment analysis, DNA is separated by electrophoresis, with each probe represented by a peak of a defined length. The decrease of relative size (height)

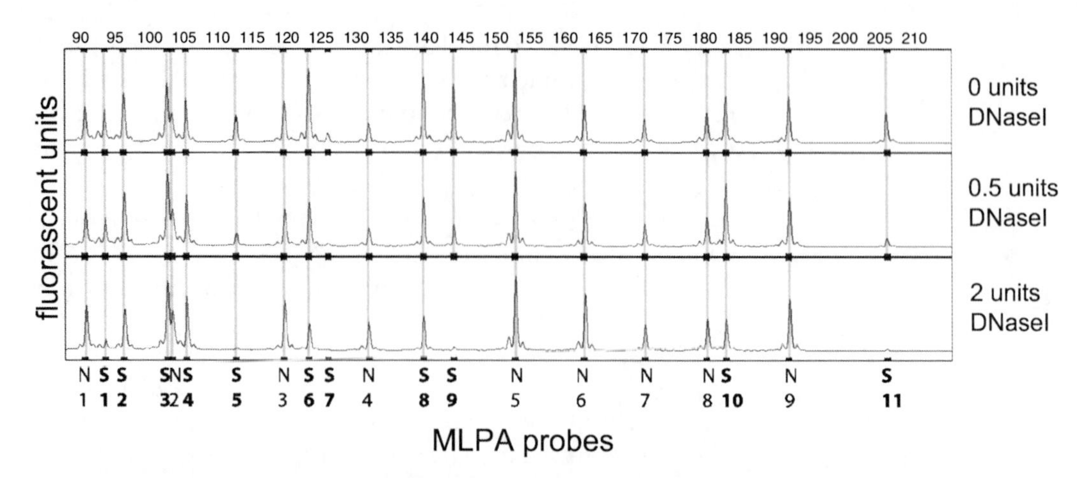

Fig. 1. An example of DNaseI-MLPA output. Each peak represents signal from a specific MLPA probe after nuclei were isolated from HeLa cells and treated with 0, 0.5, and 2 U of DNaseI. The vertical axis shows the amount of fluorescent signal produced by each probe. On the lower horizontal axis, numbers starting with N indicate probes designed to bind to DNaseI nonsensitive regions, numbers starting with S indicate probes binding to DNaseI sensitive regions for HeLa cells (S) as published by the ENCODE consortium (15). The numbers along the upper horizontal axis are the product length in base pairs. Figure derived from ref. 9.

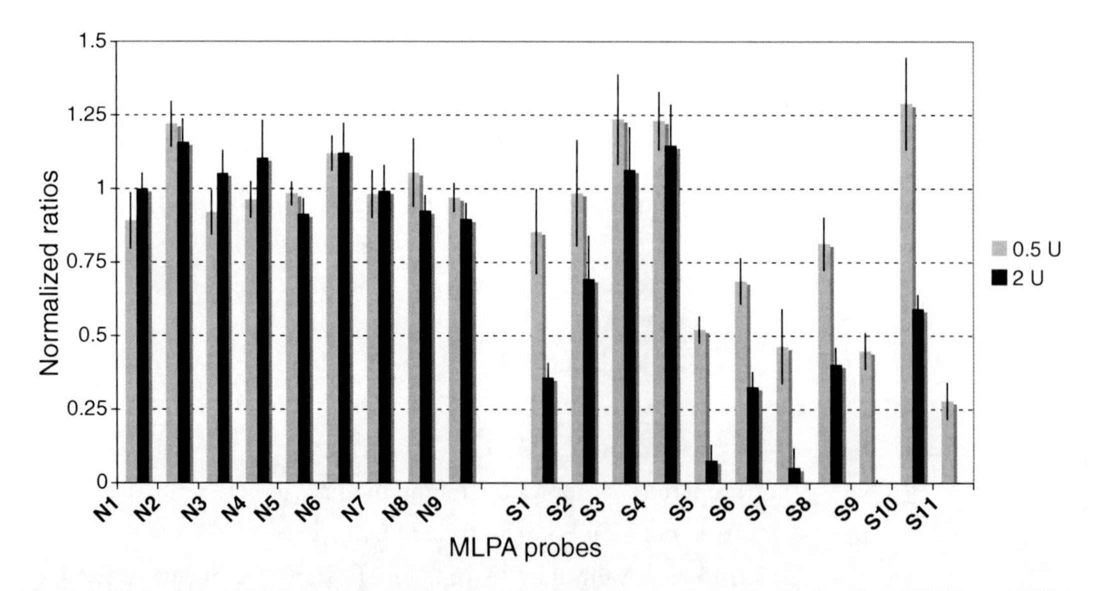

Fig. 2. Analysis of the peaks shown in Fig. 1. Each peak is normalized by dividing the height of that peak by the sum of the heights of the nonsensitive probes. In this analysis, a reduction to less than 75% of the nondigested probe is considered to show DNaseI-hypersensitivity. Data for each probe is shown as mean ± SD. Figure derived from (9). Note that two loci (S3 and S4) identified by the ENCODE consortium as showing DNaseI-hypersensitivity (15) did not show evidence of this with DNaseI-MLPA.

of each peak when comparing undigested with DNaseI digested DNA is then used to measure DNaseI (hyper)sensitivity (Fig. 1). Different normalization approaches have been described, based on the number of probes within the probe mix that may be affected. Depending on the probe set used, many probes may recognize DNaseI-hypersensitive sites. It is therefore recommended to normalize each peak against several control probes from diverse loci that are unlikely to be DNaseI-hypersensitive in any of the samples being screened.

Once normalized ratios have been calculated, it is necessary to define thresholds for DNaseI (hyper)sensitivity. Assuming a normalized value of 1 corresponds to DNaseI insensitivity, a probe peak should drop below a value of 0.75 to be considered DNaseI sensitive (Fig. 2).

4. Notes

1. For analysis, it is necessary to include reliable control probes, however, it can be difficult to find suitable genomic regions that are insensitive to DNaseI. Depending on the cell line used, it can be very helpful to use data available from the ENCODE project and other data from the UCSC genome browser (http://genome.ucsc.edu).
2. To minimize changes in chromatin structure, nuclei should be kept on ice and the procedure performed as quickly as possible.
3. Reduce the amount of solutions if using less than 10^6 cells, however, do not use less than 50 μl of CER I. Depending on the cell line, isolated nuclei can be sticky and very hard or impossible to resuspend. In this case, double amounts of NE-PER solution can be used. For some cell lines, reducing the incubation time to 6–7 min can facilitate subsequent resuspending of the nuclei. If unsuccessful, harvested cells can also be aliquoted before nuclei isolation.
4. The integrity of the nuclei can be checked at this point by observation through a light microscope.
5. It is recommended to use two different DNaseI concentrations (e.g. 0.5, 2 U) and an undigested control. If more concentrations are required, increase the amount of reaction buffer.
6. It is important to efficiently purify and recover very long as well as very short fragments of DNA. Phenol–Chloroform extraction is not suitable for this task, as traces of phenol are known to interfere with MLPA. Alternatively, the DNeasy Blood & Tissue Kit (Qiagen) can be used according to the

manufacturer's instructions, except for the elution step, which should be performed with only 50 µl elution buffer. Based on our experience, the HighPure PCR purification kit gives better results for shorter fragments, the DNeasy Blood & Tissue kit is better suited for longer DNA fragments (less digested samples).

7. Purified DNA fragments can be stored at 4°C for several weeks.
8. A critical requirement for a successful MLPA reaction is good-quality template DNA, as impurities such as organic solvents can lead to inaccurate results. For best results, only DNA samples that were isolated using the same method should be normalized together.
9. Incomplete denaturation of the genomic template can mean that some probes do not have complete access to their target sequences. It may be necessary to increase the duration of denaturation to improve probe accessibility.
10. Use of filter tips is recommended throughout the procedure.
11. Although the protocol uses an overnight (~16 h) hybridization time to allow the probes to anneal to the genomic template, there have been reports showing that 2–3 h is sufficient (13, 14). As probes may differ in annealing efficiency, shorter hybridization times should be tested for each probe mix before being routinely implemented.
12. The ligated products are stable, and can be stored for several months at –20°C.
13. To save on reagents, it is possible to reduce the volume of the PCR.
14. When using small amounts of starting material, we have successfully increased the number of PCR cycles up to 40.
15. This assumes the use of a capillary sequencer such as the ABI3700 (Applied Biosystems). Other machines may have different requirements.
16. The optimal amount of PCR product may need to be determined, depending on the sensitivity of the specific sequencer.

Acknowledgments

We would like to thank Professor Andrew Sinclair for helpful discussions. SE is supported by a University of Melbourne Postgraduate Research Scholarship. S.J.W. is funded by NHMRC Australia (grant numbers 491293 and 546478).

References

1. Gross, D. S., and Garrard, W. T. (1988) Nuclease hypersensitive sites in chromatin. *Annual review of biochemistry* **57**, 159–197.
2. Wu, C., Wong, Y. C., and Elgin, S. C. (1979) The chromatin structure of specific genes: II. Disruption of chromatin structure during gene activity. *Cell* **16**, 807–814.
3. Burgess-Beusse, B., Farrell, C., Gaszner, M., Litt, M., Mutskov, V., Recillas-Targa, F., Simpson, M., West, A., and Felsenfeld, G. (2002) The insulation of genes from external enhancers and silencing chromatin. *Proceedings of the National Academy of Sciences of the United States of America* **99 Suppl 4**, 16433–16437.
4. Follows, G. A., Janes, M. E., Vallier, L., Green, A. R., and Gottgens, B. (2007) Real-time PCR mapping of DNaseI-hypersensitive sites using a novel ligation-mediated amplification technique. *Nucleic acids research* **35**, e56.
5. Dorschner, M. O., Hawrylycz, M., Humbert, R., Wallace, J. C., Shafer, A., Kawamoto, J., Mack, J., Hall, R., Goldy, J., Sabo, P. J., Kohli, A., Li, Q., McArthur, M., and Stamatoyannopoulos, J. A. (2004) High-throughput localization of functional elements by quantitative chromatin profiling. *Nature methods* **1**, 219–225.
6. Crawford, G. E., Davis, S., Scacheri, P. C., Renaud, G., Halawi, M. J., Erdos, M. R., Green, R., Meltzer, P. S., Wolfsberg, T. G., and Collins, F. S. (2006) DNase-chip: a high-resolution method to identify DNase I hypersensitive sites using tiled microarrays. *Nature methods* **3**, 503–509.
7. Sabo, P. J., Kuehn, M. S., Thurman, R., Johnson, B. E., Johnson, E. M., Cao, H., Yu, M., Rosenzweig, E., Goldy, J., Haydock, A., Weaver, M., Shafer, A., Lee, K., Neri, F., Humbert, R., Singer, M. A., Richmond, T. A., Dorschner, M. O., McArthur, M., Hawrylycz, M., Green, R. D., Navas, P. A., Noble, W. S., and Stamatoyannopoulos, J. A. (2006) Genome-scale mapping of DNase I sensitivity in vivo using tiling DNA microarrays. *Nature methods* **3**, 511–518.
8. Crawford, G. E., Holt, I. E., Whittle, J., Webb, B. D., Tai, D., Davis, S., Margulies, E. H., Chen, Y., Bernat, J. A., Ginsburg, D., Zhou, D., Luo, S., Vasicek, T. J., Daly, M. J., Wolfsberg, T. G., and Collins, F. S. (2006) Genome-wide mapping of DNase hypersensitive sites using massively parallel signature sequencing (MPSS). *Genome research* **16**, 123–131.
9. Ohnesorg, T., Eggers, S., Leonhard, W. N., Sinclair, A. H., and White, S. J. (2009) Rapid high-throughput analysis of DNaseI hypersensitive sites using a modified MLPA approach. *BMC Genomics* **10**, 412.
10. Schouten, J. P., McElgunn, C. J., Waaijer, R., Zwijnenburg, D., Diepvens, F., and Pals, G. (2002) Relative quantification of 40 nucleic acid sequences by multiplex ligation-dependent probe amplification. *Nucleic acids research* **30**, e57.
11. Lalic, T., Vossen, R. H., Coffa, J., Schouten, J. P., Guc-Scekic, M., Radivojevic, D., Djurisic, M., Breuning, M. H., White, S. J., and den Dunnen, J. T. (2005) Deletion and duplication screening in the DMD gene using MLPA. *Eur J Hum Genet* **13**, 1231–1234.
12. Zeng, F., Ren, Z. R., Huang, S. Z., Kalf, M., Mommersteeg, M., Smit, M., White, S., Jin, C. L., Xu, M., Zhou, D. W., Yan, J. B., Chen, M. J., van Beuningen, R., Huang, S. Z., den Dunnen, J., Zeng, Y. T., and Wu, Y. (2008) Array-MLPA: comprehensive detection of deletions and duplications and its application to DMD patients. *Human mutation* **29**, 190–197.
13. Aten, E., White, S. J., Kalf, M. E., Vossen, R. H., Thygesen, H. H., Ruivenkamp, C. A., Kriek, M., Breuning, M. H., and den Dunnen, J. T. (2008) Methods to detect CNVs in the human genome. *Cytogenetic and genome research* **123**, 313–321.
14. Bayley, J. P., Grimbergen, A. E., van Bunderen, P. A., van der Wielen, M., Kunst, H. P., Lenders, J. W., Jansen, J. C., Dullaart, R. P., Devilee, P., Corssmit, E. P., Vriends, A. H., Losekoot, M., and Weiss, M. M. (2009) The first Dutch SDHB founder deletion in paraganglioma-pheochromocytoma patients. *BMC Med Genet* **10**, 34.
15. Birney, E., Stamatoyannopoulos, J. A., Dutta, A., Guigo, R., Gingeras, T. R., Margulies, E. H., Weng, Z., Snyder, M., Dermitzakis, E. T., Thurman, R. E., Kuehn, M. S., Taylor, C. M., Neph, S., Koch, C. M., Asthana, S., Malhotra, A., Adzhubei, I., Greenbaum, J. A., Andrews, R. M., Flicek, P., Boyle, P. J., Cao, H., Carter, N. P., Clelland, G. K., Davis, S., Day, N., Dhami, P., Dillon, S. C., Dorschner, M. O., Fiegler, H., Giresi, P. G., Goldy, J., Hawrylycz, M., Haydock, A., Humbert, R., James, K. D., Johnson, B. E., Johnson, E. M., Frum, T. T., Rosenzweig, E. R., Karnani, N., Lee, K., Lefebvre, G. C., Navas, P. A., Neri, F., Parker, S. C., Sabo, P. J., Sandstrom, R., Shafer, A., Vetrie, D., Weaver, M., Wilcox, S., Yu, M., Collins, F. S., Dekker, J., Lieb, J. D., Tullius, T. D., Crawford, G. E., Sunyaev, S., Noble, W. S.,

Dunham, I., Denoeud, F., Reymond, A., Kapranov, P., Rozowsky, J., Zheng, D., Castelo, R., Frankish, A., Harrow, J., Ghosh, S., Sandelin, A., Hofacker, I. L., Baertsch, R., Keefe, D., Dike, S., Cheng, J., Hirsch, H. A., Sekinger, E. A., Lagarde, J., Abril, J. F., Shahab, A., Flamm, C., Fried, C., Hackermuller, J., Hertel, J., Lindemeyer, M., Missal, K., Tanzer, A., Washietl, S., Korbel, J., Emanuelsson, O., Pedersen, J. S., Holroyd, N., Taylor, R., Swarbreck, D., Matthews, N., Dickson, M. C., Thomas, D. J., Weirauch, M. T., Gilbert, J., Drenkow, J., Bell, I., Zhao, X., Srinivasan, K. G., Sung, W. K., Ooi, H. S., Chiu, K. P., Foissac, S., Alioto, T., Brent, M., Pachter, L., Tress, M. L., Valencia, A., Choo, S. W., Choo, C. Y., Ucla, C., Manzano, C., Wyss, C., Cheung, E., Clark, T. G., Brown, J. B., Ganesh, M., Patel, S., Tammana, H., Chrast, J., Henrichsen, C. N., Kai, C., Kawai, J., Nagalakshmi, U., Wu, J., Lian, Z., Lian, J., Newburger, P., Zhang, X., Bickel, P., Mattick, J. S., Carninci, P., Hayashizaki, Y., Weissman, S., Hubbard, T., Myers, R. M., Rogers, J., Stadler, P. F., Lowe, T. M., Wei, C. L., Ruan, Y., Struhl, K., Gerstein, M., Antonarakis, S. E., Fu, Y., Green, E. D., Karaoz, U., Siepel, A., Taylor, J., Liefer, L. A., Wetterstrand, K. A., Good, P. J., Feingold, E. A., Guyer, M. S., Cooper, G. M., Asimenos, G., Dewey, C. N., Hou, M., Nikolaev, S., Montoya-Burgos, J. I., Loytynoja, A., Whelan, S., Pardi, F., Massingham, T., Huang, H., Zhang, N. R., Holmes, I., Mullikin, J. C., Ureta-Vidal, A., Paten, B., Seringhaus, M., Church, D., Rosenbloom, K., Kent, W. J., Stone, E. A., Batzoglou, S., Goldman, N., Hardison, R. C., Haussler, D., Miller, W., Sidow, A., Trinklein, N. D., Zhang, Z. D., Barrera, L., Stuart, R., King, D. C., Ameur, A., Enroth, S., Bieda, M. C., Kim, J., Bhinge, A. A., Jiang, N., Liu, J., Yao, F., Vega, V. B., Lee, C. W., Ng, P., Shahab, A., Yang, A., Moqtaderi, Z., Zhu, Z., Xu, X., Squazzo, S., Oberley, M. J., Inman, D., Singer, M. A., Richmond, T. A., Munn, K. J., Rada-Iglesias, A., Wallerman, O., Komorowski, J., Fowler, J. C., Couttet, P., Bruce, A. W., Dovey, O. M., Ellis, P. D., Langford, C. F., Nix, D. A., Euskirchen, G., Hartman, S., Urban, A. E., Kraus, P., Van Calcar, S., Heintzman, N., Kim, T. H., Wang, K., Qu, C., Hon, G., Luna, R., Glass, C. K., Rosenfeld, M. G., Aldred, S. F., Cooper, S. J., Halees, A., Lin, J. M., Shulha, H. P., Zhang, X., Xu, M., Haidar, J. N., Yu, Y., Ruan, Y., Iyer, V. R., Green, R. D., Wadelius, C., Farnham, P. J., Ren, B., Harte, R. A., Hinrichs, A. S., Trumbower, H., Clawson, H., Hillman-Jackson, J., Zweig, A. S., Smith, K., Thakkapallayil, A., Barber, G., Kuhn, R. M., Karolchik, D., Armengol, L., Bird, C. P., de Bakker, P. I., Kern, A. D., Lopez-Bigas, N., Martin, J. D., Stranger, B. E., Woodroffe, A., Davydov, E., Dimas, A., Eyras, E., Hallgrimsdottir, I. B., Huppert, J., Zody, M. C., Abecasis, G. R., Estivill, X., Bouffard, G. G., Guan, X., Hansen, N. F., Idol, J. R., Maduro, V. V., Maskeri, B., McDowell, J. C., Park, M., Thomas, P. J., Young, A. C., Blakesley, R. W., Muzny, D. M., Sodergren, E., Wheeler, D. A., Worley, K. C., Jiang, H., Weinstock, G. M., Gibbs, R. A., Graves, T., Fulton, R., Mardis, E. R., Wilson, R. K., Clamp, M., Cuff, J., Gnerre, S., Jaffe, D. B., Chang, J. L., Lindblad-Toh, K., Lander, E. S., Koriabine, M., Nefedov, M., Osoegawa, K., Yoshinaga, Y., Zhu, B., and de Jong, P. J. (2007) Identification and analysis of functional elements in 1% of the human genome by the ENCODE pilot project. *Nature* *447*, 799–816.

Chapter 13

Detecting Long-Range Chromatin Interactions Using the Chromosome Conformation Capture Sequencing (4C-seq) Method*

Nele Gheldof, Marion Leleu, Daan Noordermeer, Jacques Rougemont, and Alexandre Reymond

Abstract

Eukaryotic transcription is tightly regulated by transcriptional regulatory elements, even though these elements may be located far away from their target genes. It is now widely recognized that these regulatory elements can be brought in close proximity through the formation of chromatin loops, and that these loops are crucial for transcriptional regulation of their target genes. The chromosome conformation capture (3C) technique presents a snapshot of long-range interactions, by fixing physically interacting elements with formaldehyde, digestion of the DNA, and ligation to obtain a library of unique ligation products. Recently, several large-scale modifications to the 3C technique have been presented. Here, we describe chromosome conformation capture sequencing (4C-seq), a high-throughput version of the 3C technique that combines the 3C-on-chip (4C) protocol with next-generation Illumina sequencing. The method is presented for use in mammalian cell lines, but can be adapted to use in mammalian tissues and any other eukaryotic genome.

Key words: Chromosome conformation capture, Looping interactions, Interactome, Regulatory elements

1. Introduction

On top of the linear sequence of the genome lay more levels of chromatin organization that play a crucial role in transcriptional regulation. One important aspect of transcriptional regulation is the physical proximity of regulatory elements, for example,

**Author Contributions*: Nele Gheldof and Alexandre Reymond wrote the paper. Marion Leleu, Daan Noordermeer, and Jacques Rougemont contributed the part about the data analysis.

Bart Deplancke and Nele Gheldof (eds.), *Gene Regulatory Networks: Methods and Protocols*,
Methods in Molecular Biology, vol. 786, DOI 10.1007/978-1-61779-292-2_13, © Springer Science+Business Media, LLC 2012

looping of a distant enhancer element to its gene of interest. Hence, many studies have recently focused on investigating the three-dimensional structure of the genome, and specifically looping mechanisms of long-range regulatory elements. It has been proposed that the genome is structured as a network of interactions between genomic elements. Until about a decade ago, the structure of a genome was mostly studied by using cytological methods. However, this method is limited by its resolution: even current advances in the technique have only been able to reach a resolution of ~250 nm. In addition, cytological studies do not allow ab initio detection of novel regulatory elements. The development of the chromosome conformation capture (3C) technique in 2002 by Job Dekker (1) and its adaptation for use in mammalian cells (2, 3) have proved a major advancement in the field. It has since been widely used to detect physical looping interactions, as evidenced by its citation in more than 340 research articles as of July 2010.

The original 3C technique is however limited by the fact that it can detect only few interactions as it relies on locus-specific PCR primers. Therefore, many efforts have been made to allow both ab initio detection of looping interactions and increase the throughput of the technique (4–10).

Here, we describe the 4C-sequencing methodology (4C-seq) that we developed in our laboratory, which is based on the 4C (3C-on-chip) protocol developed by the de Laat group (9, 11). This technique allows identification of genome-wide interactors with one or a few elements ("baits") without prior knowledge of the interactors. Figure 1 depicts the outline of the methodology.

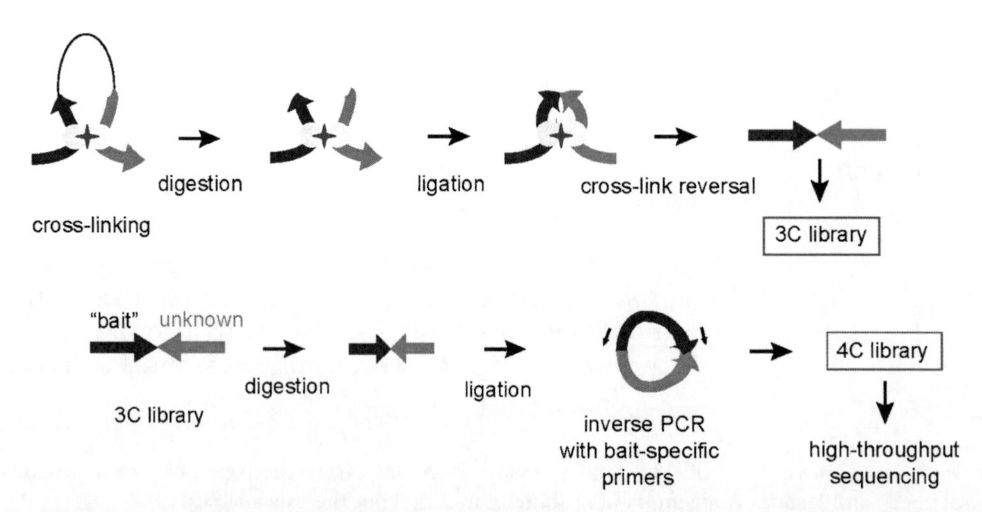

Fig. 1. Schematic representation of the Chromosome Conformation Capture sequencing (4C-seq) technique. Chromatin contacts are crosslinked with formaldehyde, digested with a primary restriction enzyme and ligated under dilute conditions to favor intramolecular binding. After reversal of the crosslinks, the purified DNA represents the 3C library. This library is subjected to a second digestion, after which the DNA is circularized and using bait-specific primers, the 4C library can be amplified and detected by high-throughput sequencing.

Briefly, formaldehyde is added to cells to crosslink the DNA and DNA-bound proteins, after which the cells are digested and ligated under dilute conditions to favor intramolecular binding. After reversal of the crosslinks, the DNA is purified to obtain a 3C library. This 3C library is subjected to a second round of digestion and ligation, generating a 4C library. Subsequently, the 4C library is circularized and amplified in a 4C-seq PCR amplification reaction by using bait-specific primers. Using next-generation Illumina sequencing, the composition of the 4C-seq PCR products can be determined. By using a tailored data analysis pipeline, the sequencing results can then be mapped to the genome and translated into a genome-wide pattern of interactions for the initially chosen bait(s).

2. Materials

2.1. Selection of Appropriate Restriction Enzymes

For optimal capture of DNA interactions during the generation of the 3C library, it is advised to choose a six-cutter as primary restriction enzyme (such as *Bgl*II, *Hin*dIII, or *Eco*RI). Subsequently, the fragment of interest is digested into a smaller fragment by a four-cutter before circularization of the 4C library (see Note 1).

2.2. Primer Design

Optimal primer design is crucial for successful 4C-seq experiments. Obtaining good primers may take several trial assays (see Note 2).

For the 4C-seq primers, we added the Illumina adaptors directly onto the bait-specific primer sequences to avoid inefficient ligation afterwards. Order of these sequences is the following (see Notes 3–5):

4C-seq forward primer:

5′–Illumina P5 sequence–Illumina sequencing primer–bait-specific primer sequence–3′

corresponding to: 5′–AATGATACGGCGACCACCGA–ACACT CTTTCCCTACACGACGCTCTTCCGATCT–bait-specific forward primer sequence–3′

4C-seq reverse primer:

5′–Illumina P7 sequence–bait-specific primer sequence–3′

corresponding to: 5′–CAAGCAGAAGACGGCATACGA–bait-specific reverse primer sequence–3′

2.3. Generation of 3C and 4C Template

1. 37% Formaldehyde.
2. 2.5 M Glycine.
3. Lysis buffer: 10 mM Tris–HCl, pH 8.0, 10 mM NaCl, 0.2% NP-40.
4. Protease inhibitor cocktail (Sigma-Aldrich).

5. 1% and 10% (w/v) Sodium dodecyl sulfate (SDS).
6. Selected restriction enzymes and corresponding 10× restriction enzyme buffers.
7. 10% Triton X-100.
8. 10× ligation buffer: 500 mM Tris–HCl, pH 7.5, 100 mM $MgCl_2$, 100 mM dithiothreitol (DTT).
9. 10 mg/mL Bovine serum albumin (BSA).
10. 100 mM Adenosine triphosphate (ATP).
11. T4 DNA ligase, 400,000 U/mL (NEB).
12. 10 mg/mL Proteinase K in TE buffer, pH 8.0.
13. Phenol/chloroform (1/1) (see Note 6).
14. 3 M Sodium acetate, pH 5.2.
15. 70 and 100% Ethanol.
16. 10× TE buffer, pH 8.0.
17. Glycogen (20 mg/mL).
18. 10 mM Tris-HCl, pH 7.5.
19. 4C-seq PCR: we obtained optimal results using the Expand Long Template System (Roche).

2.4. Data Analysis Software

1. Any global alignment algorithm (e.g., Align0 or Exonerate) for the separation of the reads (in case of multiplexing) and the filtering of noninformative reads.
2. Perl or any scripting program handling the processing of strings for the creation of the virtual library. The *coverage* functionality of bedTools should help in evaluating the coverage in repeats of the generated fragments.
3. Any short-read alignment algorithm (e.g., Bowtie or BWA) for mapping of reads to a reference genome.
4. Scripting and bedTools for quantitation and visualization of interactions at the fragment level.
5. Scripting and any statistical analysis platform (e.g., R or Matlab) for profile-analysis.

3. Methods

3.1. Cell Culture and Crosslinking of Cells

1. Grow 5×10^7 cells in their appropriate medium according to standard conditions. Harvest exponentially growing cells.
2. For suspension cells: spin down the cells and redissolve in 22.5 mL fresh medium.
   For monolayer cells: remove the medium and add 10 mL fresh medium per dish.

3. Add formaldehyde to a concentration of 1% or 2%, incubate 10 min at room temperature on a shaking platform. Fix the cells by adding 2.5 M glycine to a final concentration of 0.125 M, incubate 5 min at room temperature on a shaking platform. Store at 4°C for at least 15 min.
4. For suspension cells: centrifuge the cells 10 min at 400 × *g*, 4°C.

   For monolayer cells: scrape off the cells, transfer to a conical tube and centrifuge 10 min at 400 × *g*, 4°C.
5. Remove the medium, and either store the pellet at −80°C after quick freezing until further use, or proceed immediately to Subheading 3.2.

3.2. Generation of 3C Template

1. Redissolve the pellet in 1 mL ice-cold lysis buffer, supplemented with 0.3 mL protease inhibitor cocktail, and incubate 15 min on ice.
2. Dounce homogenize the cells on ice with a tight pestle, by gently scraping 15 times, incubating 1 min on ice, and scraping 15 more times.
3. Transfer to a microfuge tube and centrifuge the cells 5 min at 2,000 × *g*, 4°C.
4. Remove the medium and wash the pellet with 0.5 mL 1× appropriate restriction enzyme buffer. Centrifuge again and resuspend in 0.5 mL 1× restriction enzyme buffer.
5. Distribute 50 μL of cells to individual microfuge tubes (so you have ten tubes of 0.5×10^7 cells per tube).
6. Add 312 μL 1× restriction enzyme buffer to each tube.
7. Add 38 μL 1% SDS and incubate for 10 min at 65°C.
8. Add 44 μL 10% Triton X-100. Mix by pipetting up and down.
9. Keep one tube separate to determine cutting efficiency (see Subheading 3.3). Add 400 U restriction enzyme per tube. Mix and incubate overnight at 37°C in a shaking incubator (see Note 7).
10. Add 86 μL 10% SDS and incubate for 30 min at 65°C.
11. Transfer each reaction to a 50 mL conical tube. Make a mastermix and add to each tube: 745 μL 10% Triton X-100, 745 μL 10× ligation buffer, 80 μL 10 mg/mL BSA, 80 μL 100 mM ATP, 5,960 μL dH_2O. Add 7.5 μL T4 DNA ligase (3,000 cohesive end units) per tube and incubate 2 h at 16°C.
12. Add 50 μl 10 mg/mL proteinase K in 1× TE buffer, pH 8.0, and incubate overnight at 65°C.

13. Add an additional 50 μL 10 mg/mL proteinase K in 1× TE buffer, pH 8.0, and incubate at 42°C for 2 h.
14. Add 8 mL of phenol to each tube, vortex for 2 min, and centrifuge for 5 min at 2,500 × *g* at room temperature.
15. Remove the top aqueous phase (see Note 8), add phenol/chloroform to it, vortex for 2 min, and centrifuge for 5 min at 2,500 × *g* at room temperature.
16. Transfer the top aqueous phase to a high-speed centrifuge tube, add one-tenth of the volume of 3 M sodium acetate, pH 5.2, vortex briefly, add 2.5× volume of 100% ethanol, and mix.
17. Incubate at least 30 min at –80°C (or overnight) and centrifuge for 20 min at 12,000 × *g* at 4°C.
18. Remove supernatant and resuspend each pellet in 400 μL 1× TE buffer, pH 8.0. Transfer DNA to microfuge tubes.
19. Add 400 μL phenol/chloroform (1/1), vortex 1 min, and centrifuge 5 min at 16,000 × *g*.
20. Remove the aqueous top phase and add one-tenth of the volume of 3 M sodium acetate, pH 5.2, vortex briefly, add 2.5× volume of 100% ethanol.
21. Incubate at least 30 min at –80°C and centrifuge for 20 min at 16,000× *g* at 4°C.
22. Remove the supernatant and wash the obtained pellet up to five times with 70% ethanol (or until the size of the pellet remains the same).
23. Briefly air dry the pellet and redissolve in 50 μL 1× TE buffer, pH 8.0, pool the samples and add 1 μl 10 mg/mL RNase A.
24. Incubate for 15 min at 37°C. This is the 3C library that can be stored at –20°C.
25. To check the quality and quantity of 3C template, run the 3C library on a 0.8% agarose gel next to a 1 kb ladder (Fig. 2). We typically obtain ~100 μg of 3C library (~200 ng/μL). Also, a 3C PCR titration is done starting with 200 ng/μL as highest concentration. Set up a twofold serial dilution series of the 3C library and use standard PCR conditions (12) with two sets of 3C primers, one between close fragments (~5 kb), and one between two fragments further away (~20 kb). Analyze the titration results by plotting the amount of PCR product obtained versus the amount of 3C library used. The PCR products can be quantified by using the ImageQuant software. Use this plot to determine the linear range of PCR product formed by various amounts of the template.

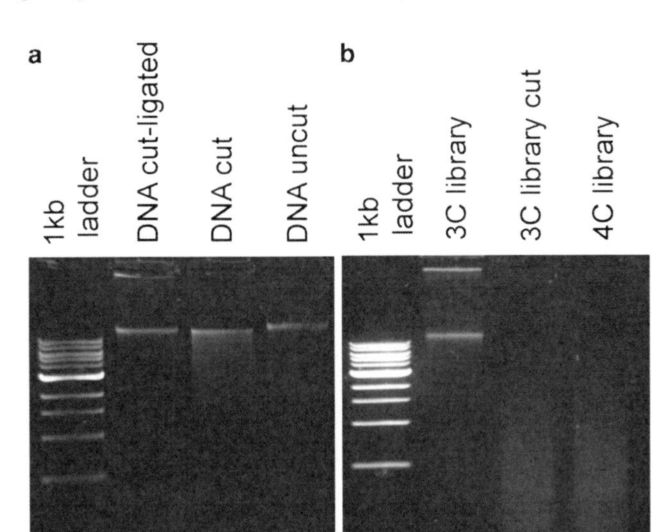

Fig. 2. Visualization of the different stages involved in generation of the 3C and 4C library on 0.8% agarose gels. (**a**) *Lane 2* shows the 3C library (DNA cut and ligated) which migrates above the 10-kb marker. *Lane 3* shows the digested, unligated DNA, which runs lower and as a smear. *Lane 3* is the undigested and unligated DNA, which migrates as a single band above the 10-kb marker. (**b**) *Lane 2* shows the 3C library, *lane 3* the 3C library cut with the secondary restriction enzyme, which digests the library into smaller fragments and shows a smear. *Lane 4* is the 4C library (3C library cut and ligated) which also shows a smear.

3.3. Determination of Cutting Efficiency (see Note 9)

1. Keep one tube separate after crosslinking and lysing of the cells and split in two tubes (see step 9 in Subheading 3.2). Add 200 U restriction enzyme to one tube, but not to the other. Mix and incubate overnight at 37°C in a shaking incubator.
2. Add 43 μL 10% SDS and incubate for 30 min at 65°C.
3. Add 1.5 μL proteinase K (10 mg/mL) and reverse the cross-links overnight at 65°C.
4. Perform a phenol/chloroform (1/1) extraction.
5. Remove the top aqueous phase, add one-tenth of the volume of 3 M sodium acetate, pH 5.2, to it, vortex briefly, add 2.5× volume of 100% ethanol, and mix.
6. Incubate at least 30 min at −80°C and centrifuge for 20 min at 16,000 × g at 4°C.
7. Remove the supernatant and resuspend in 25 μL 1× TE buffer, pH 8.0.
8. Run 1 μL on a 0.8% agarose gel (Fig. 2). In addition, by performing a PCR using primers across the restriction site at various loci, one can determine the amount of undigested versus digested material, which typically amounts to about 30%.

3.4. Generation of 4C Template

1. Digest the 3C library overnight at 37°C in a shaking incubator with the secondary restriction enzyme (see above) at a concentration in the linear range as determined in step 25 in Subheading 3.2 (typically ~100 ng/μL).
2. Heat inactivate for 20 min at 65°C.
3. Add an equal amount of phenol/chloroform (1/1), vortex 2 min, and spin 5 min at 16,000 × *g*, room temperature.
4. Remove the aqueous top phase and add one-tenth of the volume of 3 M sodium acetate, pH 5.2, vortex briefly, add 2.5× volume of 100% ethanol.
5. Incubate at least 30 min at −80°C and centrifuge for 20 min at 16,000 × *g* at 4°C.
6. Remove the supernatant and wash the obtained pellet with 70% ethanol.
7. Centrifuge for 5 min at 16,000 × *g* at 4°C.
8. Remove the supernatant, briefly air dry the pellet and redissolve in 100 μL dH_2O.
9. Keep an aliquot separate (5 μL) to check digestion efficiency (Fig. 2).
10. Transfer the digested sample to a 50 mL conical tube. Add 12.46 mL dH_2O, 1.4 mL 10× ligation buffer, 140 μL 100 μM ATP and 5 μL T4 DNA ligase (2,000 U). Incubate 4 h at 16°C.
11. Add 14 mL phenol/chloroform (1/1), vortex 2 min, and spin 10 min at 2,500 × *g* at room temperature.
12. Transfer the aqueous top phase into two high-speed centrifuge tubes (7 mL each), add 7 mL water and add 14 μL glycogen (20 mg/mL), one-tenth of the volume of 3 M sodium acetate, pH 5.2, vortex briefly, add 2.5× volume of 100% ethanol.
13. Incubate at least 30 min at −80°C (or overnight) and centrifuge for 20 min at 12,000 × *g* at 4°C.
14. Remove the supernatant and wash the obtained pellet with 5 mL 70% ethanol.
15. Centrifuge for 15 min at 12,000 × *g* at 4°C.
16. Remove the supernatant, briefly air dry the pellet, and redissolve in 100 μL 10 mM Tris-HCl, pH 7.5. Incubate for 30 min at 37°C and pool the appropriate samples.

3.5. 4C-seq PCR Amplification and Sequencing

1. Perform an inverse 4C-seq PCR titration starting with 200 ng 4C sample as highest input with the 4C-seq primers (see above for the primer design). Set up a twofold serial dilution series and use the following PCR conditions: 2.5 μL 10× buffer I (Expand Long Template system), 0.35 μL 25 mM dNTPs,

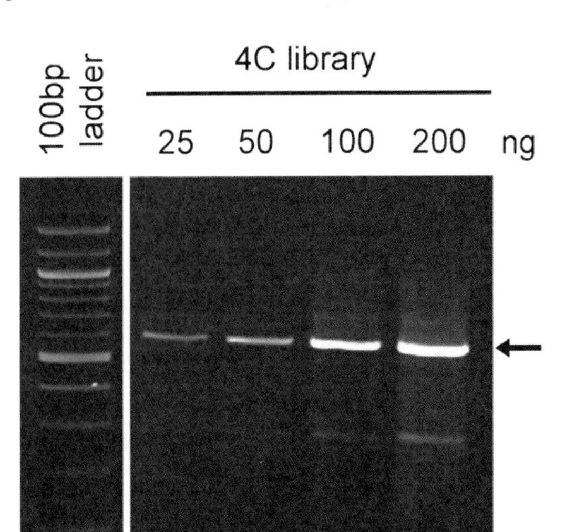

Fig. 3. Titration of a 4C PCR visualized on a 1.5% agarose gel. The *arrow* indicates the size of the undigested self-ligated fragment, which is the most prominent band. A smear of PCR products and some distinct bands are observed in every lane, which linearly increase with increasing amount of template.

2.5 μL 20 μM 4C-seq forward primer, 2.5 μL 20 μM 4C-seq reverse primer, 0.375 μL DNA polymerase mix, template and dH_2O up to 25 μL. Amplify the reactions using the following cycles: 1 cycle for 2 min at 94°C, followed by 30 cycles of: 15 s at 94°C, 1 min at 55°C, 3 min at 68°C, followed by 1 cycle for 7 min at 68°C.

2. Analyze the titration results on a 1.5% agarose gel (Fig. 3). The 4C-seq reaction is successful when a smear on the gel is observed with a most distinct band in the ~30% undigested "bait" restriction site that is self-ligated. Verify the correct size of this undigested self-ligated band. The smear should also be linear and represents the amplification of all fragments that physically interact with the bait.
3. Plot the amount of PCR product of one of the most distinct bands versus the amount of 4C library used. The PCR products can be quantified by using the ImageQuant software. Use this plot to determine the linear range of PCR product formed by various amounts of the template. Choose a template concentration toward the higher side of the linear range. Because of the complexity of the sample, it is important to perform enough PCRs to capture a good representation of all possible interactions. It is therefore recommended to amplify up to a total amount of 3.2 μg of 4C input material (e.g., if 200 ng 4C template was chosen, perform 16 inverse PCRs). Perform the inverse PCR as described in step 1 in Subheading 3.5.

4. Pool and purify the PCR products using the Qiaquick PCR purification kit. Check the sample on a 1.5% agarose gel and measure the amount of sample by comparing to a reference sample of genomic DNA of known concentration, or by fluorometric quantification using PicoGreen.
5. If multiplexing is desired, mix the samples in equimolar ratios (see Note 10).
6. Run the entire 4C-seq sample without size selection on the Illumina Genome Analyzer II using a 76 single end cycle.

3.6. Data Analysis

Analysis of a 4C-seq dataset requires a string of steps in which the list of unsorted Illumina sequences is translated into a genome-wide pattern of interaction partners (see Note 11). In Fig. 4, the main steps of a typical analysis are illustrated, including the separation of reads in case of multiplexed samples.

It is important to realize that, even though both 4C-seq and ChIP-seq intend to visualize genome-wide interaction profiles, analysis of data between the two techniques is fundamentally different both at a biological and experimental level. Biologically, ChIP-seq aims to identify discrete binding sites for DNA-binding factors. Due to the fibrous nature of the chromatin template, 4C-seq peak analysis should always identify clusters of neighboring restriction fragments. Experimentally, ChIP-seq reads can be found anywhere in the genome, while 4C-seq reads can only be found directly bordering the recognition sites of the restriction enzymes.

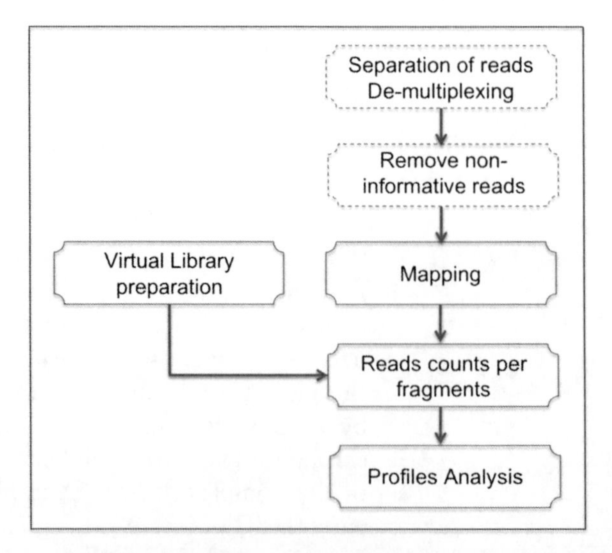

Fig. 4. Main steps of a 4C-seq data analysis. *Dashed boxes* indicate optional steps. The virtual library preparation has to be done only once for each combination of genome and restriction enzymes. Profiles analysis can be followed by any downstream analysis (e.g., comparison between several conditions, tissues, and so on).

1. *Separation of multiplexed samples*: When multiplexed samples are analyzed, the reads can be separated by identifying and removing the barcode and/or different bait-specific forward primer sequences at the beginning of the 5′-end of the Illumina sequences. Any global alignment algorithm (e.g., Align0 or Exonerate) can be used for this purpose, with sufficient stringency to balance between correct classification and number of discarded reads (dependent on the dissimilarity between the primers).
2. *Removal of noninformative reads*: We suggest removing reads representing undigested and self-ligated fragments before the mapping process. This can be done by comparing reads with corresponding genomic sequences (using global alignment software or simple text matching).
3. *Mapping*: Informative reads can be mapped to the reference genome by using any short-reads aligners (we recommend Bowtie or BWA). Ideally, reads of sufficient length should be considered (generally 25 bps or more) and it is preferable to allow for mismatches and multiple hits to compensate for sequencing errors (in our experience, allowing up to two mismatches and ten hits in reads of 30–35 bps results in a good balance between specificity and sensitivity).
4. *Virtual library preparation*: Due to the nature of 3C-like techniques, the occurrence of reads at specific locations in the genome can be predicted to either stem from bona fide ligation events, or to represent technical artifacts and/or random background reads. Similar to the microarray-design in a 4C-experiment, for 4C-seq, a virtual library of all analyzable restriction fragments can be created to exclude reads mapping to regions that cannot be amplified. The library should contain the restriction fragments that are defined by the sequences delimited by two consecutive occurrences of the first restriction enzyme. Here additional quality filters can also be included: presence of at least one occurrence of the second restriction site, absence of annotated repeat sequences and other technical issues (see Note 12).
5. *From reads to fragments*: Interactions at the fragment level are quantified by binning reads falling into the previously defined fragments. Since reads should only map to the outer edges of each fragment, we first evaluate the number of reads mapping to the two extreme parts of the fragments (the first 30–35 bps of the sequence, depending on the size of the reads considered), and then aggregate those numbers at the fragment level (e.g., by taking the average). If multiple hits have been allowed during the mapping step, the resulting counting bias can be corrected by weighting each read to 1/(number of hits) (see Fig. 5).
6. *Profiles analysis*: As mentioned above, due to structural constraints of the chromatin fiber, analysis of 4C-seq data should

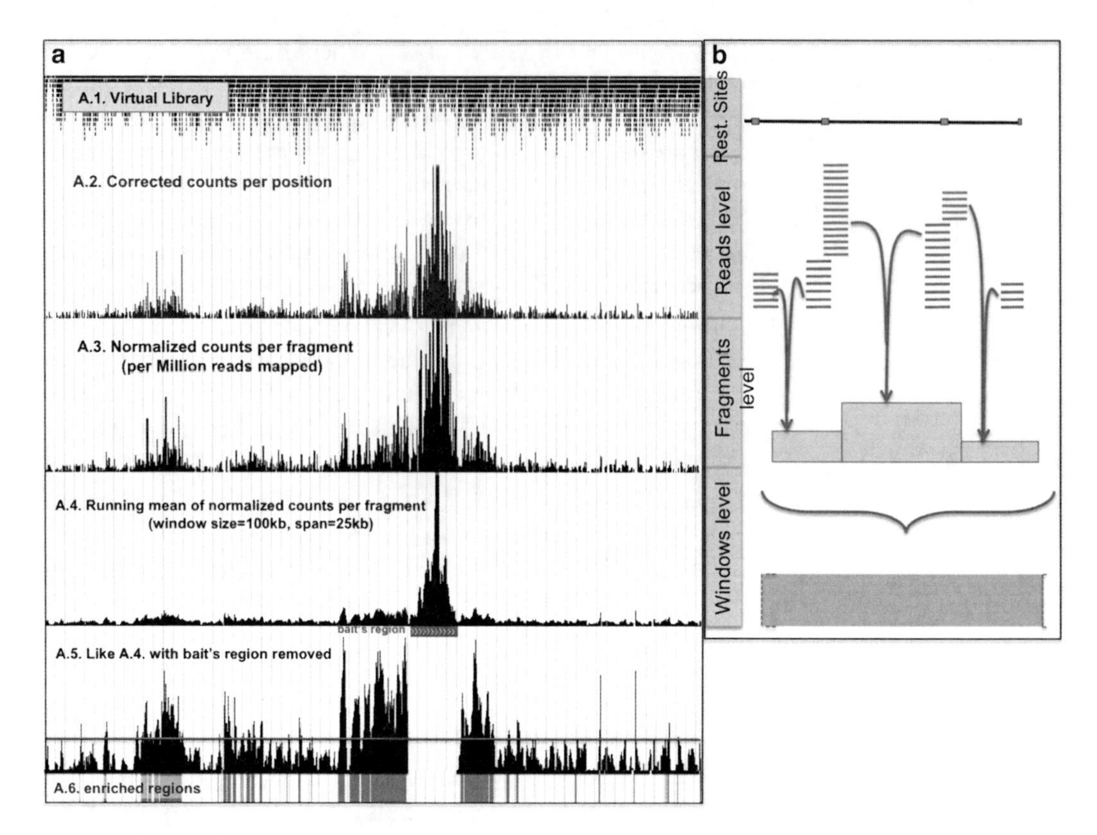

Fig. 5. Visualization of individual steps in the processing of 4C-seq data. (**a**) 10 Mb region showing (1) the coverage of a typical library of analyzable restriction fragments, (2) the number of reads per position (corrected by weighting each read to 1/nb of hits), (3) the counts per fragments (normalized to the number of reads per million reads mapped), (4) the running mean of (3) (window size = 100 kb and span = 25 kb), (5) same view as (**a** 4) but with the region neighboring the bait excluded (notice the change in the vertical scale), the *vertical line* represents the threshold as determined by randomization of the data, and (6) resulting interacting regions. (**b**) Schematic view of different levels of visualization: Restriction sites level (represented by *boxes*), the reads level (reads represented in full lines (resp. *dashed lines*) map on the forward (resp. reversed) strand), the fragments level and the Windows level.

be aimed at identifying clusters of over-represented restriction fragments, rather than identifying a single high-frequency fragment. Previous analyses of 4C data (9) used running mean and running median algorithms to identify clusters of interacting fragments and to minimize the influence of isolated peaks (generally, a running median analysis is less sensitive to high isolated "peaks" and thus more robust for identifying less frequent, but still significant interactions). A typical window size covers a sufficient number of fragments (e.g., 100 kb or around 25 adjacent fragments). Statistically significant regions of interactions can subsequently be determined by comparing the resulting windows to randomly shuffled data allowing a certain preset false discovery rate (FDR of, e.g., 5%).

Importantly, in 4C-seq experiments a very large proportion of reads maps to the region directly neighboring the bait (up to a few Mb's, see Fig. 5 for a typical profile), which will strongly bias any genome-wide statistical method aiming at identifying long-range interactions. For the identification of interacting regions at long distance, it may therefore be required to discard this region from the analysis.

4. Notes

1. It is important that the distance between the six-cutter and the four-cutter is larger than 250 bp, as shorter stretches of DNA are too rigid to circularize (11). Successful libraries have been obtained using as four-cutter *Nla*III, *Dpn*II, and *Mse*I. Generally, enzymes that are blocked by CpG methylation work less well and are advised to be avoided.
2. It is advised to first test primers without addition of Illumina adapter sequences, as 4C-seq PCR efficiency in the majority of cases is not considerably influenced by the presence of these adapter sequences. After obtaining functional primers, high-quality (HPLC or PAGE purified) primers with the Illumina adaptor tails should be obtained.
3. The 4C-seq bait-specific primers can be designed using the Primer3 software, with following parameters for optimal primer design: 20–22 mers with a melting temperature between 55 and 65°C and a maximum difference of 2°C between the primers. Importantly, primers should not be designed in repetitive sequences in the genome.
4. For multiplexing of different templates with the same primer-set, it is possible to add a barcode sequence as follows: 5′–Illumina P5 sequence–Illumina sequencing primer–BARCODE–bait-specific primer sequence–3′. It is important to keep in mind that the barcode will be included in the 76 bp Illumina reads.
5. It is important to choose the primer with the sequencing tail as close as possible to the restriction site, such as few bases as possible of the 76 bp Illumina reads are spent on the bait-specific sequence. Choose the best primer combination from the four possible locations to design the primer with sequencing tail: 5′ of your bait at the 3′ end of the primary restriction enzyme, or at the 5′end of your secondary restriction enzyme, or 3′ of your bait at the 5′ end of the primary restriction enzyme, or at the 3′ end of your secondary restriction enzyme.
6. Make sure to use phenol that is buffer-equilibrated around neutral pH 6–8.

7. The cell lysate will not be completely dissolved at this point, so for more efficient digestion, agitate the digestion reaction overnight in the incubator.
8. The DNA tends to accumulate close to the phenol interface so make sure to pipette as close as possible to the interface without touching it.
9. The cutting efficiency needs to be determined every time you start working with a new cell line or new restriction enzyme.
10. Considering the number of mappable reads that can be obtained from one Solexa lane and the number of reads that are required to faithfully resolve a genome-wide interaction pattern, we recommend multiplexing a limited number of samples per Illumina lane (generally in the order of four samples).
11. As mentioned in the data analysis section, one has to be very careful about interpretation of the data. Only clustered increased interaction frequencies will reveal significantly interacting regions. In addition, it is recommended to perform at least two biological replicates because many random chromatin interactions occur, so only double positive regions are reliable interacting regions.

    If possible, it is also advised to confirm looping interactions found by 3C-based methods with a complementary method such as microscopy-based techniques like fluorescence in situ hybridization (FISH).
12. We strongly advise against mapping the reads directly to the virtual library, as most of the aligners are not designed to efficiently handle references with a large number of sequences.

Acknowledgments

We are grateful for the help from Keith Harshman, Jérôme Thomas, Floriane Consales, and Emmanuel Beaudoing from the Lausanne Genomic Technologies Facility (GTF). This work was supported by the Marie Heim-Vögtlin fellowship to N.G.

References

1. Dekker, J., Rippe, K., Dekker, M., and Kleckner, N. (2002) Capturing chromosome conformation. *Science 295*, 1306–1311.
2. Gheldof, N., Tabuchi, T. M., and Dekker, J. (2006) The active FMR1 promoter is associated with a large domain of altered chromatin conformation with embedded local histone modifications. *Proceedings of the National Academy of Sciences of the United States of America 103*, 12463–12468.
3. Tolhuis, B., Palstra, R. J., Splinter, E., Grosveld, F., and de Laat, W. (2002) Looping and interaction between hypersensitive sites in the active beta-globin locus. *Molecular Cell* ***10***, 1453–1465.
4. Dostie, J., Richmond, T. A., Arnaout, R. A., Selzer, R. R., Lee, W. L., Honan, T. A., Rubio, E. D., Krumm, A., Lamb, J., Nusbaum, C., Green, R. D., and Dekker, J. (2006) Chromosome Conformation Capture Carbon

Copy (5C): a massively parallel solution for mapping interactions between genomic elements. *Genome Research* ***16***, 1299–1309.

5. Duan, Z., Andronescu, M., Schutz, K., McIlwain, S., Kim, Y. J., Lee, C., Shendure, J., Fields, S., Blau, C. A., and Noble, W. S. (2010) A three-dimensional model of the yeast genome. *Nature* ***465***, 363–367.
6. Fullwood, M. J., Liu, M. H., Pan, Y. F., Liu, J., Xu, H., Mohamed, Y. B., Orlov, Y. L., Velkov, S., Ho, A., Mei, P. H., Chew, E. G. Y., Huang, P. Y. H., Welboren, W.-J., Han, Y., Ooi, H. S., Ariyaratne, P. N., Vega, V. B., Luo, Y., Tan, P. Y., Choy, P. Y., Wansa, K. D. S. A., Zhao, B., Lim, K. S., Leow, S. C., Yow, J. S., Joseph, R., Li, H., Desai, K. V., Thomsen, J. S., Lee, Y. K., Karuturi, R. K. M., Herve, T., Bourque, G., Stunnenberg, H. G., Ruan, X., Cacheux-Rataboul, V., Sung, W.-K., Liu, E. T., Wei, C.-L., Cheung, E., and Ruan, Y. (2009) An oestrogen-receptor-[agr]-bound human chromatin interactome. *Nature* ***462***, 58–64.
7. Lieberman-Aiden, E., van Berkum, N. L., Williams, L., Imakaev, M., Ragoczy, T., Telling, A., Amit, I., Lajoie, B. R., Sabo, P. J., Dorschner, M. O., Sandstrom, R., Bernstein, B., Bender, M. A., Groudine, M., Gnirke, A., Stamatoyannopoulos, J., Mirny, L. A., Lander, E. S., and Dekker, J. (2009) Comprehensive Mapping of Long-Range Interactions Reveals Folding Principles of the Human Genome. *Science* ***326***, 289–293.
8. Schoenfelder, S., Sexton, T., Chakalova, L., Cope, N. F., Horton, A., Andrews, S., Kurukuti, S., Mitchell, J. A., Umlauf, D., Dimitrova, D. S., Eskiw, C. H., Luo, Y., Wei, C.-L., Ruan, Y., Bieker, J. J., and Fraser, P. Preferential associations between co-regulated genes reveal a transcriptional interactome in erythroid cells. *Nature Genet.* ***42***, 53–61.
9. Simonis, M., Klous, P., Splinter, E., Moshkin, Y., Willemsen, R., de Wit, E., van Steensel, B., and de Laat, W. (2006) Nuclear organization of active and inactive chromatin domains uncovered by chromosome conformation capture-on-chip (4C). *Nature Genet.* ***38***, 1348–1354.
10. Zhao, Z., Tavoosidana, G., Sjolinder, M., Gondor, A., Mariano, P., Wang, S., Kanduri, C., Lezcano, M., Sandhu, K. S., Singh, U., Pant, V., Tiwari, V., Kurukuti, S., and Ohlsson, R. (2006) Circular chromosome conformation capture (4C) uncovers extensive networks of epigenetically regulated intra- and interchromosomal interactions. *Nature Genet.* ***38***, 1341–1347.
11. Simonis, M., Kooren, J., and de Laat, W. (2007) An evaluation of 3C-based methods to capture DNA interactions. *Nature Methods* ***4***, 895–901.
12. Miele, A., Gheldof, N., Tabuchi, T. M., Dostie, J., and Dekker, J. (2006) Mapping chromatin interactions by chromosome conformation capture. *Current Protocols in Molecular Biology / edited by Frederick M. Ausubel et al.* Chapter ***21***, Unit 21 11.

Part III

Mapping Protein–DNA Interactions

Chapter 14

Analyzing Transcription Factor Occupancy During Embryo Development Using ChIP-seq

Yad Ghavi-Helm and Eileen E.M. Furlong

Abstract

Accurately assessing the binding of transcription factors to *cis*-regulatory elements in vivo is an essential step toward understanding the mechanisms that govern embryonic development. Genome-wide transcription factor location analysis has been facilitated by the development of high-density tiling arrays (ChIP-on-chip), and more recently by next-generation sequencing technologies, which are used to sequence the DNA fragments obtained from chromatin immunoprecipitation experiments (ChIP-seq). This chapter provides a detailed protocol of the different steps required to generate a successful ChIP-seq library, starting from embryo collection and fixation to chromatin preparation, immunoprecipitation, and finally library preparation. The protocol is optimized for *Drosophila* embryos, but can be adapted to any organism. The obtained library is suitable for sequencing on an Illumina GAIIx platform.

Key words: *Drosophila*, Transcription, Development, Chromatin immunoprecipitation, ChIP-seq, Next-generation sequencing

1. Introduction

An extensive number of nuclear proteins, involved in processes ranging from the regulation of DNA repair and RNA transcription to chromatin modifications, interact with chromatin or with chromatin binding partners. Mapping these protein–chromatin interactions in vivo is a crucial first step to understand their mechanism of action.

In multicellular organisms, embryonic development is governed by highly orchestrated patterns of gene expression in both a temporal- and tissue-specific manner. Understanding the regulatory networks that control these expression patterns is thus an essential step to understanding metazoan development. *Cis*-regulatory networks are defined by a combination of

Bart Deplancke and Nele Gheldof (eds.), *Gene Regulatory Networks: Methods and Protocols*, Methods in Molecular Biology, vol. 786, DOI 10.1007/978-1-61779-292-2_14, © Springer Science+Business Media, LLC 2012

sequence-specific transcription factors, which bind to enhancer element or *cis*-regulatory modules (CRMs) (1, 2). Recent years has seen substantial progress in dissecting transcriptional networks during embryonic development at a genome-wide scale, revealing extensive and dynamic patterns of transcription factor occupancy and combinatorial binding (3–10).

A powerful method to analyze transcription factors occupancy in vivo is chromatin immunoprecipitation (ChIP) experiments (11) followed by quantitative real-time PCR or by hybridization of the immunoprecipitated DNA to microarrays (ChIP-on-chip) (12). Briefly, ChIP experiments are based on the in vivo cross-linking of proteins to the chromatin using formaldehyde. The chromatin is then extracted and sheared by sonication into short fragments of approximately 500 bp. The DNA fragments specifically bound by the protein of interest are then isolated by immunoprecipitation using antibody directed against the protein of interest. After this "pull down" step, the cross-links are reversed and the DNA is purified. ChIP-on-chip experiments have been successfully used to identify the binding sites of a wide range of proteins. High-density tiling arrays have been extensively applied for organisms with a smaller genome such as yeast and *Drosophila* where the entire genome can fit on a single microarray slide. For organisms with a larger genome, such as mice and humans, ChIP-on-chip analysis has mainly been restricted to promoter analysis due to the prohibitory cost and logistics of using many tiling arrays for a single sample. However, the recent development of next-generation sequencing technologies has enabled researchers to accurately define the binding sites of a protein of interest on the entire genome for any organism using high-throughput sequencing of the immunoprecipitated DNA (ChIP-seq). ChIP-seq technology thereby provides an obvious advantage for the analysis of transcription factor occupancy in organisms with large genomes, as the genome-wide location of a protein can be assessed in a single sequencing run. The method also provides the added advantage of increased resolution around the transcription factors binding site; where ChIP-on-chip typically has a resolution of ±100 bp (10), ChIP-seq allows the location of binding sites within ±25 bp in some cases (13). To facilitate this increase in resolution, and also for the library preparation, the chromatin used for ChIP-seq experiments should be sheared into fragments of approximately 200 bp.

This chapter provides a detailed description of the different steps needed to prepare a ChIP-seq library, including many required quality controls. The initial steps of the protocol are tailored for preparing chromatin from *Drosophila* embryos, but the protocol could be adapted to any organism with minor modifications. The obtained library is suited for high-throughput sequencing using the Illumina GAIIx platform.

2. Materials

2.1. Embryo Collection and Cross-linking

1. Staged *Drosophila* embryos grown in population cages as described in (14) (see Notes 1 and 2).
2. Collection sieve apparatus (112 μm/355 μm/710 μm bottom to top; Retsch).
3. Nitex membranes (approximately 5×5 cm each), 120 μm (Sefar).
4. 37% Formaldehyde solution (w/w).
5. Cross-linking solution (store at room temperature): 1 mM EDTA, 0.5 mM EGTA, 100 mM NaCl, 50 mM HEPES, pH 8.0. Filter (0.45 μm) before storage at room temperature.
6. Glycine solution: 125 mM Glycine, 0.1% Triton X-100 (v/v) in PBS. Store at room temperature.
7. Heptane.
8. Methanol.
9. Dechorionating solution: 3% sodium hypochlorite (v/v) (or 50% commercial bleach). Prepare fresh.
10. PBT solution: 0.1% Triton X-100 (v/v) in PBS. Store at room temperature.

2.2. Evaluation of Embryos Developmental Stages Distribution Within a Collection

1. PBT solution: 0.1% Triton X-100 (v/v) in PBS. Store at room temperature.
2. Methanol.
3. Glycerol.
4. Mounting medium: Prolong Gold antifade reagent with DAPI (Invitrogen).
5. Microscope slides.

2.3. Chromatin Preparation

1. Protease inhibitor stock solutions (1,000× stock): 10 mg/mL aprotinin (in water), 10 mg/mL leupeptin and pepstatin (in DMSO). Store in small aliquots at −20°C.
2. Phenylmethylsulfonyl fluoride (PMSF): 100 mM stock (100×) in 2-propanol. Store at room temperature.
3. PBT solution: 0.1% Triton X-100 (v/v) in PBS. Store at room temperature.
4. 15-mL Dounce homogenizer (Wheaton).
5. Cell lysis buffer: 85 mM KCl, 0.5% IGEPAL CA-630 (v/v) (Sigma), 5 mM HEPES, pH 8.0. Autoclave without IGEPAL and then add appropriate amount of IGEPAL from 10% stock. Store at 4°C. Add protease inhibitors and PMSF just prior to use.

6. Nuclear lysis buffer: 10 mM EDTA, 0.5% *N*-lauroylsarcosine (w/v), 50 mM HEPES, pH 8.0. Store at 4°C. Add protease inhibitors and PMSF just prior to use.
7. Bioruptor sonicator (Diagenode).
8. 1.6-mL low-binding reaction tubes (Biozym).
9. TE buffer: 1 mMEDTA, 10 mM Tris–HCl, pH 8.0. Filter (0.45 μm) before storage at 4°C.
10. RNase A (QIAGEN): Prepare 1 mg/mL aliquot and store at –20°C.
11. Proteinase K (Roche). Prepare 10 mg/mL aliquots and store at –80°C.
12. 10% SDS solution (w/v). Store at room temperature.
13. Phenol–Chloroform–IAA, 25:24:1, pH 7.9.
14. Chloroform.
15. Glycogen (Roche). Prepare 5 mg/mL aliquots and store at –20°C.
16. NanoDrop ND-1000 Spectrophotometer (Thermo Fisher Scientific).

2.4. Chromatin Immunoprecipitation

1. 1.6-mL low-binding reaction tubes (Biozym).
2. Protein A-Sepharose 4B, Fast Flow (Sigma) (use as 50% (v/v) slurry).
3. RIPA buffer: 140 mM NaCl, 1 mM EDTA, 1% Triton X-100 (v/v), 0.1% SDS (w/v), 0.1% sodium deoxycholate (w/v), 10 mM Tris–HCl, pH 8.0. Filter (0.45 μm) before storage at 4°C. Add protease inhibitors and PMSF just prior to use for the immunoprecipitation, but not to the wash buffers.
4. RIPA-BSA buffer: RIPA buffer supplemented with 1 mg/mL BSA. Store at 4°C.
5. IP dilution buffer: 0.35 M NaCl, 2.5% Triton X-100 (v/v), 0.25% SDS (w/v), 0.25% sodium deoxycholate (w/v). Store at 4°C. Add protease inhibitors and PMSF just prior to use.
6. RIPA500 buffer: RIPA buffer adjusted to 500 mM NaCl. Filter (0.45 μm) before storage at 4°C. Add protease inhibitors and PMSF just prior to use for the immunoprecipitation, but not to the wash buffers.
7. LiCl buffer: 250 mM LiCl, 1 mM EDTA, 0.5% IGEPAL CA-630 (v/v) (Sigma), 0.5% sodium deoxycholate (w/v), 10 mM Tris–HCl, pH 8.0. Filter (0.45 μm) before storage at 4°C.
8. TE buffer: 1 mM EDTA, 10 mM Tris–HCl, pH 8.0. Filter (0.45 μm) before storage at 4°C.
9. RNase A (QIAGEN): Prepare 1 mg/mL aliquot and store at –20°C.

10. Proteinase K (Roche). Prepare 10 mg/mL aliquots and store at –80°C.
11. 10% SDS solution (w/v). Store at room temperature.
12. 2-mL phase-lock heavy gel tubes (5 Prime).
13. Phenol–Chloroform–IAA, 25:24:1, pH 7.9.
14. Chloroform.
15. Glycogen (Roche). Prepare 5 mg/mL aliquots and store at –20°C.
16. Qubit fluorometer (Invitrogen).
17. Quant-iT dsDNA HS Assay Kit (Invitrogen).

2.5. Library Preparation

1. NEBNext™ DNA Sample Prep Reagent Set 1.
2. QIAquick PCR purification kit and MinElute gel extraction kit (QIAGEN).
3. 1.6-mL low-binding reaction tubes (Biozym).
4. 6× orange G loading dye.
5. Qubit fluorometer (Invitrogen).
6. Quant-iT dsDNA HS Assay Kit (Invitrogen).
7. Agilent 2100 Bioanalyzer (Agilent).
8. Agilent High Sensitivity DNA kit and Agilent DNA 1000 kit (Agilent).
9. GelGreen Nucleic Acid Gel Stain, 10,000× in water (Biotium).
10. Safe Imager transilluminator (Invitrogen).
11. Gene Catcher Disposable Gel Excision Tips 6.5 mm (Diamed).
12. Paired-End Adapter Oligo Mix (Illumina) (see Notes 3 and 4).
13. Multiplexing Sample Preparation Oligonucleotide Kit (Illumina) (see Notes 3 and 5).
14. Paired-End PCR primers 1.0 and 2.0 (Illumina) (see Notes 3 and 6).

3. Methods

3.1. Preparation of Cross-linked Embryos

1. At time points appropriate for the specific study, collect embryos into a collection sieve apparatus by washing the embryos from agar plates with tap water and a paintbrush (see Note 1).
2. Transfer embryos to 400 mL fresh dechorionating solution and incubate for 2.5 min at room temperature with stirring to dechorionate.

3. Poor the embryos into the 112-μm size sieve and wash them extensively with tap water (see Note 7). Remove as much liquid as possible by blotting the sieve with a paper towel.
4. Transfer embryos to 10 mL of stirred PBT solution in a stirred glass beaker per 1.5 g of embryos.
5. Place a sufficient number of Nitex membranes onto a pile of tissue paper and pipette 10 mL of embryo suspension (corresponding to approximately 1.5 g of embryos) onto each membrane. Fold the membrane over to cover the embryos and gently blot the embryos dry through the membrane with a paper towel.
6. Transfer each membrane into a separate 50-mL Falcon tube containing 10 mL of cross-linking solution and 30 mL of heptane. Shake off the embryos, recover the membrane, and add 485 μL of Formaldehyde solution 37%. Shake the tube vigorously at room temperature (20–25°C) for 15 min (see Note 8).
7. Pellet the embryos in each tube by centrifugation at 500 × *g* for 1 min. Replace the supernatant with 30 mL of Glycine solution and shake vigorously at room temperature for at least 1 min to stop the cross-linking reaction.
8. Pellet the embryos by centrifugation at 500 × *g* for 1 min. Carefully decant the supernatant and wash the pellet with 50 mL of ice-cold PBT solution.
9. Pellet the embryos by centrifugation at 500 × *g* for 1 min, decant the supernatant, and resuspend the embryos in approximately 10 mL of PBT solution per tube. Transfer the embryos onto separate Nitex membranes as in step 5, fold the membrane over to cover the embryos, and blot them dry with a paper towel.
10. Transfer a small number of embryos (100–200) from any of the membranes into a microfuge tube containing 0.5 mL of heptane and 0.5 mL of methanol. Shake vigorously to devitellinize the embryos, let them settle and then remove as much liquid as possible. Wash the embryos with methanol twice and store them at −20°C. This sample from the collection is set aside to evaluate if the collected embryos are at the correct developmental stage (Subheading 3.2).
11. Transfer the remaining aliquots of dry cross-linked embryos from the Nitex membranes into separate 12-mL cryotubes and freeze them in liquid nitrogen. Cross-linked embryos can be stored at −80°C for at least 1 year.

3.2. Evaluation of Embryos Developmental Stages Distribution Within a Collection

1. For staging purposes, rehydrate the sample set aside in step 10 in Subheading 3.1 by rinsing the embryos once in 1 mL of each of the following solutions (v/v): 70% of methanol/30% of PBT, 50% of methanol/50% of PBT, and 30% of methanol/70% of PBT.
2. Finally, wash the embryos once in 1 mL of PBT for 5 min.

3. Replace the supernatant with 85% of glycerol and mount the embryos on a microscope slide.
4. The developmental stages present in the collection can now be examined microscopically using the morphological features defined by Campos-Ortega and Hartenstein (15). Compare the distribution of developmental stages between repeated collections and exclude those containing inappropriate stages from further analysis, if necessary.

3.3. Chromatin Preparation

1. Thaw embryos quickly at room temperature and resuspend them in 15 mL of ice-cold PBT solution containing protease inhibitors and PMSF. Transfer the suspension to a 15-mL Dounce homogenizer.
2. Homogenize each 1.5 g aliquot of embryos in a 15-mL Dounce homogenizer on ice by applying 20 strokes with the loose-fitting pestle.
3. Transfer the lysate to a 50-mL Falcon tube and centrifuge at $400\times g$ at 4°C for 1 min to precipitate the vitelline membranes and large debris.
4. Decant the supernatant into a fresh 50-mL Falcon tube and centrifuge at $1,100\times g$ at 4°C for 10 min. Decant the supernatant and discard it.
5. Resuspend the cell pellet in 15 mL ice-cold cell lysis buffer containing protease inhibitors and PMSF.
6. Homogenize the cells in a 15-mL Dounce homogenizer on ice by applying 20 strokes with the tight-fitting pestle. Split the sample into two approximately 8 mL aliquots in two separate 15-mL Falcon tubes.
7. Centrifuge the samples at $2,000\times g$ at 4°C for 4 min to pellet the nuclei. Discard the supernatant (see Note 9).
8. Resuspend each pellet in 1 mL of ice-cold nuclear lysis buffer containing protease inhibitors and PMSF and incubate at room temperature for 20 min.
9. Add 1 mL of ice-cold nuclear lysis buffer containing protease inhibitors and PMSF to each sample and sonicate using a Bioruptor sonicator (15 cycles 30 s on/30 s off, high-energy settings) (see Note 10).
10. Transfer the chromatin to 1.5-mL Eppendorf tubes and centrifuge at $20,000\times g$ at 4°C for 10 min.
11. Pool the supernatants to ensure a homogenous sample, keep 50 μL for quality assessment, and freeze the remaining chromatin in 200 μL aliquots in liquid nitrogen. Chromatin can be stored at −80°C for at least 1 year.
12. To determine the yield and average fragment length of the chromatin preparations, dilute the 50 μl of chromatin set aside

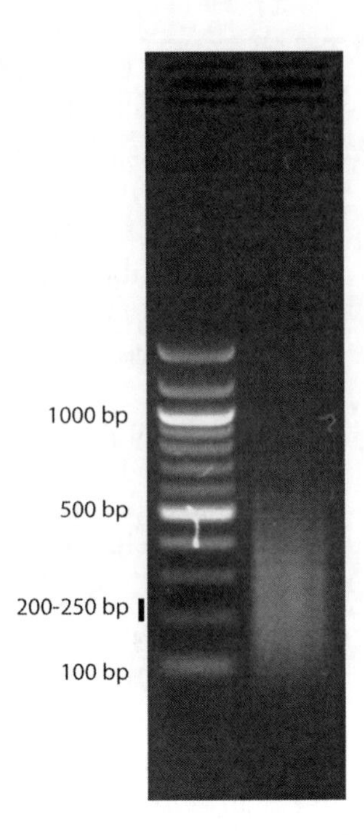

Fig. 1. Verification of sonicated fragments length. A 50 μL aliquot of chromatin is treated with RNAse A and proteinase K, reverse cross-linked, purified by phenol–chloroform extraction, and ethanol-precipitated. The average length of sonicated chromatin is assessed by electrophoresis on a 1.5% TAE agarose gel. The sonication settings should produce fragments with an average size of 200–250 bp.

in step 11 with 50 μL of TE buffer. Add 50 μg/mL of RNase A and incubate at 37°C for 30 min. Treat the sample as described in steps 9–13 in Subheading 3.4 and resuspend the purified DNA in 30 μL of TE buffer. Determine the concentration of sheared DNA using a NanoDrop ND-1000 Spectrophotometer and verify its size distribution by gel electrophoresis using a 1.5% agarose gel (see Note 11). The Bioruptor settings used in this protocol give rise to an average fragment length of approximately 200–250 bp. An example is shown in Fig. 1.

3.4. Chromatin Immunoprecipitation

1. For each immunoprecipitation, wash 25 μl of 50% Protein A-Sepharose suspension with 1 mL of RIPA-BSA buffer for 10 min on a rotating wheel at 4°C. Pellet the beads by centrifugation at 1,000 × *g* for 2 min and incubate in 1 mL of RIPA-BSA buffer on a rotating wheel at 4°C overnight (see Notes 12 and 13).
2. Thaw an aliquot of chromatin on ice. Transfer 30–50 μg of chromatin to a 1.6-mL low-binding reaction tube and adjust the final volume to 500 μL with ice-cold TE buffer. Add 400 μL of ice-cold IP dilution buffer containing protease

inhibitors and PMSF. Retain 10 μL of the sample (1% input) in a separate tube and store at 4°C until step 8.

3. Add 3–20 μl of serum or a suitable amount of purified antibody to each tube. Incubate at 4°C overnight on a rotating wheel (see Note 14).
4. Centrifuge the preblocked Protein A-Sepharose beads at 1,000 × *g* for 2 min, discard the supernatant and resuspend the beads in 100 μl of RIPA buffer per reaction.
5. Add 100 μl of bead suspension to each chromatin sample from step 3 and incubate at 4°C on a rotating wheel for 3 h.
6. To purify the antigen–antibody complexes, pellet the beads by centrifugation at 1,000 × *g* for 2 min, discard the supernatant, and rinse the beads once with 1 mL of ice-cold RIPA buffer.
7. Pellet the antigen–antibody complexes again by centrifugation at 1,000 × *g* for 2 min and wash them with 1 mL of each of the following buffers at 4°C on a rotating wheel for 10 min: 1× with RIPA buffer, 4× with RIPA500 buffer, 1× with LiCl buffer, and 2× with TE buffer.
8. Resuspend the beads in 100 μL of TE buffer supplemented with 50 mg/mL of RNase A and incubate at 37°C for 30 min. From this point on, include also the 1% input sample retained in step 2. Add 90 μL of TE buffer to yield a final volume of 100 μL before RNase A addition.
9. Add SDS to a final concentration of 0.5% (w/v) from a 10% stock and incubate with 0.5 mg/mL of proteinase K at 37°C overnight.
10. Transfer the samples to 65°C for at least 6 h to reverse the cross-links.
11. Adjust the samples to 200 μL with TE buffer. Extract the DNA by combining the sample with 200 μL of phenol–chloroform–isoamylalcohol in a prespun phase-lock tube. Mix briefly and centrifuge at 15,000 × *g* at room temperature for 5 min. Add 200 μL of chloroform, mix briefly, and centrifuge again at 15,000 × *g* at room temperature for 5 min. Transfer the aqueous sample to a fresh 1.6-mL low-binding reaction tube.
12. Supplement the samples with 0.25 mg/mL of glycogen, add 20 μL of 3 M sodium acetate solution, pH 5.2, and 550 μL of ethanol, vortex briefly, and incubate the sample at −80°C for at least 1 h.
13. Centrifuge the sample at 4°C at 15,000 × *g* for 30 min to precipitate the DNA, wash the pellet once with 1 mL of 70% ethanol, and centrifuge again at 15,000 × *g* at 4°C for 10 min.
14. Resuspend the purified DNA in 30 μL of TE buffer.

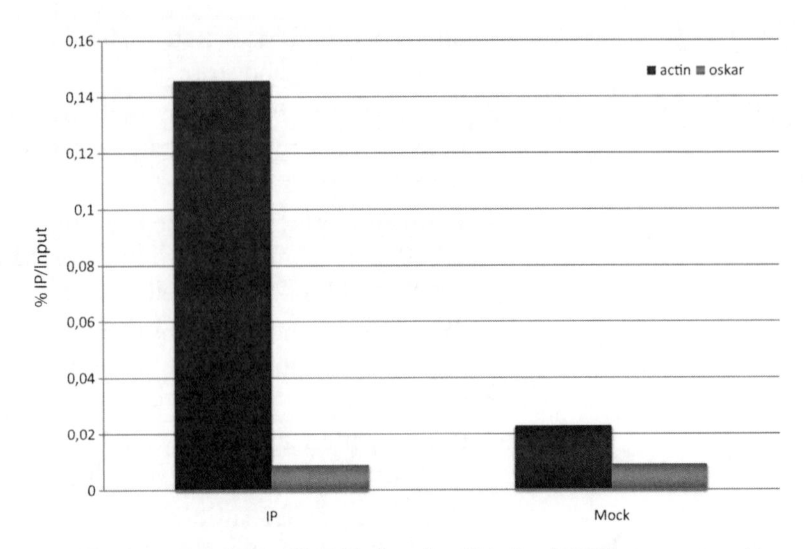

Fig. 2. Enrichment of a known Mef2-binding site within the *Act57B* locus compared to a mock control. A chromatin imunoprecipitation experiment was realized using 3 μL of anti-Mef2 antiserum (IP) or 3 μL of preimmune serum (Mock), and assayed by quantitative real-time PCR. A specific enrichment can be observed within the Act57B enhancer (actin), a known binding site of Mef2, compared to a negative control region (oskar). Note that enrichments of 0.01–3% are typical for tissue-specific transcription factors, while enrichments of 10–50% can be obtained for chromatin remodeling factors or histone methylation marks.

15. If sequence-specific binding sites for the protein of interest are known, evaluate the efficiency of enrichment by quantitative real-time PCR (see Note 15). An example of result obtained by immunoprecipitation of the Mef2 transcription factor is shown in Fig. 2.
16. Determine the concentration of ChIP DNA using a Qubit fluorometer and a Quant-iT dsDNA HS Assay Kit following the manufacturer's instructions.

3.5. Library Preparation

1. Dilute Klenow DNA polymerase (Large Fragment) 1:5 with deionized water for a final Klenow concentration of 1 U/μL.
2. Combine and mix the following components in a 0.5 mL PCR tube: 2–5 ng of ChIP (or input) DNA to be end-repaired (see Notes 16 and 17), 5 μL of 10× phosphorylation buffer, 2 μL of 10 mM dNTP mix, 1 μL of T4 DNA Polymerase, 1 μL of Klenow DNA Polymerase (LF), and 1 μL of T4 polynucleotide kinase. Adjust final volume to 50 μL with water.
3. Incubate the reaction for 30 min at 20°C in a thermal cycler.
4. Purify the reaction using a QIAquick PCR purification kit. Elute the end-repaired DNA in 34 μL of EB buffer into a 1.6-mL low-binding reaction tube (see Note 18).

5. Add and mix the following components to the end-repaired DNA: 5 μL of 10× NEB buffer 2, 10 μL of 1 mM dATP, 1 μL of Klenow exo-.
6. Incubate the reaction for 30 min at 37°C.
7. Purify the reaction using a PCR purification kit and MinElute columns. Elute the A-tailed DNA in 10 μL of EB buffer into a 1.6-mL low-binding reaction tube (see Note 18).
8. Dilute the Illumina Paired-End Adapter oligo mix 1:10 with water (see Note 19).
9. Add and mix the following components to the A-tailed DNA: 15 μL of 2× quick ligation buffer, 1 μL of diluted adapters or 1 μL of 2 μM multiplexing oligonucleotide adapters, and 1 μL of T4 quick ligase.
10. Incubate the reaction for 15 min at room temperature.
11. Purify the reaction using a PCR purification kit and MinElute columns. Elute the adapter-ligated DNA in 16 μL of EB buffer into a 1.6-mL low-binding reaction tube (see Note 18).
12. Prepare a 2% agarose TAE gel prestained with GelGreen.
13. Mix the adapter-ligated DNA sample with 4 μL of 6× orange G loading dye.
14. Load and run the gel at 80 V for 15 min, then at 110 V for approximately 2 h. Include a 100-bp DNA ladder to two lanes of the gel (see Note 20).
15. View the gel on a Safe Imager transilluminator to avoid being exposed to UV light. Extract gel slice with a Gel Excision Tip at appropriate size (see Note 21).
16. Purify the DNA from the agarose slice using a gel extraction kit and QIAquick columns (see Note 22). Elute in 36 μL of EB buffer into a 1.6-mL low-binding reaction tube (see Note 18).
17. Combine and mix the following components in a 0.5-mL PCR tube: 36 μL of size-selected DNA, 10 μL of 5× Phusion HF buffer, 1.5 μL of 10 mM dNTP mix, 1 μL of Paired-End PCR primers 1.0, 1 μL of Paired-End PCR primers 2.0, and 0.5 μL of Phusion polymerase.
18. Amplify using the following PCR protocol:

step 1	40 s at 98°C
step 2	10 s at 98°C
step 3	30 s at 65°C
step 4	30 s at 72°C
step 5	GO TO step 2 for 17–20 times (see Note 23)
step 6	5 min at 72°C
step 7	Hold at 4°C

19. Prepare a tick 2% agarose TAE gel prestained with GelGreen.
20. Mix the amplified DNA sample with 10 μL of 6× orange G loading dye.
21. Load and run the gel at 80 V for 15 min, then at 110 V for approximately 2 h. Include a 100 bp DNA ladder to two lanes of the gel (see Note 20).
22. View the gel on a Safe Imager transilluminator to avoid being exposed to UV light. Extract gel slice with a scalpel (see Note 24).
23. Purify the DNA from the agarose slice using a MinElute gel extraction kit (see Note 22). Elute in 15 μL of EB buffer into a 1.6-mL low-binding reaction tube (see Note 18).
24. Determine the concentration of the library using a Qubit fluorometer and a Quant-iT dsDNA HS Assay Kit following the manufacturer's instructions.
25. Verify the size, purity, and concentration of the library using an Agilent 2100 Bioanalyzer. Depending on the concentration of the library (as determined in step 24), use either an Agilent High Sensitivity DNA kit (5–500 pg/μL) or an Agilent DNA 1000 kit (0.1–50 ng/μL) following manufacturer's instructions (Fig. 3).
26. If sequence-specific binding sites for the protein of interest are known, verify that the library has maintained a specific enrichment by quantitative real-time PCR (i.e., repeat step 15 in Subheading 3.4).
27. If using multiplexing oligonucleotide adapters, prepare an equimolar mix of the different multiplexed libraries and repeat step 25 again (see Note 25).
28. The library is now ready for sequencing on the Illumina GAIIx platform. Follow instructions from your sequencing facility for sample submission.

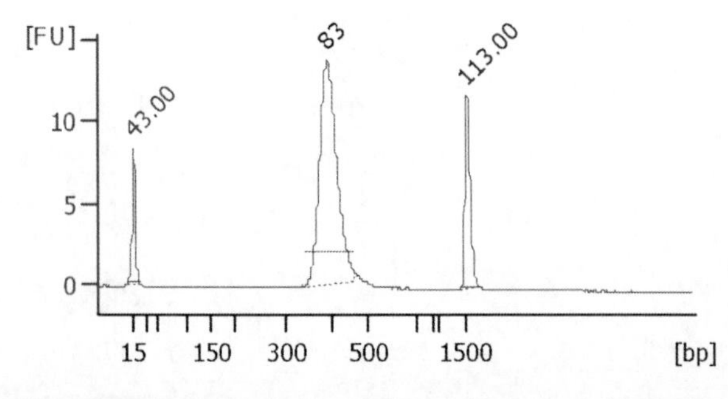

Fig. 3. Agilent 2100 Bioanalyzer analysis of a successful library before Illumina sequencing. A successful library is characterized by a single intense peak with an expected size between 350 and 400 bp.

29. Approximately 20 million uniquely mapped reads (corresponds to one lane of a flow cell using the Illumina GAIIx) should be sufficient to identify all the binding sites of a given transcription factor. For organisms with a smaller genome, this is likely to be a huge excess; therefore, multiplexing of different transcription factor libraries will significantly reduce costs.

4. Notes

1. For staged embryo collections, it is advisable to perform three embryo "pre-lays" of 1 h each before taking any embryos for experiments.
2. Depending on the amount of embryos to process, we start collecting the embryos from 15 to 20 min before they reach the desired time point, which is approximately the required time to reach the fixation step (step 6).
3. The sequences of the adapters and primers are available from Illumina upon request.
4. Paired-end oligos generate libraries compatible with both single and paired-end flow cells. We, thus, always use the Paired-End Adapter oligo mix, even though we do not use the paired-end option for classical ChIP-seq libraries.
5. Multiplexing oligonucleotide adapters can be purchased from Illumina; however, we routinely use our own adapters. We purchase single-stranded HPLC purified oligonucleotides from Sigma. The 3′ thymine overhang is protected from digestion with a phosphorothioate and the 5′ extremity of the nonoverhanging oligonucleotide is phosphorylated. The oligonucleotides are annealed as described in (16). Briefly, the oligonucleotides are resupended and mixed to a final concentration of 50 μM in annealing buffer (10 mM Tris–HCl pH 8.0, 50 mM NaCl, and 0.1 mM EDTA) and then incubated in a thermal cycler using the following protocol:

   2 min at 95°C

   Ramp from 95°C to 75°C at 0.1°C/s, hold 2 min at 75°C

   Ramp from 75°C to 65°C at 0.1°C/s, hold 2 min at 65°C

   Ramp from 65°C to 50°C at 0.1°C/s, hold 2 min at 50°C

   Ramp from 50°C to 37°C at 0.1°C/s, hold 2 min at 37°C

   Ramp from 37°C to 20°C at 0.1°C/s, hold 2 min at 20°C

   Ramp from 20°C to 4°C at 0.1°C/s, hold at 4°C

   The multiplexing oligonucleotide adapters are further diluted to 0.2 μM in EB buffer and stored at −20°C.

6. Paired-End PCR primers 1.0 and 2.0 can be purchased from Illumina; however, we routinely use our own primers. HPLC-purified primers are purchased from MWG, diluted to 10 μM in EB buffer, and stored at –20°C.
7. Wash embryos until bleach smell is gone. Removal of the chorion will cause the embryos to float and clump together.
8. This step cross-links proteins to chromatin as well as to other proteins. The time required for this step should be kept constant between repeated collections. Importantly, as some proteins are more easily cross-linked to chromatin than others, the formaldehyde concentration/length of cross-linking reaction might require optimization for different proteins of interest. It is important not to over-cross-link, as this will result in what looks like spreading of the ChIP signal from the actual site of binding.
9. Steps 5–7 allow isolating the nuclei, which increases the signal-to-noise ratio.
10. Wear ear protection. We routinely sonicate in 15-mL polystyrene Falcon tubes. However, 1.5-mL and 0.5-mL tubes may also be used by adjusting the sample volume and the number of cycles. Sonication conditions should be optimized for each sonicator and may vary depending on the extent of cross-linking and sample type. Ensure that the water bath remains cold by changing water (always keep several bottles of water in the cold room for sonication purpose) every five sonication cycles, or use a cooling system.
11. Load different amounts of reverse cross-linked DNA (we routinely run between 200 and 600 ng of DNA) to assess the size range accurately. Migration of too large quantities of DNA on agarose gel might not reflect the real DNA fragmentation distribution.
12. Depending on the species in which the antibodies have been raised, Protein A or Protein G-Sepharose beads can be used. Always use a cutoff pipette tip when pipetting Sepharose beads. With appropriate protocol modifications, magnetic beads may be substituted for Sepharose beads.
13. Always include a mock control. The mock control is done by performing all the steps of the chromatin immunoprecipitation protocol but using the equivalent amount of preimmune serum instead of antibody.
14. The optimal amount of serum/antibody is difficult to predict, as it depends on the abundance of the protein of interest, the concentration of specific IgGs in the serum, etc. We recommend to test a range of different serum/antibody amounts (e.g., 3, 5, and 10 μL) while keeping the amount of chromatin and beads constant, and to monitor the enrichment of a known binding site using quantitative real-time PCR.

15. If high background is observed, additional washes may be needed. Alternatively, a preclearing step may be added by incubating the chromatin with Protein A/G-Sepharose beads for 1 h prior to step 3. Any nonspecific binding of chromatin to Protein A/G-Sepharose beads will be removed during this additional step. Transfer the supernatant to a new 1.6-mL low-binding tube, retain 10 μL of the sample as 1% input control, and add the appropriate amount of antibody as described in step 3.
16. Although the library preparation could be performed with as low as 1 ng of ChIP DNA, we recommend using 3–5 ng, if possible, to avoid the requirement of too many amplification cycles during step 18. If the amount of ChIP DNA obtained from one immunoprecipitation is too low, up to three replicates can be pooled together.
17. We routinely sequence at least two immunoprecipitation replicates (independent chromatin preparation, independent immunoprecipitations with different antibodies if available, and independent library preparation) as well as one input library for each condition to be analyzed.
18. Incubate the column for 3 min at room temperature after addition of EB buffer to increase the yield.
19. The amount of adapters to be used may have to be titrated relative to starting material. If starting from less than 2 ng of ChIP DNA, consider further diluting the adapters, as the presence of nonligated adapters will promote the formation of adapter dimers during the amplification step. Adapter dimers will migrate at approximately 120 bp and, if present in the final library, yield useless reads.
20. If making multiple libraries, be very careful to avoid cross-contamination either by leaving many empty lanes between samples or by using one gel per sample.
21. At this step, the adapter-ligated DNA is not visible on the gel. Excise the gel slice in the 250–300 bp range (expected size after ligation of the 33 bp adapter to each side of the A-tailed DNA).
22. Melt the gel slice at room temperature, as it has been shown that heating to 50°C in chaotropic buffer induces an under-representation of A+T rich sequences (17).
23. We recommend keeping the number of amplification cycles as low as possible to avoid amplification artifacts. Too many amplification cycles will result in a low complexity library. The complexity of a library can be defined as the number of independent DNA molecules that compose it. This complexity should be significantly higher than the total number of sequenced reads, as a low complexity library will increase the

likelihood of generating duplicated reads. However, when starting with very low amount of ChIP DNA, up to 21 cycles of amplification might be needed to obtain enough material for Illumina cluster generation.

24. Gel purification removes the 120-bp adapter dimer.

25. Depending on the desired sequencing coverage, up to four libraries can be multiplexed. The coverage required to identify all the binding sites of a given transcription factor depends on the transcription factor itself and on the size of the genome to be analyzed. It can thus not be predicted in advance. A saturation analysis should be performed, a posteriori, to verify whether the sequencing depth was sufficient enough to identify all the binding sites of the transcription factor of interest. Briefly, a saturation analysis consists in sampling the data and analyzing how the number of predicted binding sites changes when only a subset of the data is used for prediction. By sampling increasing fractions of the data, the number of identified binding sites should reach a plateau corresponding to the number of binding sites identified from the complete dataset. See also (18) for more information on sequencing depth.

References

1. Bonn, S., and Furlong, E. E. (2008) cis-Regulatory networks during development: a view of Drosophila *Curr Opin Genet Dev* **18**, 513–20.
2. Davidson, E. H. (2006) The Regulatory Genome: Gene Regulatory Networks In Development And Evolution *Academic Press.*
3. Sandmann, T., Jensen, L. J., Jakobsen, J. S., Karzynski, M. M., Eichenlaub, M. P., Bork, P., and Furlong, E. E. (2006) A temporal map of transcription factor activity: mef2 directly regulates target genes at all stages of muscle development *Dev Cell* **10**, 797–807.
4. Sandmann, T., Girardot, C., Brehme, M., Tongprasit, W., Stolc, V., and Furlong, E. E. (2007) *Genes Dev* **21**, 436–49.
5. Li, X. Y., MacArthur, S., Bourgon, R., Nix, D., Pollard, D. A., Iyer, V. N., Hechmer, A., Simirenko, L., Stapleton, M., Luengo Hendriks, C. L., Chu, H. C., Ogawa, N., Inwood, W., Sementchenko, V., Beaton, A., Weiszmann, R., Celniker, S. E., Knowles, D. W., Gingeras, T., Speed, T. P., Eisen, M. B., and Biggin, M. D. (2008) Transcription factors bind thousands of active and inactive regions in the Drosophila blastoderm *PLoS Biol* **6**, e27.
6. Liu, Y. H., Jakobsen, J. S., Valentin, G., Amarantos, I., Gilmour, D. T., and Furlong, E. E. (2009) A systematic analysis of Tinman function reveals Eya and JAK-STAT signaling as essential regulators of muscle development *Dev Cell* **16**, 280–91.
7. Zeitlinger, J., Zinzen, R. P., Stark, A., Kellis, M., Zhang, H., Young, R. A., and Levine, M. (2007) Whole-genome ChIP-chip analysis of Dorsal, Twist, and Snail suggests integration of diverse patterning processes in the Drosophila embryo *Genes Dev* **21**, 385–90.
8. Jakobsen, J. S., Braun, M., Astorga, J., Gustafson, E. H., Sandmann, T., Karzynski, M., Carlsson, P., and Furlong, E. E. (2007) Temporal ChIP-on-chip reveals Biniou as a universal regulator of the visceral muscle transcriptional network *Genes Dev* **21**, 2448–60.
9. MacArthur, S., Li, X. Y., Li, J., Brown, J. B., Chu, H. C., Zeng, L., Grondona, B. P., Hechmer, A., Simirenko, L., Keranen, S. V., Knowles, D. W., Stapleton, M., Bickel, P., Biggin, M. D., and Eisen, M. B. (2009) Developmental roles of 21 Drosophila transcription factors are determined by quantitative differences in binding to an overlapping set of thousands of genomic regions *Genome Biol* **10**, R80.

10. Zinzen, R. P., Girardot, C., Gagneur, J., Braun, M., and Furlong, E. E. (2009) Combinatorial binding predicts spatio-temporal cis-regulatory activity *Nature* **462**, 65–70.
11. Aparicio, O., Geisberg, J. V., and Struhl, K. (2004) Chromatin immunoprecipitation for determining the association of proteins with specific genomic sequences in vivo *Curr Protoc Cell Biol* **17**, Unit 17 7.
12. Sandmann, T., Jakobsen, J. S., and Furlong, E. E. (2006) ChIP-on-chip protocol for genome-wide analysis of transcription factor binding in Drosophila melanogaster embryos *Nat Protoc* **1**, 2839–55.
13. Johnson, D. S., Mortazavi, A., Myers, R. M., and Wold, B. (2007) Genome-wide mapping of in vivo protein-DNA interactions *Science* **316**, 1497–502.
14. Sisson, J. C. (2000) Culturing Large Populations Of *Drosophila* for Protein Biochemistry *Drosophila Protocols*.
15. Campos-Ortega, J. A., and Hartenstein, V. (1985) The Embryonic Development of *Drosophila melanogaster Springer*.
16. Ng, P., Wei, C. L., and Ruan, Y. (2007) Paired-end diTagging for transcriptome and genome analysis *Curr Protoc Mol Biol* **21**, Unit 21 12.
17. Quail, M. A., Kozarewa, I., Smith, F., Scally, A., Stephens, P. J., Durbin, R., Swerdlow, H., and Turner, D. J. (2008) A large genome center's improvements to the Illumina sequencing system *Nat Methods* **5**, 1005–10.
18. Park, P. J. (2009) ChIP-seq: advantages and challenges of a maturing technology *Nat Rev Genet* **10**, 669–80.

Chapter 15

Genome-Wide Profiling of DNA-Binding Proteins Using Barcode-Based Multiplex Solexa Sequencing

Sunil Kumar Raghav and Bart Deplancke

Abstract

Chromatin immunoprecipitation (ChIP) is a commonly used technique to detect the in vivo binding of proteins to DNA. ChIP is now routinely paired to microarray analysis (ChIP-chip) or next-generation sequencing (ChIP-Seq) to profile the DNA occupancy of proteins of interest on a genome-wide level. Because ChIP-chip introduces several biases, most notably due to the use of a fixed number of probes, ChIP-Seq has quickly become the method of choice as, depending on the sequencing depth, it is more sensitive, quantitative, and provides a greater binding site location resolution. With the ever increasing number of reads that can be generated per sequencing run, it has now become possible to analyze several samples simultaneously while maintaining sufficient sequence coverage, thus significantly reducing the cost per ChIP-Seq experiment. In this chapter, we provide a step-by-step guide on how to perform multiplexed ChIP-Seq analyses. As a proof-of-concept, we focus on the genome-wide profiling of RNA Polymerase II as measuring its DNA occupancy at different stages of any biological process can provide insights into the gene regulatory mechanisms involved. However, the protocol can also be used to perform multiplexed ChIP-Seq analyses of other DNA-binding proteins such as chromatin modifiers and transcription factors.

Key words: RNA polymerase II, Barcode multiplexing, ChIP-sequencing, Protein–DNA interaction, ChIP-chip

1. Introduction

Spatiotemporal gene expression is controlled by gene regulatory networks (1). These networks are composed of genomic components such as genes and their respective regulatory elements and regulatory state components including transcriptional complexes which bind the latter elements, and as such can activate or repress gene expression (2). Thus, to understand how differential gene expression programs underlying a specific biological process are enacted, we need to identify for each implicated gene which regulatory state components bind to which gene-specific

Bart Deplancke and Nele Gheldof (eds.), *Gene Regulatory Networks: Methods and Protocols*,
Methods in Molecular Biology, vol. 786, DOI 10.1007/978-1-61779-292-2_15, © Springer Science+Business Media, LLC 2012

genomic components at a certain time and space. There are two complementary approaches to map such protein–DNA interactions: the first is DNA-centered and inquires which proteins can bind to a specific regulatory element of interest (3); the second is protein-centered and reveals where in the genome a specific protein of interest binds (2). A prominent example of the latter is chromatin immunoprecipitation (ChIP) which, when coupled to microarray analysis (ChIP-chip) (4) or next-generation sequencing (ChIP-Seq) (5), enables the genome-wide profiling of DNA-binding proteins.

ChIP-Seq has quickly become the preferred method as depending on the sequence coverage, it tends to be more comprehensive, less costly, more quantitative, and provides higher binding site resolution than ChIP-chip (2, 5). However, while the final read-out has improved significantly due to sequencing advances, the actual ChIP procedure is still cumbersome and success of the assay depends heavily on multiple factors including the quality of the primary antibody and how well the chromatin can be isolated and fragmented. It has therefore been recommended to re-optimize the ChIP protocol for each new ChIP antibody, specifically focusing on three critical steps (6). The first is the fixation time during which protein–DNA interactions are crosslinked typically using formaldehyde. This time should be kept short as over-crosslinking may reduce the accessibility of protein epitopes, increase the number of false-positive interactions, and reduce the chromatin fragmentation efficiency. The latter is important since the resolution of ChIP data correlates inversely with the final length of chromatin fragments. However, when dealing with proteins that only bind weakly to DNA or indirectly such as co-factors (2), longer fixation times are required. The second step is the amount of input material, which typically inversely correlates with the expression level of the protein of interest. Too much material may increase the background and obscure relevant protein–DNA interactions, while too little material may limit the ability to detect any interaction. The third step is the amount of antibody. Since the titer and specificity of antibody aliquots tends to be quite variable, the most optimal antibody amount needs to be carefully titrated using positive control protein–DNA interactions. However, the latter are not always available, especially when dealing with a novel DNA interacting protein, making the optimization of the ChIP procedure a lengthy if not impossible undertaking. With the advances in high-throughput sequencing, these difficulties can now be easily circumvented by preparing ChIP samples based on distinct protocols and analyzing which protocol is most optimal based on the best genome-wide enrichment of protein-specific binding sites compared to the input. To reduce sequencing costs, these distinct samples can be sequenced simultaneously using the barcode-based multiplexing sequencing procedure described in this chapter. Thus, not only can multiplexing be effective to analyze distinct biological samples, it is also useful to evaluate the efficiency of a ChIP protocol in unbiased fashion.

As indicated, this chapter provides a detailed protocol on how to perform genome-wide profiling of DNA-binding proteins using barcode-based multiplex Solexa sequencing. As a proof-of-concept, we multiplex ChIP samples from a genome-wide DNA occupancy assay of RNA polymerase II (RNAPII) at distinct time points during 3T3-L1 preadipocyte differentiation with the aim of understanding how the transcriptional activity of genes changes during this process as described in ref. 7. An outline of this procedure is shown in Fig. 1. It is clear however that the same procedure can be used for other DNA-binding proteins of interest.

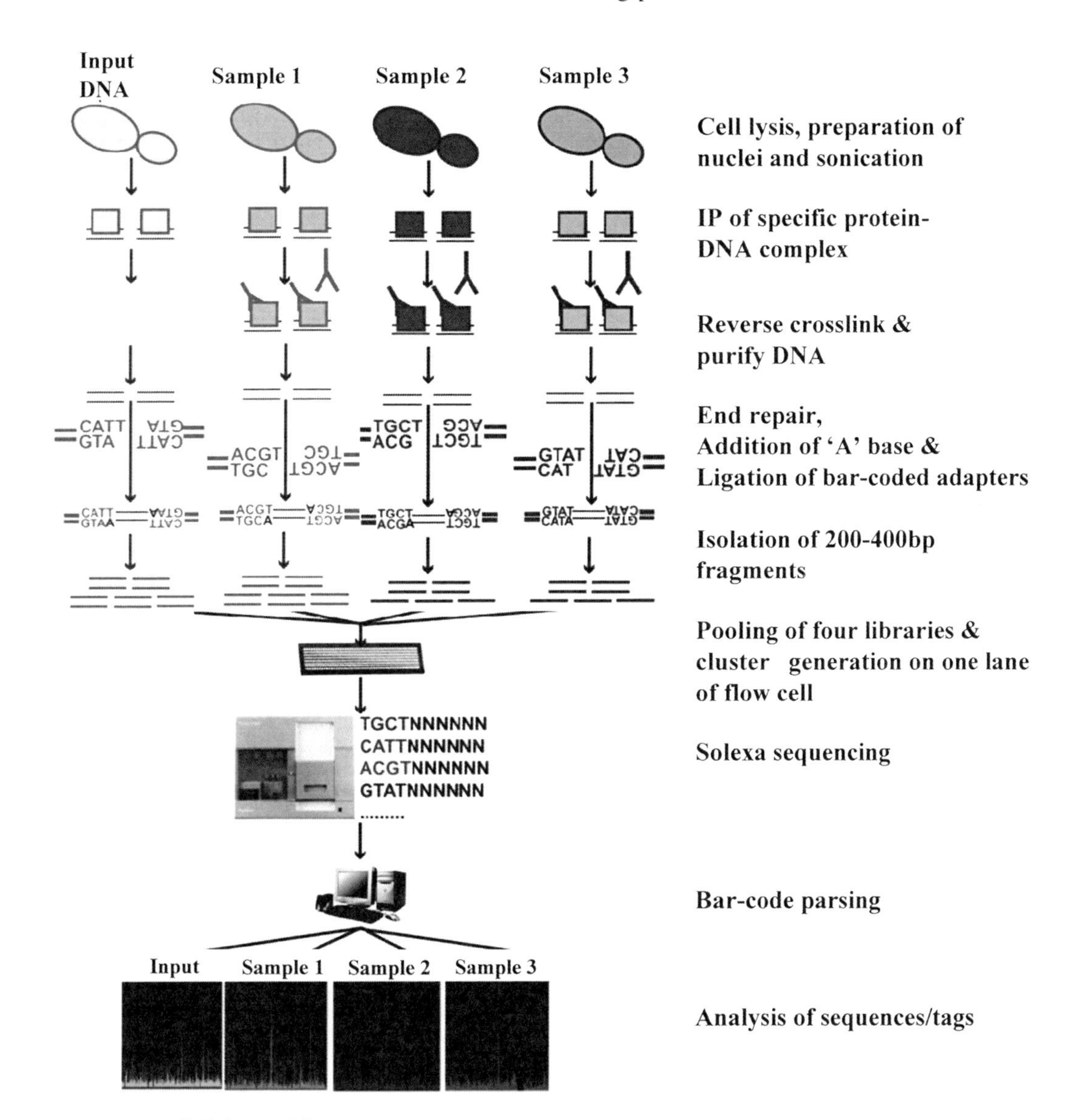

Fig. 1. Multiplex ChIP-Seq workflow.

2. Materials

2.1. Cell Culture and Differentiation of 3T3-L1 Preadipocytes

1. Dulbecco's modified Eagle's medium (DMEM).
2. Fetal calf serum (FCS).
3. Antibiotic solution with glutamine:penicillin/streptomycin (100×) with l-glutamine (200 mM).
4. Solution of trypsin (0.25%) with ethylenediamine tetra-acetic acid (EDTA, 1 mM).
5. Iso-butyl-methyl-xanthine (IBMX): (a) Prepare IBMX stock solution by dissolving 0.0115 g/mL (w/v) IBMX in 0.5 N KOH, (b) Filter sterilize through a 0.22 mm syringe filter (make fresh).
6. Dexamethasone stock solution: 10 mM of dexamethasone in 100% ethanol (store at –20°C). Dilute stock to 1 mM using 1× PBS (tissue culture grade) prior to use.
7. Insulin stock solution (Bovine): 167 μM in 0.02 M HCl. Filter sterilize through 0.22 mm filter and store at –20°C in small aliquots.
8. Differentiation medium: DMEM supplemented with 10% FCS and 1× antibiotic solution with glutamine, –1:100 IBMX stock, 1:1,000 dexamethasone diluted stock, 1:1,000 insulin working stock.
9. Insulin medium: 1:1,000 Insulin working stock in DMEM supplemented with 10% FCS and 1× antibiotic solution with glutamine.
10. Cell scrapers.

2.2. Preparation of Cells for ChIP

1. Formaldehyde solution (1×): 1% formaldehyde (36.5%) in 1× PBS.
2. Quenching solution: 2.5 M glycine.
3. RNAPII antibody (Cat. No. sc-9001, Santa Cruz Biotechnology, Santa Cruz, CA).
4. Normal rabbit IgG (Cat. No. 12-370, Millipore, USA).

2.3. Preparation of Protein-A Sepharose Beads for ChIP

1. Recombinant Protein-A Sepharose (Cat. No. -17-5138-6, GE Healthcare).
2. Fragmented Salmon sperm DNA, 10 mg/mL (Sigma, USA).
3. Bovine serum albumin, 50 mg/mL (BSA).
4. ChIP dilution buffer: 1% Triton X-100, 1.2 mM EDTA, 16.7 mM Tris–HCl, pH 8.0, 167 mM NaCl.

2.4. Lysis of the Cells and Preparation of the Chromatin for IP

1. Nuclei extraction buffer (Buffer LI): 50 mM HEPES–NaOH, pH 7.5, 140 mM NaCl, 1 mM EDTA, pH 8.0, 10% glycerol, 0.5% NP-40, 0.25% Triton X-100, Complete Protease

Inhibitors (Roche), Phosphatase Inhibitors (5 mM NaF, 1 mM β-glycerol phosphate, and 1 mM sodium orthovanadate).

2. Protein extraction buffer (Buffer LII): 200 mM NaCl, 1 mM EDTA, pH 8.0, 0.5 mM EGTA, pH 8.0, 10 mM Tris–HCl, pH 8.0, Complete Protease Inhibitors (Roche), Phosphatase Inhibitors (5 mM NaF, 1 mM β-glycerol phosphate, and 1 mM sodium orthovanadate).
3. Chromatin extraction buffer (Buffer LIII): 1 mM EDTA, pH 8.0, 0.5 mM EGTA, pH 8.0, 10 mM Tris–HCl, pH 8.0, 1% Triton X-100, Complete Protease Inhibitor (Roche), Phosphatase Inhibitor (Roche).
4. 50× Protease inhibitor tablets-EDTA free (Roche).
5. 10× Protease inhibitor tablets-EDTA free (Roche).
6. 10× Phosphatase inhibitor tablets-EDTA free (Roche).
7. Sodium fluoride (IM).
8. Sodium orthovanadate (IM).
9. Beta glycerol phosphate (IM).

2.5. Washing of Beads, Elution, and Reversal of the Chromatin Crosslinks

1. Low-salt immune complex wash buffer: 0.1% sodium dodecyl sulfate (SDS), 1% Triton X-100, 2 mM EDTA, 20 mM Tris–HCl, pH 8.0, 150 mM NaCl.
2. High-salt immune complex wash buffer: 0.1% SDS, 1% Triton X-100, 2 mM EDTA, 20 mM Tris–HCl, pH 8.0, 500 mM NaCl.
3. LiCl immune complex wash buffer: 0.25 M LiCl, 1% NP40, 1% deoxycholate, 1 mM EDTA, 10 mM Tris–HCl, pH 8.0.
4. TE buffer: 10 mM Tris–Cl, pH 8.0, and 1 mM EDTA, pH 8.0.
5. Elution buffer: 1% SDS, 100 mM sodium bicarbonate.
6. 5 M NaCl.

2.6. RNase and Proteinase-K Treatment and Purification of ChIP-DNA

1. DNase-free RNase-A (10 mg/mL).
2. Proteinase-K (20 mg/mL).
3. Qiagen PCR purification kit.

2.7. Verification of Enrichment in ChIP-DNA Before Library Preparation for Solexa Sequencing

1. Real-time PCR instrument (e.g., ABI 7900).
2. Oligonucleotide primers to genomic regions of interest.
3. Sybr-Green PCR master mix (Applied Biosystems, Foster City, CA).
4. Optical PCR plates and adhesive covers compatible with the real-time PCR instrument.

2.8. Preparation of DNA Library for Solexa Sequencing

1. End-it DNA End-repair kit (Epicentre Technologies, Madison, WI).
2. Klenow (3′–5′ exo-) (5 U/μL, NEB).
3. SYBR Gold 10,000× (Invitrogen).
4. Dark Reader transilluminator.
5. Quick ligation kit (New England Biolabs, Ipswich, MA).
6. HPLC purified barcoded adapter oligos: to maintain equal representation of all four nucleotides, the number of barcodes per sequencing run needs to be a multiple of four whereby each barcode should start with a different nucleotide (underlined):

Oligo name	Barcode	Modification	Sequence (5′–3′)
Seq_1F	GTAT	None	ACACTCTTTCCCTAC ACGACGCTCTTCC GATCT**GTAT**
Seq_1R	GTAT	5′ phosphate	TACAGATCGGAAGAG CTCGTATGCCGTC TTCTGCTTG
Seq_2F	CATT	None	ACACTCTTTCCCTAC ACGACGCTCTTCC GATCT**CATT**
Seq_2R	CATT	5′ phosphate	ATGAGATCGGAAGAG CTCGTATGCCGTC TTCTGCTTG
Seq_3F	ACGT	None	ACACTCTTTCCCTAC ACGACGCTCTTCC GATCT**ACGT**
Seq_3R	ACGT	5′ phosphate	CGTAGATCGGAAGAG CTCGTATGCCGTC TTCTGCTTG
Seq_4F	TGCT	None	ACACTCTTTCCCTAC ACGACGCTCTTCC GATCT**TGCT**
Seq_4R	TGCT	5′ phosphate	GCAAGATCGGAAGAG CTCGTATGCCGTC TTCTGCTTG

7. PCRprimer 1.1: 5′ AATGATACGGCGACCACCGAGATCTACACTCTTTCCCTACACGACGCTCTTCCGATCT 3′.

   PCRprimer2.1: 5′ CAAGCAGAAGACGGCATACGAGCTCTTCCGATCT 3′.
8. Phusion Hot Start High-fidelity DNA Polymerase (New England Biolabs).
9. QIAGEN Minelute purification kit.

10. QIAGEN gel extraction kit.
11. Certified Low Range Ultra Agarose.
12. 50 bp DNA ladder.
13. 1× TBE buffer.
14. 6× DNA loading buffer.

3. Methods

The method outlined below is as indicated in the introduction used for ChIP-Seq of RNAPII at different stages of mouse 3T3-L1 preadipocyte differentiation to mature fat cells. The same protocol can however be used to perform ChIP of chromatin modifiers, transcription factors, as well as indirect DNA-binding proteins such as co-factors. The control regions selected for verification of the enrichment of ChIP are the promoter regions of the housekeeping genes β-actin and GAPDH and proximal promoter and exonic regions of genes that are well known to be expressed during 3T3-L1 differentiation such as Ebf1 or Rarγ (7).

3.1. Differentiation of 3T3-L1 Preadipocytes to Mature Adipocytes

1. Seed cells in 150-mm tissue culture petri-plates at a density of 1×10^6 cells/plate.
2. Change the cell culture medium every alternate day and let the cells grow to absolute confluency.
3. When the cells are completely confluent, leave them for 2 more days before starting differentiation.
4. Add the differentiation medium (see Subheading 2.1) and incubate the cells for 2 days at 37°C and 10% CO_2 in incubator.
5. Aspirate the differentiation medium, wash the cells using 1× PBS, and add insulin medium (see Subheading 2.1).
6. Add fresh DMEM supplemented with FCS and antibiotics after removing insulin medium and incubate further for 2 days.
7. Observe the differentiation of cells into fat droplet containing mature adipocytes using phase contrast microscope.

3.2. Preparation of Cells for ChIP

1. Add fresh 1% formaldehyde solution (e.g., 15 mL for a 15 cm plate) to the plates or flasks after washing the cells two times with 1× PBS. Perform this step in a fume hood to avoid toxic formaldehyde fumes.
2. Swirl the plates or flasks briefly and let them sit at room temperature for 10 min.

3. Add 1/20 volume of 2.5 M glycine to the plates or flasks to quench the formaldehyde. Swirl to mix glycine and then incubate for 5 min at room temperature.
4. Rinse the cells twice with 5-mL cold PBS on the plate. Place the plates on ice and scrape the cells in 5 mL PBS and collect in 50-mL falcon tubes and fill the tubes with cold PBS (see Note 1).
5. Centrifuge at 600 × *g* for 5 min at 4°C. Discard the supernatant. Wash the cell pellets three times with ice-cold PBS to remove any traces of formaldehyde and glycine.
6. Flash freeze the cell pellets in liquid nitrogen and store at −80°C.

3.3. Preparation of Protein-A Sepharose Beads for ChIP (see Note 7)

1. Aliquot an adequate amount (130 μL/ChIP) of 50% suspension of Recombinant Protein-A Sepharose beads from the stock bottle in a fresh 15 mL falcon tube.
2. Centrifuge the beads at 200 × *g* for 1 min and remove the supernatant. Mark the tube at double the volume of beads for making the 50% re-suspension again with ChIP dilution buffer.
3. Wash the beads thrice using 10 mL of ChIP dilution buffer. For each wash, re-suspend the beads gently in 10 mL of buffer and then centrifuge at 200 × *g* for 1 min to pellet them.
4. After the third wash, add 1.1 mL of 50 mg/mL BSA per mL of beads and 250 μg of 10 mg/mL fragmented salmon sperm DNA and make up the volume to 10 mL using ChIP dilution buffer.
5. To block the beads, rotate them overnight at 4°C.
6. Wash the beads thrice with ChIP dilution buffer as in step 3 to remove unbound BSA and salmon sperm DNA.
7. Re-suspend the beads in an adequate amount of ChIP dilution buffer as indicated by the mark made during step 2.

3.4. Lysis of the Cells and Preparation of the Chromatin for IP (see Note 2)

1. Place and thaw the frozen cells on ice (10–15 min). Re-suspend gently by pipetting up and down each pellet of approximately 40×10^6 cells in 40 mL of nuclei extraction buffer. Rock at 4°C for 10 min. Centrifuge at 850 × *g* for 10 min at 4°C. Pour off the supernatant and discard.
2. Re-suspend by pipetting each pellet in 20 mL of protein extraction buffer. Rock gently at room temperature for 10 min. Pellet nuclei by spinning at 850 × *g* for 10 min at 4°C.
3. Discard the supernatant with vacuum aspiration.
4. Re-suspend each pellet of 40×10^6 cells nuclei per 1.6 mL of Buffer LIII. Avoid making bubbles. Transfer nuclei to 15 mL polypropylene tubes (800 μL/tube).

5. Clean the sonicator probe (Bioruptor) specific for 15 mL falcon tubes with 70% EtOH before putting into the sample tubes. Samples should be kept in ice-cold water during sonication. Sonicate for 65 min with cycles of 30 s ON and 30 s OFF at high frequency. Change the ice and cold water every 10 min to avoid over-heating of the samples (change of cold water is not required if attached cooling unit is used). Check the fragment size on gel and if needed, continue the sonication (see Notes 3 and 4 and Fig. 2a).
6. Remove the samples to the new tubes and spin them at maximum speed for 10 min at 4°C to pellet the debris.
7. Transfer the supernatant to 15-mL polypropylene tubes.

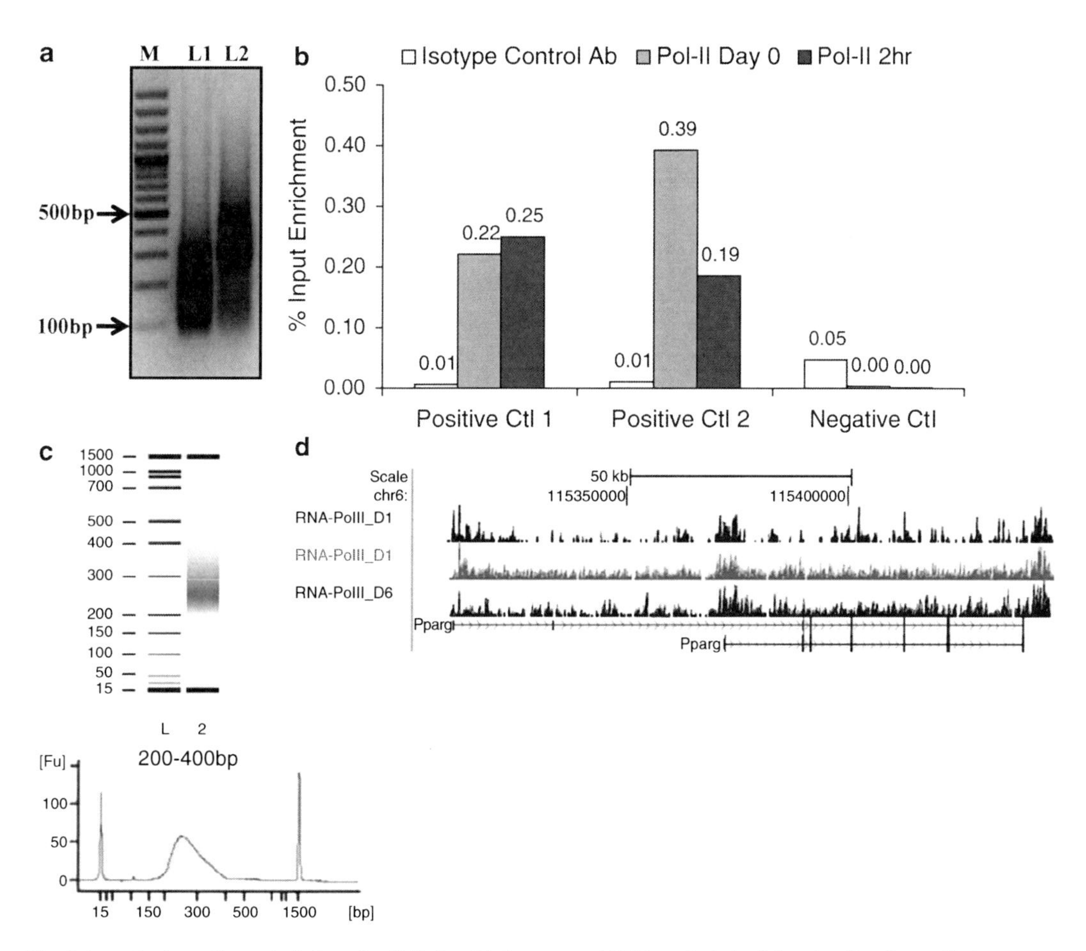

Fig. 2. Important quality control steps for ChIP-Seq (**a**) Agarose gel (1%) analysis to determine the chromatin fragmentation efficiency; *M*: 100 bp DNA ladder, *L1*: sonicated chromatin after decrosslinking, *L2*: sonicated chromatin without decrosslinking. (**b**) Validation of RNA Pol-II ChIP-DNA enrichment using positive and negative control genomic regions at 0 and 2 h time point of 3T3L1 differentiation. (**c**) Validation of the ChIP-DNA library using a Bioanalyzer prior to Solexa sequencing; *L*: ladder, 2: representative sample. (**d**) UCSC genome browser snap-shot showing the RNAPII occupancy across the *PPARγ* gene.

8. Estimate roughly the DNA concentration by using a Nanodrop. You should get approximately 1 mg of chromatin from 50×10^6 3T3-L1 cells. For RNA Pol-II, 200 μg of chromatin is sufficient for performing one IP.
9. Take an equal amount of chromatin DNA for a control IP (rabbit control IgG) and specific antibody IP. Adjust the chromatin volume to 1 mL with ChIP dilution buffer. Save 1% of diluted chromatin from each sample as input DNA. Sonicated chromatin can be stored at −80°C if the ChIP is to be performed later or to preserve the left over chromatin (see Notes 5 and 6).
10. Add 80 μL of blocked Protein-A Sepharose beads and rotate at 4°C for 2 h to remove the nonspecifically bound chromatin from the samples.
11. Centrifuge the samples at $200 \times g$ for 2 min at 4°C to pellet the beads. Transfer the precleared chromatin supernatant to new tubes.
12. Add 10 μg of RNAPII antibody or Isotype control antibody for a control IP. Incubate overnight at 4°C in an Eppendorf rotator.

3.5. Washing of Beads to Remove Nonspecific Complexes (see Notes 2 and 8)

1. Add 50 μL of Protein-A Sepharose to each IP sample and place the tubes in an Eppendorf rotator for 1 h 30 min at 4°C. This serves to collect the antibody–protein–DNA complex.
2. Pellet Protein-A Sepharose by brief centrifugation at $400 \times g$ for 1 min and remove the supernatant fraction by vacuum aspiration. The supernatant should be aspirated carefully without disturbing or aspirating the bead pellet.
3. Wash the Protein-A Sepharose–antibody–chromatin complex-containing beads by resuspending the beads in 1 mL each of the cold buffers in the order listed below (a–d). Incubate for 2 min on a rotating platform followed by brief centrifugation at $400 \times g$ for 1 min. Carefully remove the supernatant using vacuum aspiration.
   (a) Low-salt immune complex wash buffer, four washes.
   (b) High-salt immune complex wash buffer, two washes.
   (c) LiCl immune complex wash buffer, two washes.
   (d) TE buffer, two washes.

3.6. Elution of Protein–DNA Complexes and Reverse Crosslinking

1. Bring 1 M $NaHCO_3$ to room temperature. A precipitate may be observed but will go into solution once room temperature is achieved. Set heat block to 65°C.
2. Make elution buffer for all IP tubes as well as for Input tubes. For each tube, prepare 200 μL of elution buffer as follows: 20 μL 10% SDS, 20 μL 1 M $NaHCO_3$, and 160 μL dH_2O.

Alternatively, make a large volume to accommodate all tubes. For example, if there are nine tubes, mix together 200 μL 10% SDS, 200 μL 1 M $NaHCO_3$, and 1.6 mL dH_2O.

3. For input tubes (see Subheading 3.3, step 9), add 190 μL elution buffer and set aside at room temperature.
4. Add 110 μL elution buffer to each tube containing the antibody/agarose complex. Mix by flicking the tube gently.
5. Incubate at room temperature for 15 min with gentle agitation.
6. Pellet the Sepharose by brief centrifugation at 400 × *g* for 1 min and transfer 100 μL of supernatant into new microfuge tubes.
7. Repeat steps 4–6 and combine elutes (total volume = 200 μL).
8. To reverse crosslink the protein–DNA complexes, add 8 μL 5 M NaCl to all tubes (IPs and Inputs) and incubate at 65°C overnight.

3.7. RNA, Protein Digestion, and DNA Purification

1. To all tubes, add 2 μL of RNase-A (10 mg/mL) and incubate for 1 h at 37°C.
2. Add 4 μL 0.5 M EDTA, 8 μL 1 M Tris–HCl, and 2 μL Proteinase-K (20 mg/mL) and incubate at 45°C for 2 h.
3. Purify the ChIP-DNA using Qiagen PCR purification columns and elute the DNA in 50 μL elution buffer. Store the purified DNA at −20°C till verification of ChIP enrichment followed by library preparation for high-throughput sequencing.

3.8. Verification of ChIP-DNA Enrichment Using q-PCR (See Note 9)

1. Mix 0.3 μM final concentration of each primer with water and Sybr-Green Master mix to a 1× final concentration.
2. Aliquot 10 μL per well of Master mix into an optical q-PCR plate in duplicate or triplicate depending on the availability of template.
3. Add 1 μL of ChIP-DNA in each well containing Master mix, seal plates with adhesive covers, mix, and centrifuge briefly.
4. Use the PCR conditions suggested for use with the Sybr-Green mix: 2 min at 50°C, 10 min at 95°C, and thereafter 40 cycles of 15 s at 95°C and 1 min at 60°C. Enabling the generation of a dissociation curve at the end of the PCR allows an assessment of the specificity of the primers.
5. Specific enrichment of ChIP-DNA is assessed as follows:

   Extrapolate the input Ct from 1 to 100%, i.e., the corrected input Ct value = $Ct_{Input} - 6.64$ as we are taking 1% input chromatin (see Subheading 3.3, step 9).

$$\% \text{ Input enrichment} = 100 \times 2^{(Ct_{\text{Corrected Input}} - Ct_{\text{positive control or negative control}})}$$

$$\text{Background enrichment} = 100 \times 2^{(Ct_{\text{Corrected Input}} - Ct_{\text{Control antibody}})}$$

$$\% \text{ ChIP enrichment} = \% \text{ Input enrichment (positive control)} - \% \text{ Input enrichment (negative control)}$$

If the ChIP enrichment for the positive control genomic region is significantly greater than the negative control one, then it implicates that the ChIP protocol is working fine and the DNA obtained from ChIP can be used for library preparation for deep sequencing (see Fig. 2b).

3.9. Preparation of ChIP-DNA Library for Solexa Sequencing

3.9.1. End-Repair of Input Control and ChIP-DNA

1. Use 10 ng of Input DNA and 1–10 ng of ChIP-DNA as starting material for DNA end-repair. During this step, DNA ends are repaired to blunt ends by T4 DNA polymerase and phosphorylated at the 3′ ends by T4 Polynucleotide Kinase.
2. Prepare the following reaction mix:
   (a) DNA sample (34 μL)
   (b) 10× End-repair buffer (5 μL)
   (c) 2.5 mM dNTP mix (5 μL)
   (d) 10 mM ATP (5 μL)
   (e) END-it Enzyme mix (1 μL).
   Total volume should be 50 μL per reaction
3. Incubate the reaction mix for 45 min at room temperature.
4. Purify the end-repaired DNA using a Qiagen Minelute PCR purification kit. Elute the DNA first in 20 μL of elution buffer, and then in 12 μL second time. Total elution volume is thus 32 μL.

3.9.2. Addition of an "A" Base to the 3′ End of the DNA Fragments

1. Prepare the following reaction mix:
   (a) DNA from step 1 in Subheading 3.9 (32 μL)
   (b) 10× NEB buffer 2 (5 μL)
   (c) 1 mM ATP (10 μL)
   (d) Klenow exo-polymerase (5 U/μL) enzyme (3 μL).
   Total volume should be 50 μL per reaction
2. Incubate for 30 min at 37°C.
3. Purify the DNA using a Minelute PCR purification kit in 12.5 μL of elution buffer.

3.9.3. Ligation of Adapters to the Ends of the DNA Fragments

1. Prepare the following reaction mix:
   (a) DNA sample from step 2 in Subheading 3.9 (12.5 μL)
   (b) 2× DNA ligase buffer (15 μL)
   (c) Barcoded 15 μM DNA oligo mix diluted 1:10 (1 μL)
   (d) DNA quick ligase enzyme (1.5 μL).
   Total volume should be 30 μL per reaction

2. Incubate for 15 min at room temperature.
3. Minelute purify the DNA in 10 μL of elution buffer.
4. At this stage, DNA can be stored at −80°C for several days.

3.9.4. Gel Purification of the Ligated DNA for Fragment Size Selection

1. Prepare a 30 mL (volume depends on the size of gel cast), 2% agarose gel in 1× TBE buffer.
2. Load 1 μg of 50 bp DNA ladder into one lane of the gel.
3. Add 3 μL of 6× loading buffer to 10 μL of the DNA from the purified ligation reaction.
4. Load the entire sample into another lane of the gel, leaving one empty lane between ladder and sample.
5. Run the gel at 120 V for 60 min.
6. Stain the gel using 1× SYBR Gold dye in 1× TBE for 20 min while shaking. Cover the gel container with foil to avoid exposure to light.
7. View the gel on a Dark Reader transilluminator to avoid being exposed to UV light.
8. Use a clean scalpel to excise a gel slice containing DNA in the 200–400 bp range. Be sure to image the gel before and after the slice is excised.
9. Use a QIAGEN gel extraction kit to purify the DNA from the agarose slices and elute DNA in 36 μL of elution buffer.

3.9.5. Enrichment of Adapter-Modified, Size-Selected DNA Fragments by PCR

1. Prepare the following PCR mix:
   (a) DNA (33.5 μL)
   (b) 5× Phusion buffer (10 μL)
   (c) 10 mM dNTP mix (1 μL)
   (d) 10 μM PCR primer 1.1 (2.5 μL)
   (e) 10 μM PCR primer 2.1 (2.5 μL)
   (f) Phusion Hot Start DNA polymerase (0.5 μL)
   The total volume should be 50 μL per reaction
2. Amplify using the following PCR protocol:
   (a) Step 1: 98°C for 30 s
   (b) Step 2: 98°C for 10 s
   (c) Step 3: 65°C for 30 s
   (d) Step 4: 72°C for 30 s
   (e) Step 5: go to step 2, 17 cycles
   (f) Step 6: 72°C for 5 min
   (g) Step 7: 4°C forever
3. Purify the DNA using a Minelute PCR purification kit and elute in 20 μL of elution buffer.

4. Quantify the DNA concentration of sequencing library using Qubit and check the quality using Bioanalyzer (see Note 10 and Fig. 2c).
5. Store the amplified DNA at −80°C till preparation for sequencing reaction.

4. Notes

1. At later stages of 3T3L1 differentiation, the cell layer is easily detached during fixing of the cells using formaldehyde or quenching afterwards by glycine, if the cells are detached from the plates then they can be scraped in glycine solution after putting the plates on ice.
2. Perform all the steps on ice until the chromatin decrosslinking step to avoid the denaturation of protein–DNA complexes and add protease inhibitor and phosphatase inhibitors to the lysis buffers/ChIP dilution buffer right before use. Always chill the buffers on ice for at least 20 min before use to prevent the decrosslinking of protein–DNA complexes.
3. The efficiency and resolution of ChIP-Seq depends in large part on the chromatin fragment size. It is therefore very important to standardize the sonication conditions according to the cell material before starting the real experiment. Use the lowest settings that result in sheared DNA ranging from 100 to 500 bp in size (Fig. 2a). Shearing varies greatly depending on cell type, quantity, volume, crosslinking, and equipment.
4. To verify the chromatin length:
   (a) Transfer a 10 μL aliquot of sonicated chromatin sample to a microfuge tube. Keep the remaining chromatin samples on ice.
   (b) Denature for 10 min at 95°C in a heat block.
   (c) Add 2 μL Proteinase-K and incubate for 20 min at 45°C.
   (d) Purify the decrosslinked DNA using a Qiagen PCR purification column and elute in 20 μL of elution buffer.
   (e) Load 10 μL of purified DNA after adding 4 μL of 6× loading buffer into a 1% agarose gel. In an adjacent well, load a 100 bp DNA ladder to verify the fragment size of purified DNA.
5. Chromatin samples after fragmentation check can be stored at −80°C if processing will be performed later on. However, make sure to add 10% glycerol (final concentration). Addition of glycerol to the chromatin reduces the chances of protein–DNA decrosslinking during the freeze thaw.

6. Quantitation of chromatin DNA using Nanodrop after sonication helps to determine equal amounts of chromatin for different conditions or time points of culture to be used for ChIP.
7. For pull down of the protein–DNA complexes after fragmentation, Protein-A Sepharose, Protein-G Sepharose or magnetic beads conjugated with protein-G/Protein-A/secondary antibody can be used. It is thereby important to optimize the ChIP protocol for each bead type, especially the washing steps. We have observed that the enrichment with magnetic beads tends to be lower than that with Sepharose beads, although the latter tend to produce more noise and therefore need to be washed more extensively.
8. The washing step after pull down of chromatin by the specific antibody needs to be standardized for each protein or antibody of interest.
9. The validation of ChIP-DNA enrichment is very important prior to starting the library preparation and sequencing. Enrichment of ChIP-DNA can be measured by using positive and negative control genomic regions or by measuring the fold enrichment over control IgG antibody samples (Fig. 2b). Typically, these experiments are performed using real-time PCR thereby employing primers that yield a linear amplification curve when analyzing serially diluted input samples (1/2, 1/4, 1/8, 1/16, 1/32). The resulting data provide an estimate of the signal-to-noise ratio in the ChIP samples, which may be helpful when establishing a positive hit threshold upon analyzing the sequencing results.
10. The amount of ChIP-DNA starting material used to prepare the library is very low (<10 ng), and even after 18 PCR cycles, the yield may be too low to be visible on a regular agarose gel. Therefore, it is recommended to check the quality of the library before sequencing using a sensitive technique such as Bioanalyzer (Agilent Technologies 2100). This also allows the precise determination of the fragment size, which is important to establish the molar concentration of DNA, which will be used for downstream sequencing (Fig. 2c). In addition, the DNA concentration needs to be determined using a Qubit fluorometer (Invitrogen) because if the concentration is less than 4 ng/μL, then the cluster estimation for Solexa sequencing will not be very accurate.

References

1. Davidson, E. H. (2006) *The Regulatory Genome: Gene Regulatory Networks In Development And Evolution*, Academic press.
2. Deplancke, B. (2009) Experimental advances in the characterization of metazoan gene regulatory networks, *Briefings in Functional Genomics and Proteomics* **8**, 12–27.
3. Simicevic, J., and Deplancke, B. (2010) DNA-centered approaches to characterize regulatory protein-DNA interaction complexes, *Mol Biosyst* **6**, 462–468.
4. Ren, B., Robert, F., Wyrick, J., Aparicio, O., Jennings, E., Simon, I., Zeitlinger, J., Schreiber, J., Hannett, N., Kanin, E., Volkert, T., Wilson, C.,

Bell, S., and Young, R. (2000) Genome-wide location and function of DNA binding proteins, *Science* **290**, 2306–2309.

5. Johnson, D. S., Mortazavi, A., Myers, R. M., and Wold, B. (2007) Genome-Wide Mapping of in Vivo Protein-DNA Interactions, *Science* **316**, 1497–1502.
6. Mukhopadhyay, A., Deplancke, B., Walhout, A. J., and Tissenbaum, H. A. (2008) Chromatin immunoprecipitation (ChIP) coupled to detection by quantitative real-time PCR to study transcription factor binding to DNA in Caenorhabditis elegans, *Nature Protocols* **3**, 698–709.
7. Nielsen, R., Pedersen, T. A., Hagenbeek, D., Moulos, P., Siersbaek, R., Megens, E., Denissov, S., Borgesen, M., Francoijs, K.-J., Mandrup, S., and Stunnenberg, H. G. (2008) Genome-wide profiling of PPARγ:RXRα and RNA polymerase II occupancy reveals temporal activation of distinct metabolic pathways and changes in RXR dimer composition during adipogenesis, *Genes & Development* **22**, 2953–2967.

Chapter 16

Computational Analysis of Protein–DNA Interactions from ChIP-seq Data

Jacques Rougemont and Felix Naef

Abstract

Chromatin immunoprecipitation experiments followed by ultra-high-throughput sequencing (ChIP-seq) is becoming the method of choice to identify transcription factor binding sites in prokaryotes and eukaryotes in vivo. Here, we review the computational steps that are necessary for analyzing the sequenced chromatin fragments, including mapping of short reads onto reference genomes, normalization of multiple conditions, detection of *bona fide* peaks or binding sites, annotation of sites, characterization of sequence-specific binding affinities, and relationships with biophysical models for protein–DNA interactions. The goal is that following the indicated steps will help the discovery of novel mechanisms underlying transcription regulation in a broad range of experimental systems.

Key words: Chromatin immunoprecipitation, Ultra-high-throughput sequencing, Transcriptional factor binding, Transcription regulation, Bioinformatics

1. Introduction

Our current understanding of transcription regulatory networks hinges on our ability to accurately measure and predict specific interactions between transcription regulatory proteins and their cognate sites on the DNA, and to study the consequence of such interaction sites on gene products, i.e., cytosolic mRNA and protein abundances. These questions can now be tackled using modern analytical methods, notably through the genome-wide analysis of immunoprecipitated chromatin samples (CHiP).

Chromatin immunoprecipitation and sequencing (ChIP-seq) has become the most popular technique to investigate genome-wide in vivo binding patterns of transcription factors and thereby delineate their potential targets and mechanisms of action. Proteins and DNA are crosslinked, then DNA is fragmented by sonication and

Bart Deplancke and Nele Gheldof (eds.), *Gene Regulatory Networks: Methods and Protocols*,
Methods in Molecular Biology, vol. 786, DOI 10.1007/978-1-61779-292-2_16, © Springer Science+Business Media, LLC 2012

immunoprecipitated with an antibody specific to one particular transcription factor. Short reads (30–50 bases) from either ends of size-selected (200–600 bp) DNA fragments are then sequenced by a high-throughput technology. The number of reads associated with a genomic locus will be proportional to the occupancy of, or residency time at, that locus by the transcription factor.

Beyond the crucial role of antibodies in this assay, the accessible sensitivity will be limited by the total number of mappable reads (1), while the read length needs only be sufficient to unambiguously map back to the genome (2). This makes current ultra-high-throughput short read sequencers particularly adapted to this application.

We discuss here prevalently the situation corresponding to sequence-specific transcription factors with well-defined and narrow binding locations on the genome. Other cases, such as histone modifications or RNA polymerases profiles, have been reviewed extensively in refs. 2, 3. In our context, the following properties of ChIP-seq technology are the cornerstones of the data processing:

- Protein-bound sites generate two mirror-symmetric read density profiles, one on each strand with an upstream shift that reflects the size of sonication fragments. This proves essential for successful peak search algorithms (4).
- The overwhelming majority of sequencing reads are not associated with an IP-enriched locus but reflect a random background expected to be similarly distributed in independent IP and control samples. This property justifies the use of normalization schemes between samples that rely on simple scaling, even linear scaling in the simplest form (1).
- The statistics for the local sequencing depth or coverage are relatively well approximated by a generalization of the Poisson distribution with overdispersion (typically negative binomial distributions). Such statistics are then used to evaluate the significance of enriched regions (5).

2. Materials

2.1. Software

The following pieces of software are freely available and will run on most Unix-like systems:

- Bowtie (6) or BWA (7) to map short reads onto reference genomes.
- Samtools (8) to manipulate and store read alignments.
- MACS (9), QuEST (4) to extract enriched regions from the read density on the genome.

- Meme (10) and MatrixReduce (11) for searching for over-represented sequence motifs in enriched regions.
- Selected R packages, e.g., Biostrings, ShortRead (12), ChipPeakAnno (13), biomaRt for general data manipulation, statistics and fetching genome annotations.

2.2. Databases

Several resources are recommended to perform a complete analysis of ChIP-seq data (see Note 1):

- Reference genomic sequences from NCBI, e.g., mouse chromosome 2, is found at this URL:
  - ftp://ftp.ncbi.nih.gov/genomes/M_musculus/Assembled_chromosomes/mm_ref_chr2.fa.gz
- Precomputed genome indexes, using *bowtie-build* or *bwa index*, e.g.,
  - bowtie-build mm_ref_chr2.fa mm9_chr2
- Genomic annotations: UCSC (14) and Ensembl (15) provide comprehensive collections of gene annotations, conservation scores, locations of regulatory elements, etc. The following two links are good starting points:
  - http://hgdownload.cse.ucsc.edu/goldenPath/
  - http://www.ensembl.org/info/data/ftp/index.html
- Many transcription factors have known or putative position-specific scoring matrices that can be used to compare to ChIP-seq data, see, e.g., Jaspar (16) and Transfac (17).

3. Methods

3.1. Mapping ChIP-seq Reads

The sequencer produces a collection of short reads with individual nucleotide quality scores (for the Illumina sequencer, this comes in the *fastq* format). Typical yields for the Illumina GAII technology are 20 million (M) reads of 38 bases.

These short sequences are mapped onto a preindexed reference genome using a CPU and memory-efficient tool, e.g., Bowtie or BWA (see Note 2). It is important to allow two to three mismatches between the read (or the alignment seed in typical aligner options) and the genome to accommodate polymorphisms and sequencing errors. With such short sequences, ungapped alignments are normally sufficient. Allowing multiple genome matches for the same read is useful to evaluate the abundance of repetitive sequences in the library. These can later be discarded or normalized. It is also important to keep track of the strand of the match to compute separate densities for each strand, see below. After mapping, duplicate reads (namely, any additional read matching the same genomic position and orientation as a previously aligned read, see Note 3) are typically discarded, as they most likely reflect an amplification bias (called also PCR duplicate).

We recommend storing read alignments as a sorted and indexed binary BAM file (8). There exist many tools and programming libraries to manipulate and interrogate such files. Moreover, BAM files can be directly visualized on the UCSC genome browser (14). In Note 4, we describe two alternative ways of doing this, using *samtools* and a perl script on the command line, or using the python API (8). C and Java programmers can use the respective APIs to do the same operations, and other tasks described below can follow the same general pattern.

3.2. Normalization

Having mapped reads and removed duplicates, we next compute the sequencing depth at each genomic position. In UCSC terminology, we process the BAM file into either a wiggle or a bedGraph description: run through the bam file (via a call to *samtools view*, or using one of the APIs mentioned above) and fill a hash table with genomic positions as keys and number of reads as values. Each read will add one count to every position it covers. Several normalizations and filters can be applied during this operation:

- The number of bases included downstream of the read start position can help visualize lowly covered sites. It may be taken as one single base if one is interested in recording only the ends of the ChIP fragments. This choice is preferred for subsequent statistical analyses as it makes statistical models easier to formulate. Otherwise it is often taken as the read length, another possible choice is to extend reads to the average sonication size, both give nicely smoothed visual representations, without actually adding information.
- Multiple alternative alignments of the same read can either each contribute 1/(number of matches) to their respective genomic positions or (equivalently but not as straightforward to implement) one of them is selected randomly.
- We like to report counts per position normalized by 10^{-7} times the total number of mapped reads (this factor is here to keep the numbers within a convenient range and simply reflects the typical throughput of a sequencing run). Typically, we find that a strong site for a specific transcription factor has a few hundred tags per 10^{7} sequenced tags. This normalization factor is computed while the hash table is being filled and applied to each position when printing out the result.

Two different samples (sequencing runs from independent libraries, whether they are biological replicates or not) can yield significantly different numbers of unique mapped reads depending on the amount of starting material, the sequencing error rate, the severity of amplification biases, etc. Following the above procedure including the normalization by total mapped reads before pooling replicates or comparing conditions makes sure that one library does not outweigh the others and efficiently decreases the effect of library-specific artifacts.

Visualization of these read densities is best done using UCSC's bigWig file format (14): generate a bedGraph formatted file from the hash table described above and then convert it using *wigToBigWig* from http://hgdownload.cse.ucsc.edu/admin/exe/. One option is to generate two separate tracks for the forward and reverse strands, another possibility is a merged track which sums the contributions from both strands after shifting their coordinates downstream by half the sonication size. See Fig. 1 for an illustration of these steps.

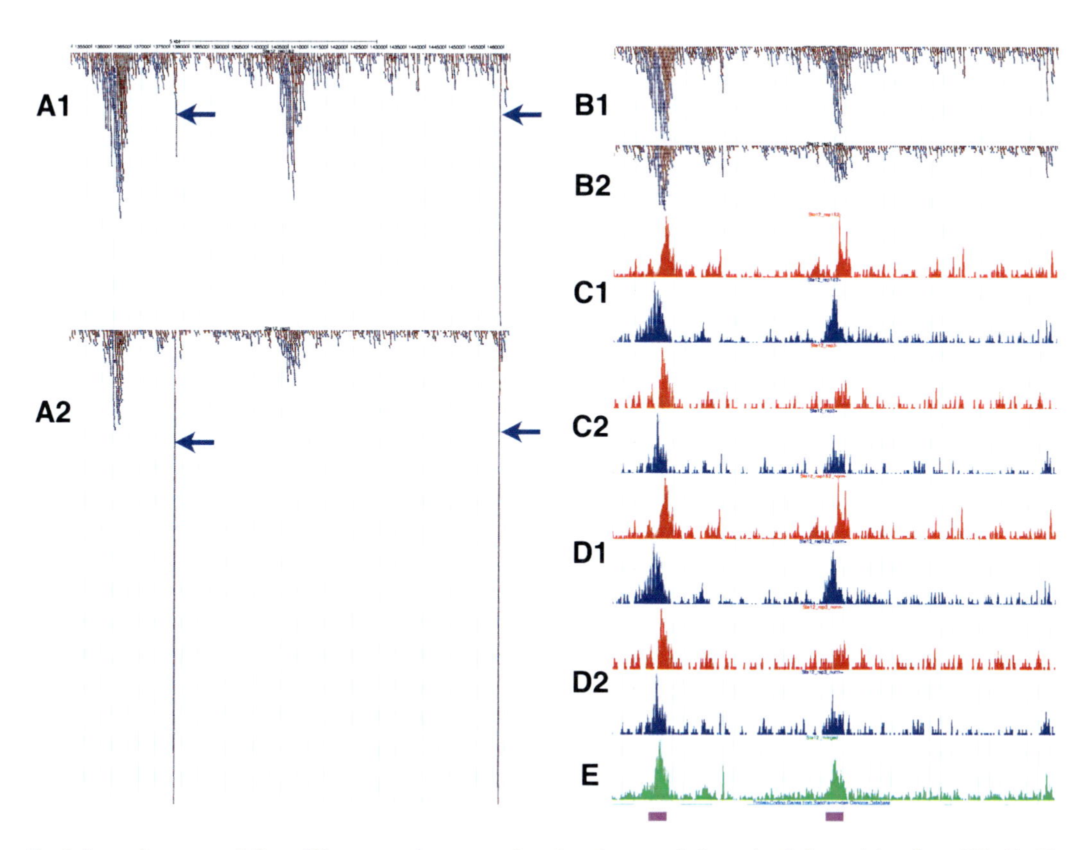

Fig. 1. Example representation of the successive processing steps from reads to peaks. Data are taken from (22). A1–A2: Two replicates of the same ChIP experiment with different throughput, showing large amplification biases (*blue arrow*). B1–B2: Removing duplicate reads largely removes the artifacts. C1–C2: Strand-specific read densities (*red tracks*: reverse strand, *blue tracks*: forward strand) show the upstream shift of the peaks. D1–D2: Same as C1–C2 but after normalization by total number of mapped reads: peaks in replicate 2 are now of comparable size to those in replicate 1. E: Merged densities, the four tracks from C1–C2 have been shifted downstream by 80 bases and summed (vertical scale is different). Clearly enriched regions emerge (*pink bars*). A1–A2 represent the bam files produced by Bowtie (Subheading 3.1), B1–B2 are the reduced bam files after PCR duplicate removal (**see** Note 4). C1–E are densities such as described in Subheading 3.2 (uploaded to the UCSC browser as bigWig files).

3.3. Peak Search and Binding Site Identification

The next step in the analysis consists in searching for genomic regions displaying enriched IP signal (putative transcription factor binding sites). These procedures are commonly referred to as "peak calling". A peak corresponding to a real binding site must meet several conditions to be considered trustable: (1) consistent spatial spread with the expected symmetry between strands, (2) significant enrichment in read density compared to a local background and to control if available (such as total input or unspecific IP). False positives are pervasive in ChIP-seq data (18) and several algorithms have been proposed to control the false-discovery rate (FDR). Here, we shortly describe the MACS algorithm (9) which we found to be useful for this task. The most sensitive parameters seem to be the *p*-value cutoff and mfold parameter, and it is also important to check the default value for genome size (valid mostly for the human genome). It uses the following steps:

- Evaluate the peak shift d by computing the maximum of the cross-correlation between forward and reverse read densities restricted to a set of regions with high coverage compared to the background. We have experienced that this quantity is not always optimally estimated and it may be beneficial to independently estimate it and specify it as an input.
- Compute the merged density as described above using the value of d previously estimated.
- Evaluate the significance of the enrichment of each window of size $2d$ based on Poisson distribution with a local estimate of its mean.

FDR is typically estimated empirically by running the same peak search on the control sample or a set of shuffled reads from the IP sample and counting the number of peaks thus identified as the rate of false positives. Remark that it is important to understand the meaning of false positive in this case: when comparing two different IP conditions, a peak present in both will be flagged as false positive even if significantly reduced in the second condition relative to the first.

MACS tends to return regions of putative ChIP enrichment which are often much larger than the expected binding site, and even larger then the apparent spread of the ChIP signal. One way of increasing the spatial resolution of peak detection is to smooth the data, for example, as implemented in QuEST (4), which uses convolution of the read starts with downstream shifted Gaussian kernels and then searches local maxima of these smoothed densities as peak centers. We have experienced that while Quest is currently more accurate at identifying the optimal shift of the strands, its FDR control is very sensitive to having balanced numbers of reads. Another possibility is to infer the binding site from the detailed symmetry between the two strands, see CSDeconv (19):

- Assume that each strand profile is generated by the convolution of a binding profile and a strand-specific density kernel, then infer the binding profile from the data.

The difficulty here is to choose a kernel that will minimize the uncertainty in the deconvolved profile. CSDeconv uses a Gaussian kernel, which is convenient but does not necessarily reflect the expected profile from the sample preparation. It is often beneficial to run several peak callers, a fast one to delineate broad regions of putative enrichment and a second more expensive algorithm to evaluate the precise peak profiles.

3.4. Annotation

Binding sites identified by a genome-wide search are likely to be proximal to their target genes and are also often expected to be evolutionary conserved. To further classify peaks we therefore cross the BED file containing the list of peak regions with several other sources of data:

- Search for genes in the immediate neighborhood of sites (e.g., in windows of ±10 kb of the peak for a mammalian genome) and report the fraction of sites that fall in gene promoters, gene body, downstream or intergenic regions. This is straightforward using ChipPeakAnno from Bioconductor (13).
- Compute the distribution of conservation scores (e.g., by intersection with the UCSC PhastCons conservation scores) within peaks and compare to control regions of the same size taken randomly.
- Scan peak regions with known or inferred position-weight matrices to estimate the co-occurrence of binding sites with sequence motifs. This is conveniently done in R with the Biostrings package or using the FIMO tool from the Meme suite (10).

3.5. Motif Search and Biophysical Modeling

A positive binding site corresponds to a sequence in the genome with high occupancy, which can mean that the site has high affinity for the transcription factor, or that co-factors contribute to recruiting this factor to the DNA. The high spatial resolution of ChIP-seq make it well suited to identify over-represented DNA sequence motifs characteristic of bound regions. Typically, we extract relatively short regions (about 100 bp) around the peak centers in a fasta file (use fastaFromBed (20)) and use programs such as Meme (10). A strategy to increase the sensitivity of detecting sequence motifs associated with binding sites is to use the ChIP profiles as position-specific priors, which can be done conveniently as in ref. 21.

Another approach aims at estimating a binding energy profile for the sequence-specific protein–DNA interaction. For convenience, such models often use the approximation that the individual genomic positions (typically 6–10) contribute additively to the binding energy. The MatrixReduce (11) algorithm offers one possibility to infer position-specific affinity matrices (PSAM). It is based on the assumption that a quantitative relationship exists between sequence occupancy as measured in ChIP and site strength, and uses the further approximation that sites are lowly occupied (Boltzmann limit) in order to reduce the number of free parameters to fit.

4. Notes

1. When data are obtained from several different sources, it is important to check that they are consistently associated with the same version of the genome assembly.
2. All our examples use bowtie. BWA and Soap provide similar options and give comparable results. While BWA might be a better option for longer reads as it permits gapped alignments, in our experience even 75 bp reads align well without gaps.
3. The expected (strand-specific) coverage is the total number of reads divided by (twice) the genome size. In an IP-enriched region, this can be multiplied by the IP enrichment ratio (typically 10–100). We suggest restricting the number of reads allowed to map at the same genomic position to a threshold as a function of this number. For example, a human Chip-seq with 40 million reads gives an expected coverage of $40/6{,}000 = 0.007$. With an IP-enrichment factor of 100, even enriched regions will have an average number of 0.7 reads per position. Therefore, only the first read mapping at each position will be kept. In yeast, the genome size is 24 Mb, therefore a similar sequencing throughput will result in a coverage of 0.8, and with a tenfold enrichment, the Poisson probability of having more than 12 reads at the same (enriched) position is under 5%.
4. How to remove reads exceeding NMAX at each position:
   (a) PERL programmers use the command line

- *samtools view -h mybam.bam | counter.pl | samtools view -Sb -o cleaned.bam -*

where counter.pl is the following script:

```
while (<>) {
    if (/^@/) {
        print;
    } else {
        @r=split"\t";
        $str = ($r[1] & 0x10) ? 1 : 0;
        $pos0 = "$r[2]:$r[3]";
        if ($pos0 ne $pos[$str]) {
            $pos[$str] = $pos0;
            $n[$str]=0;
        }
        print if $n[$str]<NMAX;
        $n[$str]++;
    }
}
```

(b) PYTHON programmers use the script

```
import pysam
infile = pysam.Samfile("mybam.bam", "rb")
outfile = pysam.Samfile("cleanbam.bam", "wb",
template=infile)
counts = {'0':0,'1':0}
pos = {'0':'','1':''}
for read in infile.fetch():
    strnd = (read.is_reverse and '1' or '0')
    pos0 = "%s:%d" % (read.rname, read.pos)
    if (pos0 != pos[strnd]):
        pos[strnd]=pos
        counts[strnd]=0
    if (counts[strnd] < NMAX:
        outfile.write(read)
    counts[strnd]+=1
outfile.close()
infile.close()
```

Acknowledgments

F.N. and J.R. are supported by the SystemsX.ch Swiss initiative in systems biology and the Ecole Polytechnqiue Federale de Lausanne, F.N. is supported by the Swiss National Foundation (SNF grant 31-130714).

References

1. Barski, A., and Zhao, K. (2009) Genomic location analysis by ChIP-Seq, J Cell Biochem **107**, 11–18.
2. Park, P. J. (2009) ChIP-seq: advantages and challenges of a maturing technology, Nat Rev Genet **10**, 669–680.
3. Leleu, M., Lefebvre, G., and Rougemont, J. (2010) Processing and analyzing ChIP-seq data: from short reads to regulatory interactions, Briefings in functional genomics.
4. Valouev, A., Johnson, D. S., Sundquist, A., Medina, C., Anton, E., Batzoglou, S., Myers, R. M., and Sidow, A. (2008) Genome-wide analysis of transcription factor binding sites based on ChIP-Seq data, Nat Methods **5**, 829–834.
5. Kharchenko, P., Tolstorukov, M., and Park, P. (2008) Design and analysis of ChIP-seq experiments for DNA-binding proteins, Nat Biotechnol **26**, 1351–1359.
6. Langmead, B., Trapnell, C., Pop, M., and Salzberg, S. (2009) Ultrafast and memory-efficient alignment of short DNA sequences to the human genome, Genome Biol **10**, R25.
7. Li, H., and Durbin, R. (2009) Fast and accurate short read alignment with Burrows-Wheeler transform, Bioinformatics **25**, 1754–1760.
8. Li, H., Handsaker, B., Wysoker, A., Fennell, T., Ruan, J., Homer, N., Marth, G., Abecasis, G. R., Durbin, R., and Subgroup, G. P. D. P. (2009) The Sequence Alignment/Map format and SAMtools, Bioinformatics **25**, 2078–2079.
9. Hesselberth, J. R., Chen, X., Zhang, Z., Sabo, P. J., Sandstrom, R., Reynolds, A. P., Thurman, R. E., Neph, S., Kuehn, M. S., Noble, W. S., Fields, S., and Stamatoyannopoulos, J. A. (2009) Global mapping of protein-DNA interactions in vivo by digital genomic footprinting, Nat Methods **6**, 283–289.
10. Bailey, T. L., Boden, M., Buske, F. A., Frith, M., Grant, C. E., Clementi, L., Ren, J., Li, W. W., and Noble, W. S. (2009) MEME SUITE: tools for motif discovery and searching, Nucleic Acids Research **37**, W202–208.
11. Foat, B. C., Morozov, A. V., and Bussemaker, H. J. (2006) Statistical mechanical modeling of genome-wide transcription factor occupancy data by MatrixREDUCE, Bioinformatics **22**, e141–149.
12. Morgan, M., Anders, S., Lawrence, M., Aboyoun, P., Pagès, H., and Gentleman, R. (2009) ShortRead: a bioconductor package for input, quality assessment and exploration of high-throughput sequence data, Bioinformatics (Oxford, England) **25**, 2607–2608.
13. Zhu, L. J., Gazin, C., Lawson, N. D., Pages, H., Lin, S. M., Lapointe, D. S., and Green, M. R. (2010) ChIPpeakAnno: a Bioconductor package to annotate ChIP-seq and ChIP-chip data, BMC Bioinformatics **11**, 237.
14. Rhead, B., Karolchik, D., Kuhn, R. M., Hinrichs, A. S., Zweig, A. S., Fujita, P. A., Diekhans, M., Smith, K. E., Rosenbloom, K. R., Raney, B. J., Pohl, A., Pheasant, M., Meyer, L. R., Learned, K., Hsu, F., Hillman-Jackson, J., Harte, R. A., Giardine, B., Dreszer, T. R., Clawson, H., Barber, G. P., Haussler, D., and Kent, W. J. (2010) The UCSC Genome Browser database: update 2010, Nucleic Acids Research **38**, D613–619.
15. Hubbard, T. J. P., Aken, B. L., Ayling, S., Ballester, B., Beal, K., Bragin, E., Brent, S., Chen, Y., Clapham, P., Clarke, L., Coates, G., Fairley, S., Fitzgerald, S., Fernandez-Banet, J., Gordon, L., Graf, S., Haider, S., Hammond, M., Holland, R., Howe, K., Jenkinson, A., Johnson, N., Kahari, A., Keefe, D., Keenan, S., Kinsella, R., Kokocinski, F., Kulesha, E., Lawson, D., Longden, I., Megy, K., Meidl, P., Overduin, B., Parker, A., Pritchard, B., Rios, D., Schuster, M., Slater, G., Smedley, D., Spooner, W., Spudich, G., Trevanion, S., Vilella, A., Vogel, J., White, S., Wilder, S., Zadissa, A., Birney*, E., Cunningham, F., Curwen, V., Durbin, R., Fernandez-Suarez, X. M., Herrero, J., Kasprzyk, A., Proctor, G., Smith, J., Searle, S., and Flicek, P. (2009) Ensembl 2009, Nucleic Acids Research **37**, D690–697.

16. Portales-Casamar, E., Thongjuea, S., Kwon, A. T., Arenillas, D., Zhao, X., Valen, E., Yusuf, D., Lenhard, B., Wasserman, W. W., and Sandelin, A. (2010) JASPAR 2010: the greatly expanded open-access database of transcription factor binding profiles, Nucleic Acids Research **38**, D105–110.

17. Matys, V., Fricke, E., Geffers, R., Gössling, E., Haubrock, M., Hehl, R., Hornischer, K., Karas, D., Kel, A. E., Kel-Margoulis, O. V., Kloos, D.-U., Land, S., Lewicki-Potapov, B., Michael, H., Münch, R., Reuter, I., Rotert, S., Saxel, H., Scheer, M., Thiele, S., and Wingender, E. (2003) TRANSFAC: transcriptional regulation, from patterns to profiles, Nucleic Acids Research **31**, 374–378.

18. Wilbanks, E., and Facciotti, M. (2010) Evaluation of Algorithm Performance in ChIP-Seq Peak Detection, PLoS ONE **5**, e11471 EP -.

19. Lun, D. S., Sherrid, A., Weiner, B., Sherman, D. R., and Galagan, J. E. (2009) A blind deconvolution approach to high-resolution mapping of transcription factor binding sites from ChIP-seq data, Genome Biol **10**, R142.

20. Quinlan, A. R., and Hall, I. M. (2010) BEDTools: a flexible suite of utilities for comparing genomic features, Bioinformatics (Oxford, England) **26,** 841–842.

21. Bailey, T. L., Boden, M., Whitington, T., and Machanick, P. (2010) The value of position-specific priors in motif discovery using MEME, BMC Bioinformatics **11**, 179.

22. Lefrançois, P., Euskirchen, G. M., Auerbach, R. K., Rozowsky, J., Gibson, T., Yellman, C. M., Gerstein, M., and Snyder, M. (2009) Efficient yeast ChIP-Seq using multiplex short-read DNA sequencing, BMC Genomics **10**, 37.

Chapter 17

Using a Yeast Inverse One-Hybrid System to Identify Functional Binding Sites of Transcription Factors

Jizhou Yan and Shawn M. Burgess

Abstract

Binding of transcription factors to promoters is a necessary step to initiate transcription. From an evolutionary standpoint, the regulatory proteins and their binding sites are considered to have molecularly coevolved. We developed an efficient yeast strategy, an "inverse one-hybrid system", to identify binding targets of transcription factors globally in a genome of interest. The technique consists of a yeast strain expressing a transcription factor of interest mated to yeast containing a library of random genomic fragments cloned upstream of a reporter gene (*URA3*). Positive growth on media without uracil denotes a fragment being bound by the transcription factor, e.g., zebrafish FoxI1. The bound fragments in hundreds of positive clones are sequenced and retested for their binding activities using a colony PCR and sequencing strategy. The resulting tools allow for rapid and genomic-wide identification of transcriptional binding targets.

Key words: Yeast, Inverse one-hybrid, DNA binding sites, Transcription factor, Protein–DNA interaction, Genome-wide screening

1. Introduction

Identification of transcription factor DNA binding sites is a difficult task, particularly because they typically consist of short and degenerate DNA sequences. Traditionally, protein-DNA binding elements were identified by SELEX, electrophoretic mobility shift assay (EMSA), and/or DnaseI footprinting assays (1). These approaches can identify the particular sequences that a protein will bind in vitro, but does not address the in vivo activity of a particular binding site. In addition, traditional methods have failed to create high-resolution, genome-wide maps of the interaction between a DNA-binding protein and DNA. To determine binding sites in the context of chromatin, recent emphasis has been on the powerful ChIP-chip and ChIP-Seq approaches (2). This is a combination of

Bart Deplancke and Nele Gheldof (eds.), *Gene Regulatory Networks: Methods and Protocols*,
Methods in Molecular Biology, vol. 786, DOI 10.1007/978-1-61779-292-2_17, © Springer Science+Business Media, LLC 2012

chromatin immunoprecipitation (ChIP) with whole-genome DNA microarrays (chips) or high-volume sequencing. However, both these approaches require antibody precipitation from the sample. For many reasons, this may not be available to all researchers for all transcription factors.

The yeast one-hybrid technique takes a known promoter element cloned upstream of a reporter gene (such as *URA3*) and screens a library of cDNAs for a protein that will bind to the promoter and activate the reporter gene. A potential strategy to reverse this approach and characterize DNA–protein interactions of a known transcription factor is a modification of the yeast one-hybrid screen first developed by the Voglestein lab to identify targets of p53 binding (3). At that time, this was an effective strategy, but downstream analysis was difficult without the availability of the entire human genome sequence to give the fragments a broader context for analysis. With the advent of whole genomes being sequenced and new techniques for multispecies comparisons to identify conserved, noncoding elements, the potential power of this inverse one-hybrid approach has significantly increased. Genomic fragments can, in principle, be sequenced in high throughput, mapped unambiguously to a genomic location, and more than one species can be processed in parallel to determine common map positions or potential regulatory targets. We have created tools and protocols to perform yeast inverse one-hybrid screening on both zebrafish and mouse genomic DNA libraries (4). The library is maintained in yeast of opposite mating type to the strain expressing the desired transcription factor, allowing for efficient screening of 10^7 genomic fragments or essentially full coverage of the genome. The plasmid expressing the transcription factor contains the *ADE2* gene, allowing us to efficiently identify color colonies that require the target transcription factor for growth on -uracil plates. Colonies not requiring the plasmid carrying the transcription factor for growth, can lose the *ADE2* carrying plasmid, and the resulting *ade2* yeast will turn red (5). In addition, we have adapted a colony PCR and sequencing strategy that will allow for hundreds of genomic fragments to be both sequenced and retested for binding activity. The resulting tools will allow for rapid identification and multispecies comparisons of transcriptional binding targets in cases where ChIP-chip or other available techniques are not viable options.

2. Materials

2.1. Reagents and Kits for Cloning

1. Plasmids: pYoh1 (modified from pACT2 AD vector, Clontech, Mountain View, CA, USA), and pHQ366, a URA3 reporter plasmid (6).

2. Oligonucleotides (oligos) and primers.
   (a) ADE2-primers: (forward 5′-AAT GCA ATC GAT TAA CGC CGT ATC GTG ATT AAC-3′ and reverse 5′-ACG TAA GCG GCC GCC GCT ATC CTC GGT TCT GC-3′).
   (b) pYoh366-primers: P1(5′-GCG CTT TAA GAG AAA ATA TTT GTC CTG-3′) and P2 (5′-GTA GCA GCA CGT TCC TTA TAT GTA GC-3′).
   (c) pYoh366 sequencing primer: P3 (5′-CTC AAT ATA CTC CTA ATT AAT AC-3′).

   All oligos and primers were synthesized by Integrated DNA Technologies (IDT)
3. Enzymes and buffers for cloning:

   EcoR I, Tsp509I, T4 DNA ligase, Shrimp Alkaline Phosphatase (SAP). Restriction enzymes and T4 DNA ligase with NEBuffers were purchased from New England BioLabs (Ipswich, MA, USA); SAP was purchased from Promega (San Luis Obispo, CA).
4. PCR reagents
   (a) Platinum Taq DNA polymerase and buffer (Invitrogen, Carlsbad, CA, USA).
   (b) dNTP mix (2.5 mM) (Invitrogen).
   (c) UltraPure agarose (Invitrogen).
5. Competent cells (Invitrogen)

   One Shot® TOP10 Chemically Competent *E. coli*, Electro-MAX™ DH5α-E™ Cells.
6. DNA purification kits (Qiagen, Valencia, CA, USA)

   QIAprep Spin Miniprep Kit, QIAGEN Plasmid Midi Kit, QIAquick Gel Extraction Kit, QIAquick PCR Purification Kit.

2.2. Yeast Manipulations

1. Yeast strains: W303 (MATα and MATa).
2. YPDA medium: 20 g/L Difco peptone,10 g/L Yeast extract, 20 g/L glucose, 20 mg/L Adenine hemisulfate, 20 g/L Agar (for plates only). Add H_2O to 1,000 mL. Adjust the pH to 6.5 and then autoclave for 15 min at 121°C. See Note 1.
3. SD-Ade-Trp-Ura Dropout (DO) medium: 6.7 g Yeast nitrogen base without amino acids (Difco), 20 g Agar (for plates only), 0.71 g CSM-Ade-Trp-Ura dropout mixture (Sunrise Science Product, San Diego, CA, USA), to 1 L of deionized water. To prepare other SD-DO media, such as SD-Ade, SD-Trp, SD-Ura, SD-Ade-Trp, and SD-Trp-Ura, supplement SD-Ade-Trp-Ura medium with the appropriate amino acids as follows: 0.02 g L-Adenine hemisulfate (Sigma), 0.02 g L-Tryptophan (Sigma), or 0.02 g L-Uracil (Sigma). Mix and autoclave at 121°C for 15 min.

4. 5-FOA plates: Autoclave 20 g of agar in 750 mL deionized water. Add 3 g SC+1 mg/mL 5-FOA mixture (Sunrise) to 250 mL of deionized water. Warm up 5-FOA mixture to 55–65°C on a heat plate and stir until it dissolves (this takes ~30 min to 1 h). After the 5-FOA is completely dissolved, sterilize it using a 2-μm filter. Place both melted agar and the 5-FOA containing solution in a water bath (50–60°C) for 1 h. Finally, add the 5-FOA solution into the melted agar and pour plates.
5. LiOAc (lithium acetate) mix: 100 mM LiOAc, 10 mM Tris–HCl (pH 7.5), 1 mM EDTA. Make fresh before use, filter, and store on ice.
6. PEG mix: 40 mL 50% PEG (molecular weight 3,350), 5 mL 1 M LiOAC, 500 μL 1 M Tris–HCl (pH 7.5), 100 μL 0.5 M EDTA, and 4.4 mL H_2O. Make fresh, filter, and store on ice.
7. SOS: 25 mL YPD, 700 μL 1 M $CaCL_2$, 270 μL 1% uracil, 18.2 g sorbitol, and H_2O to 100 mL. Dissolve and filter-sterilize. Store at 4°C.
8. Carrier DNA (Salmon sperm DNA, 10 mg/mL). Store at –20°C.
9. (Optional) Alkali cation yeast transformation kit (Qbiogene, Carlsbad, CA), contains TE, pH 7.5, Lithium/Cesium acetate, carrier DNA (10 mg/mL), Histamine, PEG, TE/cation mix, and SOS.

2.3. DNA Sequencing Reagents and Materials

1. Shrimp alkaline phosphatase (SAP) (Amersham Pharmacia Biotech, Piscataway, NJ, USA).
2. Exonuclease I (USB, Cleveland, OH, USA).
3. Big Dye Terminator (BDT) mix: 50 μL of BDT v3.1 reaction mix (Applied Biosystems), 175 μL of 5× BDT v 1.1/3.1 sequencing buffer (AB), add ultrapure water to 400 μL.
4. Formamide/Crystal violet loading dye (1,000×): 0.5–1 g crystal violet, 50 mL ultrapure water. Dilute 1:1,000 with deionized water to make 1× dye. Store at –20°C.
5. (Optional) ABI BigDye Terminator cycle sequencing kit.
6. DyeEx96 kit (Qiagen).

2.4. Zebrafish Genomic DNA Extraction Buffer

100 mM NaCl, 10 mM Tris–HCL pH 8.0, 10 mM EDTA pH 8.0, and 0.4% SDS, before use add 200 μg/mL Proteinase K.

2.5. Other Equipment

1. Cell-Porator *E. coli* Electroporation System (Life Technologies).
2. Applied Biosystems ABI 377 DNA sequencer.

3. Methods

We modified the yeast one-hybrid system by using the ADE2-color selection in yeast (Fig. 1) to identify genomic sequences that can activate transcription in the presence of a known transcription factor. The new system makes use of two plasmids. The first plasmid, pYoh366, contains a *URA3* reporter downstream of the *SPO13* promoter and uses *TRP1* as a selective marker (6) (Fig. 2). Random genomic DNA sequences are cloned into the *SPO13* promoter (*SPO13* is actively repressed in growing yeast; thus, the *SPO13* promoter reduces background *URA3* expression). URA3 is not expressed unless the inserted genomic DNA fragment contains a promoter sequence bound by the tested transcription factor. The second plasmid, pYoh1 (marked with *ADE2* as the selective gene), allows for fusion of the transcription factor of

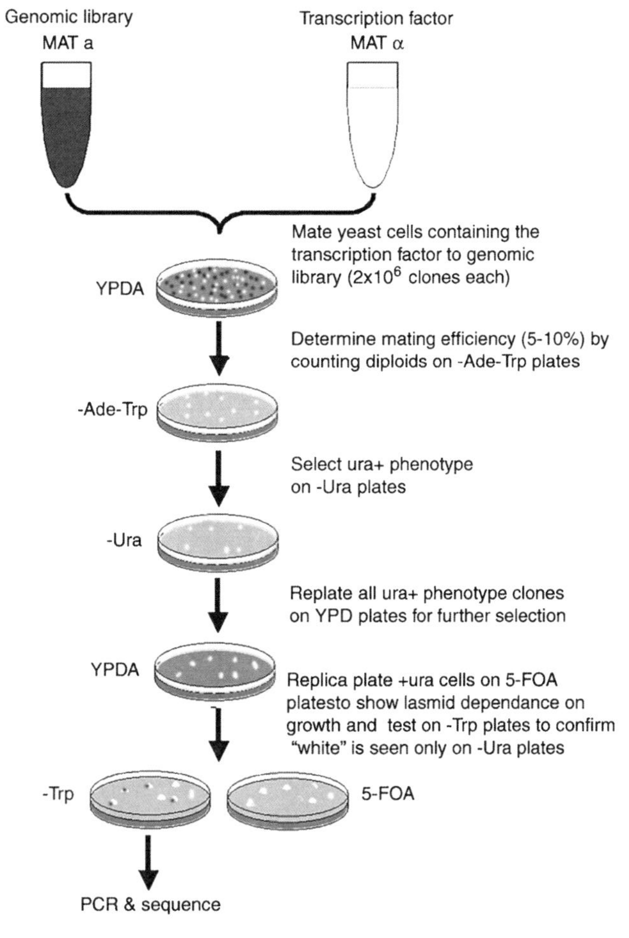

Fig. 1. Strategy for identifying tanscription factor-DNA binding sites from a genomic library in yeast using the "inverse one-hybrid" approach.

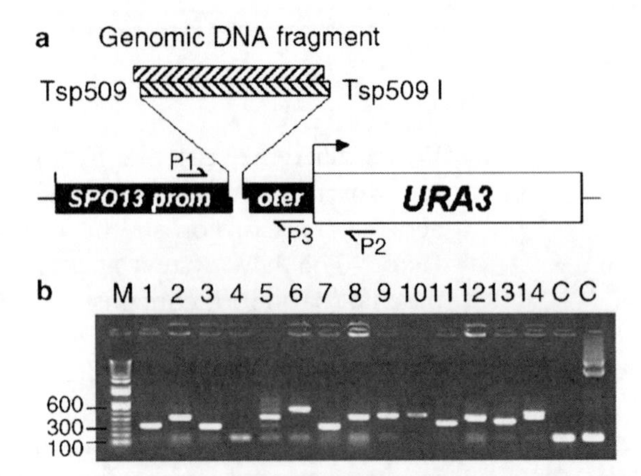

Fig. 2. Generation of the genomic libraries. (**a**) The zebrafish genomic DNA library was constructed in the *URA3* reporter vector, pYoh366. pYoh366 was modified from pHQ366 (*TRP1* as a selective marker) by replacing EcoRI-Pst I insert with pYoh366-linker. P1, P2, and P3 indicate the locations of the primers used for amplification and sequencing. (**b**) The library quality was assessed by PCR using primers P1 and P2 to amplify 14 random colonies (1–14) and two empty vector samples as a control (C) The zebrafish genomic DNA library contains 3×10^7 colonies (average size of 300 bp). P1, P2, and P3 indicate the locations of the primers.

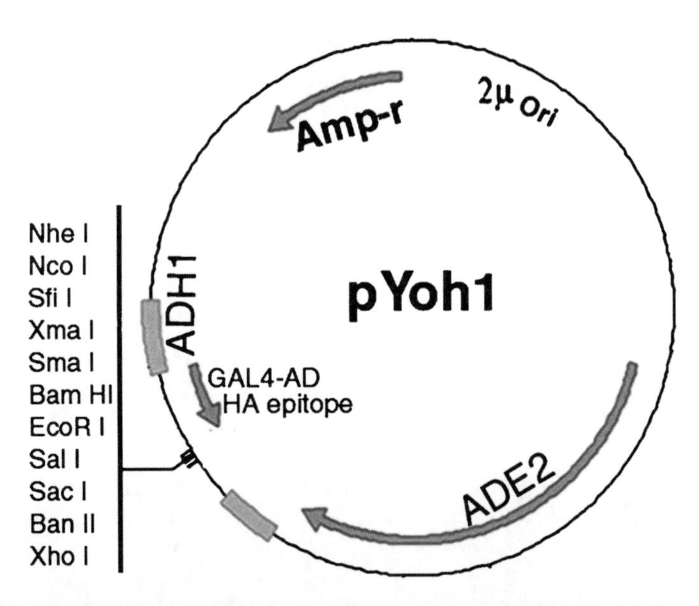

Fig. 3. Construction of transcription factor expressing vector (pYoh1). pYoh1 was modified from pACT2 by inserting the wild-type *ADE2* gene at the ClaI-NotI sites and adding pYoh1-MCS1. Shown is a vector map and the multiple cloning site of the pYoh1 vector.

interest with the *GAL4* activation domain (AD). By fusing the AD to the transcription factor, it ensures transcriptional activation of *URA3* regardless of the normal cellular function (activator or repressor) of the transcription factor. In addition, we introduced the *ADE2* gene into the pYoh1 plasmid (Fig. 3) for reasons that are described later.

To efficiently screen for protein–DNA interactions, we employ a yeast mating strategy (7). Mating two haploid yeast strains of opposite mating type, each harboring one half of the screening system, results in the formation of doubly transformed diploid zygotes. This results in a much more efficient screening of millions of library plasmids than transformation does. Selection for diploids is accomplished by using a combination-Trp-Ade plate. Cells are then washed off and replated on a -Ura plate. Only cells in which there is an interaction between the transcription factor expressed from pYoh1 and the genomic DNA fragment will survive on the -Ura-selective plates. If by chance there is a genomic DNA sequence that can be activated by an endogenous yeast transcription factor, then the presence of pYoh1 is not required and will be lost at some frequency (roughly 1% per cell division). The loss of the *ADE2* gene causes the resulting *ade2* yeast to turn red on plates with low adenine supplement. Therefore, it is a simple visual test to identify colonies that require the test transcription factor to grow; positive colonies are entirely white, while a false positive will have "sectors" of red. The all white, ura+ clones are duplicated on SD-Trp plates and 5-FOA selective plates to demonstrate that the growth on SD-Ura plates is not a gene conversion event (these colonies would not be viable on 5-FOA). Individual positive clones that display sectoring pink on SD-Trp plates and grow on 5-FOA as well are PCR-amplified, and the PCR fragments are directly sequenced in 96-well format (Fig. 1). "Sectoring" is caused by the randomly loss of the *ADE2* carrying plasmid. *ADE2* plasmid is lost in one cell in roughly 10% of all cell divisions. A colony that is randomly losing the *ADE2* marker on a SD -Trp plate will have pie-shaped wedges of pink-colored cells as well as other wedges of white where the plasmid has been maintained (see the first panel of Fig. 4 for an example). If the cells can grow on 5-FOA plates, this indicates that the growth on SD-Ura

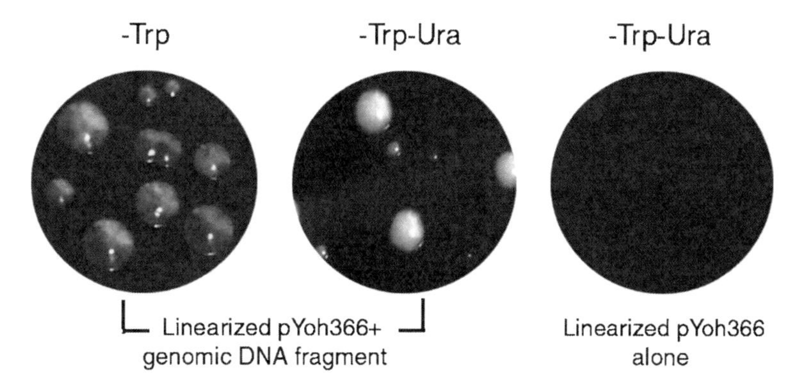

Fig. 4. Clone validation. The figure shows the confirmation of a PCR'ed genomic fragment cotransformed with the linearized pYoh366 vector. The homologous ends of the PCR fragment recombine with the plasmid ends allowing for recircularization to occur. Growth on -trp plates shows plasmid rescue, the color sectoring is evident when the TF is not required for growth. On the -ura plate, both the genomic fragment and the TF are needed for growth (no growth without the fragment and white colonies with the genomic fragment).

plates is controlled by the transcription factor. Colonies that grow on 5-FOA should be completely pink (no white cells), as only the cells that have lost the transcription factor expressing *ADE2* plasmid will be able to grow in the presence of 5-FOA.

3.1. Generation of the Genomic Library

This section includes extracting genomic DNA, constructing a genomic library (pYoh366-g), and transforming this genomic library into yeast cells.

3.1.1. Preparation of Genomic DNA

1. Dissect a small piece of tissue or collect cells from the organism of interest, mince tissue (if necessary) and transfer into a 15-mL centrifuge tube containing 4 mL of extraction buffer. Incubate at 50°C overnight with occasional gentle swirling.
2. Cool solution to room temperature and extract with 2 mL of equilibrated phenol. Mix gently until an emulsion has formed; separate phases by centrifugation at 3,000 × *g* for 10 min.
3. Remove the aqueous phase carefully using 1 mL pipette to a fresh tube. Overlay with an equal volume of isopropanol by slowly letting it run down the side of the tube. Swirl the tube gently until the solution is thoroughly mixed. The DNA will precipitate immediately and should be easily visible.
4. Remove the precipitated DNA using a Pasteur pipette and transfer it to a tube containing 70% ethanol. Let DNA stand in 70% ethanol for about 5 min and gently move it around from time to time using the Pasteur pipette.
5. Remove as much of the 70% ethanol as possible. Let the DNA air-dry until the last visible traces of ethanol have evaporated.
6. Resuspend the DNA in 0.5 mL TE (pH 8.0) and 100 μg/mL DNase-free RNase A.
7. Extract the DNA solution with phenol–chloroform–isoamyl alcohol (as in step 3) and transfer the aqueous phase to a fresh tube.
8. Add 0.1 volume of 7.5 M ammonium chloride and overlay with two volumes of ethanol. Precipitate DNA by slowly inverting the tube until the solution is thoroughly mixed. Rinse twice in 70% ethanol as in steps 5 and 6.
9. Resuspend DNA in 200 μL of nuclease-free water for further enzyme digestion. See Note 2.

3.1.2. Constructing a Genomic Library: pYoh366-g

1. EcoRI (5 U) digest pYoh366 (2 μg of vector in 50 μL digestion reaction) at 37°C for 2 h, add 1 μL of SAP for 1 h at 37°C to dephosphorylate the vector. Stop the reaction by incubating at 80°C for 15 min.
2. Compare digested and undigested pYoh366 on a 1% agarose gel. Undigested pYoh366 should run faster than the

digested version. Rescue and purify the digested pYoh366 using QIAEX II purification.

3. Rapidly mix 50 μg of zebrafish genomic DNA in a 50 μL digestion reaction containing 0.01 U Tsp509I in NEBuffer 1. Incubate at 65°C for 10 min.
4. Run digested genomic DNA on a 1% agarose gel and using a razor blade, cut out the genomic DNA ranging from 200 to 800 bp in length, cut gel into pieces, and split gel fragments into three or more tubes. Use QIAEX II gel purification to isolate genomic DNA fragments.
5. Mix the digested genomic DNA (about 10 μg), 2 μg of the digested pYoh366, and 20 μL of T4 DNA ligase in 0.5 mL of ligation reaction, and incubate at 16°C overnight with occasional mixing.
6. Phenol–chloroform-extract and ethanol-precipitate the ligated DNA. Dissolve in 20 μL of 0.5×TE.

3.1.3. Electroporation

The number of clones needed for the genomic library depends directly on the size of the genome. For good representation of the genome, a target of 5–10× coverage should be attempted. For example, with a genome size of 1 GB, the target number of clones would be 2×10^7 (500 bp inserts $\times 2 \times 10^7 = 10$ GB).

1. Place 0.5 mL S.O.C medium in each culture tube on ice.
2. Aliquot 20 μL of pYoh366-g plasmids into ten Eppendorf tubes, 1–2 μL in each tube on ice.
3. Thaw ELECTROMAX DH10B cell on wet ice.
4. When cells are thawed, mix cells by tapping gently. Add 19 μL of cells to each chilled Eppendorf tube containing DNA plasmid mix.
5. Pipette the cell–DNA mixture into a chilled disposable microelectroporation chamber.
6. Electroporate at 2.2 kV, 2 μF, and 4 kΩ. The time should be set at ~4–5″ (Cell-Porator, Life Technologies).
7. Remove the cells from the microelectroporation chamber and immediately transfer to the culture tube containing S.O.C.
8. Repeat steps 5–7 till all electroporations are finished.
9. Shake at 225 rpm at 37°C for 30 min.
10. Pool all transformants (5 mL), mix, dilute 5 μL of the transformants into 495 μL of LB medium, and plate 10 μL onto five 100 mm LB-ampicillin plates each to calculate the electroporation efficiency. Plate the remaining transformants onto 50 large plates and incubate overnight at 37°C.
11. It is possible to preserve unused transformants by adding an equal volume of 80% glycerol–PBS and store at –80°C.

12. To calculate the library size, count the total number of colonies on small five plates, and multiply by the dilution factor, i.e., 5×10^3). For example, if the total number of colonies on five small plates is 6,000, then the total number of colonies of the library is 3×10^7.
13. Harvest all colonies by rinsing the plates with 10 mL of LB media and rubbing colonies off with a culture spreader. Remove LB media from all plates and use to inoculate 500 mL of LB media. Let grow overnight and purify the plasmid library using a Qiagen Maxiprep column according to the manufacturer's instructions. Resuspend pYoh366-g in 20 μL of TE, pH 7.5. Preserve 10–20 mL of culture media by adding an equal volume of 80% glycerol–PBS, after which the resulting mix is aliquotted into Eppendorf tubes and stored at −80°C.

3.1.4. Transformation of pYoh366-g into Yeast Strain W303 (MATa)

1. Inoculate 50 mL YPDA with a single colony of W303 (*MATa*) and grow with shaking at 30°C until cells reach early stationary phase (~$0.6–2 \times 10^8$ cells/mL).
2. Collect the cells in a 50-mL sterile conical tube, spin 5 min at 400 × *g* at 4°C, and keep the cells on ice throughout the procedure.
3. Wash the cells with 40 mL ice-cold sterile dH_2O and pellet for 5 min at 400 × *g* at 4°C.
4. Repeat wash with 20 mL sterile dH_2O (ice-cold) and pellet the cells.
5. Resuspend cells in 5 mL 1 M Sorbitol (ice-cold) and pellet for 5 min at 400 × *g* at 4°C.
6. Resuspend cells with 190 μl 1 M Sorbitol (ice-cold), aliquot 19 μL into ten microcentrifuge tubes that contain 1–2 μL (1 μg) of pYoh366-g and keep on ice.
7. Pipette the cell–plasmid mixture into a prechilled microelectroporation chamber (20 μL). Tap contents to the bottom, making sure that the sample is in contact with both sides of the aluminum chamber.
8. Electroporate at 2.2 kV, 2 μF, and 4 kΩ. The time should be set at ~4–5″ (Cell-Porator, Life Technologies).
9. Immediately transfer the cells to a sterile Eppendorf tube containing 0.5 mL 1 M Sorbitol (ice-cold). Spread onto five SD-trp plates.
10. Repeat steps 7–9 until all ten cell–DNA mixtures are electroporated.

11. Calculate the transformation efficiency (cfu/μg DNA) by counting colonies (cfu) growing on plates from serial dilutions: cfu × total suspension volume (μL) × fold dilution/μg DNA.
12. Plate remaining cells on ten large, square culture dishes SD-trp, +5-FOA (Thermo Scientific), grow at 30°C for 72 h. Wash colonies off plates with 10 mL SD media. See Note 3.

3.2. Ectopic Expression of Target Transcription Factors in Yeast Cells

3.2.1. Introducing the Target Transcription Factor into pYoh1

1. PCR-amplify 0.1 μg of the transcription factor cDNA with restriction sites designed to preserve the open reading frame and allowing for a fusion between the transcription factor of interest and the activation domain of Gal4 (Gal4-AD).
2. Subclone the amplified fragment into pYoh1 using traditional molecular biology techniques. See Notes 4–6.
3. Midi-prep pYoh1-TF.

3.2.2. Transformation of pYoh1-TF into Yeast Cells

1. Thaw yeast strain W303 (*MATα*) cells, streak on a YPDA plate, incubate at 30°C overnight.
2. Inoculate 100 mL of YPDA from a single colony, vortex vigorously for 5 min to disperse any clumps, incubate overnight (16–18 h) at 30°C with shaking at 250 rpm to $OD_{600} > 1.5$.
3. Split the overnight culture into two 50 mL conical tubes, spin to pellet cells at 400 × *g* for 5 min at room temperature, discard supernatant.
4. Resuspend cells in 15 mL LiOAc mix, spin to pellet cells and discard supernatant.
5. Repeat step 4.
6. Gently resuspend cells in 0.5 mL LiOAc mix. Cells are ready for transformation.
7. Set up transformations. To each transformation add:

   5 μL of 5 mg/mL carrier DNA (boiled and cool in ice)

   5 μL histamine solution (optional)

   0.1–1 μg DNA of pYoh1-TF and pRS313-*HIS3* (see Note 7) (10 μL maximum volume), vortex

   100 μL yeast cells, vortex

   0.7 mL PEG Mix, vortex.
8. Incubate at 30°C for 60 min.
9. Heat shock for 20 min in a 42°C water bath; cool to 30°C.
10. Centrifuge cells for 5 s at 13,000 × *g* at room temperature. Remove the supernatant.
11. Resuspend cells in 150 μL of SOS media and plate directly on an SD-His-Ade plate. Incubate at 30°C for 2–3 days.

12. Culture a W303 (*MAT*α)/pYoh1-TF positive clone in 5 mL of SD-His-Ade medium at 30°C with shaking at 250 rpm for 2 days.
13. Inoculate a larger scale culture of W303 (*MAT*α)/pYoh1-TF expressing cells in 200 mL of selective medium until the culture reaches 10^9 cells. See Note 8.
14. Spin down the culture, dissolve the pellet in 3 mL of fresh selective medium. Aliquot the cells in 2×10^8 per tube and add an equal volume of 50% glycerol. Mix and store at −80°C.

3.3. Genome-Wide Screening Transcription Factor Binding Sites

3.3.1. Yeast a/α Mating

1. In order to ensure mating and also allow for loss from cell death, thaw one vial of W303(*MAT*α)/pYoh1-TF (2×10^8cells) and inoculate in 100 mL SD-Ade-His medium; thaw 1 vial of W303 (*MATa*)/pYoh366-g (2×10^8 cells) and inoculate in a separate 100 mL SD-Trp culture overnight. Spin cells down at 400 × *g* for 5 min. Resuspend each in a separate 1,000 mL YAPD and shake in a shaker incubator for 2 h at 250 rpm and 30°C. Pellet cells at 400 × *g* for 5 min. Resuspend in 25 mL each.
2. Mix *MATa* and *MAT*α cells in 1:1 ratio, pellet cells at 400 × *g* for 5 min.
3. Resuspend the pellet in YPDA medium at a cell density of 10^8/mL.
4. Transfer 0.5 mL of mixed cell solution into a 50 mL tube; incubate with shaking at 30°C for 15 min.
5. Plate on a YPDA agar plate, and incubate at 30°C for 4.5 h.
6. Harvest all colonies by washing the plate with 4 mL SD-His-Trp medium, spin at 400 × *g* for 5 min. See Note 9.
7. Resuspend the pellet in 1 mL of SD-his-trp medium, plate 10 μL on a YPDA plate, 10 μL on an SD-His-Trp plate, and the remaining cells on ten SD-His-Ura plates and incubate at 30°C for 2 days. Determine mating efficiency by counting diploids on SD-His-Trp plate/per total colony number on YPDA plates. See Note 10.
8. Using a toothpick, transfer ura + clones to YPDA plates and incubate at 30°C overnight.
9. Replica-plate the clones on SD-Trp plates SD-ura, and 5-FOA plates (Fig. 4). Incubate at 30°C for 2–3 days. Colonies that "sector" on SD-Trp, grow on 5-FOA, and are white on SD-Ura plates are true positives.

3.3.2. Large Scale Sequencing of Positive Clones

1. In each well of a 96-well PCR plate, add 20 μL of PCR reaction mixture containing the pYoh366 primers.
2. Use a toothpick or a 10-μL pipette tip to pick a tiny (a small amount is important for success, too much will inhibit the

reaction) spot of fresh-growth cells (the cells become increasingly refractory to PCR the longer they have been in stationary phase) from the individual clones and rinse into 20 μL of a PCR reaction cocktail (one clone, one well) and PCR-amplify the positive clones using primers P1 and P2 (5 min 95°C, 30 × (30 s 95°C, 30 s 55°C, 1 min 72°C)).

3. Treat the PCR product with 0.3 U shrimp alkaline phosphatase (Promega) and 3 U EXO I (USB) for 1 h at 37°C and 80°C for 15 min.
4. Add 95 μL of ddH_2O to dilute the reaction.
5. In a fresh 96-well PCR plate, add 5 μl of presequencing mix: 4 μL of BDT mix and 1 μL of pYoh366 sequencing primer P3 (3.2 pmol).
6. Transfer 5 μL of the diluted PCR reaction from step 4 to the sequencing plate in step 5 using a multichannel pipette. Mix and spin at 3,000 × *g* 4°C for 5 min.
7. Place the plate in a PCR machine using the program: 96°C 10 s, 52°C 10 s, and 60°C 4 min for 30 cycles.
8. Purify the sequencing reaction using a Qiagen DyeEx™ kit following the manufacturer's instructions. See Note 11.
9. After purification, dry the 96-well plate in a vacuum dryer for 20–30 min.
10. Add 7 μL of formamide solution and spin at 3,000 × *g* at 4°C for 5 min. Transfer products immediately to an ABI 377 DNA synthesizer or store at −20°C until needed. See Note 11.

3.3.3. Validation of Positive Genomic Fragments

1. Use primers P1 and P3 to PCR amplify positive clones and a negative clone in Subheading 3.3.1. See Note 12.
2. Purify the PCR products using a QIAquick PCR purification kit following the manufacturer's instructions.
3. EcoRI digest 1 μg of pYoh366 at 37°C for 2 h, add 1 μL of SAP and incubate at 37°C for 1 h. Inactivate SAP at 80°C for 15 min.
4. Purify the linearized pYoh366 using the QIAquick PCR purification kit.
5. Cotransform 0.5 μg of the PCR products in step 2 and 0.1 μg of the linearized pYoh366 in step 4 into W303(*MATα*)/pYoh1-TF cells. Fragments will homologously recombine with the linearized pYoh366 plasmid to create a functional plasmid.
6. Plate on SD-Trp-Ura plates and incubate at 30°C for 2–4 days. Colonies should grow and remain white.

4. Notes

1. Autoclaving for a longer time can darken the color of glucose-containing YPDA medium and decrease performance. Alternatively, allow medium to cool to ~55°C and then add dextrose (glucose) to 2% (50 mL of a sterile 40% stock solution). YPD medium is the same as YPDA, but without addition of an adenine hemisulfate supplement. Both can be used as enriched medium, but *ade2* mutant strains will grow more slowly in YPD. Store all yeast culture media at room temperature and plates at 4°C.
2. Genomic DNA is difficult to resuspend, and this may take several hours. Place DNA solution at 37°C and gently mix and swirl from time to time. Resuspending can be facilitated by warming the DNA solution to 50°C. For long-term storage, dissolve genomic DNA in TE, pH 8.0 and store at 4°C or −80°C; do not store DNA frozen at −20°C.
3. Growing the library on 5-FOA-containing plates eliminates "auto-activating" fragments, i.e., some genomic sequences might contain binding sites for endogenous yeast transcription factors, and 5-FOA will suppress the growth of yeast colonies containing these fragments.
4. It is important to functionally verify the sequence of the amplified wild-type *ADE2* gene. Yeast cells with a mutated *ade2* gene become pink on YPDA plates.
5. The PCR program will vary depending on the required conditions. The annealing temperature depends on the melting temperature of the primers used, while the extension time depends on the length of the target DNA fragment. We generally consider 1 min per 1 kb of target DNA extension.
6. It is required to confirm compatible NEBuffers at the NEB Web site before performing double digestions. If it is not possible to perform double digestion in a single reaction buffer, then digest 20 μL in the lower salt buffer containing the first enzyme, and then increase the digestion volume to 30 μL with the second enzyme.
7. Transforming the *HIS3* marker into the same cells containing pYoh1-TF (*ADE2*) allows for the selection of successfully mated cells using-His-Trp-Ura media. Because *ADE2* is not specifically required for growth, it allows for the pYoh1-TF to sector on the -ura library plates if the fragment generates a false positive (i.e., *URA3* expression independent of the expressed TF), while not allowing growth of any unmated haploid cells that will not have all the necessary markers.

8. Before large-scale screening, it is important to test that the transcription factor is expressed and correctly fused with the AD and HA. In our case, we tested 1 mL of W303 (*MATα*)/pYoh1-FoxI1 by Western blot analysis to confirm that the Foxi1-fusion with HA epitope was being properly expressed. Detailed in reference (8).
9. Calculate the mating efficiency using the formula: Number of diploid colonies on the double selective plate/½(the total number of colonies on a YPDA plate + the number of observed diploid colonies).
10. Alternatively, the sequencing reaction can be cleaned up by isopropanol (IPA) precipitation:
    (a) Spin the sequencing plate briefly. Add 30 μL 75% Isopropanol and gently shake to mix. Incubate at room temp for 15 min.
    (b) Centrifuge at 3,000 × *g* for 30 min. Invert onto a paper towel on a bench top until most of the IPA leaks out. Do not bang/slap/slam/tap the inverted plate or pellet could be lost.
    (c) Add 50 μL 70% IPA. Spin at 3,000 × *g* for 10 min. Invert onto fresh paper towels.
    (d) With plate inverted on paper towel, centrifuge at 200 × *g* for 1 min to remove the residual IPA. Air-dry for 20 min.
    (e) Resuspend in 7 μL formamide loading dye.
11. Different sequencers may use different sequencing reagent kits, refer to the specific manual. Adaptations to the PCR primers are possible for analyzing PCR products through next-generation sequencing technologies.
12. For the most stringent control, mutate the transcription factor binding sequence and cotransform with linearized pYoh366 into W303 (MATα)/pYoh1-TF.

Acknowledgments

The authors would like to thank Deborah M. Davis for her technical assistance, Rainer K. Brachmann for providing the pHQ366 plasmid, Kj Myung for plasmid pRS313, Hong Liu for providing wild-type yeast genomic DNA samples, and Carl Wu for providing the W303 strains *MATa* and α. JY was partially supported by the Shanghai Leading Academic Discipline Project- (S30701). This research was supported by the Intramural Research Program of the National Human Genome Research Institute, National Institutes of Health (SB).

References

1. Kim, S., Ip, H. S., Lu, M. M., Clendenin, C., and Parmacek, M. S. (1997) A serum response factor-dependent transcriptional regulatory program identifies distinct smooth muscle cell sublineages, *Mol Cell Biol 17*, 2266–2278.
2. Buck, M. J., and Lieb, J. D. (2004) ChIP-chip: considerations for the design, analysis, and application of genome-wide chromatin immunoprecipitation experiments, *Genomics 83*, 349–360.
3. Tokino, T., Thiagalingam, S., el-Deiry, W. S., Waldman, T., Kinzler, K. W., and Vogelstein, B. (1994) p53 tagged sites from human genomic DNA, *Hum Mol Genet 3*, 1537–1542.
4. Zeng, J., Yan, J., Wang, T., Mosbrook-Davis, D., Dolan, K. T., Christensen, R., Stormo, G. D., Haussler, D., Lathrop, R. H., Brachmann, R. K., and Burgess, S. M. (2008) Genome wide screens in yeast to identify potential binding sites and target genes of DNA-binding proteins, *Nucleic Acids Res 36*, e8.
5. Hieter, P., Mann, C., Snyder, M., and Davis, R. W. (1985) Mitotic stability of yeast chromosomes: a colony color assay that measures nondisjunction and chromosome loss, *Cell* ***40***, 381–392.
6. Qian, H., Wang, T., Naumovski, L., Lopez, C. D., and Brachmann, R. K. (2002) Groups of p53 target genes involved in specific p53 downstream effects cluster into different classes of DNA binding sites, *Oncogene 21*, 7901–7911.
7. Soellick, T. R., and Uhrig, J. F. (2001) Development of an optimized interaction-mating protocol for large-scale yeast two-hybrid analyses, *Genome Biol 2*, RESEARCH0052.
8. Yan, J., Xu, L., Crawford, G., Wang, Z., and Burgess, S. M. (2006) The forkhead transcription factor FoxI1 remains bound to condensed mitotic chromosomes and stably remodels chromatin structure, *Mol Cell Biol 26*, 155–168.

Chapter 18

Using cisTargetX to Predict Transcriptional Targets and Networks in *Drosophila*

Delphine Potier, Zeynep Kalender Atak, Marina Naval Sanchez, Carl Herrmann, and Stein Aerts

Abstract

Gene expression regulation is a fundamental biological process leading to complete organism development by controlling processes like cell type specification and differentiation. The accuracy of this process is governed by transcription factors (TFs) acting within a complex gene regulatory network. CisTargetX has been developed to enable a user to predict TFs, enhancers, and target genes involved in the regulation of co-expressed genes. It uses a strategy that incorporates the genome-wide prediction of clusters of transcription factor binding sites (TFBSs), starting from a large, unbiased collection of position weight matrices (PWMs) and uses comparative genomics criteria to filter potential TFBS. We describe in this chapter, step-by-step, how to use cisTargetX starting from a set of genes or TF(s) to predict transcriptional targets with their putative binding sites and networks in *Drosophila*. Next, we illustrate this approach on a particular developmental system, namely, sensory organ development, and identify relevant TFs, DNA regions regulating gene expression, and TF/target gene interactions. CisTargetX is available at http://med.kuleuven.be/lcb/cisTargetX.

Key words: *cis*-Regulatory module, Gene regulatory network, Motif discovery, Drosophila, cisTargetX, Transcriptional targets

1. Introduction

Mapping the gene regulatory network (GRN) underlying the developmental and physiological programs of each and every cell type in an organism will require the joint forces of high-throughput technologies, genetic assays, and computational methods. Here, we focus on how computational methods contribute to GRN mapping. We present a bioinformatics application, called cisTargetX, and we show how cisTargetX can be used to predict enhancers and target genes in the genome for specific TFs.

Bart Deplancke and Nele Gheldof (eds.), *Gene Regulatory Networks: Methods and Protocols*,
Methods in Molecular Biology, vol. 786, DOI 10.1007/978-1-61779-292-2_18, © Springer Science+Business Media, LLC 2012

High-confidence TF to target gene predictions either directly represent candidate regulatory interactions in a GRN under study that can be taken to further analysis or validation; or alternatively they can be used as additional filters for TF/target gene interactions that are identified by experimental techniques as illustrated in several other chapters in this book.

The computational prediction of transcription factor binding site (TFBSs) in a genome remains a serious challenge, given the large amounts of false-positive predictions and the yet limited knowledge about the *cis*-regulatory logic. Recent progress in the field has improved the predictions of regulatory elements by taking advantage of the clustering behavior of TFBSs in *cis*-regulatory modules (CRMs) (reviewed in ref. 1); the evolutionary conservation of binding sites (2–5); and the presence of particular chromatin marks in CRMs (6, 7). Generally, CRM scoring methods predict clusters of TFBSs in a genomic region using a TF consensus motif or position weight matrix (PWM) as input. These methods can also take multiple PWMs, representing different TFs, as input and are then used to predict heterotypic CRMs. Ideally, the prediction performance of TFBSs, and hence TF/target gene interactions needs to be high enough to allow direct assimilation of the predicted interactions into the GRN under study. However, although these methods are very useful to score a candidate genomic region under study, using them for a full-genome scoring, even when all these filters are applied, does usually not yield satisfactory prediction accuracies (8), except for well-known TF combinations such as in the *Drosophila* segmentation enhancers (9). Moreover, additional challenges arise when the query TFs and their recognition motifs are not known a priori, when multiple PWMs exist for the same TF, or when a TF has in vivo binding properties that are not represented by in vitro-determined PWMs. Solutions to these problem have traditionally be presented by ***motif discovery*** algorithms that aim to identify over-represented motifs across a set of co-expressed genes. Although advanced statistical solutions to this problem have been presented (10), they are only applicable to small search spaces, such as the 1 kb promoter sequences of the co-expressed genes (11–13). Even under these simplified conditions, the prediction accuracies are often not high enough to allow immediate follow-up by experimental validations.

Recently, two types of approaches have proven to be accurate enough to allow immediate follow-up of the predictions by experimental validations. The first approach uses high-quality training sets of similar CRMs to construct well-defined enhancer models (14, 15). The second approach is independent of previously known enhancers but combines gene expression data with genome-wide scoring to simultaneously perform motif discovery and CRM prediction. Examples of these methods are PhylCRM/Lever (16) and ModuleMiner (17) for vertebrates, and cisTargetX (18) for *Drosophila*. These methods combine motif clustering in CRMs, comparative genomics, genome-wide CRM scoring, gene co-expression, and use large collections of candidate PWMs. They allow to include much larger sequence search spaces for motif discovery and are no longer limited to proximal promoter sequences. From the computational and experimental validations performed for these methods, the performance of correct identification of TF/target gene interactions is high enough (usually above 50%) to allow resource-effective experimental validation.

In this chapter, we describe the different components of the cisTargetX method (Fig. 1 and Subheading 2.2) and we illustrate its usage on a case study, using various sets of co-expressed genes extracted from online databases. This case study (referred to as the PNS case in this chapter) involves gene regulation underlying proneural cell fate specification in the *Drosophila* wing imaginal disc. In this process, the basic Helix-Loop-Helix (bHLH) proneural genes, such as Achaete/Scute and Atonal, operate at the top of the hierarchy and involve lateral inhibition through the Notch signaling pathway. Central to this regulation is the control of gene expression by the Notch effector Suppressor of Hairless (Su(H)). In this case study, we ask whether we can identify proneural and Su(H) target genes, and target genes of other TFs involved in this network, such as pointed (Pnt) and Enhancer of Split (E(spl)).

2. Materials and Methods

2.1. Materials

2.1.1. Sequences

DNA sequences used by cisTargetX are extracted from the April 2004 *Drosophila melanogaster* assembly (referred to as *dm2*) and the April 2006 assembly (referred to as *dm3*). Both versions are available within cisTargetX. For each *D. melanogaster* gene, all introns and the 5 kb upstream of the transcription start site (TSS) are extracted. If the neighboring gene is less than 5 kb away, we extract the largest intergenic upstream region. For cross-species comparisons, orthologous regions are obtained using the LiftOver standalone program from the Kent software suite available from the UCSC Web site. The following assemblies are used for the 11 *Drosophila* species: *Dp3*, *DroAna2*, *DroEre1*, *DroGri1*, *DroMoj2*, *DroPer1*, *DroSec1*, *DroSim1*, *DroVir2*, *DroWil1*, *DroYak1* for the *dm2* release and *Dp4*, *DroAna3*, *DroEre2*, *DroGri2*, *DroMoj3*, *DroPer1*, *DroSec1*, *DroSim1*, *DroVir3*, *DroWil1*, *DroYak2* for the *dm3* release.

2.1.2. Motifs

We compiled various motif collections from several data sources. Currently, two PWM collections can be used in the online cisTargetX application. The first collection, termed PWM_LIB1, contains 1981 PWMs representing TFBS from different taxa such as mammal, insect, yeast, or plants which were used in the original cisTargetX

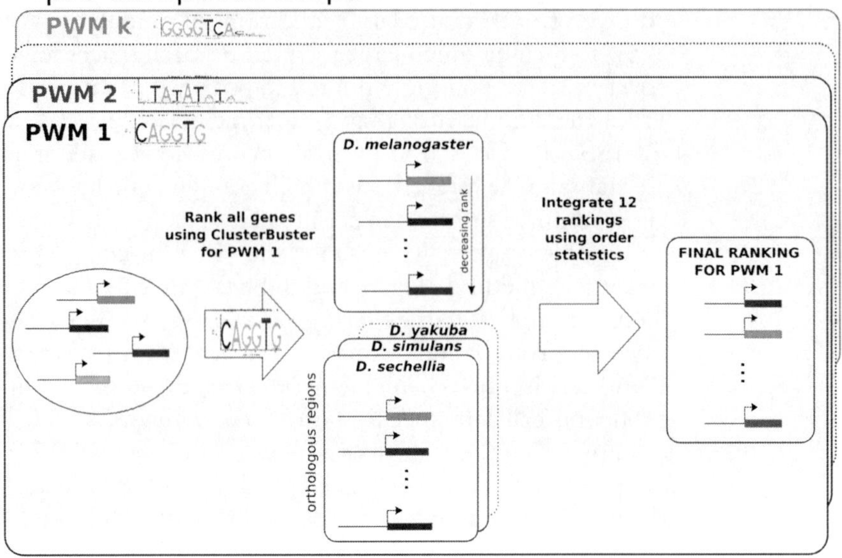

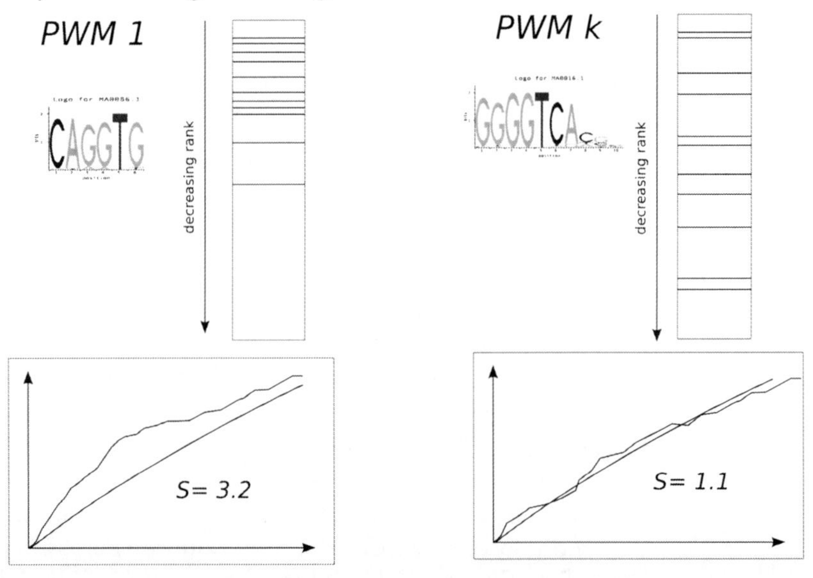

Fig. 1. A general scheme of the cisTargetX procedure; (**a**) Precomputed steps: 5 kb upstream regions and introns for all *Drosophila melanogaster* genes are extracted, and scored using either SWAN or Cluster-Buster for each PWM in the library. For each PWM, all genes are ranked according to the score of the highest scoring cluster in its surrounding region. Simultaneously, orthologous regions in all 11 other *Drosophila* species are scored, and the corresponding genes are ranked as in the melanogaster case. For each PWM, the 12 rankings are combined into a final rank using order statistics. (**b**) Scoring a user gene set: for each PWM, the ranks of the genes (the *vertical rectangles* represent all the genes ranked as explained in (**a**), and the *black lines* indicate the position of the user's genes in this ranking) in the set given by the user are represented as ROC-curves, for which the significance is evaluated using an appropriate score, which measures the deviation from random rankings. The PWMs that achieve a score above a given threshold are displayed in the output screen.

publication (18). This library contains known PWMs from Jaspar3 (19), from the TRANSFAC Professional release (20), and PWMs from curated binding sites from the literature (21). In addition, it contains unknown PWMs derived from motif discovery and comparative genomics (2, 22–24). PWM_LIB2 contains 3338 PWMs derived from the same sources as PWM_LIB1 but now using Jaspar Release 4 which includes PWMs from UNIPROBE (25).

2.1.3. Motif Clustering

To simplify the interpretation of cisTargetX results, motifs represented as PWMs are clustered using STAMP (26) to take into account the redundancy in the library used. We use STAMP with the SSD (sum of squared distances) metric and the -chp option which implements the Calinski and Harabasz statistic to determine the optimal number of clusters. Each motif cluster is assigned a distinct color on the cisTargetX output page. For example, PWM_LIB2 contains four insect PWMs for Su(H), two vertebrate PWMs corresponding to Su(H) homologues, and an unknown motif which appears to be similar to the Su(H) PWM (see Fig. 2). STAMP allows to systematically recognize such clusters of similar matrices and attributes a common color code to this set of PWMs in the cisTargetX output.

2.1.4. CRM Finding Algorithms

For motif scoring, the Cluster-Buster (27) or SWAN (28) algorithms are used; kindly provided by Martin Frith and Saurabh Sinha, respectively. Because these alternative implementations of Hidden Markov Models to score CRMs yield slightly different results, we allow the user to choose between both methods in the online cisTargetX application. In the Procedure Section, we provide results from both methods.

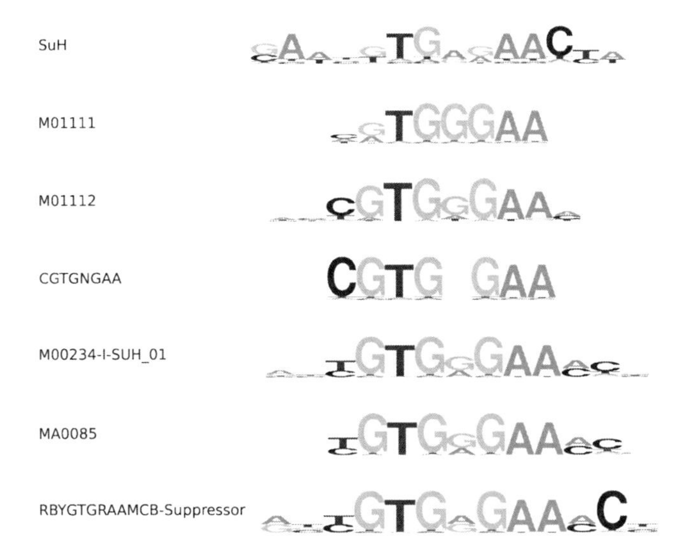

Fig. 2. An example of 7 Su(H)-like matrices that are represented in the PWM_LIB2; STAMP cluster these PWMs together and assigns a common color code in the cisTargetX output.

2.1.5. Gene Sets

Gene sets can be obtained in different ways from different types of sources. In the results, we illustrate various configurations, with gene sets of different sizes (from a few dozen to several hundred genes) extracted from high-throughput data or curated from the literature. Particularly, for the PNS case, we extract gene sets from microarray data sets, ChIP-chip data sets, Gene Ontology, FlyBase, and STRING.

1. High-throughput gene sets:
   (a) We use microarray data from Reeves and Posakony (29) (GEO reference: GDS1068) in which the proneural cluster cells are marked with GFP and specifically selected by a cell-sorting assay. 209 genes are reported by the authors to be specifically enriched in GFP-positive cells with respect to GFP-negative cells (these 209 genes can be found in Supplementary Table 1 of the Reeves and Posakony publication).
2. Annotation-based gene sets
   We use the "TermLink" tool from FlyBase to extract genes:
   (b) With reported expression in a particular anatomical structure, based on the fly anatomy ontology:
      - *Proneural cluster* (FBbt:00001135): 33 genes.
   (c) With a particular Gene Ontology (GO) annotation:
      - *Sensory organ precursor cell fate determination* (GO:0016360): 18 genes.
3. TF-centered gene sets.

STRING (http://string-db.org/) (30) is a database of known and predicted protein–protein associations, including genetic interactions, physical interactions, and associations based on literature co-occurrence or gene co-expression. We query STRING using a TF of interest and extract a network around it. We use medium confidence interactions and set the maximum number of neighbors for each node to 200. We use the query as TF for which we wish to identify target genes, which is Su(H) in our case. We extended the core network by clicking four times on "More" to obtain a larger network containing 114 genes.

2.2. Methods

2.2.1. Principle of the Method

The detailed algorithm of cisTargetX has been presented elsewhere (18). We refer to this publication for details, and will outline the most important aspects of the method in what follows. The key specificities of cisTargetX are (Fig. 1):

- An unbiased, comprehensive collection of PWMs compiled from a wide variety of available data sources (see previous section for a description of this dataset).

Table 1
Result for the PNS case with cisTargetX using PWM_LIB1 with a Cluster-Buster CRM scoring and a cutoff at 2.5

Motif identifier	Motif rank	Cluster	Candidate TF	Enrichment score	Family binding profile consensus
M01111-V-RBPJK_Q4	1[b]	1	*Su*(*H*)	5.7526	GTGGGAA
CGTGNGAA	2[a]	1		4.5729	
M00234-I-SUH_01	3[a]	1		4.4525	
M01112-V-RBPJK_01	4[b]	1		4.3846	
RBYGTGRGAAMCB-Suppressor	15	1		3.1664	
MA0085	17[a]	1		3.0959	
M00497-V-STAT3_02	5[a]	2	Pnt/aop	3.9107	GGAA
M00086-V-IK1_01	11[a]	2		3.3304	
M00498-V-STAT4_01	21[a]	2		2.9167	
M00655-V-PEA3_Q6	6[b]	3	Pnt/aop	3.6967	MGGAWGT
M00032-V-CETS1P54_01	7[a]	3		3.6224	
MA0080	9[a]	3		3.5238	
MA0026	16[a]	3		3.1574	
Eip74EF	23[b]	3		2.7898	
MA0076	24[a]	3		2.7245	
MA0062	26[a]	3		2.695	
M00108-V-NRF2_01	27[a]	3		2.695	
GATCCTC	8	4	New candidate	3.543	GAGGATCT
AGATCCT	19	4		2.9384	
M00025-V-ELK1_02	10[a]	5	pnt/aop	3.3855	YTTCCGG
M00074-V-CETS1P54_02	13[a]	5		3.2343	
MA0098	20[a]	5		2.9294	
MA0028	28[a]	5		2.6669	
TCCTGCA	12	6	New candidate	3.2676	TCCTGCA
MA0081	14[a]	7	pnt/aop	3.2228	ASMGGAA
CAGCTTC	18	8	New candidate	3.0243	CAGCTTC
M00312-V-BEL1_B	22[a]	9	Proneural	2.9064	KNGRNAGT
GCAGSTGK-scute	25	9	bHLH	2.7091	NMNCKCW
M00262-V-STAF_01	29[a]	9		2.6438	GYRTAGCA
M00804-V-E2A_Q2	30[a]	9		2.5925	GNT

[a]Present in one other cisTargetX scoring
[b]Present in all the cisTargetX scoring

- An algorithm to identify clusters of binding sites, like Cluster-Buster (27) or SWAN (28).
- An integration of the 12 *Drosophila* genomes not based on alignments, therefore allowing for binding site reshufflings. In fact, studies have revealed a high rate of binding site turnover in *Drosophila*, meaning that binding sites are frequently gained or lost between *Drosophila* species, often resulting in substantial changes in the organization of the functional elements (31). Alignment-based approaches could therefore miss such elements.
- A statistical analysis based on ranks in the form of a receiver-operating curve (ROC curve) from which (1) area under the curve (AUC) values are computed and transformed into a motif enrichment score, allowing to find over-represented motifs; and (2) the optimal gene cutoff is determined to identify candidate target genes for the enriched motifs.

In order to speed-up the online processing, several computational steps of the algorithm are precomputed offline and stored. Processing of the user gene list is therefore fast: an average gene list of a few dozen genes is processed within 5 min depending on server load and on the number of PWMs in the selected PWM library.

2.2.2. Precomputed Steps

For each PWM, the regions surrounding all the *D. melanogaster* genes (for *dm2* and *dm3*) are scored using either SWAN or Cluster-Buster (see previous section for details on the selection of sequences); for a given gene, if several clusters of binding sites are identified, the top scoring cluster is used, and its score is attributed to the whole region.

This is repeated for all orthologous regions in the 11 other *Drosophila* species: for each *D. melanogaster* gene, the surrounding regions are mapped to the genome of the other *Drosophila* species using the LiftOver tool from the Kent tool suite (UCSC genome browser Web site). Each orthologous region is then scored similarly to step 1 (Fig. 1).

The result of steps 1 and 2 are, for each PWM, 12 rankings of all the *D. melanogaster* genes (13,992 in the dm3 version). In order to combine these rankings, order statistics is used as in ref. 32; the result is a single ranking of all the genes, for each PWM. We will call these integrated motif-based rankings (IMbRs). IMbRs are precomputed offline and stored. If new matrices are added in the library, all regions for all *Drosophila* species need to be scored using these new matrices, which results in a new ranking. Advanced users willing to get insight into the raw rankings might obtain these files on request. The next steps detail what happens when the user inputs a set of genes.

2.2.3. Processing of User Input

Once the user has uploaded or pasted a list of CG identifiers, cisTargetX determines the rankings of the user genes in the IMbR. These rankings are displayed as ROC curves, in which, for each PWM, the x-axis represents the ordered list of all genes, and the y-axis represents the number of user genes encountered above a particular rank. The relevance of the particular PWM for the user gene set is quantified by computing the AUC value of the ROC curve, over a fraction of the total genome set (default: 3%). The rationale behind this choice is that we want to identify PWMs that rank the user-genes best in the top 3% of the genome (which represents about 420 genes) in order to maximize the specificity of the method. In parallel, the mean AUC value over all PWMs is computed together with the standard deviation and will serve as a reference to compute the score as follows

$$S(m) = \frac{a(m) - \bar{a}}{s},$$

where m represents a PWM, a the AUC value, $\bar{a}$ the mean AUC over all PWMs, and σ the standard deviation of AUC values over all PWMs.

The optimal score threshold depends on the size of the input list; simulations based on random sets of various sizes have shown that the maximal score of the gene set decreases with larger set sizes (unpublished data). Therefore, we determined for each set-size an optimal threshold based on random simulations and apply this threshold by default in the online application, indicated as "determine threshold automatically." Note however that this automatically determined cutoff is rather conservative and that the user has the possibility to apply a less stringent cutoff of 2 or 2.5. These ad hoc thresholds were determined empirically based on various test sets and appear to give reasonable results for a wide range of cases.

Once a threshold has been selected (either automatically or by the user), all motifs for which $S(m)$ is greater than this threshold are selected, and clustered using STAMP (26) in order to take into account the redundancy in the library of PWM. The clustering is done as described in the previous section, and each cluster is assigned a color.

On the results page, the enriched PWMs are displayed using their color code, together with their ROC-curve and logo; a link is provided displaying the genes from the user list that are ranked better than an empirically determined rank ("target genes"). This rank is based on the maximum distance between the observed ROC and the average ROC curve over all PWMs. The genes ranked in the top 1,000 genes are also displayed.

3. Procedure

In this section, we present a detailed example, based on a gene set obtained from the microarray analysis of Reeves and Posakony, containing genes that are preferentially expressed in proneural cluster cells (29) (available as Supplementary material at http://med.kuleuven.be/lcb/resources). Each step of the analysis is described below.

Steps 1–3: How to submit a job? (Fig. 3)

Steps 4 and 5: How to interpret and use the results?

Step 6: How to define and extract a CRM?

Step 7: How to reconstruct a gene regulatory network?

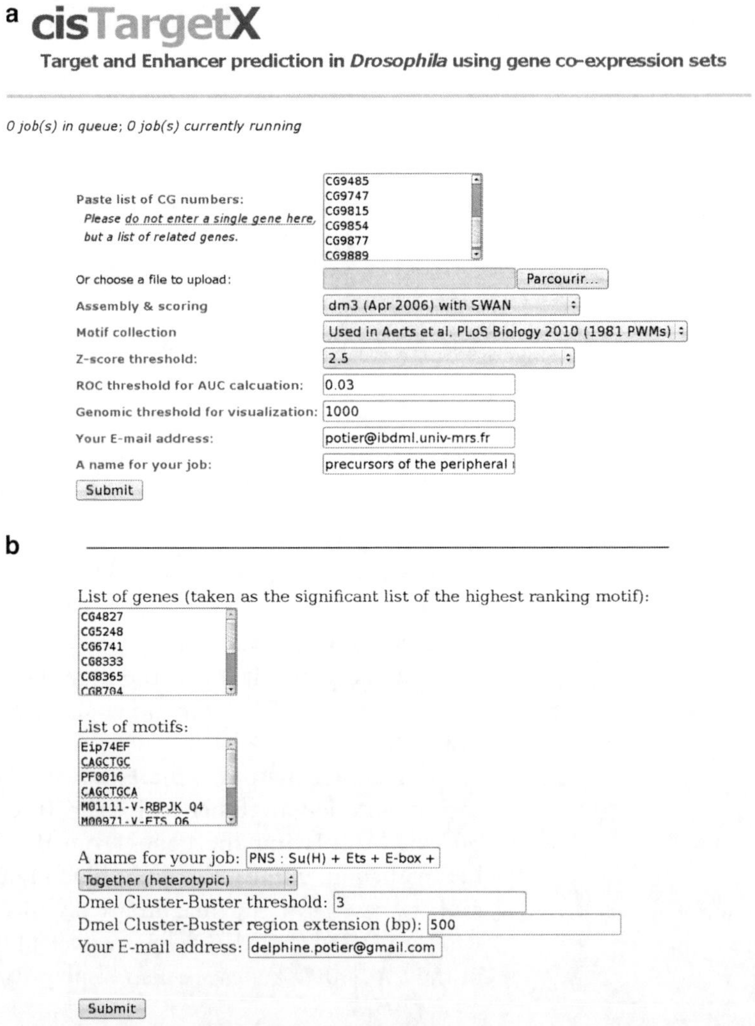

a **cisTargetX**

Target and Enhancer prediction in *Drosophila* using gene co-expression sets

0 job(s) in queue; 0 job(s) currently running

Paste list of CG numbers: CG9485 CG9747 CG9815 CG9854 CG9877 CG9889
Please do not enter a single gene here, but a list of related genes.

Or choose a file to upload: Parcourir...

Assembly & scoring: dm3 (Apr 2006) with SWAN

Motif collection: Used in Aerts et al. PLoS Biology 2010 (1981 PWMs)

Z-score threshold: 2.5

ROC threshold for AUC calcuation: 0.03

Genomic threshold for visualization: 1000

Your E-mail address: potier@ibdml.univ-mrs.fr

A name for your job: precursors of the peripheral

Submit

b

List of genes (taken as the significant list of the highest ranking motif): CG4827 CG5248 CG6741 CG8333 CG8365 CG8704

List of motifs: Eip74EF CAGCTGC PF0016 CAGCTGCA M01111-V-RBPJK_04 M00971-V-ETS_06

A name for your job: PNS : Su(H) + Ets + E-box +

Together (heterotypic)

Dmel Cluster-Buster threshold: 3

Dmel Cluster-Buster region extension (bp): 500

Your E-mail address: delphine.potier@gmail.com

Submit

Fig. 3. (**a**) cisTargetX submission form filled-in for the PNS test case. (**b**) Proceed form to identify CRMs and visualize them on the UCSC genome browser.

3.1. Define a Gene Set and Upload It

Paste or upload the list of *D. melanogaster* genes using Computed Gene identifiers (CGxxxx). See Note 1 for an explanation on how to obtain CG identifiers for *Drosophila* genes and Note 4 concerning the minimal size of the set.

3.2. Choose cisTargetX Parameters

3.2.1. Assembly and Scoring

Choose the genome assembly release and the method of CRM scoring. Two genome releases are available, *dm2* and *dm3* (*dm3* being the more recent; we recommend using this one). Two algorithms for binding site cluster identification can be used: Cluster-Buster (27) and SWAN (28) (see Subheading 2). We recommend using Cluster-Buster in a first approach because individual motif locations cannot be obtained from SWAN, which impedes running the second step of cisTargetX.

3.2.2. Motif Collection

Choose the motif collection. Two motif collections are available (see Subheading 2 for a detailed description) to score the sequences around this set of co-expressed genes. Choose PWM_LIB1 to use the same PWMs as in Aerts et al. (18) or PWM_LIB2 to use an updated PWM library, which represents the most complete collection to date.

3.2.3. Score Threshold

Choose the score threshold: choose the minimal score threshold to consider a motif as being relevant (see Subheading 2 for a detailed description of the score). Based on simulations using random sets of genes of various sizes, we have determined an optimal threshold which depends of the size of the gene list. If this threshold appears to be too stringent (i.e., no or few results are returned), the user can select a different threshold (2 or 2.5).

3.2.4. ROC Threshold for AUC Calculation

As described in Subheading 2, AUC values are not computed on the whole ROC curve, but on a fraction, in which we expect to find genes from our gene set. By default, this value is set to 3%, which corresponds to the first 420 genes, but this value can be changed, although we do not recommend so.

3.2.5. Genomic Threshold for Visualization

The *x*-axis cutoff for visualization of the ROC curve. When set to 1,000, the ROC curve of the top 1,000 genes is shown on the results page.

3.2.6. Your E-mail Address

Provide an e-mail address. The results will be e-mailed to the user at this address once the job is completed.

3.2.7. A Name for Your Job (Optional)

Provide a name for the job.

3.3. Submit the Job

Once the submission form is filled-in, click on the "Submit" button. A page with the job ID, the name of the job, and its status will appear. Once completed (which can take several minutes) a link to the result page will appear, and a message with the URL will be sent by e-mail. Note that the result will stay on the server for 1 month.

For the PNS case, we run cisTargetX on the set of 209 genes from the Reeves and Posakony dataset using the *dm3* release and SWAN scoring with a manual cutoff of 2.5.

3.4. Interpret Motif Results: Select Relevant Motifs

3.4.1. Output Presentation

In the result page, there is at the top of the Web page:

- A link to the input gene list to check which CG identifiers were entered in the submission form (click on "input gene list").
- The enrichment score cutoff you have chosen (in this case 2.5).
- A link to an archive with a file for each displayed (i.e., enriched) motif containing the optimal subset of genes that are predicted as direct targets (click on "download").
- A link to a summary table (click on "view summary" table that provides a survey of the results (Fig. 4)).
- The list of the significant motifs for the entered gene set, with, for each significant motif:
  - In column 1, the name of the PWM.
  - In column 2, the score of your gene set for this particular PWM.
  - In column 3, the logo corresponding to the PWM.

cisTargetX AUC results

Motif	Enrichment score	Logo	ROC	Candidate targets	All genes in top 1000
Eip74EF	4.6235021858339	LOGO	ROC	link	link
CAGCTGC	3.7267001106303	LOGO	ROC	link	link
GGTCACC	3.5907607963140	LOGO	ROC	link	link
PF0016	3.5463278960966	LOGO	ROC	link	link
CAGCTGCA	3.5084573130045	LOGO	ROC	link	link
M01111-V-RBPJK_Q4	3.4744299993046	LOGO	ROC	link	link
M00971-V-ETS_Q6	3.3998914888486	LOGO	ROC	link	link
HLHm5	3.3462948748764	LOGO	ROC	link	link
M00771-V-ETS_Q4	3.2678608056488	LOGO	ROC	link	link
TIFDMEM0000033	3.2103424882153	LOGO	ROC	link	link
M00655-V-PEA3_Q6	3.1606806500493	LOGO	ROC	link	link
SelexConsensus_sna	3.1593734155622	LOGO	ROC	link	link
CGTGNGAA	3.1463402877255	LOGO	ROC	link	link
M01112-V-RBPJK_01	3.1175811290087	LOGO	ROC	link	link
M00339-V-ETS1_B	3.0966392325250	LOGO	ROC	link	link
M00433-V-HMX1_01	2.7998447145677	LOGO	ROC	link	link
M01009-V-HES1_Q2	2.7541437968978	LOGO	ROC	link	link
PF0035	2.7175281589133	LOGO	ROC	link	link
[illegible]	2.6822197554160	LOGO	ROC	link	link
GCANGTCC	2.6796575758213	LOGO	ROC	link	link
M00184-V-MYOD_Q6	2.6442968829445	LOGO	ROC	link	link
MA0054	2.6390418003063	LOGO	ROC	link	link

Fig. 4. cisTargetX Summary Table. Summary view of the results with a score threshold above 2.5 obtained by SWAN scoring on dm3. Nine clusters of motifs are identified, the first related to the ETS family (*yellow*), a second to the bHLH family (proneural factors, *green*), a third to Su(H) (*purple*), and a fourth to Enhancer of split family factors (*brown*). These four families were expected to be involved in the PNS case. The remaining clusters with unknown motifs are new candidates. We added a *red line* showing the motifs that pass the automatic threshold.

- In column 4, the ROC curve, which is drawn using the PWM-based ranking (on the *x*-axis) and the recovery of the input gene list (on the *y*-axis) in blue. The red curve represents the average for the PWMs present in the used library. The green curve represents 1.96 times the standard deviation above the red curve.
- In column 5, the list of candidate target genes based on the ROC curve optimal cutoff.
- In column 6, all genes of the user gene set present in the top 1,000 target genes for this motif, which allows the user to examine possible target genes beyond the optimal cutoff.
- In column 7, a checkbox to select this motif for downstream analysis (see Subheading 3.5).

In this list (as in the summary table), each motif appears with a background color which helps recognizing the STAMP-derived cluster of similar motifs, as described in Subheading 2.

3.4.2. Interpretation of the Results

The motif background color can be used to interpret the results, although the interpretation of the motif clusters will depend on the expert knowledge of the user regarding the system under study. In the PNS case, nine motif families appeared significant (Fig. 4). The yellow cluster contains PWMs for an ETS-related motif (GGAA) (note that most are derived from vertebrate binding sites as indicated by the "-V-" prefix in the TRANSFAC nomenclature), and one motif is from a *Drosophila* TF, namely *Eip74EF*. However, current knowledge of sensory organ precursor (SOP) and PNC development would rather suggest *Pnt* as the binding TF. The green cluster contains E-box motifs (CANNTG), typically bound by bHLH factors that play an important role in SOP specification (e.g., *Achaete*, *Scute*, *Atonal*). The blue cluster contains *Su*(*H*) motifs. Finally, the brown cluster contains the *HLHm5* motif. All these TFs are known to be involved in the specification or differentiation of SOP or PNC cells. The other clusters are new candidate motifs for yet unidentified TFs involved in SOP or/and PNS development. For the results of the extended analysis of the PNS case, including more gene sets and parameter settings we refer to the Results section.

3.5. Obtain a Subset of the Co-expressed Genes as Predicted Direct Targets for One or More Motifs

For significant motifs, cisTargetX determines the optimal subset of genes that are predicted as direct targets (see Subheading 2 for details). To obtain this subset for one motif, click on the "candidate targets" link. The result is displayed as a list of genes with the rank of the gene, its CG identifier, its name, and its associated FBgn identifier. To obtain the candidate target genes for a set of motifs that are in the same cluster (e.g., the yellow cluster above), one should compute the union of the list for all motifs in this cluster. Potential common targets of TFs represented by motifs in different motif clusters can be obtained by computing the intersection of these lists (see Note 2 to know how to make a list union/intersection). Note that all the lists of "candidate targets" for the displayed motifs are available in an archive by clicking on the "Download" link at the top of the page.

For the PNS test case, the union of the candidate targets of *Su(H)*-like motifs yields 72 candidate target genes; the union of the candidate targets of E-box motifs yields 81 proneural target genes; the union of the candidate targets of the ETS motif yields 92 genes; and the union of the candidate targets of *HLHm5*-like motifs yields 30 genes. 16 genes are common to these four lists. These target gene predictions can immediately lead to a predicted GRN (see Subheading 3.7).

3.6. Scoring Candidate Target Genes with One or More Motifs to Obtain CRM Locations: Visualize the Results in the UCSC Genome Browser

Once a set of interesting motifs and corresponding candidate target genes are identified, the location of putative binding sites and predicted CRMs can be obtained by checking the motif in the right-most column, and clicking on the "Proceed" button on the top of the page. A new prefilled submission form appears (Fig. 3b).

3.6.1. List of Genes

By default, the input gene list contains the candidate targets of the highest ranked motif. This list could be replaced by a user defined list, for example, as the union or intersection of the lists of each individual motif.

List of motifs: contains all the motifs ticked.

In our example, we have selected motifs for Su(H) (M0111-RBPJK_Q4), Achaete/Scute (GCAGSTGK-scute), and the E(spl)-complex (motifs Espl and HLHm5).

3.6.2. Type of Clusters to Look for

Use the "separate" option for parallel homotypic CRMs or the "together" option for heterotypic CRMs. The parallel homotypic option means that each motif yields independent CRM predictions, while the heterotypic CRM option yields CRM binding sites for the different input PWMs. Technically, Cluster-Buster is run multiple times with one PWM under the parallel homotypic option, while Cluster-Buster is run only once with all PWMs as input under the heterotypic option.

Dmel Cluster-Buster Threshold corresponds to the score threshold for which the motif clusters with scores on *D. melanogaster* higher than this value will be reported. Based on our experience, we recommend using a conservative threshold of 3.

3.6.3. Region Extension

Cluster-Buster is used to score the *D. melanogaster* sequences first. Each high-scoring CRM (score > threshold) is selected. The genomic locations for the selected CRMs are extended with 500 bp by default (this is the region extension parameter that is adjustable by the user). For the extended locations, the orthologous sequences are obtained from 11 other *Drosophila* species and scored with

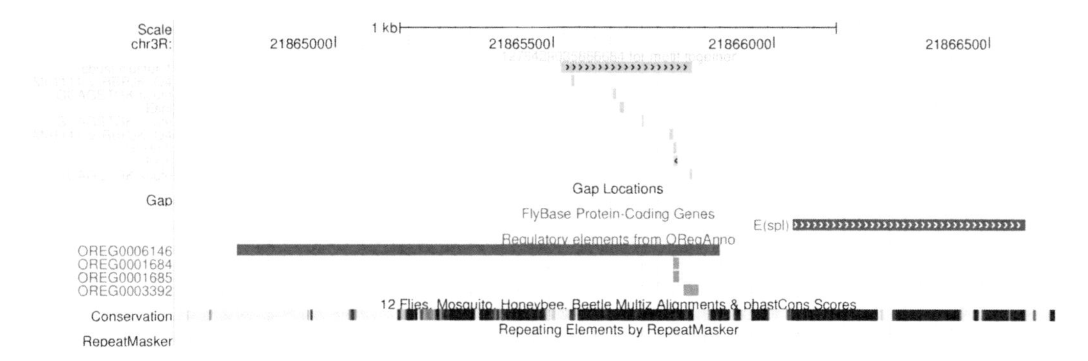

Fig. 5. UCSC view of a predicted CRM in the upstream region of E(spl). The predicted CRM and its individual binding sites are shown in *light gray*, a RedFly validated CRM and individual binding sites from OregAnno are visualized in *dark gray*.

Cluster-Buster to obtain a cross-species ranking of CRM predictions across the selected candidate target genes.

3.6.4. Your E-mail Address

Provide an e-mail address where to send the link to the results and click "Submit."

A new page with the status of the job will open. When the job is finished, two links are available: one displays the result directly in the UCSC genome browser as a custom track (see Note 3 on using the UCSC genome browser to visualize predicted CRMs), the other displays the result as a BED file that can be saved. An e-mail with the link will be sent to the user.

The custom track displays the CRM as predicted by Cluster-Buster as well as the individual predicted binding sites, with a score above the threshold defined in the previous screen. Figure 5 shows a predicted CRM in the upstream region of the E(spl) gene; this CRM contains binding sites for Su(H), Ac-Sc, E(spl), and HLHm5. Interestingly, the CRM coincides with an OregAnno CRM from Redfly, which has been shown to drive expression in the PNCs (33). Individual binding sites for E(spl) and HLHm5 are also supported by OregAnno annotations. This view can be used to define putative CRMs to be confirmed experimentally.

3.7. Reconstructing a GRN

Using the predicted target genes, we can partly reconstruct the underlying GRN. On the result page, one can download all sets of predicted target genes for all relevant motifs identified by cisTargetX. This is presented as an archive file (.tar.gz format) which can be opened with any unzip tool. The archive contains one file per motif, and each file contains the list of predicted target genes for this motif. The target genes are presented with (1) their rank, (2) their CG identifier, (3) their gene symbol, (4) their Flybase identifier (FBgn), (5) possible synonyms (all fields are tab delimited). These files can be used to reconstruct a partial GRN, provided one can attribute each relevant motif identified by cisTargetX to a *Drosophila* TF. This is the bottleneck of the reconstruction procedure, and we will present a possible partial solution to this.

1. Download and unzip the archive containing the lists of target genes.
2. Open each file individually in a spreadsheet application.
   We now need to add a column in the spreadsheet corresponding to the regulator, i.e., the candidate TF matching the motif. There are various alternatives for this, which we will discuss in order of increasing complexity.
   (a) If the motif corresponds to a fly TF, we only need to identify the corresponding CG identifier in Flybase. This is the case for the motifs from the FlyReg collection, the JASPAR insect collection, a small set of TRANSFAC motifs (recognizable from the -I- in their identifier), and a subset of motifs from the Stark et al. Collection (2).
   (b) If the motif corresponds to a TF from another species, then orthology relationships can be used to identify the corresponding fly protein. For this, use the UniProt database which provides links to orthology databases like InParanoid and PhylomeDB; or use the orthology predictions provided by Ensembl. However, in certain cases, such a prediction is missing and there is no simple way to identify a *Drosophila* TF.
   (c) If the motif does not correspond to any known TF, the TF family may be recognized (e.g., a typical E-box motif, such as CAGCTGC in Fig. 6 represents bHLH binding sites)

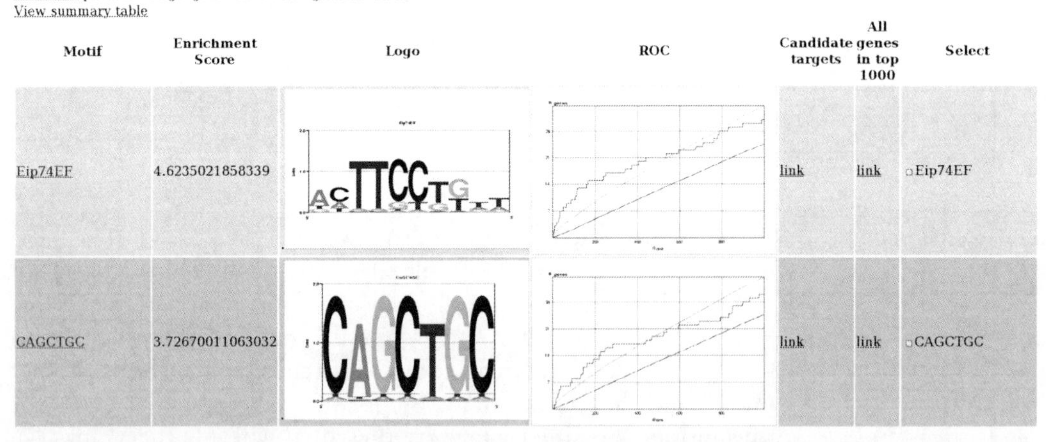

Fig. 6. cisTargetX Results. Screenshot of the top part of the cisTargetX results page showing the two first-predicted motifs. The first one, Eip74EF represent an ETS motif (GGAA) that could be bound by Pnt and the second motif is an E-box motif (CANNTG) that could be bound by the proneural factors Scute or Atonal. The ROC curve in the fourth column shows the recovery of the input genes in a cumulative fashion along the ranked list of genes (lmBR, *x*-axis). The steeper this curve (toward the upper left corner), and the more it differs from the average curve in *red*, the higher the enrichment of targets in the input set. The *green curve* represents 1.96 times the standard deviation above the *red curve*. By clicking on the link in the "candidate Targets" column, the predicted target genes for this particular motif are shown.

or it may be related to a *Drosophila* TF if the motif is found in a cluster together with a motif that corresponds to one of the two situations above.

Once a candidate *Drosophila* TF with its CG identifier is obtained, one or several columns can be added to the spreadsheet containing its potential target genes, for example, the gene symbol and the CG identifier. If the fly TF could not be identified, one can choose to ignore the motif and its target genes.

Once this is done, the various spreadsheets can be merged into one large sheet and adjusted, for example, to the following format:

1. CG identifier of the fly TF
2. Gene symbol of the fly TF
3. CG identifier of the target gene
4. Gene symbol of the target gene
5. FBgn of the target gene

Note that only columns 1 and 3 are required to build the network. Once this is done, save the file, and visualize the network using, for example, Cytoscape (http://www.cytoscape.org/) or any other network visualization tool. In Cytoscape use "File > Import > Network from table (text/MS excel)" and choose the CG identifier of the fly TF column as source interaction and the CG identifier of the target gene column as target interaction. It is convenient to distinguish the TFs from the other genes in the network. Using Cytoscape, this can be achieved by importing a "Node attributes" file, allowing the distinction of TFs from other kinds of proteins by highlighting them in the network. Such a file can be built by downloading the list of CGs identified as potential TFs in the FlyTF project (http://www.flytf.org/, a link to all putative *D. melanogaster* TFs is provided on the FlyTF homepage). This file should have the format (one line for each TF):

TF

CGxxx = 1

CGyyy = 1

Using "File > Import > Node attribute" in Cytoscape, load this file, and use this attribute to change the shape, color, size, etc. of the TF-associated nodes in the network. For help on how to use Cytoscape, we refer to the tutorial at http://www.cytoscape.org/tut/tutorial.php.

The reconstructed GRN using only the Reeves and Posakony dataset is shown in Fig. 7.

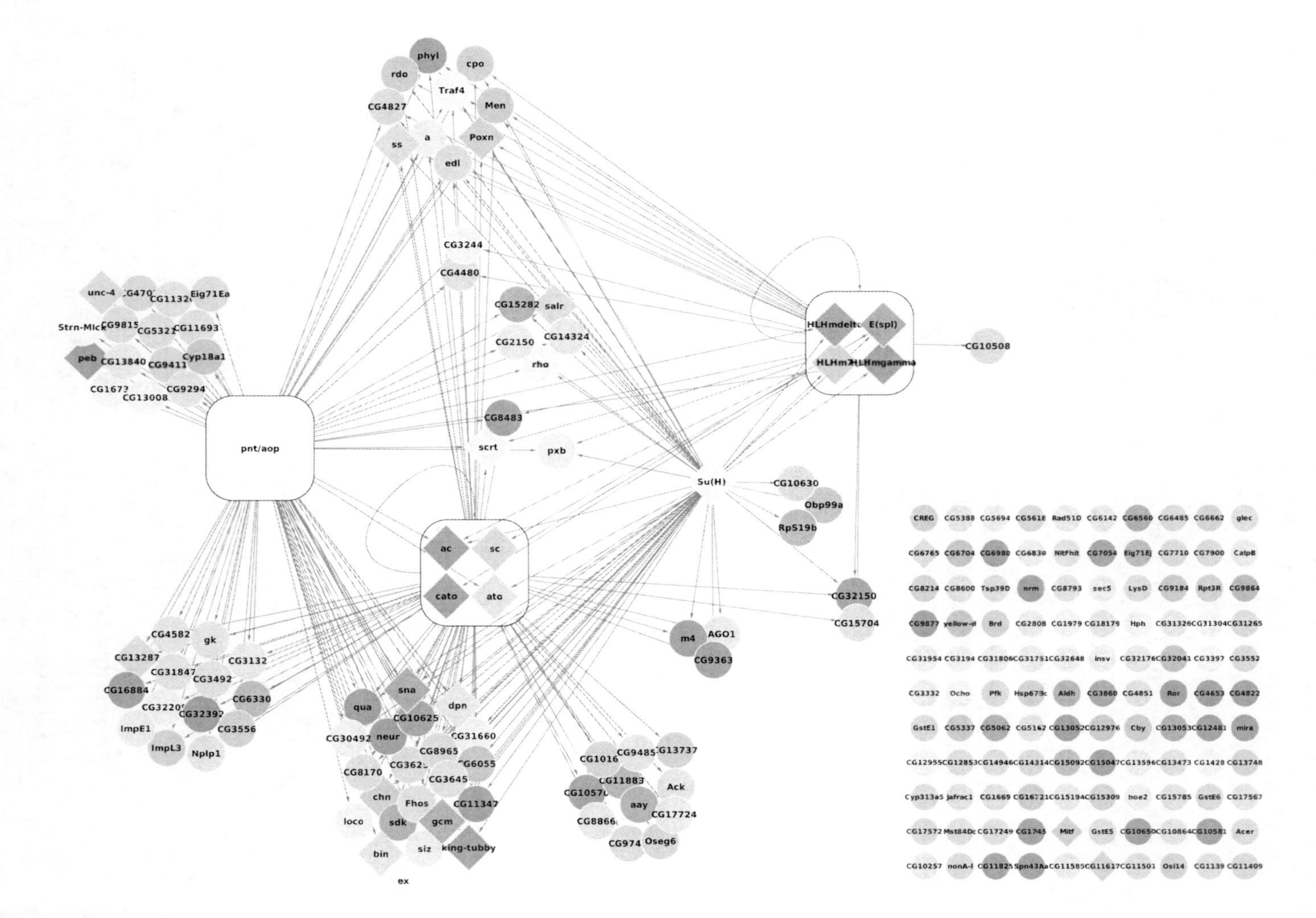

Fig. 7. Reconstructed GRN derived from all predicted TF/target gene interactions in the Reeves and Posakony gene set. Diamond nodes represent TFs. *Shades of gray* represent the level of over-expression in the GFP+ vs. GFP− cells, where *darkest gray* is more than fourfold upregulation. Metanodes (*gray boxes*) represent groups of TFs that are all equally likely to be attributed to the same motif cluster (e.g., members of the E(spl) complex, the proneural genes, or pnt/aop).

3.8. Combining the Results from Multiple Analyses and Gene Sets into One GRN

We have applied cisTargetX on related gene sets obtained from various other sources besides microarrays, namely Gene Ontology, FlyBase, and STRING (see Subheading 2.2). The different sets of genes are co-expressed during, or jointly involved in, the specification of sensory organs in *Drosophila*, with a focus on sensory organ development in the wing imaginal disc during third instar larval development. The ultimate goal of this analysis is to use cisTargetX to unravel these sets into high-confidence predicted TF/target gene interactions from which a larger GRN can be drawn. All gene sets were used with the three scoring possibilities of cisTargetX: (1) using Cluster-Buster as the CRM scoring algorithm, with the set of 1981 matrices used in ref. 18; (2) using SWAN as CRM scoring algorithm with the same set of 1981 matrices; (3) using Cluster-Buster with a larger selection of 3,335 matrices. Cluster-Buster and SWAN often result in different top-scoring motifs, yet both usually contain biologically meaningful motifs, and can be considered complementary. The first gene set is the Reeves and Posakony gene set that was used in the step-by-step procedure above. This set contains 209 genes that are co-expressed in the proneural clusters of the wing imaginal disc. Table 1 summarizes the results of cisTargetX on this geneset, this time using Cluster-Buster as scoring algorithm (Figs. 4 and 6 show the results obtained with SWAN scoring; Supplementary Table 1, available at http://med.kuleuven.be/lcb/resources, contains all detailed results). Similarly as for SWAN, the motifs identified represent TFs that are known to be involved in sensory organ development, such as the proneural bHLH factors (e.g., Achaete/Scute, Atonal, Daughterless), the Notch pathway effector TF Su(H), and the EGFR pathway effector TF Pointed. In contrast to the SWAN-based analysis, motifs representing the E(spl)-bHLH repressors are not found. This might be due to the different statistical approach used by SWAN in contrast to Cluster-Buster. The predicted target genes of these TFs can now be derived from the union of the targets of the motifs in the respective motif clusters. For example, the six Su(H) motifs in the first cluster in Table 1 result in 51 unique target genes. A third analysis on the same gene set was performed using the large PWM collection, generating similar results (Table S1 available at http://med.kuleuven.be/lcb/resources). Using all the TF/target gene predictions from these three analyses, a first GRN is drawn, following the procedure described in Subheading 2 (Fig. 7). This GRN connects 48% of the input genes (101 genes are connected) with 213 interactions in total. Next, we expanded this network further using the cisTargetX results on three other gene sets that we extracted from public resources, namely, Gene Ontology (GO), FlyBase TermLink, and STRING. The GO set represents all genes known to be involved in SOP cell fate commitment (GO:0016360); the FlyBase set represents all genes reported to be expressed in proneural clusters (Fbbt:00001135); and the STRING set represents all genes from a predicted interaction network around Su(H).

Table 2
Summary of the results for the PNS case

Test case 1: sensory organ precursors

Origin of the data	Data set	Expected TF: Su(H)	espl/HLHm5	Pros	Proneural (scute/ato/...)	EGF sig (pnt/aop)
High-throughput data	Reeves and Posakony dataset (209 genes)	Top3	Present	Absent	Top3	Top3
	GFP-negative cells (3 random gene sets of 209 genes)	Absent	Absent	Absent	Absent	Absent
Curated expression profiles (FlyBase)	FBbt:00001135 proneural cluster (33 genes)	Top3	Present	Absent	Present	Absent
Gene ontology annotated genes	GO:0016360 Sensory organ precursor cell fate determination (18 genes)	Top3	Absent	Absent	Present	Top3
TF-centered query (STRING)	Su(H) query (114 genes)	Top3	Absent	Present	Absent	Absent

For each TF, we indicate if a corresponding motif is found in the three first motif clusters (Top3), if it is not, then we indicate whether it is present or absent in cisTargetX data generated by one of the CRM scoring algorithms and PWM libraries (Cluster-Buster/PWM_LIB1, Cluster-Buster/PWM_LIB2, SWAN/PWM_LIB1)

From all cisTargetX results on these sets (Table 2 and Tables S1–S4 available at http://med.kuleuven.be/lcb/resources), a total of 394 interactions are predicted for five different TFs (see Fig. S1, available at http://med.kuleuven.be/lcb/resources, for the entire network), from which the vast majority are novel predictions.

Although the vast majority of the interactions in these networks are novel predictions, and although these networks show already a high degree of combinatorial regulation, it is obvious that they represent the tip of the iceberg. Indeed, many target genes in these networks are themselves transcription factors (the diamond-shaped nodes in Figs. 7 and S1 for which no target genes could be predicted from the used input sets). Additional cisTargetX runs with more co-expressed gene sets focused on these factors (e.g., derived experimentally by genetic perturbations or derived from public sources) may further extend these networks. We believe this procedure is useful to generate hypotheses that can be directly tested experimentally (as done in ref. 18), or that can represent the input of further computational analyses integrating additional knowledge of chromatin states, gene expression, protein function, or regulatory variation.

4. Notes

1. Converting gene IDs to CG identifiers: converting identifiers into CG identifiers can be done using the BioMart tool (http://metazoa.ensembl.org/biomart/martview/) or the FlyBase "Batch Download" tool. The different steps in BioMart are:
   (a) Choose "Metazoa Mart5 database" in the drop-down menu.
   (b) Choose the *D. melanogaster* dataset in the next drop-down menu.
   (c) On the left-hand side, click on "Filters," then on the "+" symbol left of "GENES" in the main panel.
   (d) Check "ID list limit" and paste or upload your list of genes, and select the type of identifiers used in your list.
   (e) Click on "Attributes" in the left panel, then on the "+" symbol next to "EXTERNAL" in the main panel, and select "FlyBase CGID Gene" under "External references."
   (f) On the topleft, click on "results," and then on "Go" next to "Export all results to...," after selecting the XLS format to import the list into Excel.
   (g) You can now select the column containing the CG identifiers, and paste it into cisTargetX.

2. Making unions/intersections of gene lists.

   Creating the union or intersection from two gene lists can be done by iterative use of the Web tool "Compare two lists" (http://jura.wi.mit.edu/bioc/tools/compare.php).

3. Visualize predicted CRMs in the UCSC Genome Browser.

   After clicking on the "Custom Track" in the UCSC Genome Browser the user can:

   (a) Move to another genomic region by typing the coordinates in the position/search box (in the format chrX:start_position-end_position).

   (b) Move to a gene by typing its name in the gene search box (see http://genome.ucsc.edu/goldenPath/help/geneSearchBox.html).

   (c) Create hyperlinks to all predicted CRMs using the Table Browser, and selecting as group: "Custom Tracks"; as track: "the prediction track for which you want the hyperlinks to CRMs"; as output format: "hyperlinks to Genome Browser," followed by clicking on "get output." For more informations on using the UCSC Genome Browser, we refer to the tutorial at http://www.openhelix.com/ucsc.

4. Particular cautions.

   Do not use cisTargetX with too small gene sets (e.g., less than 10), and certainly not with only a single gene, in fact random simulations have shown that small data sets result in more false positive than larger sets.

   Some of the results in this chapter are obtained using TRANSFAC Professional database. Because a license is required for this database, some of the presented examples may not be visible or repeatable for users without such a license.

Acknowledgments

We thank P.A. Salmand, A. Aubry, C. Oliva for their advice as cisTargetX users. This work is supported by a PhD fellowship from FWO (to M.N.S.) and KULeuven CREA grant 3 M100189 (to S.A.).

Supplementary material is available online at the following URL: http://med.kuleuven.be/lcb/resources/.

References

1. Van Loo, P., and Marynen, P. (2009) Computational methods for the detection of cis-regulatory modules, *Briefings in Bioinformatics* 10, 509–524.
2. Stark, A., Lin, M. F., Kheradpour, P., Pedersen, J. S., Parts, L., Carlson, J. W., Crosby, M. A., Rasmussen, M. D., Roy, S., Deoras, A. N., Ruby, J. G., Brennecke, J., Hodges, E., Hinrichs, A. S., Caspi, A., Paten, B., Park, S., Han, M. V., Maeder, M. L., Polansky, B. J., Robson, B. E., Aerts, S., van Helden, J., Hassan, B., Gilbert, D. G., Eastman, D. A., Rice, M., Weir, M., Hahn, M. W., Park, Y., Dewey, C. N., Pachter, L., Kent, W. J., Haussler, D., Lai, E. C., Bartel, D. P., Hannon, G. J., Kaufman, T. C., Eisen, M. B., Clark, A. G., Smith, D., Celniker, S. E., Gelbart, W. M., and Kellis, M. (2007) Discovery of functional elements in 12 Drosophila genomes using evolutionary signatures, *Nature* 450, 219–232.
3. Li, L., Zhu, Q., He, X., Sinha, S., and Halfon, M. S. (2007) Large-scale analysis of transcriptional cis-regulatory modules reveals both common features and distinct subclasses, *Genome Biol* 8, R101.
4. He, X., Ling, X., and Sinha, S. (2009) Alignment and Prediction of cis-Regulatory Modules Based on a Probabilistic Model of Evolution, *PLoS Comput Biol* 5, e1000299.
5. Sinha, S., Schroeder, M., Unnerstall, U., Gaul, U., and Siggia, E. (2004) Cross-species comparison significantly improves genome-wide prediction of cis-regulatory modules in Drosophila, *BMC Bioinformatics* 5, 129.
6. Whitington, T., Perkins, A. C., and Bailey, T. L. (2009) High-throughput chromatin information enables accurate tissue-specific prediction of transcription factor binding sites, *Nucleic Acids Research* 37, 14–25.
7. Ernst, J., Plasterer, H. L., Simon, I., and Bar-Joseph, Z. (2010) Integrating multiple evidence sources to predict transcription factor binding in the human genome, *Genome Res* 20, 526–536.
8. Aerts, S., van Helden, J., Sand, O., and Hassan, B. A. (2007) Fine-Tuning Enhancer Models to Predict Transcriptional Targets across Multiple Genomes, *PLoS ONE* 2, e1115.
9. Schroeder, M. D., Pearce, M., Fak, J., Fan, H., Unnerstall, U., Emberly, E., Rajewsky, N., Siggia, E. D., and Gaul, U. (2004) Transcriptional Control in the Segmentation Gene Network of Drosophila, *PLoS Biol* 2, e271.
10. Tompa, M., Li, N., Bailey, T. L., Church, G. M., De Moor, B., Eskin, E., Favorov, A. V., Frith, M. C., Fu, Y., Kent, W. J., Makeev, V. J., Mironov, A. A., Noble, W. S., Pavesi, G., Pesole, G., Regnier, M., Simonis, N., Sinha, S., Thijs, G., van Helden, J., Vandenbogaert, M., Weng, Z., Workman, C., Ye, C., and Zhu, Z. (2005) Assessing computational tools for the discovery of transcription factor binding sites, *Nat Biotech* 23, 137–144.
11. Frith, M. C., Fu, Y., Yu, L., Chen, J., Hansen, U., and Weng, Z. (2004) Detection of functional DNA motifs via statistical over-representation, *Nucleic Acids Research* 32, 1372–1381.
12. Ho Sui, S. J., Fulton, D. L., Arenillas, D. J., Kwon, A. T., and Wasserman, W. W. (2007) oPOSSUM: integrated tools for analysis of regulatory motif over-representation, *Nucleic Acids Research* 35, W245–W252.
13. Roider, H. G., Manke, T., O'Keeffe, S., Vingron, M., and Haas, S. A. (2009) PASTAA: identifying transcription factors associated with sets of co-regulated genes, *Bioinformatics* 25, 435–442.
14. Kantorovitz, M. R., Kazemian, M., Kinston, S., Miranda-Saavedra, D., Zhu, Q., Robinson, G. E., Göttgens, B., Halfon, M. S., and Sinha, S. (2009) Motif-Blind, Genome-Wide Discovery of cis-Regulatory Modules in Drosophila and Mouse, *Developmental Cell* 17, 568–579.
15. Rouault, H., Mazouni, K., Couturier, L., Hakim, V., and Schweisguth, F. (2010) Genome-wide identification of cis-regulatory motifs and modules underlying gene coregulation using statistics and phylogeny, *Proceedings of the National Academy of Sciences* 107, 14615–14620.
16. Warner, J. B., Philippakis, A. A., Jaeger, S. A., He, F. S., Lin, J., and Bulyk, M. L. (2008) Systematic identification of mammalian regulatory motifs' target genes and functions, *Nat Meth* 5, 347–353.
17. Van Loo, P., Aerts, S., Thienpont, B., De Moor, B., Moreau, Y., and Marynen, P. (2008) ModuleMiner – improved computational detection of cis-regulatory modules: are there different modes of gene regulation in embryonic development and adult tissues? *Genome Biol* 9, R66.
18. Aerts, S., Quan, X., Claeys, A., Naval Sanchez, M., Tate, P., Yan, J., and Hassan, B. A. (2010) Robust Target Gene Discovery through Transcriptome Perturbations and Genome-Wide Enhancer Predictions in Drosophila Uncovers a Regulatory Basis for Sensory Specification, *PLoS Biol* 8, e1000435.

19. Portales-Casamar, E., Thongjuea, S., Kwon, A. T., Arenillas, D., Zhao, X., Valen, E., Yusuf, D., Lenhard, B., Wasserman, W. W., and Sandelin, A. (2009) JASPAR 2010: the greatly expanded open-access database of transcription factor binding profiles, *Nucleic Acids Research 38*, D105–D110.
20. Matys, V. (2006) TRANSFAC(R) and its module TRANSCompel(R): transcriptional gene regulation in eukaryotes, *Nucleic Acids Research 34*, D108–D110.
21. Bergman, C. M., Carlson, J. W., and Celniker, S. E. Drosophila DNase I footprint database: a systematic genome annotation of transcription factor binding sites in the fruitfly, Drosophila melanogaster, *Bioinformatics 21*, 1747–1749.
22. Elemento, O., and Tavazoie, S. (2005) Fast and systematic genome-wide discovery of conserved regulatory elements using a non-alignment based approach, *Genome Biol 6*, R18.
23. Xie, X., Lu, J., Kulbokas, E. J., Golub, T. R., Mootha, V., Lindblad-Toh, K., Lander, E. S., and Kellis, M. (2005) Systematic discovery of regulatory motifs in human promoters and 3(prime) UTRs by comparison of several mammals, *Nature 434*, 338–345.
24. Down, T. A., Bergman, C. M., Su, J., and Hubbard, T. J. P. (2007) Large-Scale Discovery of Promoter Motifs in Drosophila melanogaster, *PLoS Comput Biol 3*, e7.
25. Newburger, D. E., and Bulyk, M. L. (2009) UniPROBE: an online database of protein binding microarray data on protein-DNA interactions, *Nucleic Acids Research 37*, D77–D82.
26. Mahony, S., and Benos, P. V. (2007) STAMP: a web tool for exploring DNA-binding motif similarities, *Nucleic Acids Research 35*, W253–W258.
27. Frith, M. C., Li, M. C., and Weng, Z. (2003) Cluster-Buster: finding dense clusters of motifs in DNA sequences, *Nucleic Acids Research 31*, 3666–3668.
28. Kim, J., Cunningham, R., James, B., Wyder, S., Gibson, J. D., Niehuis, O., Zdobnov, E. M., Robertson, H. M., Robinson, G. E., Werren, J. H., and Sinha, S. (2010) Functional Characterization of Transcription Factor Motifs Using Cross-species Comparison across Large Evolutionary Distances, *PLoS Comput Biol 6*, e1000652.
29. Reeves, N., and Posakony, J. W. (2005) Genetic Programs Activated by Proneural Proteins in the Developing Drosophila PNS, *Developmental Cell 8*, 413–425.
30. Snel, B., Lehmann, G., Bork, P., and Huynen, M. A. (2000) STRING: a web-server to retrieve and display the repeatedly occurring neighbourhood of a gene, *Nucleic Acids Research 28*, 3442–3444.
31. Hare, E. E., Peterson, B. K., Iyer, V. N., Meier, R., and Eisen, M. B. (2008) Sepsid even-skipped Enhancers Are Functionally Conserved in Drosophila Despite Lack of Sequence Conservation, *PLoS Genet 4*, e1000106.
32. Aerts, S., Lambrechts, D., Maity, S., Van Loo, P., Coessens, B., De Smet, F., Tranchevent, L., De Moor, B., Marynen, P., Hassan, B., Carmeliet, P., and Moreau, Y. (2006) Gene prioritization through genomic data fusion, *Nat Biotech 24*, 537–544.
33. Castro, B., Barolo, S., Bailey, A. M., and Posakony, J. W. (2005) Lateral inhibition in proneural clusters: cis-regulatory logic and default repression by Suppressor of Hairless, *Development 132*, 3333–3344.

Chapter 19

Proteomic Methodologies to Study Transcription Factor Function

Harry W. Jarrett

Abstract

Transcription factors regulate transcription by binding to regulatory regions of genes including the promoter. Few of the transcription factors are well characterized, and few promoters have been described in detail. New methods have been developed to improve both transcription factor and promoter characterization, some of which are discussed here. Trapping methodology applicable to both individual transcription factors and intact transcription complexes are described, as well as 2D gel electrophoresis, Southwestern blotting, and basic liquid chromatography/tandem mass spectrometry methodology. These methods have proved useful in the study of transcriptional regulation.

Key words: Transcription, Regulation, DNA, Oligonucleotide, Promoter, Mass spectrometry, Southwestern blot

1. Introduction

Transcription factors are the activator and repressor proteins that bind to the promoter region of genes to regulate expression. One class of transcription factors is the specific transcription factors that bind to response elements, individual DNA binding sites present throughout the promoter. Another group is often referred to as the general transcription factors, which for genes is generally the RNA polymerase II transcription (TFII) complex, which assemble a group of approximately 60 proteins and subunits near the initiation region where transcription will begin. Current models propose that the specific factors bind early and recruit the TFII complex, RNA polymerase II, and other proteins to initiate transcription (1).

Bart Deplancke and Nele Gheldof (eds.), *Gene Regulatory Networks: Methods and Protocols*,
Methods in Molecular Biology, vol. 786, DOI 10.1007/978-1-61779-292-2_19,

In humans, there are approximately 23,000 genes (2), many of which have alternate promoters used at different stages of development or in different tissues. There are at least 1,500 specific human transcription factors (3) (http://dbd.mrc-lmb.cam.ac.uk/DBD/index.cgi?Home). Very few of these promoters have ever been characterized in detail and only a few percent of the transcription factors are well characterized. Thus, the task of characterizing the transcription factor proteome and clearly understanding genetic regulation is monumental. We have developed new methods, many of which are described here, which may be useful to scientists who also aim to comprehend transcriptional regulation. Here, we focus on those techniques that may be unfamiliar to the reader and do not discuss reporter assays, chromatin immunoprecipitation (ChIP), promoter footprinting methodology, overexpression, or silencing methods, which are also commonly used in transcription factor characterization.

Rather, we focus on a basic set of techniques that can be used to purify and characterize proteins that bind to DNA. These include the isolation of nuclear extract from cultured cells, oligonucleotide purification and labeling, and the electrophoretic mobility shift assay for DNA-binding activity, which utilizes nondenaturing gel electrophoresis to detect the DNA-binding protein complexes. Methods for purifying individual transcription factors (oligonucleotide trapping) or complete transcription complexes (promoter trapping) are then presented. Further purification by 2D gel electrophoresis and the use of Southwestern blotting techniques to detect specific, high-affinity DNA-binding proteins are presented next. Finally, we describe methods to digest the detected proteins with trypsin and their characterization by liquid chromatography-electrospray ionization-mass spectrometry.

2. Materials

2.1. Cell Culture and Nuclear Extract

1. Cell culture flasks of 182 cm^2 (Celltreat, Shirley, MA, USA).
2. Dulbecco's Modification of Eagle's medium (DMEM, Cellgro, Manassas, VA).
3. Heat-inactivated adult bovine serum (Sigma Chemical Co., St. Louis, MO, USA). The serum is inactivated by incubation at 56°C for 30 min and stored at 4°C prior to use.
4. PBS (10×): 80 g NaCl, 2 g KCl, 27.2 g Na_2HPO_4 7 H_2O, and 2.4 g KH_2PO_4 made up to 1 L with H_2O (see Note 1). The PBS is autoclaved prior to use.
5. Trypsin-EDTA (1×) (Sigma).
6. 15-mL and 50-mL high-density polyethylene, sterile, graduated culture tubes (Fisher Scientific, Pittsburgh, PA, USA).

7. Nuclear extract hypotonic buffer: 10 mM HEPES, pH 7.9, 1.5 mM $MgCl_2$, 10 mM KCl, 0.5 mM EDTA, and freshly added 0.5 mM dithiothreitol (DTT) and 0.2 mM phenylmethylsufonylfluoride (PMSF).
8. Nuclear extract low-salt buffer: 20 mM HEPES, pH 7.9, 1.5 mM $MgCl_2$, 20 mM KCl, 0.2 mM EDTA, 25% glycerol, and freshly added 0.5 mM DTT and 0.2 mM PMSF.
9. Nuclear extract high-salt buffer: 20 mM HEPES, pH 7.9, 1.5 mM $MgCl_2$, 1.6 M KCl, 0.2 mM EDTA, 25% glycerol, and freshly added 0.5 mM DTT and 0.2 mM PMSF.
10. Nuclear extract dialysis buffer: 20 mM HEPES, pH 7.9, 100 mM KCl, 20% glycerol, and freshly added 0.5 mM DTT and 0.2 mM PMSF.

2.2. Oligonucleotide Preparation (see Note 2)

1. TE buffer: 10 mM Tris, 1 mM EDTA, pH 7.5. TE0.1 is TE containing 0.1 M NaCl. Similarly, TE0.4 and TE1.2 contain 0.4 and 1.2 M NaCl, respectively.
2. 0.5 M EDTA (free acid), titrated to pH 8 with 5 M NaOH.
3. 3 M sodium acetate solution: 40.8 g sodium acetate, 0.2 mL 0.5 M EDTA, 5 mL glacial acetic acid, and 66 mL water.

2.3. Oligonucleotide Labeling

1. 10 μCi/μL, 6,000 Ci/mmol γ-^{32}P-ATP (PerkinElmer, Boston, MA), stored at −20°C.
2. T4 Polynucleotide kinase (10 units/μL, New England BioLabs, Ipswich, MA, USA), stored at −20°C.
3. 1 mg/mL salmon sperm DNA (Sigma) in TE stored at −20°C.
4. 10% trichloroacetic acid (TCA): 10 g TCA dissolved to 100 mL in H_2O.
5. Supelco silanized glass wool (Sigma).
6. 17 × 100 mm sterile culture tubes (VWR, Sugar Land, TX, USA).
7. Bio-Gel P-6 fine resin (BioRad laboratories, Hercules, CA, USA). 10 g is suspended in 250 mL of TE in an autoclavable glass bottle and autoclaved for 45 min. After cooling and the resin has settled down, excess liquid is removed to give a 1:1 slurry.

2.4. Electrophoretic Mobility Shift Assay

1. Acrylamide/Bis (29:1): dissolve 29 g acrylamide, 1 g *N,N′*-methylene-bis-acrylamide (Bis) to a final volume of 100 mL.
2. 5× TBE: 30.03 g Tris base, 15.25 g boric acid, 10 mL 0.5 M EDTA, H_2O to 1,000 mL, autoclave for 45 min, and store at room temperature.
3. 1 M Tris/HCl, pH 7.5: Dissolve 121.1 g Tris-free base in 800 mL H_2O. Titrate to pH 7.5 with concentrated HCl. Adjust volume to 1 L and autoclave for 45 min.

4. 5 M NaCl: Dissolve 292.2 g NaCl to a total volume of 1 L with H_2O. Autoclave for 45 min.
5. 5× Electrophoretic mobility shift assay (EMSA) buffer (8 mL): 400 μL 1 M Tris–HCl (pH 7.5), 80 μL 0.5 M EDTA, 320 μL 5 M NaCl, 1.6 mL glycerol, 3.2 μL 2-mercaptoethanol, and 5.6 mL H_2O. Stable for 1 week at room temperature.
6. Poly dI:dC (Sigma).
7. Bromophenol blue: 0.1 g bromophenol blue, 50 mL glycerol, adjust volume to 100 mL with water.

2.5. DNA-Sepharose Preparation

1. Cyanogen Bromide (CNBr)-Sepharose coupling buffer. 0.1 M $NaHCO_3$, pH 8.3, 0.5 M NaCl.
2. CNBr-Sepharose blocking buffer. 0.1 M Tris, pH 8, 0.5 M NaCl.
3. 1.5-mL columns with polypropylene frits (Alltech).

2.6. Oligonucleotide Trapping Chromatography

1. EP24 (5′-GCTGCAG**ATTG**CG**CAAT**CTGCAGC-3′).
2. ACEP24$(GT)_5$ (5′-GCTGCAG**ATTG**CG**CAAT**CTGCAG CGTGTGTGTGT-3′). The bold text highlights the binding site.
3. Heparin (Sigma). Please note that many heparins are available, but we prefer the one defined by catalog number H-3393.
4. $(dT)_{18}$ (5′-TTTTTTTTTTTTTTTTTT-3′).
5. Tween-20 (Fisher Chemical).

2.7. Promoter Trapping Chromatography

1. Taq polymerase (5 units/μL, New England BioLabs).
2. Lambda exonuclease (5 units/μL, New England BioLabs).

2.8. Two-Dimensional Gel Electrophoresis

1. Two-dimensional gel electrophoresis (2DGE) IEF rehydration buffer: 7 M urea, 2 M thiourea, 2% CHAPS, 65 mM DTT, 0.8% pH 3–10 Ampholytes (BioRad), 1% Zwittergent 3–10 (Sigma), and 0.01% bromophenol blue.
2. 2DGE equilibration buffer: 50 mM Tris–HCl, pH 6.8, 6 M urea, 2% SDS, 30% glycerol, and 0.001% bromophenol blue.
3. Blotting buffer: 10% methanol in 25 mM Tris, 192 mM glycine.
4. Blotting membrane is Sequi-blot PVDF membrane (0.2 μm) (BioRad Laboratories). Prior to use, after cutting to size, the membrane is placed in methanol for 10 min to hydrate and then transferred to blotting buffer.

2.9. 2DGE-Southwestern Blotting

1. Southwestern Blotting binding buffer is 10 mM HEPES/NaOH, pH 7.9, 50 mM NaCl, 10 mM $MgCl_2$, 0.1 mM EDTA, 1 mM DTT, 50 μM $ZnSO_4$, and 0.1% Tween-20.

2.10. On-Blot Trypsin Digestion for ESI-MS Analysis

1. SDS-stripping buffer: 62.5 mM Tris–HCl, pH 6.8, 2% SDS, 100 mM 2-mercaptoethanol.

2.11. Capillary HPLC

1. Mobile phase A: 0.5% acetic acid (HAc)–0.005% trifluoroacetic acid (TFA), from Fischer.
2. Mobile phase B: 90% acetonitrile (ACN)–0.5% HAc–0.005% TFA.

3. Methods

3.1. HEK293 Nuclear Extract

1. The method used is a minor modification of (4). Human Embyonic Kidney -293 (HEK293, American Type Culture Collection) cells are cultured in a NuAire IR Autoflow CO_2 water-jacketed incubator at 37°C with 5% CO_2 and 95% atmospheric air. The 182 cm^2 cell culture flasks are seeded with 5×10^6 HEK293 cells and grown in 60 mL/flask of DMEM containing 10% heat-inactivated adult bovine serum. Cells are grown to over 90% confluence, and eight flasks are harvested for a typical nuclear extract preparation.
2. Cell harvesting: the cells are first rinsed with PBS and harvested by adding 10 mL of trypsin-EDTA solution for 5 min at 37°C. The cells are then immediately suspended in 40 mL 4°C DMEM containing 10% inactivated adult bovine serum and removed from the flask.
3. All subsequent steps are on ice or at 4°C. The cells are centrifuged (1,850 $\times g$ for 5 min, 4°C) in 50-mL disposable, sterile plastic conical tubes, one tube per flask, and the cells are washed with 50 mL PBS. The cells are resuspended in sterile PBS, combined, and transferred to a 15-mL sterile, graduated conical tube and again centrifuged. The final pellet will be approximately 2 mL and contain $2–3 \times 10^8$ cells. Carefully note the packed cell volume (pcv) and remove the supernatant.
4. Quickly resuspend the cells in 5 pcv of ice-cold nuclear extract hypotonic buffer, centrifuge the cells (1,850 $\times g$ for 5 min, 4°C), and discard the supernatant.
5. Resuspend the cells in 2 pcv of nuclear extract hypotonic buffer and allow to swell for 10 min on ice.
6. Transfer the cells to an ice-cold Dounce homogenizer and homogenize with the type B pestle using ten slow up-and-down strokes.
7. Collect the nuclei by centrifugation (3,300 $\times g$, 15 min) in a graduated conical tube. Carefully observe the packed nuclear volume (pnv). Discard the supernatant.

8. Add 0.5 pnv of nuclear extract low-salt buffer and transfer the nuclei to a clean, ice-cold Dounce homogenizer and homogenize with the type B pestle for two slow strokes.
9. While gently mixing, add drop-wise 0.5 pnv of nuclear extract high-salt buffer. The nuclei are then homogenized with two slow strokes with the pestle. Allow the nuclei to extract for 30 min on ice with gentle stirring. Transfer to a JA-20 centrifuge tube.
10. Pellet the extracted nuclei by centrifugation (25,000 × *g*, 30 min). Save the supernatant and discard the pellet.
11. Dialyze the supernatant nuclear extract three times versus 50 volumes of nuclear extract dialysis buffer, allowing 4 h between each buffer change, 12 h total (see Note 3).
12. Centrifuge the dialyzed nuclear extract (25,000 × *g*, 20 min) and carefully remove the supernatant. Determine the protein concentration (5) and dilute as needed with fresh dialysis buffer to prepare 5 mg/mL protein. The yield is usually 1–2 mL. Prepare 50–100 μL aliquots and store frozen at –85°C. The nuclear extract can be stored for at least 1 year.

3.2. Oligonucleotide Cleanup

1. Oligonucleotides (see Note 2) arrive dry after deblocking in NH_4OH, which can interfere with labeling and coupling. They are ethanol-precipitated.
2. Dissolve in 300 μL of TE buffer.
3. Add 30 μL of 3 M NaOAc and 1 mL of ethanol (absolute), mix, and allow to precipitate in the –85°C freezer for 1 h (see Note 4).
4. Centrifuge the tube in the cold room at 14,000 × *g* for 12 min at 4°C. Carefully observe tube orientation so that the position of the pelleted DNA, which may not be visible, is known and can be avoided. Carefully remove the supernatant with a Pasteur pipette. Wash the pellet with 500 μL of ice-cold 70% ethanol by vortexing. Again centrifuge and carefully remove the supernatant. The tube is left open, covered with a tissue, and allowed to air-dry for 2 h at room temperature.
5. If used for coupling to Sepharose, dissolve the pellet in water. For long-term storage for other uses such as EMSA, dissolve the pellet in 500 μL of TE. Allow 30 min, with occasional vortexing, for the pellet to dissolve.
6. Determine the absorption at 260 nm using approximately 1 μL diluted to 1 mL with TE and determine the concentration (see Note 5).

3.3. Oligonucleotide Labeling

1. Mix 20 μL of 0.1 μM oligonucleotide, 5 μL polynucleotide kinase buffer (10×, supplied with enzyme), 2 μL γ-^{32}P-ATP (10 μCi/μL, 6,000 Ci/mmol), and 21 μL water to make a

total volume of 48 μL and centrifuge briefly. Add 2 μL T4 Polynucleotide Kinase (10 unit/μL) and mix by gentle tapping.

2. Incubate at 37°C for 60 min.
3. Stop reaction by adding 2 μL 0.5 M EDTA.
4. TCA analysis:
   (a) Remove 1 μL from the reaction mixture to a 13 × 100 mm test tube containing 100 μL of 100 μg/mL salmon sperm DNA in TE. Mix well.
   (b) Spot 1 μL of this mixture directly onto a Whatman GF/C 47-mm filter disk.
   (c) To remaining 100 μL, add 5 mL ice-cold 10% TCA, vortex, and leave on ice 15 min.
   (d) Collect precipitate by vacuum filtration through a GF/C filter. Wash the tube and filter 5 × 5 mL ice-cold TCA, then 2 × 5 mL ice-cold ethanol.
5. Analysis. Count both filters either by Cerenkov radiation (without scintillation fluid) or with 5 mL scintillation fluid. Total counts are obtained from the directly spotted 1 μL by:

   $\text{Total} = \text{C.P.M.} \times (101\ \mu\text{L}/1\ \mu\text{L}) \times (52\ \mu\text{L}/1\ \mu\text{L})$

   And that precipitated:

   $\text{TCA} = \text{C.P.M.} \times (100\ \mu\text{L}/101\ \mu\text{L}) \times (52\ \mu\text{L}/1\ \mu\text{L})$

   The efficiency of labeling is determined from the percent of total counts incorporated into TCA precipitable DNA. The specific activity is just the TCA precipitate counts divided by the 2 pmol of oligonucleotide used for labeling.
6. Recovery. The oligonucleotide is desalted on a spin column. This treatment removes over 90% of the remaining γ-^{32}P-ATP. This can be done using a commercially available column. However, we make our own columns from a 1-mL tuberculin syringe barrel plugged with silanized glass wool and placed inside a 17 × 100 mm culture tube through a small hole made in the lid with a pair of scissors (Fig. 1). The column (syringe barrel) is filled with a 1:1 slurry of BioRad P-6 resin. The culture tube serves as a receptacle. The column is centrifuged until no more fluid elutes in an IEC clinical centrifuge at maximum speed for 10 min. The eluate is discarded and a 1.5-mL centrifuge tube is placed under the column outlet, all inside the culture tube. The labeled DNA (51 μL, step 3) is added to the column and the column is again centrifuged. The 50–55 μL eluate is then ready for use. The oligonucleotide is now approximately 40 nM (see Note 6).
7. The specific activity of the DNA is adjusted to 4,000 cpm/pmol, and 10 nM oligonucleotide and stored in 50 μL aliquots in the −20°C freezer. For example, if the oligonucleotide is

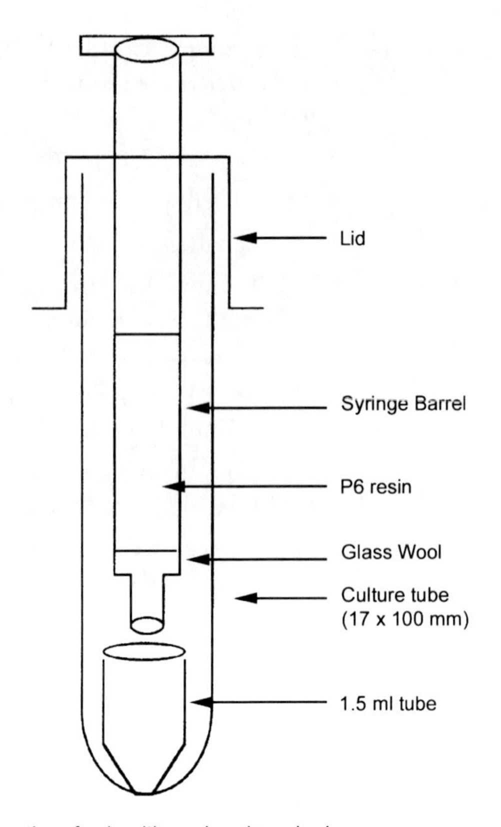

Fig. 1. A cross-section of a desalting spin column is shown.

diluted fourfold with TE, it is 10 nM. If labeling efficiency, calculated from TCA precipitable radioisotope (step 5) is 50%, the oligonucleotide is now 200 μL and 66,000 dpm/μL. Further dilution with 3.1 mL of unlabeled 10 nM oligonucleotide in TE provides the desired specific activity. Normally, only sufficient 50 μL aliquots are prepared for 2 weeks experiments and the rest discarded in liquid waste. The stored oligonucleotide is usable for 2 weeks for EMSA.

3.4. Electrophoretic Mobility Shift Assay

1. Prepare a nondenaturing polyacrylamide gel. The following procedure is for 10 mL gel (enough for two 0.75 mm × 7 × 10 cm minigels). For most purposes the 5% gel is used. In a 16 × 100 mm test tube, combine the following components:

	5%	10%	12%
H_2O	7.7 mL	6.1 mL	5.4 mL
5× TBE	0.5 mL	0.5 mL	0.5 mL
Acrylamide:bis (29:1)	1.7 mL	3.3 mL	4 mL
10% Ammonium persulfate	0.1 mL	0.1 mL	0.1 mL
TEMED	20 μL	20 μL	20 μL

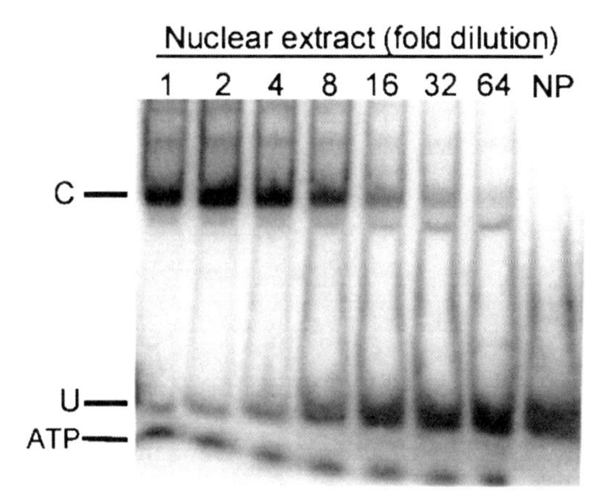

Fig. 2. Electrophoretic mobility shift assay of rat liver nuclear extract C/EBP. A twofold serial dilution of rat liver nuclear extract (220 ng/μL stock) was incubated with 1.6 nM radiolabeled ACEP24(GT)$_5$. The fold dilution is shown above the gel. Poly dI:dC was constant at 40 μg/mL. *C* specific shift complex identified by other experiments, *U* unshifted double-stranded DNA; *NP* no protein added to the gel. Reproduced with permission from Elsevier (10).

2. Mix well and quickly pour into the assembled plates and place the comb. Let the gel polymerize (approximately 20 min).
3. Fill the gel tank with 0.25× TBE (~300 mL).
4. Remove the comb, wash, and fill the wells with 0.25× TBE. Preelectrophorese for 30 min at 100 V.
5. In a microtube, combine the following components:

5× EMSA buffer	5 μL
1 μg/μL Poly (dI-dC)	1 μL
Nuclear Extract or purified fraction	2 μL
10 nM ^{32}P-labeled, annealed oligonucleotide	5 μL
H_2O	12 μL

6. Mix gently and incubate at room temperature for 20 min.
7. Add 2 μL of bromophenol blue to each sample and mix gently. Load onto the gel.
8. Gel electrophoresis. Fill the upper and lower chambers of the gel apparatus with 0.25× TBE if this has been removed. Run the gel at 100 V for 50 min or until dye front is ~1 mm from the bottom of the gel.
9. Autoradiography. Dry the gel overnight or place in a watertight zip-lock bag. Expose film overnight at −70°C using an intensifying screen. A typical EMSA readout for the CAAT enhancer binding protein transcription factor (C/EBP) is shown in Fig. 2.

3.5. DNA-Sepharose Preparation (Requires 2 Days)

1. Using the oligonucleotide cleanup protocol (Subheading 3.2), prepare the oligonucleotides and dissolve in water. For trapping methods, $(AC)_5$ (NH_2-ACACACACAC), where "NH_2" refers to the aminohexyl phosphoramidite added on the last cycle during synthesis, oligonucleotide is used. If beginning from a 1 μmol synthesis, approximately 500 nmol will be obtained. Coupling is with 50 nmol $(AC)_5$/mL Sepharose, which is 175 nmol/g lyophilized CNBr Sepharose. Assuming 500 nmol is obtained, 10 mL of oligonucleotide-Sepharose can be prepared or the recipe adjusted to fit the amount available. Keeping the oligonucleotide volume as low as possible (less than 1 mL), place 500 nmol into a 50-mL sterile, conical, graduated tube kept on ice.
2. Weigh 2.9 g of lyophilized cyanogen bromide-activated Sepharose 4B into an ice-cold 250 mL graduated cylinder.
3. Fill graduated cylinder with 250 mL ice-cold 1 mM HCl, cover with Parafilm, and invert to mix. Let the Sepharose swell for 15 min while keeping the graduated cylinder on ice, mixing by inversion every few minutes. The graduated cylinder is used here because the Sepharose settles slowly over the long dimension, keeping it well suspended over the swelling process.
4. Filter and wash the Sepharose three times, each with 100 mL ice-cold 1 mM HCl on the coarse 600-mL Kimax sintered glass filter funnel, leaving a moist cake. It is important that the funnel be fast-flowing, allowing the washes to be accomplished in 2–3 min.
5. Scrape the Sepharose into the DNA containing conical tube. Mix and add sufficient coupling buffer (approximately 20 mL) to produce an easily mixed, fairly liquid slurry.
6. Put the capped tube on a wheel rotator (Cole-Parmer Roto-Torque) overnight at room temperature. The tube is placed along the periphery of the wheel so that the solution rotates over the side wall of the tube. The rate of rotation is as slow as possible while keeping the Sepharose suspended.
7. The next day, filter the Sepharose on a coarse 30-mL sintered glass funnel using a side-arm flask of known weight and wash the tube and Sepharose four times with 5 mL of blocking buffer.
8. Reweigh the side-arm flask to determine the volume of the total filtrate plus washes (1 g = 1 mL), usually about 35–40 mL. As long as the volume is more than 10 mL, the $A_{260\ nm}$ can be determined directly without dilution.
9. The nanomole of DNA recovered is:

   nmol $(AC)_5$ NOT coupled = mLs recovered × 10^6 × $A_{260\ nm}$/111,250

10. The nanomole of DNA coupled is calculated from the results in steps 1 and 9 as:
    nmol $(AC)_5$ coupled = nmol $(AC)_5$ added − nmol $(AC)_5$ not coupled
11. Return the filtered Sepharose to the 50-mL screw-cap tube and add sufficient blocking buffer (about 20 mL) to again produce a liquid slurry. Blocking buffer reacts with any remaining activated groups on the Sepharose, rendering them inert.
12. Incubate on wheel rotator overnight at room temperature. Wash DNA-Sepharose two times with 4 mL TE0.1 (see Subheading 2.2) containing 10 mM NaN_3 on the filter funnel. Return the Sepharose to a fresh 50-mL conical screw-cap tube, and after the Sepharose has settled, add or subtract a sufficient amount of the same buffer to make an exact 1:1 slurry. Yield is about 10 mL of settled $(AC)_5$-Sepharose and about 20 mL of the 1:1 slurry. Typically, about 80–100% of the $(AC)_5$ couples, yielding about 25–50 nmol $(AC)_5$/mL Sepharose. The $(AC)_5$-Sepharose is stable at 4°C for at least 1 year.
13. To prepare a 1-mL column, a 1.5-mL column is outfitted with a frit at its outlet, a stopcock, and filled with TE0.1. Most of the TE is allowed to flow through the frit leaving approximately 0.2 mL in the column. One milliliter of well mixed 1:1 $(AC)_5$-Sepharose slurry is added, and flow is again initiated. As the column drains, a second milliliter of slurry is added, maintaining a full column until all is added. The column is then watched carefully while flowing and when the bed is no longer increasing, flow is stopped, usually leaving approximately 0.2 mL of clear buffer above the resin bed. A second frit is then placed carefully on top of the resin bed so that a flat top which cannot be easily disturbed is produced. The column is then outfitted with a stopper through which a 18 g × 1 in. needle is placed and an empty 10 mL syringe barrel is attached to the needle to serve as a reservoir. The column is then washed with 10 mL of TE0.1 to equilibrate the column and if necessary, the top frit is again adjusted to the top of the resin bed. For C/EBP trapping, the column is washed with a further 10 mL of TE0.4. For smaller columns (e.g., 0.1 mL), column barrels and frits recovered from a used QIAquick PCR purification column or Qiagen RNAeasy column are used.

3.6. Oligonucleotide Trapping Chromatography (Fig. 3)

1. The oligonucleotide used for trapping depends upon the transcription factor sought and the DNA response element it binds. For C/EBP we used EP24 (without a tail) for EMSA and ACEP24 (containing the $(GT)_5$ single-stranded tail) for trapping (6, 7).
2. Both oligonucleotides are self-complementary and are annealed in a thermocycler for 5 min at 95°C, followed by a linear decrease to 4°C over the next hour.

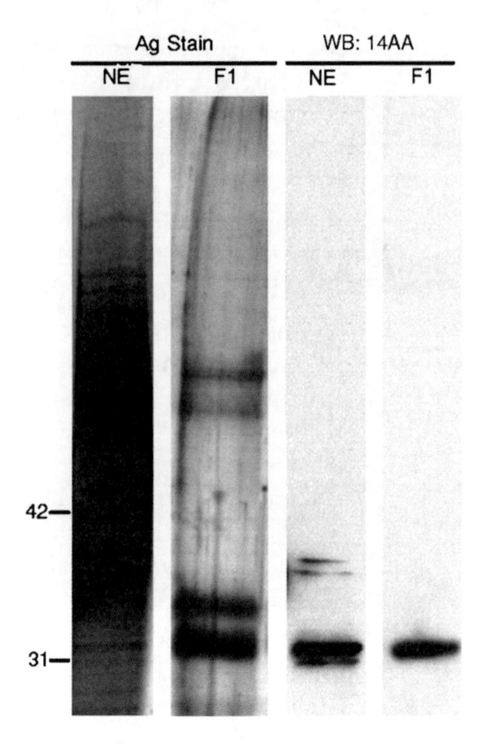

Fig. 3. Silver (Ag) stain and western blot (WB) of oligonucleotide trapping fractions of C/EBP (using ACEP24 oligonucleotide) from rat liver nuclear extract. Carbonic anhydrase (31 kDa) and ovalbumin (42 kDa) molecular masses are indicated. NE, rat liver nuclear extract. Fraction F1 was collected starting as TE1.2 eluted the column. Silver staining of the TE1.2 elution fraction (F1) shows two highly purified and isolated proteins of similar relative molecular weight as C/EBP-α isoforms. A western blot using 1:100 sc-61 (Santa Cruz Biotechnology, 14AA anti-C/EBP-α antibody) demonstrates the presence of C/EBP-α in rat liver nuclear extract, and confirms the isolation and purification of C/EBP-α using the oligonucleotide trapping method. 1 μg of NE was added in both the silver stain and western blot where indicated. For fraction F1, 1/5 of the total fraction was used for electrophoresis. For Western blot of F1, 1/20 of total fraction was used. Reproduced with permission from Elsevier (10).

3. For simple oligonucleotide trapping, trapping is modeled on conditions used for successful EMSA (Subheading 3.4) except using typically a TE based buffer for column loading and elution. For systematically optimized oligonucleotide trapping (7), EMSA assays are used to measure the DNA-binding affinity and concentration of the transcription factor in nuclear extract, and the optimal concentrations of substances such as NaCl, the single-stranded oligonucleotide $(dT)_{18}$, heparin (a competitive inhibitor of DNA-binding), the double-stranded DNA poly dI:dC, and detergents are determined. For C/EBP, the optimal conditions (see Note 7) were as follows: TE containing 0.4 M NaCl (TE0.4), 2.2 μg/mL HEK293 nuclear extract (Subheading 3.1), 1.34 nM annealed ACEP24$(GT)_5$, 930 nM $(dT)_{18}$, 50 ng/mL heparin, 50 μg/mL poly dI:dC,

and 0.1% Tween-20. This mixture (50 mL) is allowed to equilibrate on ice for 30 min to allow C/EBP to form a complex with the ACEP24$(GT)_5$ DNA. All subsequent steps are performed at 4°C.

4. The mixture is then applied to the 1 mL $(AC)_5$-Sepharose column (Subheading 3.5). After all 50 mL is applied, the column is washed with 20 mL of TE0.4, 0.1% Tween-20.
5. The column is then eluted with TE1.2 (TE containing 1.2 M NaCl) collecting 1 mL fractions.
6. The fractions containing C/EBP are then located using the EMSA assay and the purity is examined by SDS-PAGE. If necessary, the fractions can be concentrated using a Millipore Microcon Ultracell YM-10 centrifugal concentrator (see Note 8).

3.7. Promoter Trapping Chromatography

1. Oligonucleotide trapping is used when the transcription factor binding to a specific binding site is being sought. Promoter trapping allows the purification of an active transcription complex binding to a promoter DNA sequence. As a proof-of-concept, we focus in this chapter on purifying the transcriptional complex of the 281-bp core promoter of *c-jun* (8).
2. Making the DNA promoter trap: the c-jun core promoter is produced by PCR and cloned into pUC19 to yield pUC19-c-jun core promoter plasmid (8), which is used as template for PCR. To produce the DNA with single-stranded $(GT)_5$ tails for trapping, the following protocol is used: two different PCR reactions are performed with the following primers (where "Phos" denotes a 5′-phosphorylated oligonucleotide produced during synthesis).

FP	5′-cg*ggatcc*cagcggagcattacctcatc-3′
RP	5′-cg*gaattc*gctggctgtgtctgtctgtc-3′
$(AC)_5$FP	5′-Phos/acacacacacggatcccagcggagcattacc
$(AC)_5$RP	5′-Phos/acacacacacgaattcgctggctgtgtctgtc

Reaction 1 utilizes FP and $(AC)_{5R}$P while reaction 2 utilizes $(AC)_{5F}$P and RP and the reactions are performed separately. PCR (50 μL) containing 200 nM primers, 10 ng of pUC19-c-jun core promoter, 1.25 mM $MgCl_2$, 250 μM dNTP mixture, and 1 U Taq DNA polymerase in PCR buffer (New England Biolabs) is heated at 95°C for 5 min, thermocycled for 1 min at 95°C, 1 min at 50°C, and 2 min at 72°C for 35 cycles, followed by 10 min at 72°C for extension. 250 μg of each PCR product, obtained from multiple replicate reactions, is purified using the QIAquick PCR purification columns (Qiagen) and eluted in 2 mL TE buffer. The resulting 250 μg for each reaction type (1 or 2) is then digested using the supplier's protocol with 100 units of lambda exonuclease for 2 h at 37°C. Lambda exonuclease digests single strands containing a 5′-phosphoryl end to nucleotides, and since reactions one and two only have a single phosphorylated strand, the result is two single-stranded DNAs that are complementary. The two strands are then mixed and annealed (Subheading 3.6, step 2).

3. The annealed DNA contains the duplex c-jun core promoter (–200 to +81 bp) with a 3′-$(GT)_5$ single-stranded tail on each strand. The annealed DNA is purified by applying it (now approximately 500 μg duplex) to a fresh 1 mL $(AC)_5$-Sepharose column equilibrated in TE0.1 at 4°C. The column is then washed with 20 mL TE0.1 at 4°C, moved to room temperature and then eluted with 37°C TE containing 0.1% Tween-20, collecting 0.5 mL fractions (see Note 9). Fractions are analyzed by agarose gel electrophoresis and fractions containing duplex c-jun promoter DNA are combined; the concentration is determined by absorption at 260 nm (assuming 50 μg/mL DNA has an absorbance of 1.0) and stored frozen at –20°C.
4. 100 μL HEK293 nuclear extract (0.5 mg nuclear protein) is diluted to a final volume of 1 mL with TE0.1 buffer containing 0.1% Tween-20, poly dI:dC (30 μg/mL final) and incubated for 10 min at 4°C. The tailed c-jun $(GT)_5$ (calculated molecular weight 187,488) is then added to a final concentration of 60 nM and incubated to form a complex for 30 min at 4°C. At 4°C, the mixture is applied to a 0.1 mL $(AC)_5$-Sepharose column, washed with 20 column volumes of TE0.1 containing 0.1% Tween-20, and proteins bound on the column are eluted with TE0.4 buffer. Samples from TE0.4 elution are dialyzed in 50 mM NH_4HCO_3 and lyophilized (see Note 10).

3.8. Two-Dimensional Gel Electrophoresis and Blotting

1. Isoelectric focusing (IEF) is performed with ReadyStrip IPG strips (pH 3–10, linear, 7 cm) using the PROTEAN IEF cell (BioRad) according to the manufacturer's protocol. HEK293 nuclear extract (100 μg) or a similar amount obtained from oligonucleotide trapping or promoter trapping is mixed in 125 μL rehydration buffer and rehydrated at 50 V for 16 h. IEF is then performed at 40,000 Vh at 20°C.
2. The strips are equilibrated in 2.5 mL equilibration buffer containing 2% DTT at room temperature for 15 min. The strips are then removed and incubated in 2.5 mL equilibration buffer containing 2.5% iodoacetamide in the dark for 15 min. The strips are transferred to 12% SDS-PAGE gels for a second dimension of electrophoresis using the PROTEAN II xi 2-D (BioRad) cell at constant 10 mA/gel for 2 h. After electrophoresis, the gel is stained with silver nitrate or transferred to NC or PVDF membrane for Western blotting (WB) or Southwestern blotting (SW) analysis.

3. Gel blotting is performed as described (9) with minor modifications. Briefly, the protein sample, separated by SDS-PAGE or two-dimensional gel electrophoresis (2DGE), is transferred to PVDF membrane at 110 V for 1.5 h in the cold room in blotting buffer. For Southwestern blotting, PVDF gives the best performance and is used in Fig. 4.

3.9. 2DGE-Southwestern Blotting

1. The blotted proteins are denatured and renatured by immersing the blot in 10 mL 6 M guanidine HCl, which is then serially diluted to 3, 1.5, 0.75, 0.375, 0.188, 0.094 M using 1× binding buffer with incubation at 4°C for 10 min each time.
2. The blot is blocked at room temperature for 1 h in 1× binding buffer containing 0.5% polyvinyl pyrrolidone (PVP40) for 30 min. Alternatively, 5% milk in 1× binding buffer has been used, blocking overnight at 4°C (see Note 11).
3. The membrane is probed overnight for C/EBP with [γ-^{32}P] radiolabeled EP24 (1.5 nM, 10^6 cpm/mL) or with radiolabeled 2 nM 281 bp c-jun core promoter (for nuclear extract or promoter trapping fractionated nuclear extract) or with various other radiolabeled oligonucleotides (unpublished) at 4°C in 1× binding buffer containing 0.25% BSA and 10 μg/mL poly dI:dC. The washed membrane is air-dried and exposed to film for 12 h for autoradiography. Both a Western blot (upper panel) and Southwestern blot (lower panel) are shown in Fig. 4b.

3.10. On-Blot Trypsin Digestion for ESI-MS Analysis

1. The protein spots located by 2DGE-SW are excised (typically a 1–2 mm circle) and stripped at 50°C with SDS stripping buffer for 10 min, and then the membrane is washed ten times with 500 μL H_2O for 1 min/change. This stripping effectively removes the radiolabeled DNA probe and also effectively reduces interference from components used for blocking the membrane.
2. The blotted protein is reduced with 10 mM DTT at 56°C for 1 h and then alkylated by 55 mM iodoacetamide at room temperature in the dark for 1 h. After washing with 500 μL 25 mM NH_4HCO_3 five times for 1 min, the blot is immersed in 20 μL of 25 mM NH_4HCO_3 and heated at 95°C for 5 min.
3. 5 μL of 5% Zwittergent 3–16, 15 μL acetonitrile (ACN), and 10 μL of 40 μg/mL trypsin are added to a final volume of 50 μL, and the blot is digested overnight at 37°C. The blot in digestion solution is sonicated (in a bath sonicator) for 5 min, briefly centrifuged and the supernatant is collected. The peptides on the membrane are further extracted with 30 μL of 5% trifluoroacetic acid (TFA)/50% ACN at room temperature for 2 h and then 30 μL 0.5% TFA/50% ACN for another 2 h. The

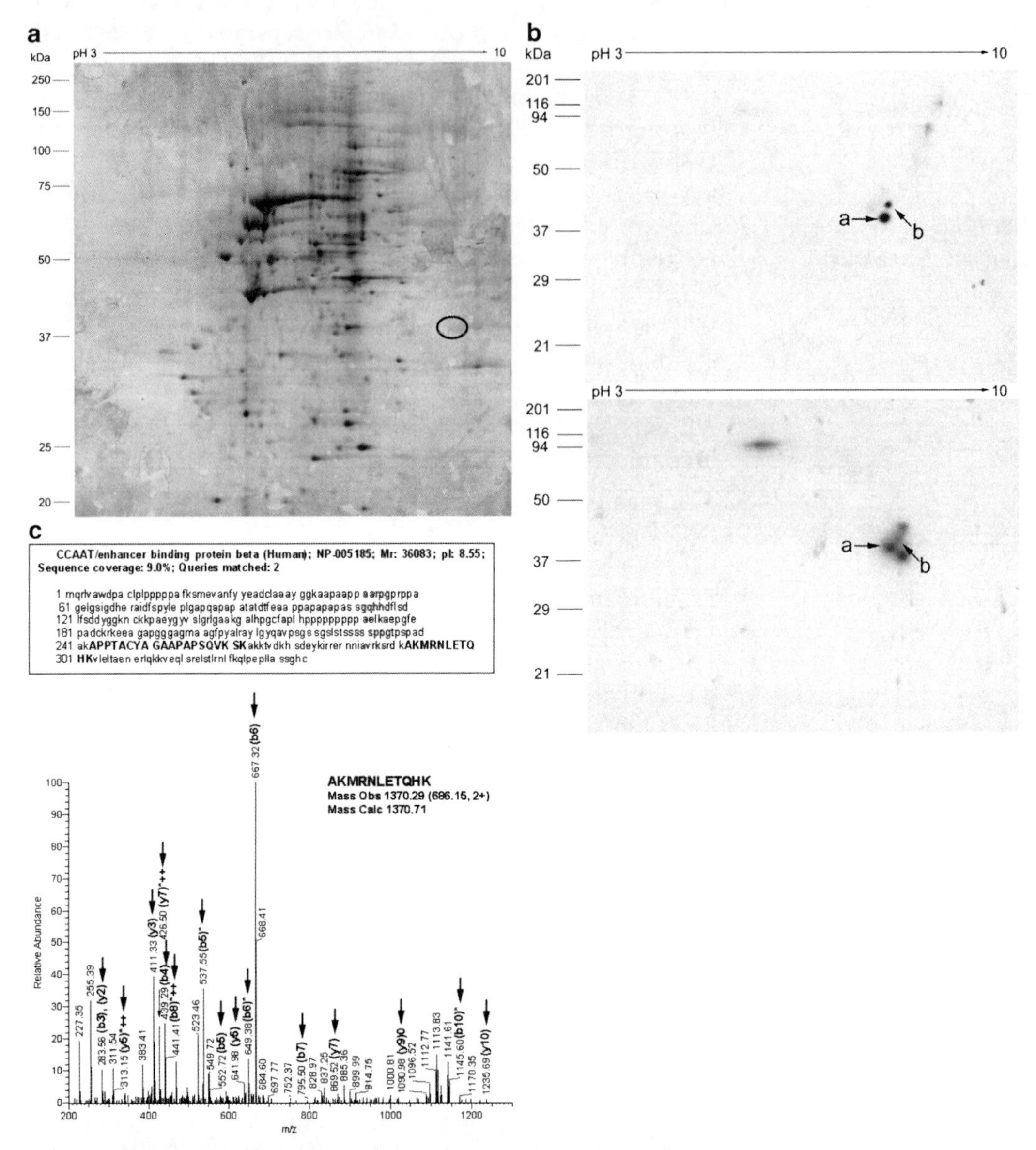

Fig. 4. 2DGE-SW analysis of HEK293 nuclear extract. (**a**) HEK293 nuclear extract was separated by 2DGE. One 2DGE gel (50 μg nuclear extract) was stained with silver nitrate in panel **a**. *Encircled* is the region of interest identified by Western and Southwestern blot as shown in the other panels. (**b**) Two 2DGE gels (100 μg nuclear extract each) were transferred to PVDF membrane for Western blot (upper panel) and Southwestern blot (lower) analysis in panel **b**. The two protein spots on the blots that reacted with the C/EBP antibody (Santa Cruz C/EBP-β antibody Δ198) and radiolabeled EP24 are indicated as "a" and "b." The 2DGE-SW spots were analyzed by on-blot digestion and HPLC-nano-ESI-MS/MS analysis. Spot "a" was successfully identified as human C/EBP-β, as shown in panel **c**. (**c**) MS/MS analysis of 2DGE-SW spot "a". MS/MS results shows two unique peptides from spot "a" matched with human C/EBP beta (NP-005185) with 9% sequence coverage after searching the local transcription factor database fused with the SwissProt database. Identification as C/EBP-β has a $P < 0.0035$. The mass spectrum shown below represent the MS/MS results of the doubly charged peptide AKMRNLETQHK at m/z 686.15 (2+). The matched b- and y-ions, derived from the peptide MS/MS, are indicated with *arrows*. The amino acid sequence of the parent peptide is shown on the top of the spectrum. Reproduced from the American Chemical Society with permission from the publisher (11).

combined eluate (110 μL) is vacuum-dried and dissolved in 30 μL of 0.1% TFA. The eluate is again vacuum-dried.

4. The eluate is dissolved in 10 μL 0.1% TFA prior to LC-nanosprayESI-MS/MS. Most scientists have available to them an excellent proteomics facility for analysis, but commercial analysis could also be an option. Therefore, this is likely to be the last step for most investigators. However, we often perform our own analysis, detailed below, which may serve as a model for analysis.

3.11. Capillary HPLC-Electrospray Tandem Mass Spectrometry (LC-nanosprayESI-MS/MS)

1. Analysis is on a Thermo Finnigan LTQ linear ion trap mass spectrometer equipped with a nano-ESI source.
2. On-line HPLC separation of the digests is accomplished with an Eksigent HPLC micro HPLC. The column is a PicoFrit™ Emitter (New Objective; 50 μm i.d., 15 μm tip) packed to 10 cm with C18 adsorbent (Alltech Altima, 5 μm, 300 Å). The gradient is from 2 to 42% mobile phase B in 30 min. The flow rate is 0.3 μL/min. MS conditions are a 2.5 kV ESI voltage, an isolation window for MS/MS of 3, 35% relative collision energy, with a scan strategy of a survey scan followed by acquisition of data dependent collision-induced dissociation (CID) spectra of the seven most intense ions in the survey scan above a set threshold.
3. Database searching is with Mascot software (Matrix Science, Boston, MA, USA, version 2.1.0) against either the Swiss-Prot 49.0 (268,833 sequences, 123,649,849 residues) or our in-house human transcription factor database. The search parameters are as follows: trypsin digestion, two possible missed cleavages, monoisotopic mass values, peptide mass tolerance of 1.0 Da, MS/MS tolerance 0.8 Da, instrument ESI-TRAP, and oxidized methionine and cysteine carboxyamidation as variable modifications. Results were scored based on the probability based Mowse score. The score threshhold to achieve $p<0.05$ is set by Mascot and is based on the size of the database searched. Proteins with a probability less than threshold and with two or more independent peptide sequences are considered as true positives. The on-blot digestion and MS/MS characterization of C/EBP-β from HEK293 nuclear extract are shown in Fig. 4c.

4. Notes

1. Water is >18 MΩ from a Millipore Synergy UV water purification unit.
2. Oligonucleotides (1 μmol) are purchased from (Integrated DNA Technologies, Coralville, IA, USA). The company

provides a molar absorptivity (E_{260nm}) used to calculate concentration.

3. Long dialysis is detrimental to activity. The dialysis procedure described gives high activity and thorough dialysis.
4. For short oligonucleotides such as $(AC)_5$, increasing ethanol to 1.2 mL and using 75% ethanol for washing will increase yield.
5. Alternatively, a nanodrop spectrophotometer can be used, but results are less accurate.
6. The centrifuge tube used for collection can be preweighed and weighed again after collection to determine the volume more accurately.
7. Determination of optimal concentrations is complex and more fully explained elsewhere (7). Essentially, the EMSA assay is used to determine the apparent dissociation constant of the protein for the DNA used (K_d) and the maximal amount of the binding protein (B_{max}) in the nuclear extract. Then, using nuclear extract containing an amount of protein equal to K_d and a concentration of DNA equal to ten times K_d is used for all subsequent EMSA assays and for the final purification. Beginning with dI:dC, different concentrations are added to the EMSA assay to determine the highest concentration that can be used without inhibiting binding. This concentration is then fixed for all subsequent assays. Similarly, the highest non-inhibiting concentration of (dT)18, heparin, and detergent is then tested and set at this concentration for all subsequent assays. The final resulting optimal concentrations are then used for oligonucleotide trapping purification.
8. Occasionally, the concentrator is leaking. It is, therefore, best to retain the filtrate and confirm by absorption (280 nm) that protein has been removed.
9. The promoter DNA binds to the column by annealing of the $(GT)_5$-tail with $(AC)_5$-Sepharose in TE0.1 at 4°C. This is disrupted in low salt (TE) at elevated temperature. The duplex promoter, because of its length, is unaffected.
10. The complex isolated has been shown to be transcriptionally active and to contain specific transcription factors, the TFII complex, and RNA polymerase 2 (8). Lyophilization may affect this activity, but not further analysis by Western blot or Southwestern blot.
11. Which blocking procedure is chosen depends on subsequent experiments. PVP-40 fragments in the mass spectrometer to give a complex spectrum but works best for renaturation and retention on the blot and gives the highest signal for Southwestern blotting. Bovine serum albumin is often not advised by commercial mass spectrometry facilities. If the

protein spots are going to be used directly after this step for mass spectrometry, some interference by either method would be expected. However, in Subheading 3.10 step 1, the blocking medium is stripped away and probably neither blocking approach would significantly interfere. We have successfully used both when the complete procedure given here is used for analysis.

Acknowledgments

We thank Dr. William Haskins and the UTSA RCMI Proteomics and Protein Biomarkers Core facility for help with analysis and the use of their equipment. We also appreciate the help of UTSA RCMI Bioinformatics and Computational Biology core for computing. The RCMI facilities are supported by grant 2G12RR013646 from the NIH. We also thank Daifang Jiang for reading the manuscript and for much helpful discussion. This work was supported by NIH grant R01GM043609 and a grant from the San Antonio Life Sciences Institute.

References

1. Hahn, S. (2004) Structure and mechanism of the RNA polymerase II transcription machinery, *Nat Struct Mol Biol.* *11*, 394–403.
2. Venter, J. C., Adams, M. D., Myers, E. W., Li, P. W., Mural, R. J., Sutton, G. G., Smith, H. O., Yandell, M., Evans, C. A., Holt, R. A., Gocayne, J. D., Amanatides, P., Ballew, R. M., Huson, D. H., Wortman, J. R., Zhang, Q., Kodira, C. D., Zheng, X. H., Chen, L., Skupski, M., Subramanian, G., Thomas, P. D., Zhang, J., Gabor Miklos, G. L., Nelson, C., Broder, S., Clark, A. G., Nadeau, J., McKusick, V. A., Zinder, N., Levine, A. J., Roberts, R. J., Simon, M., Slayman, C., Hunkapiller, M., Bolanos, R., Delcher, A., Dew, I., Fasulo, D., Flanigan, M., Florea, L., Halpern, A., Hannenhalli, S., Kravitz, S., Levy, S., Mobarry, C., Reinert, K., Remington, K., Abu-Threideh, J., Beasley, E., Biddick, K., Bonazzi, V., Brandon, R., Cargill, M., Chandramouliswaran, I., Charlab, R., Chaturvedi, K., Deng, Z., Di Francesco, V., Dunn, P., Eilbeck, K., Evangelista, C., Gabrielian, A. E., Gan, W., Ge, W., Gong, F., Gu, Z., Guan, P., Heiman, T. J., Higgins, M. E., Ji, R. R., Ke, Z., Ketchum, K. A., Lai, Z., Lei, Y., Li, Z., Li, J., Liang, Y., Lin, X., Lu, F., Merkulov, G. V., Milshina, N., Moore, H. M., Naik, A. K., Narayan, V. A., Neelam, B., Nusskern, D., Rusch, D. B., Salzberg, S., Shao, W., Shue, B., Sun, J., Wang, Z., Wang, A., Wang, X., Wang, J., Wei, M., Wides, R., Xiao, C., Yan, C., Yao, A., Ye, J., Zhan, M., Zhang, W., Zhang, H., Zhao, Q., Zheng, L., Zhong, F., Zhong, W., Zhu, S., Zhao, S., Gilbert, D., Baumhueter, S., Spier, G., Carter, C., Cravchik, A., Woodage, T., Ali, F., An, H., Awe, A., Baldwin, D., Baden, H., Barnstead, M., Barrow, I., Beeson, K., Busam, D., Carver, A., Center, A., Cheng, M. L., Curry, L., Danaher, S., Davenport, L., Desilets, R., Dietz, S., Dodson, K., Doup, L., Ferriera, S., Garg, N., Gluecksmann, A., Hart, B., Haynes, J., Haynes, C., Heiner, C., Hladun, S., Hostin, D., Houck, J., Howland, T., Ibegwam, C., Johnson, J., Kalush, F., Kline, L., Koduru, S., Love, A., Mann, F., May, D., McCawley, S., McIntosh, T., McMullen, I., Moy, M., Moy, L., Murphy, B., Nelson, K., Pfannkoch, C., Pratts, E., Puri, V., Qureshi, H., Reardon, M., Rodriguez, R., Rogers, Y. H., Romblad, D., Ruhfel, B., Scott, R., Sitter, C., Smallwood, M., Stewart, E., Strong, R., Suh, E., Thomas, R., Tint, N. N., Tse, S., Vech, C., Wang, G., Wetter, J., Williams, S., Williams, M., Windsor, S., Winn-Deen, E., Wolfe, K., Zaveri, J., Zaveri, K., Abril, J. F., Guigo, R., Campbell, M. J., Sjolander, K. V., Karlak, B., Kejariwal, A., Mi, H., Lazareva, B., Hatton, T., Narechania, A., Diemer, K., Muruganujan, A., Guo, N., Sato, S., Bafna, V., Istrail, S., Lippert, R., Schwartz, R., Walenz,

B., Yooseph, S., Allen, D., Basu, A., Baxendale, J., Blick, L., Caminha, M., Carnes-Stine, J., Caulk, P., Chiang, Y. H., Coyne, M., Dahlke, C., Mays, A., Dombroski, M., Donnelly, M., Ely, D., Esparham, S., Fosler, C., Gire, H., Glanowski, S., Glasser, K., Glodek, A., Gorokhov, M., Graham, K., Gropman, B., Harris, M., Heil, J., Henderson, S., Hoover, J., Jennings, D., Jordan, C., Jordan, J., Kasha, J., Kagan, L., Kraft, C., Levitsky, A., Lewis, M., Liu, X., Lopez, J., Ma, D., Majoros, W., McDaniel, J., Murphy, S., Newman, M., Nguyen, T., Nguyen, N., Nodell, M., Pan, S., Peck, J., Peterson, M., Rowe, W., Sanders, R., Scott, J., Simpson, M., Smith, T., Sprague, A., Stockwell, T., Turner, R., Venter, E., Wang, M., Wen, M., Wu, D., Wu, M., Xia, A., Zandieh, A., and Zhu, X. (2001) The sequence of the human genome, *Science* ***291***, 1304–1351.

3. Vaquerizas, J. M., Kummerfeld, S. K., Teichmann, S. A., and Luscombe, N. M. (2009) A census of human transcription factors: function, expression and evolution, *Nature Reviews Genetics* ***10***, 252–263.
4. Abmayr, S. M., Yao, T., Parmley, T., and Workman, J. (2006) Preparation on Nuclear and Cytoplasmic Extracts from Mammalian Cells, in *Current Protocols in Molecular Biology*, p 12.11, John Wiley & Sons.
5. Bradford, M. M. (1976) A rapid and sensitive method for the quantitation of microgram quantities of protein utilizing the principle of protein-dye binding. *Anal. Biochem.* *72*, 248–254.
6. Gadgil, H., and Jarrett, H. W. (2002) Oligonucleotide trapping method for purification of transcription factors, *J. Chromatogr. A* ***966***, 99–110.
7. Moxley, R. A., and Jarrett, H. W. (2005) Oligonucleotide trapping method for transcription factor purification systematic optimization using electrophoretic mobility shift assay, *J Chromatogr A* ***1070***, 23–34.
8. Jiang, D., Moxley, R. A., and Jarrett, H. W. (2006) Promoter trapping of c-jun promoter-binding transcription factors, *J. Chromatogr. A* ***1133***, 83–94.
9. Ohri, S., Sharma, D., and Dixit, A. (2004) Interaction of an approximately 40 kDa protein from regenerating rat liver with the −148 to −124 region of c-jun complexed with RLjunRP coincides with enhanced c-jun expression in proliferating rat liver, *Eur J Biochem* *271*, 4892–4902.
10. Moxley, R.A. and Jarrett, H.W. (2005) Oligonucleotide trapping method for transcription factor purification systematic optimization using electrophoretic mobility shift assay, *J. Chromatogr. A* ***1070***, 23–34.
11. D. Jiang, D. Jia, Y., Zhou, Y.W. and Jarrett, H.W. (2009) Two-dimensional southwestern blotting and characterization of transcription factors on-blot, *J Proteome Res* ***8***, 3693–3701.

Chapter 20

A High-throughput Gateway-Compatible Yeast One-Hybrid Screen to Detect Protein–DNA Interactions

Korneel Hens, Jean-Daniel Feuz, and Bart Deplancke

Abstract

In recent years, new techniques have spurred the discovery of *cis*-regulatory DNA elements. These stretches of noncoding DNA contain combinations of recognition sites to which transcription factors (TFs) bind, and in doing so, these TFs can activate or repress transcription. These protein–DNA interactions form the core of gene regulatory networks (GRNs) that are responsible for the differential gene expression that allow diversification of cell types, developmental programs, and responses to the environment. The yeast one-hybrid system is a genetic assay to identify direct binding of proteins to DNA elements of interest and is, therefore, instrumental in uncovering these GRNs.

Key words: Yeast one-hybrid, Transcription factor, Protein–DNA interaction, Gateway, Regulatory element, Gene regulatory element

1. Introduction

Since the sequencing of the bacteriophage phi X174 genome in 1977 (1), and later the genomes of more complex organisms such as *Caenorhabditis elegans* (2), *Drosophila melanogaster* (3), mouse (4), and human (5), genomic sequences for a large variety of organisms have become available. To date, the protein coding sequences are the best understood parts of the genome and near-complete lists of protein-coding sequences exist for several species. However, although bacteria can achieve coding densities of more than 90% of the genome (6), in complex organisms a large portion of the genomic DNA does not code for protein, for example, less than 2% of the human genome is protein coding (7). Although this noncoding DNA was sometimes referred to as "junk DNA" (8), it has become clear that a significant portion of this DNA has a defined function, and new properties are continuously being discovered. One class of noncoding sequences that has been studied extensively

Bart Deplancke and Nele Gheldof (eds.), *Gene Regulatory Networks: Methods and Protocols*,
Methods in Molecular Biology, vol. 786, DOI 10.1007/978-1-61779-292-2_20, © Springer Science+Business Media, LLC 2012

consists of *cis*-acting DNA elements that are responsible for specifying the spatial and temporal expression of genes.

Discovery of these *cis*-acting regulatory elements has long been challenging, but newly developed bioinformatics and experimental techniques as well as the vast genomic sequencing data that become available at a continuously increasing rate have made genome-wide mapping of regulatory elements feasible. For example, the cap analysis gene expression (CAGE) technique allows the precise mapping of transcriptional start sites, which in turn helps identifying the promoter driving gene expression (9). Regulatory motifs can then be found bioinformatically by correlating expression data with statistically overrepresented patterns within these promoters (10, 11). In addition, comparative genomic-based approaches can identify evolutionary conserved stretches of non-coding DNA which have a high propensity to behave as transcriptional enhancers (12). Chromatin immunoprecipitation assays coupled to microarray (ChIP-chip) or, more recently, to high-throughput sequencing techniques (ChIP-Seq) allow the precise mapping of the genomic location of enhancer-associated proteins such as the transcriptional activator p300, pointing toward active enhancers (13). Similarly, histone modifications that are enriched at active enhancers (e.g., histone H3K4 monomethylation), at active promoters (e.g., histone H3K4 trimethylation), or at repressors (histone H3K27 trimethylation) (14, 15) can be interrogated by ChIP-chip or ChIP-seq to uncover novel regulatory elements. In combination with *in vivo* reporter assays, these techniques have allowed the construction of libraries of *cis*-regulatory DNA elements that drive expression of a reporter gene in a tissue-specific manner (12, 16–20). Specific sites within these elements are bound by sequence-specific transcription factors that, together with activating or repressing cofactors, can determine the transcriptional status of a gene by respectively recruiting RNA polymerase II to or blocking it from the promoter. Several techniques have been developed to study the binding of specific TFs to regulatory elements (21). These so-called TF-centered techniques can be either aimed at determining the binding of TFs to selected regulatory elements or binding sites, e.g., electomobility shift assays, or at identifying TF binding sites genome wide, e.g., ChIP-chip or ChIP-seq. However, since regulatory elements are being discovered at an increasing rate, many biological questions arise from a gene-centered point of view, i.e., which TFs are responsible for driving expression of a particular gene or set of genes in a specific pattern. To address this type of question, a technique is needed that can map protein–DNA interactions between the full complement of TFs of an organism and a specific regulatory element or set of elements. One technique that has been used successfully to "deorphanize" regulatory elements is the yeast one-hybrid (Y1H) technique.

The Y1H technique was developed as a modification of the yeast two-hybrid (Y2H) that is used for detecting protein–protein interactions. As the name implies, Y2H uses two hybrid proteins, a bait protein that is fused to the DNA-binding domain of the yeast GAL4 TF, and a prey protein that is fused to the activation domain of GAL4. When bait and prey proteins interact, a functional GAL4 TF is reconstituted, which then drives the expression of a reporter construct that contains the GAL4 recognition sequence UAS upstream of a reporter gene (22). In the Y1H technique, the UAS sequence is replaced by a DNA bait of interest, which traditionally was a putative, small binding site in multimerized form. The hybrid protein in Y1H is a fusion between a DNA-binding protein and the GAL4 activation domain (23, 24). The system was later adapted to allow the screening of larger regulatory elements, and to also enhance the throughput by making it compatible with the Gateway cloning system that allows fast and accurate recombination-based cloning (25, 26). In addition, in contrast to the traditional Y1H which relies on cDNA libraries derived from whole organisms or specific tissues, the sensitivity of the Y1H system was significantly increased by the construction of a "TF-only" library (25).

Here, we describe in detail the generation of the prey library, the cloning in genomic integration of DNA baits and the actual Y1H itself.

2. Materials

2.1. Prey Library Preparation

1. iProof High-Fidelity DNA Polymerase (Bio-Rad, Hercules, CA).
2. dNTP Set, 100 mM (Invitrogen, Carlsbad, CA).
3. Agarose, Analytical Grade (Promega, Madison, WI).
4. 1 Kb Plus DNA Ladder (Invitrogen, Carlsbad, CA).
5. Gateway BP Clonase II enzyme mix (Invitrogen, Carlsbad, CA).
6. pDONR221 vector (100 ng/μL) (Invitrogen, Carlsbad, CA).
7. ThermoPol Reaction Buffer (New England Biolabs, Ipswich, MA).
8. Taq DNA polymerase (New England Biolabs, Ipswich, MA).
9. Glycerol solution: 40% (v/v) glycerol in ultrapure water. Autoclave before storage at room temperature.
10. NucleoSpin Multi-96 Plus Plasmid kit (Macherey Nagel, Düren, Germany).
11. Gateway LR Clonase II enzyme mix (Invitrogen, Carlsbad, CA).
12. pAD-DEST vector, 100 ng/μL.

13. pAD-DEST-2 μ vector, 100 ng/μL.
14. M13F primer: 5′-GTTGTAAAACGACGGCCAGT-3′.
15. M13R primer: 5′-CAGGAAACAGCTATGACCAT-3′.
16. AD primer: 5′-CGCGTTTGGAATCACTACAGGG-3′.
17. TERM primer: 5′- GGAGACTTGACCAAACCTCTGGCG-3′.

2.2. DNA Bait Cloning

1. iProof High-Fidelity DNA Polymerase (Bio-Rad, Hercules, CA).
2. dNTP Set, 100 mM (Invitrogen, Carlsbad, CA).
3. Agarose, Analytical Grade (Promega, Madison, WI).
4. 1 kb Plus DNA Ladder (Invitrogen, Carlsbad, CA).
5. Gateway BP Clonase II enzyme mix (Invitrogen, Carlsbad, CA).
6. pDONR221 vector (100 ng/μL) (Invitrogen, Carlsbad, CA).
7. ThermoPol Reaction Buffer (New England Biolabs, Ipswich, MA).
8. Taq DNA polymerase (New England Biolabs, Ipswich, MA).
9. Glycerol solution: 40% (v/v) glycerol in ultrapure water. Autoclave before storage at room temperature.
10. NucleoSpin Multi-96 Plus Plasmid kit (Macherey Nagel, Düren, Germany).
11. Gateway LR Clonase II enzyme mix (Invitrogen, Carlsbad, CA).
12. pAD-DEST vector, 100 ng/μL.
13. pAD-DEST-2 μ vector, 100 ng/μL.
14. M13F primer.
15. His293R primer: 5′-GGGACCACCCTTTAAAGAGA-3′.
16. 1HIFw primer: 5′-GTTCGGAGATTACCGAATCAA-3′.
17. LacZ592R primer: 5′-ATGCGCTCAGGTCAAATTCAGA-3′.

2.3. Yeast Genomic Integration

1. D-glucose (Prolabo), 40% (w/v) in ultrapure water (see Note 1). Sterilize by passing through a 22-μm filter.
2. Amino acid mix: mix 6 g of alanine, arginine, aspartic acid, asparagine, cysteine, glutamic acid, glutamine, glycine, isoleucine, lysine, phenylalanine, proline, serine, threonine, tyrosine, valine (all amino acids from Acros, Geel, Belgium), and 6 g of adenine sulfate (Applichem, Darmstadt, Germany).
3. Amino acid solutions: Histidine-HCl, 100 mM in water; Leucine, 100 mM in water; Tryptophan, 40 mM in water. Sterilize by passing through a 22-μm filter. Store the tryptophan solution at 4°C and protect from light.
4. Uracil solution: 20 mM uracil (Acros, Geel, Belgium) in water. Sterilize by passing through a 22-μm filter.

5. Yeast extract Peptone Dextrose (YPD) medium: mix 10 g yeast extract (MP Biomedicals, Irvine, CA) and 20 g bacto-peptone (Conda, Madrid, Spain). Add water to 950 mL. Sterilize by autoclaving. Add 50 mL 40% D-glucose before use.
6. YPD agar plates: mix 10 g yeast extract and 20 g bacto-peptone, add water to 450 mL, autoclave. Dissolve 20 g bacteriological agar (Conda, Madrid, Spain) in 500 mL water. Sterilize by autoclaving. Mix YPD, agar, and 50 mL 40% D-glucose and pour plates.
7. Synthetic complete Drop-Out plates (Sc -His, -Ura): Mix 2.6 g amino acid mix, 3.4 g yeast nitrogen base (BD Biosciences, Franklin Lakes, NJ), and 10 g ammonium sulfate, add water to 950 mL, adjust pH to 5.9 with 10 N NaOH. Sterilize by autoclaving. Dissolve 40 g agar in 950 mL water. Sterilize by autoclaving. Mix Sc Drop-Out medium, agar, and 100 mL 40% glucose. Add 16 mL of Leu and Trp solution. Add 3-Amino-1,2,4-triazole (3-AT) (Sigma-Aldrich) as required. For example, to make plates containing 20 mM 3-AT, add 3.36 g of 3-AT. Pour plates.
8. TE buffer 10×: 100 mM Tris–HCl, pH 8, 10 mM EDTA, autoclave before storage at room temperature.
9. LiAc solution 10×: 1 M LiAc, autoclave before storage at room temperature.
10. Polyethylene glycol (PEG) solution: 50% (w/v) PEG in ultrapure water, autoclave before storage at room temperature.
11. TE–LiAc solution: combine TE buffer 10× and LiAc solution 10× and dilute to 1× in water. Prepare fresh for each integration.
12. TE–LiAc–PEG solution: combine TE buffer 10× and LiAc solution 10× and dilute to 1× in PEG solution. Prepare fresh for each integration.

2.4. Testing for Reporter Self-activation

1. Sc -His, -Ura plates containing 3-Amino-1,2,4-triazole (3-AT). Prepare Sc -His, -Ura plates as in item 7 of Subheading 2.3. Add 3-Amino-1,2,4-triazole (3-AT) (Sigma-Aldrich, St. Louis, MO) after addition of the glucose solution as required. Pour plates.
2. Nitrocellulose membranes (Pall, Port Washington, NY).
3. Z-buffer: 60 mM $Na_2HPO_4 \cdot H_2O$, 60 mM $NaH_2PO_4 \cdot 2H_2O$, 10 mM KCl, and 1 mM $MgSO_4$. Adjust pH to 7.0. Autoclave before storage at room temperature.
4. X-Gal solution: 4% 5-bromo-4-chloro-3-indolyl- beta-D-galactopyranoside (X-Gal, MP Biomedicals, Irvine, CA) in dimethylformamide. Store at −20°C and protect from light.

5. X-Gal staining solution: Mix 6 mL z-buffer, 11 μL β-mercaptoethanol, and 100 μL X-Gal solution. Prepare fresh for each assay.
6. OmniTray (Nunc, Rochester, NY).

2.5. High-throughput Y1H Assay

1. 384-well plate (REMP, Oberdiessbach, Switzerland).
2. YPD medium: as in item 5 of Subheading 2.3.
3. YPD agar plates as in item 6 of Subheading 2.3.
4. Sc -His, -Ura, -Trp plates: Prepare plates as in item 7 of Subheading 2.3 but omit the Trp solution. For 3-AT containing plates, add 3-AT as required after addition of the glucose solution.
5. TE–LiAc solution: as in item 11 of Subheading 2.3.
6. TE–LiAc–PEG solution: as in item 12 of Subheading 2.3.
7. X-Gal staining solution: as in item 4 of Subheading 2.4.
8. Nitrocellulose membranes.
9. OmniTray (Nunc, Rochester, NY).

3. Methods

3.1. Prey Library Construction

1. The TF open reading frame (ORF) is PCR-amplified from a preexisting cDNA clone, or a cDNA library. Design primers that flank the TF open reading frame and that contain the gateway attB1 (ggggacaactttgtacaaaaaagttggcacc) and attB2 (ggggacaactttgtacaagaaagttggcaa) sites at the 5′ end of the forward and reverse primer respectively (see Note 2). The TF-specific part of the primers should be between 18 and 25 nucleotides long, have a melting temperature around 60°C and a GC content between 40 and 60%. Since the position of the primers is fixed at the beginning and the end of the ORF, these parameters may have to be relaxed to find suitable primer pairs. We obtained good results using the Primer3 primer design software (27). Removing the stop codon from the TF results in an "open-ended" ORF that can be used to create C-terminal fusions to the TF. For the Y1H system however, the GAL4-AD is fused N-terminally to the TF, so the original stop codon from the TF can be preserved if desired. An overview of the prey library construction is shown in Fig. 1.
2. Set up a 25 μL PCR reaction by mixing 5 μL iProof buffer (5×), 0.5 μL dNTP mix (10 mM of each dNTP), 0.25 μL iProof DNA polymerase, 0.5 μL of both forward and reverse primer (10 μM), and the cDNA source containing the TF cDNA.

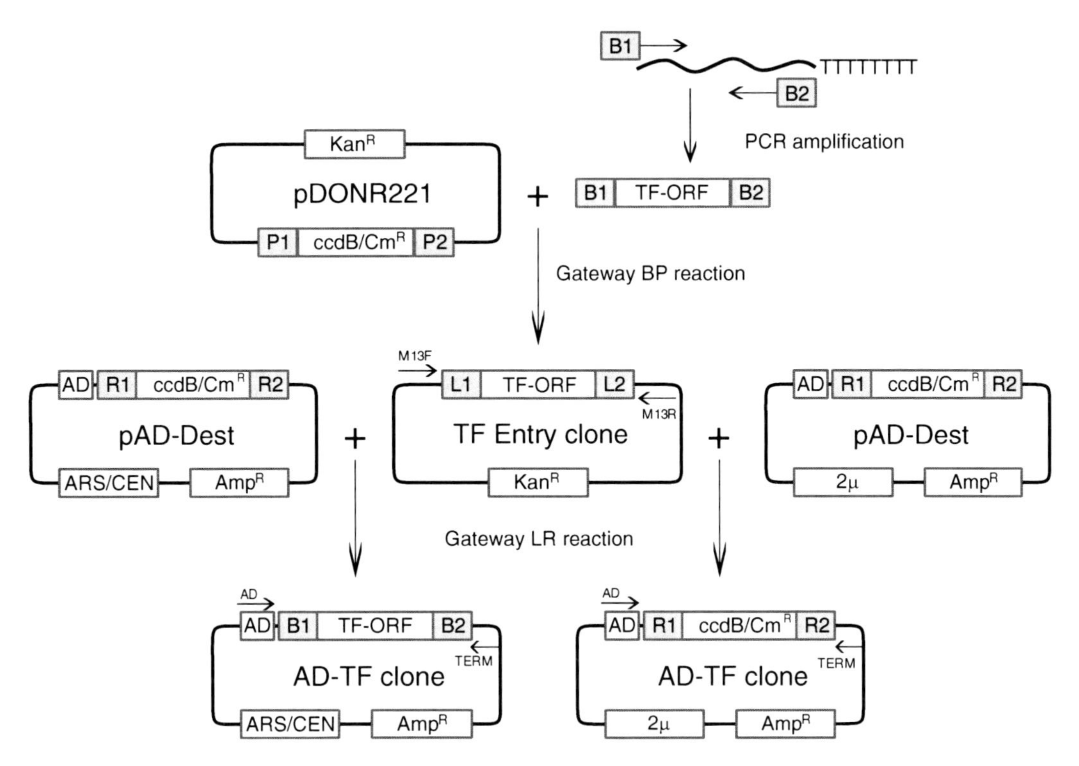

Fig. 1. Prey library construction. The TF ORF is amplified and gateway attB1 and attB2 sites are attached to the 5′ and 3′ ORF end respectively during PCR. The PCR fragment is cloned into the pDONR221 vector by performing a Gateway BP reaction, resulting in an Entry clone that can be selected for using the kanamycin resistance gene. A gateway LR reaction is subsequently performed with the Entry clone and an equimolar mixture of pAD-dest and pAD-Dest-2 μ vectors. This results in a mix of high- and low-copy number yeast expression clones coding for a hybrid protein consisting of the yeast GAL4 activation domain fused to the N-terminal end of the TF protein. Primer positions are indicated by *black arrows*.

Add up to 25 μL with ultrapure water. Perform a PCR reaction in a thermal cycler with a heated lid using the following program:

1.	98°C	60″
2.	98°C	15″
3.	65°C	30″
4.	72°C	30″ × length of the longest ORF in kilobase
5.	72°C	5′
6.	10°C	hold

Cycle between steps 4 and 2. The minimal number of cycles necessary to obtain a clear band on an agarose gel should be used to avoid the accumulation of PCR errors. Analyze 2 μL of the PCR product on an agarose gel. If multiple bands are observed, excise the correct band from the gel and purify it from the agarose gel. If a single, clear PCR fragment is obtained, no further purification is necessary.

3. Perform a Gateway BP reaction to transfer the TF ORF in the pDONR221 vector. Mix 1 μL of the pDONR221 vector (100 ng/μL), 1 μL BP clonase II (Invitrogen, Carlsbad, CA), 2 μL of the PCR product and 1 μL of ultrapure water. Incubate this mixture overnight in a 25°C incubator (see Note 3).
4. Transform the entire BP reaction mix in ultracompetent DH5α, spread the transformation mix on a LB agar plate containing 25 μg/mL kanamycin. Incubate the plate overnight at 37°C.
5. Check the size of the cloned fragment by performing colony PCR on the obtained bacterial colonies. Mix 2.5 μL thermopol buffer, 0.5 μL dNTP mix (10 mM of each dNTP), 0.5 μL of both M13F and M13R primers (10 μM), 0.25 μL Taq polymerase (New England Biolabs, Ipswich, MA), and 20.75 μL ultrapure water (see Note 4). Pick a colony by touching it with a sterile 10-μL pipette tip. Spot the bacteria on a new kanamycin containing LB-agar plate by briefly touching it with the pipette tip and then transfer the tip into the PCR mix. We typically screen six colonies per TF. Remove the pipette tip after 5 min and perform a PCR reaction in a thermal cycler with a heated lid using the following program:

1.	94°C	120″
2.	94°C	45″
3.	60°C	45″
4.	68°C	45″ × length of the longest ORF in kilobase
5.	68°C	7′
6.	10°C	hold

Cycle 30 times between steps 4 and 2. Recover the positive clones from the LB-agar plate and grow them overnight at 37°C in 3 mL LB medium containing 50 μg/mL kanamycin.

6. Prepare a glycerol stock for long-term storage by mixing 100 μL of the bacterial suspension with 100 μL of sterile 40% (v/v) glycerol. Store the glycerol stock at –80°C. Perform a plasmid isolation on the remaining bacterial suspension using the NucleoSpin Multi-96 Plus Plasmid kit (Macherey Nagel) according to the suppliers instructions (see Note 5). Determine the resulting plasmid concentration and dilute to 100 ng/μL. We strongly recommend verifying the prey sequence at this point to avoid mutations or PCR artifacts that affect the ORF (see Note 6).

7. Transfer the TFs from the Entry clone to the Y1H compatible, AD containing vectors using a Gateway LR reaction. We use an equimolar mix of pAD-Dest and pAD-Dest-2 μ. The former vector contains the low copy-number ARS/CEN yeast origin of replication, while the latter contains the high copy-number 2 μ origin of replication. By using this mix, approximately half of the yeast in the Y1H screen will contain the high copy-number vector, resulting in increased TF expression and thus increased sensitivity. The other half will contain the low copy-number vector, which may still allow detection of interactions when too high a titer of the TF proves lethal for the yeast. Set up the LR cloning reaction by mixing 1 μL of the pAD vector mix (50 ng/μL of each vector), 1 μL LR clonase II (Invitrogen, Carlsbad, CA), 2 μL TF Entry clone (100 ng/μL), and 1 μL ultrapure water. Incubate the LR reaction mix overnight at 25°C (see Note 3).
8. Transform the entire LR reaction mix in ultra-competent DH5α. Spread the transformation mix on a LB agar plate containing 50 μg/mL ampicillin and incubate the plate overnight at 37°C.
9. Check the size of the cloned fragment by performing colony PCR on the obtained bacterial colonies as described above using the AD and TERM primers (see Note 4).
10. Recover the positive clones and grow them overnight in LB medium containing 100 μg/μL ampicillin at 37°C. Prepare glycerol stocks and isolate the plasmids as described above. Dilute the plasmids to 100 ng/μL.

3.2. DNA Bait Cloning

1. Baits in the Y1H technique are DNA fragments that are suspected to contain TF binding sites. The Y1H technique described here can accommodate sizes ranging from a few base pairs (single TF binding sites) up to 1.5 kb. The DNA baits are cloned in a two-step process. First, an Entry clone is generated which is then used to transfer the bait into two reporter vector using gateway cloning. The generation of the Entry clone can be achieved in two ways: by gateway cloning for improved convenience and speed, or by traditional restriction/ligation for reduced cost. An overview of the DNA bait cloning procedure is shown in Fig. 2.
2. For the Gateway approach, design primers that flank the bait DNA fragment as described for the prey library generation in step 1 of Subheading 3.1. Add the gateway attB4 (ggggacaactt tgtatagaaaagttg) and attB1R (ggggactgctttttgtacaaacttg) sites at the 5′ end of the forward and reverse primer respectively. Set up a PCR using these primers as described in item 2 of Subheading 2.1. As DNA input for the PCR, we typically use 1 μL of genomic DNA isolated from the model organism of

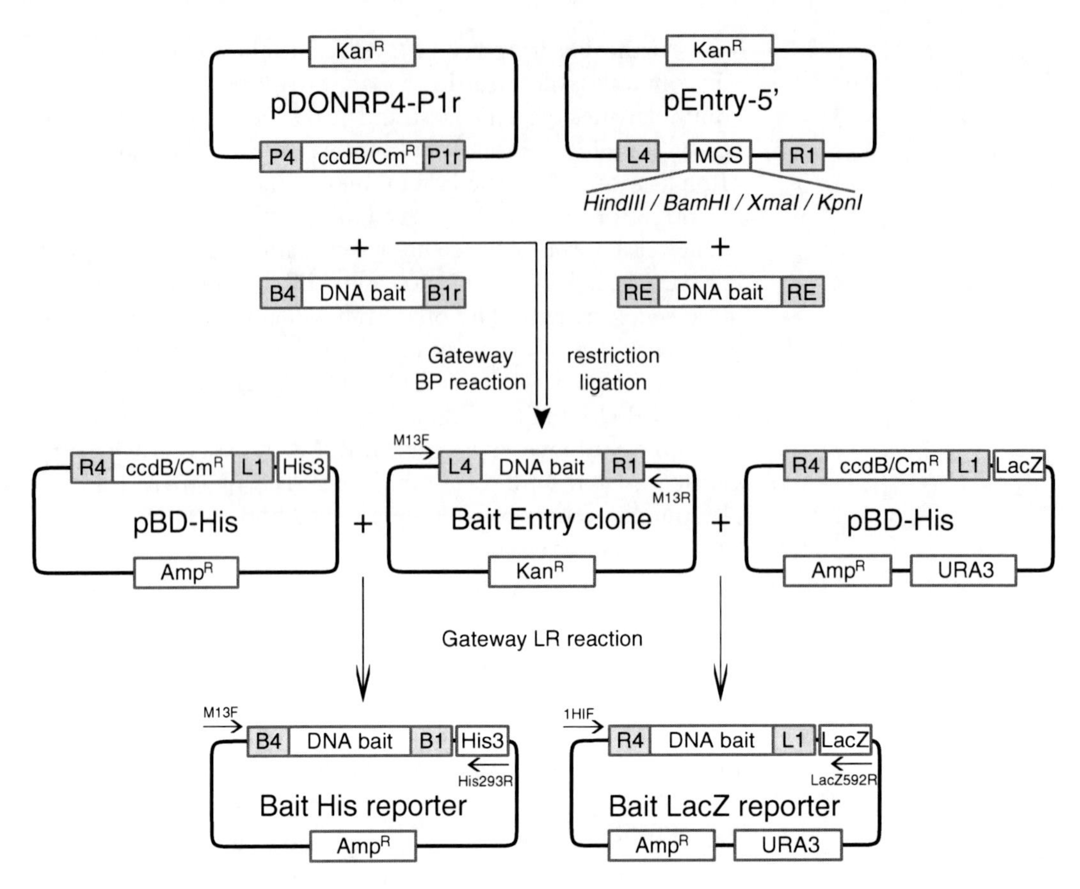

Fig. 2. DNA bait cloning strategy. DNA bait Entry clones can be generated by Gateway cloning. After PCR amplification and addition of gateway attB4 and attB1R to the 5′ and 3′ end of the bait, respectively, a Gateway BP reaction is performed to clone the bait fragment into the pDONRP4-P1R vector. Alternatively, restriction-ligation based cloning can also be used by adding appropriate restriction enzyme (RE) recognition sites to the bait fragment using PCR and cloning the bait into the pENTRY-5′ vector. Two separate gateway LR reactions are subsequently performed with the Entry clone, subcloning the bait into the pBD-His vector in one reaction and into the pBD-LAcZ in the other. Primer positions are indicated by *black arrows*.

interest at a concentration of 100 ng/μL. Set up a Gateway BP reaction as described in item 3 of Subheading 2.1 using pDONRP4-P1R as donor vector. Transform the BP reaction mix and perform a colony screen to identify colonies containing the correct Entry clone as described under steps 4 and 5 of Subheading 3.1 using M13F and M13R primers (see Note 4). Prepare glycerol stocks and isolate the plasmids as described above. Dilute the plasmids to 100 ng/μL. We strongly recommend verifying the bait sequence at this point. Continue the protocol at step 4 of Subheading 3.2.

3. For the restriction-ligation approach, design primers as above, but add restriction enzyme (RE) recognition sites at the 5′ end of the forward and reverse primer instead of gateway att sites. Any combination of the following restriction sites can be

used: *Hin*dIII (aagctt) – *Bam*HI (ggatcc) – *Xma*I (cccggg) – *Kpn*I (ggtacc). Verify that the chosen REs do not have a restriction site in the DNA bait. Add three additional nucleotides to the primer 5′ of the RE site to improve the RE cutting efficiency. Set up a PCR as described under step 2 of Subheading 3.1. Purify the PCR fragment using the QiaQuick PCR purification kit according to the manufacturer's instructions. Set up a restriction digest by mixing 6 μL of the appropriate RE buffer (according to the two enzymes that were chosen during the primer design), 50 μL of the purified PCR product and ten units of the two chosen REs. Add up to 60 μL with ultrapure water. Incubate the digestion mix for 2 h at 37°C. Set up the same digest for 5 μg of the pENTRY-5′ vector. Depending on the buffer requirements, it may be necessary to perform the digests sequentially. Purify the digested PCR fragment and the vector using the QiaQuick PCR purification kit according to the manufacturer's instructions. Set up a ligation reaction by mixing 1 μL ligation buffer, 200 cohesive end units of T4 DNA ligase, 50 ng of digested pENTRY-5′, and 50 ng of digested PCR product. Add up to 10 μL with ultrapure water. Incubate the ligation reaction overnight at 16°C. Transform ultracompetent DH5α with the ligation reaction mix, plate the bacteria on kanamycin containing LB plates and perform a colony screen to identify colonies containing the correct Entry clone as described in steps 4 and 5 of Subheading 3.1 using either the M13F or M13R primer. Prepare glycerol stocks and isolate the plasmids as described above. Dilute the plasmids to 100 ng/μL. We strongly recommend verifying the bait sequence at this point.

4. Transfer the DNA bait from the Entry clone to the Y1H compatible reporter vectors using a Gateway LR reaction. Set up two separate LR cloning reactions for each DNA bait, by mixing 1 μL of the reporter vector (pBD-His or pBD-LacZ, 100 ng/μL), 1 μL LR clonase II (Invitrogen, Carlsbad, CA), 2 μL DNA bait Entry clone (100 ng/μL), and 1 μL ultrapure water. Incubate the LR reaction mix overnight at 25°C.
5. Transform the LR reaction mix and perform a colony screen to identify colonies containing the correct DNA bait Entry clone as described items 8 and 9 of Subheading 2.1. For pBD-His, use M13F and His293R primers; for pBD-LacZ, use the 1HIFw and LacZ592R primers (see Note 4). Prepare glycerol stocks and isolate the plasmids as described above. Dilute the plasmids to 100 ng/μL.

3.3. Yeast Genomic Integration

1. The DNA baits are integrated at specific locations in the yeast genome by homologous recombination. Both reporters are integrated simultaneously, resulting in a yeast bait strain with the

His reporter vector integrated in the HIS3 locus, complementing the *his3-200* loss-of-function mutation and the LacZ reporter vector integrated in the URA3 locus, complementing the *ura3-52* loss-of-function mutation of the YM4271 yeast strain. The homologous recombination is facilitated by linearizing the reporter vectors in the homologous parts. For pBD-His, either *Xho*I or *Afl*II can be used. pBD-LacZ can be linearized with either *Nco*I or *Apa*I. It is important to choose an enzyme that does not cut in the inserted DNA bait. Set up a separate restriction digest for both the His and LAcZ reporter vector by mixing 2 μL of the appropriate RE buffer, 17.5 μL of the DNA bait reporter vector, and 0.5 μL of RE. Incubate for a minimum of 2 h at 37°C. Note: ApaI requires incubation at 25°C.

2. The yeast is made competent for transformation by the lithium acetate–poly(ethylene glycol) (PEG) method. Patch the YM4271 yeast strain on a YPD agar plate and incubate at 30°C until the yeast completely covers the plate, typically 2 days. Start a liquid yeast culture in YPD at an OD_{600} of 0.15. Incubate at 30°C with shaking until the culture reaches an OD_{600} between 0.5 and 0.6. Harvest the yeast by centrifugation for 5 min at 800×*g* and discard the supernatant (see Note 7). Resuspend the yeast in 20 mL of sterile water. Centrifuge as before and decant the supernatant. Resuspend the pellet in 20 mL TE–LiAc solution. Centrifuge as before and decant the supernatant. Resuspend the pellet in TE–LiAc solution in a volume calculated by the following formula:

$$\begin{aligned} &\mathrm{OD}_{600} \times \text{culture volume } (\mathrm{mL}) / 50 \\ &= \text{volume in which to resuspend } (\mathrm{mL}) \end{aligned}$$

For example, a 250 mL culture of $OD_{600} = 0.6$ needs to be resuspended in 3 mL TE–LiAc solution. Boil salmon sperm DNA for 5 min, cool on ice and add to the yeast suspension to a final concentration of 10% (v/v).

3. Add the entire restriction digest mix for both reporter vectors to 200 μL of competent yeast in a 2-mL microcentrifuge tube. Add 1 mL of TE–LiAc–PEG solution, and resuspend by inverting the tube ten times. Incubate the yeast suspension for 30 min at 30°C. Heat-shock the cells for exactly 20 min at 42°C in a water bath. Pellet the yeast by centrifuging for 5 s in a microcentrifuge at full speed at room temperature and remove the supernatant. Resuspend the cells in 100 μL of sterile water and spread on Sc-His, -Ura plates using glass beads. Incubate the plates at 30°C in a hot-air incubator.

3.4. Testing for Reporter Self-activation

1. Pick eight individual yeast colonies per double integration and spot them on Sc-His, -Ura plate. Incubate at 30°C overnight and replica-plate on a new Sc-His, -Ura plate and on a Sc-His, -Ura

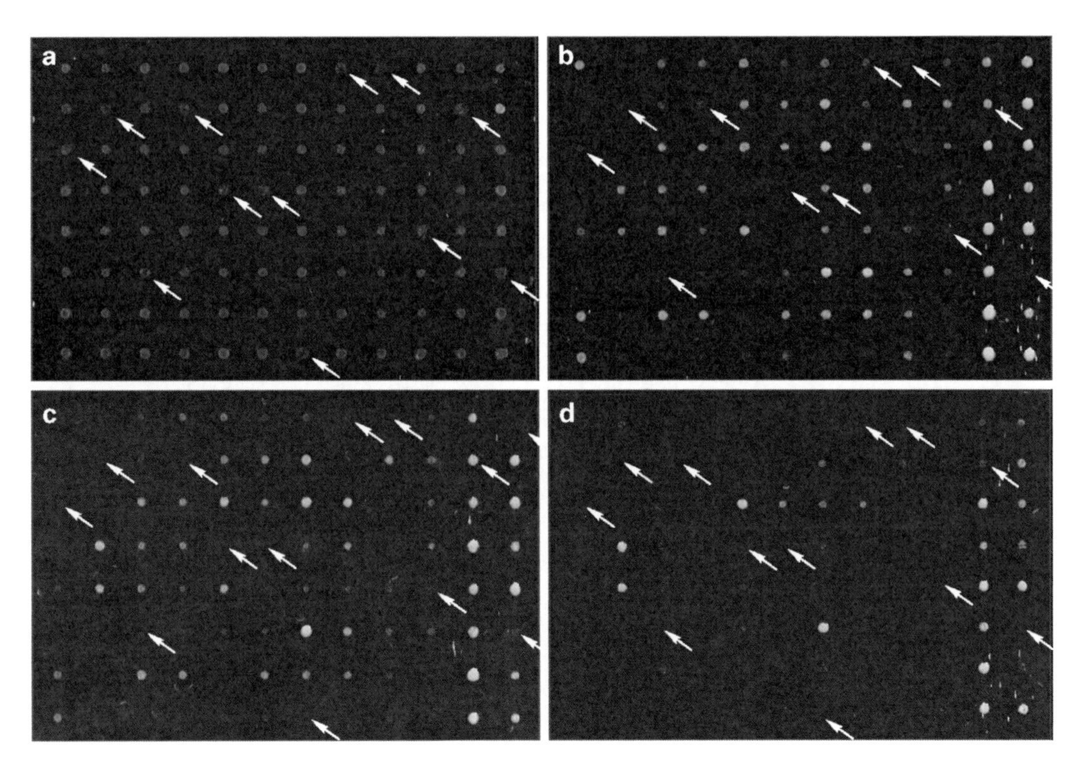

Fig. 3. Self-activation test for His reporter gene genomic insertions. Eight individual clones (columns) per bait strain for 12 different baits (*rows*) are spotted on Sc -His, -Ura plates without 3-AT and Sc -His, -Ura plates containing six different concentrations of 3-AT ranging from 10 mM to 100 mM. The plate without 3-AT (**a**) and with 10 mM (**b**), 20 mM (**c**), and 60 mM (**d**) are shown. *Arrows* indicate the colonies that were chosen for subsequent Y1H screening.

plates containing 10, 20, 40, 60, 80, and 100 mM 3-AT. Also replica-plate the colonies on a YPD plate on which a nitrocellulose membrane has been applied.

2. For testing the self-activation of the His reporter, monitor the growth of the yeast spots on the plates with different 3-AT concentrations as compared to the plate without 3-AT every day for 10 days at 30°C. Select a yeast colony that only shows growth at low 3-AT concentrations (see Notes 8 and 9). An example of a His self-activation test is shown in Fig. 3.
3. For testing the self-activation of the LacZ reporter, allow the yeast spots to grow on the nitrocellulose membrane overnight and perform a LacZ filter assay. Cut two whatman filters to size and place them in an empty OmniTray. Soak the whatman filters with X-Gal staining solution (~8 mL), remove air bubbles and decant excess staining solution. Remove the nitrocellulose membrane containing the yeast colonies from the YPD plate with forceps and submerge it in liquid nitrogen for 10 s. Thaw the membrane while holding it with the tweezers until the membrane becomes flexible again. Place the membrane on top of the whatman filters with the yeast colonies facing up. Avoid

air bubbles. Apply the lid of the OmniTray, seal the plate with parafilm and incubate at 37°C while protecting the membrane from light. Monitor the appearance of a blue coloration every 2 h for the first 4 h, then overnight. Choose the colony that shows the least coloration (see Note 10).

3.5. Y1H assay

1. The Y1H assay is performed by transforming the yeast strain containing the genomically integrated reporter vectors with the prey vectors. First, the yeast is made competent as described in step 2 of Subheading 3.3. Five microliters of competent yeast is required per prey construct (see Note 11).
2. Add 2 μL of prey plasmid to 5 μL of competent yeast in a well of a 384-microwell plate (see Note 12). Add 25 μL of TE–LiAc–PEG solution, and resuspend by pipetting five times. Incubate the yeast suspension for 30 min at 30°C. Heat-shock the cells for exactly 20 min at 42°C in a water bath. Pellet the yeast by centrifugation and remove the supernatant. Resuspend the cells in 5 μL of sterile water and spot 1 μL of this suspension on a Sc -His, -Ura, -Trp plate. We typically perform the Y1H screen using a customized robotic system (Tecan Evo) equipped with a 384 pipetting head, incubators, and a centrifuge unit capable of performing the complete transformation and spotting autonomously. Incubate the plate at 30°C for approximately 4 days. An example of a yeast transformation in 384 format is shown in Fig. 4a.
3. To test for positive interactions using the His reporter, replica-plate the colonies on a fresh Sc -His, -Ura, -Trp plate without 3-AT, and on Sc -His, -Ura, -Trp plates containing 3-AT at a level that shows no or severely reduced growth in the self-activation test (see step 2 of Subheading 3.4). Monitor the growth of the individual yeast spots every 2 days on the 3-AT containing plate as compared to the plate without 3-AT for 10 days. We typically reformat the 384 yeast colonies from the transformation plate into 1536 format by printing each yeast colony four times in a square pattern, yielding four technical replicates of each transformation and allowing easier visual detection of positives (see Note 13). We perform replica-plating and rearraying with a Singer RoToR Colony pinning robot. An example of a screen using the His reporter activity in 1536 format is shown in Fig. 4b, c.
4. To test for positive interactions using the LacZ reporter, colonies that are positive on selective (3AT) plates are picked and respotted four times in 384-well format onto a nitrocellulose filter on top of a YPD plate after which a LacZ filter assay is performed as described in step 3 of Subheading 3.4 (see Notes 13 and 14). An example of an analysis of positives from one screen using the LacZ reporter is shown in Fig. 5.

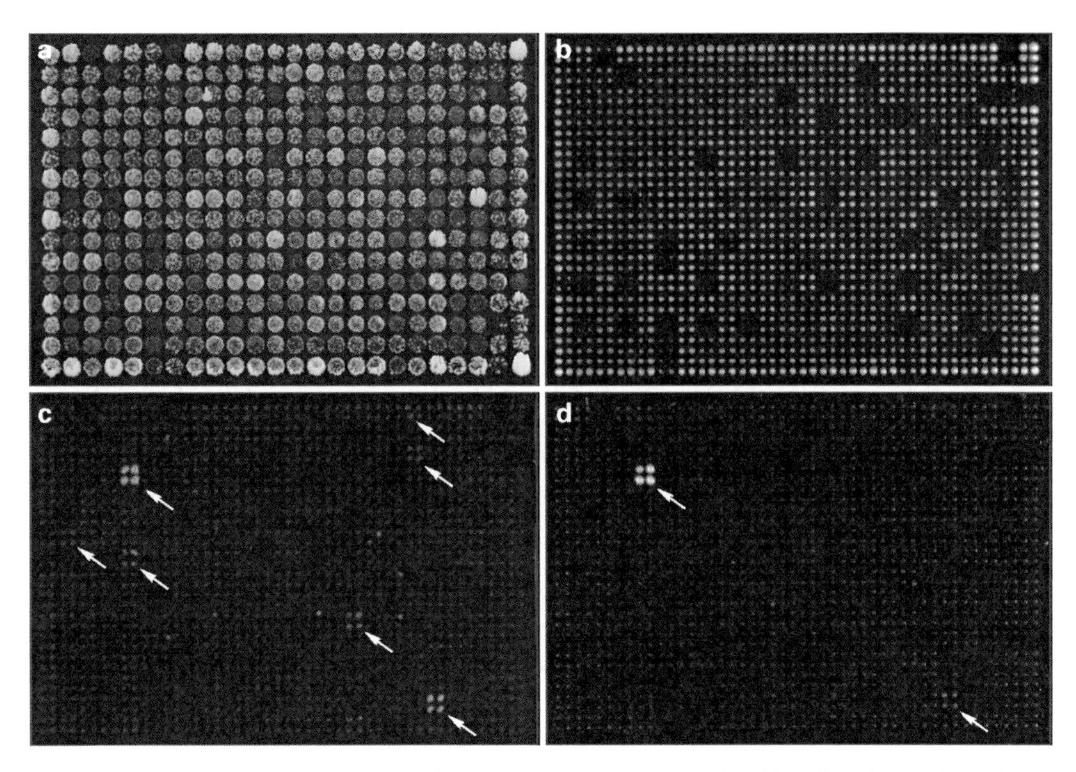

Fig. 4. The high-throughput Y1H screen. (**a**) A bait strain is transformed with 384 individual TFs and spotted on a Sc -His, -Ura, -Trp plate. Each colony is replica-plated four times in a square on a Sc -His, -Ura, -Trp plate without (**b**) and with 3-AT (respectively, 20 mM (**c**) and 40 mM (**d**) in this example). Growth on the 3-AT containing plate indicates binding of the prey TF and the DNA bait.

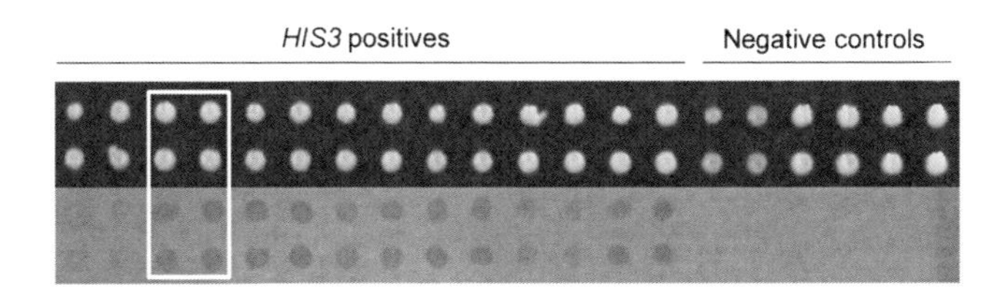

Fig. 5. Colonies that are positive on selective (3AT) plates are picked and respotted four times in 384-well format onto YPD agar plates covered by a nitrocellulose filter to perform a LacZ filter assay. Each positive is as such tested in quadruplicate. The *white rectangle* highlights one positive. Binding of the prey TF and the DNA bait results in *blue coloration* (*dark color on the image*) of the yeast colonies.

4. Notes

1. All solutions should be prepared in water that has a resistivity of 18.2 MΩ-cm and total organic content of less than five parts per billion. This standard is referred to as "ultrapure water" in this text.

2. Design the forward primer so that the ATG start codon is kept in the same reading frame as the AAA AAA triplets of the attB1 site. This will result in an in-frame fusion of the GAL4-AD sequence to the TF coding sequence in the last step of the prey library construction protocol. Additionally, when constructing an "open-ended" ORF library that can be used to create C-terminal fusions, design the reverse primer so that the last codon of the TF is kept in the same reading frame as the TTT GTA triplets of the attB2 site.

3. Both BP and LR reactions are less efficient when large inserts are being cloned (>4 kb). In this case, it might be necessary to increase the amount of PCR product or Entry vector to 3 μL for the BP and LR reaction, respectively.

4. The ccdB lethal gene present in the Gateway cassette inhibits the growth of bacterial colonies that contain vectors for which the recombination was unsuccessful. Although this system is very efficient, we sometimes observe the growth of colonies that contain a vector with an intact Gateway cassette, possibly due to a mutation in the ccdB gene. To avoid selecting these "false positive" colonies, especially when the expected insert size is roughly the same as the size of the ccdB cassette (~2.3 kb), we preferentially use one primer specific for the insert and one primer specific for the Entry or destination vector to ensure that the correct insert is cloned in the correct vector.

5. When constructing a library consisting of more than 50 clones, we perform the cloning process in 96-well plates. For smaller numbers, single-tube minipreps, e.g., the PureLink™ Quick Plasmid Miniprep Kit, can be performed.

6. Sequence verification of the TF clones can be performed by traditional Sanger sequencing. However, for large clone collections, this procedure can be very costly and time-consuming. Full-length sequencing of up to 1,000 clones of an average size of 2 kb can be achieved in a single step at reduced cost by pooling all clones together and performing a single run on a high-throughput sequencing platform. Our lab has developed an algorithm that can assemble the resulting short reads (28). A Web interface is available that combines the assembly with a decision pipeline, analyzing the presence and consequence of sequence variants (29).

7. When harvesting yeast by centrifugation, decant the medium in a glass beaker and check if it is clear. If the medium is cloudy after centrifugation, the culture may contain a bacterial contamination.

8. When growing yeast on agar plates for prolonged periods, e.g., when monitoring growth on 3-AT containing plates, protect

the plates from dehydration by incubating them in sealed plastic bags.

9. Discriminating positives from background can be problematic with yeast bait strains that show high self activation. In practice, we do not perform screens on bait strains that grow on 80 mM 3-AT or higher in the reporter self-activation assay.
10. The most critical steps in the LacZ filter assay are the length of thawing after freezing the nitrocellulose filter, and the amount of X-Gal staining solution used. If the colonies start smearing during the assay, try reducing the length of thawing and the volume of X-Gal staining solution.
11. In a traditional Y1H system, the prey library is transformed "in batch" and interacting TFs are identified by sequencing the prey vector from the resulting positive clones. This often results in strongly interacting TFs being identified multiple times, while weaker interactions may be missed (30). Therefore we describe here a protocol in which each TF from the prey library is transformed individually with the yeast reporter strain.
12. For improved throughput and reproducibility, we typically work in 384-well format using a robotic pipetting platform. However, the screen can be performed manually in 96-well format using a multichannel pipette. In this case, the volume of all solutions in steps 1 and 2 of Subheading 3.5 should be multiplied by 4, except the volume of prey plasmid, which remains 2 μL.
13. An interaction is only considered as positive if 3 out of 4 technical replicates are scored as such. In addition, we recommend to screen each bait at least twice to filter out false positives due to stochastic effects and thus to increase the overall reliability of the results. Protein preys that are positive for both reporters in both screens are considered high-confidence interactors.
14. It is also possible to directly perform a LacZ filter assay in 1536 format. However, we noticed that the overall readout is less clear and also less reproducible than when the assay is performed in 96- or 384-well format.

Acknowledgments

This work was supported by funds from the Swiss National Science Foundation, by Systems X, by a Marie Curie International Reintegration Grant (BD) from the Seventh Research Framework Programme, and by Institutional support from the Ecole Polytechnique Fédérale de Lausanne (EPFL).

References

1. Sanger, F., Air, G. M., Barrell, B. G., Brown, N. L., Coulson, A. R., Fiddes, C. A., Hutchison, C. A., Slocombe, P. M., and Smith, M. (1977) Nucleotide sequence of bacteriophage phi X174 DNA, *Nature* ***265***, 687–695.
2. (1998) Genome sequence of the nematode C. elegans: a platform for investigating biology, *Science (New York, N.Y)* ***282***, 2012–2018.
3. Adams, M. D., Celniker, S. E., Holt, R. A., Evans, C. A., Gocayne, J. D., Amanatides, P. G., Scherer, S. E., Li, P. W., Hoskins, R. A., Galle, R. F., George, R. A., Lewis, S. E., Richards, S., Ashburner, M., Henderson, S. N., Sutton, G. G., Wortman, J. R., Yandell, M. D., Zhang, Q., Chen, L. X., Brandon, R. C., Rogers, Y. H., Blazej, R. G., Champe, M., Pfeiffer, B. D., Wan, K. H., Doyle, C., Baxter, E. G., Helt, G., Nelson, C. R., Gabor, G. L., Abril, J. F., Agbayani, A., An, H. J., Andrews-Pfannkoch, C., Baldwin, D., Ballew, R. M., Basu, A., Baxendale, J., Bayraktaroglu, L., Beasley, E. M., Beeson, K. Y., Benos, P. V., Berman, B. P., Bhandari, D., Bolshakov, S., Borkova, D., Botchan, M. R., Bouck, J., Brokstein, P., Brottier, P., Burtis, K. C., Busam, D. A., Butler, H., Cadieu, E., Center, A., Chandra, I., Cherry, J. M., Cawley, S., Dahlke, C., Davenport, L. B., Davies, P., de Pablos, B., Delcher, A., Deng, Z., Mays, A. D., Dew, I., Dietz, S. M., Dodson, K., Doup, L. E., Downes, M., Dugan-Rocha, S., Dunkov, B. C., Dunn, P., Durbin, K. J., Evangelista, C. C., Ferraz, C., Ferriera, S., Fleischmann, W., Fosler, C., Gabrielian, A. E., Garg, N. S., Gelbart, W. M., Glasser, K., Glodek, A., Gong, F., Gorrell, J. H., Gu, Z., Guan, P., Harris, M., Harris, N. L., Harvey, D., Heiman, T. J., Hernandez, J. R., Houck, J., Hostin, D., Houston, K. A., Howland, T. J., Wei, M. H., Ibegwam, C., Jalali, M., Kalush, F., Karpen, G. H., Ke, Z., Kennison, J. A., Ketchum, K. A., Kimmel, B. E., Kodira, C. D., Kraft, C., Kravitz, S., Kulp, D., Lai, Z., Lasko, P., Lei, Y., Levitsky, A. A., Li, J., Li, Z., Liang, Y., Lin, X., Liu, X., Mattei, B., McIntosh, T. C., McLeod, M. P., McPherson, D., Merkulov, G., Milshina, N. V., Mobarry, C., Morris, J., Moshrefi, A., Mount, S. M., Moy, M., Murphy, B., Murphy, L., Muzny, D. M., Nelson, D. L., Nelson, D. R., Nelson, K. A., Nixon, K., Nusskern, D. R., Pacleb, J. M., Palazzolo, M., Pittman, G. S., Pan, S., Pollard, J., Puri, V., Reese, M. G., Reinert, K., Remington, K., Saunders, R. D., Scheeler, F., Shen, H., Shue, B. C., Siden-Kiamos, I., Simpson, M., Skupski, M. P., Smith, T., Spier, E., Spradling, A. C., Stapleton, M., Strong, R., Sun, E., Svirskas, R., Tector, C., Turner, R., Venter, E., Wang, A. H., Wang, X., Wang, Z. Y., Wassarman, D. A., Weinstock, G. M., Weissenbach, J., Williams, S. M., WoodageT, Worley, K. C., Wu, D., Yang, S., Yao, Q. A., Ye, J., Yeh, R. F., Zaveri, J. S., Zhan, M., Zhang, G., Zhao, Q., Zheng, L., Zheng, X. H., Zhong, F. N., Zhong, W., Zhou, X., Zhu, S., Zhu, X., Smith, H. O., Gibbs, R. A., Myers, E. W., Rubin, G. M., and Venter, J. C. (2000) The genome sequence of Drosophila melanogaster, *Science (New York, N.Y)* ***287***, 2185–2195.
4. Waterston, R. H., Lindblad-Toh, K., Birney, E., Rogers, J., Abril, J. F., Agarwal, P., Agarwala, R., Ainscough, R., Alexandersson, M., An, P., Antonarakis, S. E., Attwood, J., Baertsch, R., Bailey, J., Barlow, K., Beck, S., Berry, E., Birren, B., Bloom, T., Bork, P., Botcherby, M., Bray, N., Brent, M. R., Brown, D. G., Brown, S. D., Bult, C., Burton, J., Butler, J., Campbell, R. D., Carninci, P., Cawley, S., Chiaromonte, F., Chinwalla, A. T., Church, D. M., Clamp, M., Clee, C., Collins, F. S., Cook, L. L., Copley, R. R., Coulson, A., Couronne, O., Cuff, J., Curwen, V., Cutts, T., Daly, M., David, R., Davies, J., Delehaunty, K. D., Deri, J., Dermitzakis, E. T., Dewey, C., Dickens, N. J., Diekhans, M., Dodge, S., Dubchak, I., Dunn, D. M., Eddy, S. R., Elnitski, L., Emes, R. D., Eswara, P., Eyras, E., Felsenfeld, A., Fewell, G. A., Flicek, P., Foley, K., Frankel, W. N., Fulton, L. A., Fulton, R. S., Furey, T. S., Gage, D., Gibbs, R. A., Glusman, G., Gnerre, S., Goldman, N., Goodstadt, L., Grafham, D., Graves, T. A., Green, E. D., Gregory, S., Guigo, R., Guyer, M., Hardison, R. C., Haussler, D., Hayashizaki, Y., Hillier, L. W., Hinrichs, A., Hlavina, W., Holzer, T., Hsu, F., Hua, A., Hubbard, T., Hunt, A., Jackson, I., Jaffe, D. B., Johnson, L. S., Jones, M., Jones, T. A., Joy, A., Kamal, M., Karlsson, E. K., Karolchik, D., Kasprzyk, A., Kawai, J., Keibler, E., Kells, C., Kent, W. J., Kirby, A., Kolbe, D. L., Korf, I., Kucherlapati, R. S., Kulbokas, E. J., Kulp, D., Landers, T., Leger, J. P., Leonard, S., Letunic, I., Levine, R., Li, J., Li, M., Lloyd, C., Lucas, S., Ma, B., Maglott, D. R., Mardis, E. R., Matthews, L., Mauceli, E., Mayer, J. H., McCarthy, M., McCombie, W. R., McLaren, S., McLay, K., McPherson, J. D., Meldrim, J., Meredith, B., Mesirov, J. P., Miller, W., Miner, T. L., Mongin, E., Montgomery, K. T., Morgan, M., Mott, R., Mullikin, J. C., Muzny, D. M., Nash, W. E., Nelson, J. O., Nhan, M. N., Nicol, R., Ning, Z., Nusbaum, C., O'Connor, M. J., Okazaki, Y., Oliver, K., Overton-Larty, E., Pachter, L.,

Parra, G., Pepin, K. H., Peterson, J., Pevzner, P., Plumb, R., Pohl, C. S., Poliakov, A., Ponce, T. C., Ponting, C. P., Potter, S., Quail, M., Reymond, A., Roe, B. A., Roskin, K. M., Rubin, E. M., Rust, A. G., Santos, R., Sapojnikov, V., Schultz, B., Schultz, J., Schwartz, M. S., Schwartz, S., Scott, C., Seaman, S., Searle, S., Sharpe, T., Sheridan, A., Shownkeen, R., Sims, S., Singer, J. B., Slater, G., Smit, A., Smith, D. R., Spencer, B., Stabenau, A., Stange-Thomann, N., Sugnet, C., Suyama, M., Tesler, G., Thompson, J., Torrents, D., Trevaskis, E., Tromp, J., Ucla, C., Ureta-Vidal, A., Vinson, J. P., Von Niederhausern, A. C., Wade, C. M., Wall, M., Weber, R. J., Weiss, R. B., Wendl, M. C., West, A. P., Wetterstrand, K., Wheeler, R., Whelan, S., Wierzbowski, J., Willey, D., Williams, S., Wilson, R. K., Winter, E., Worley, K. C., Wyman, D., Yang, S., Yang, S. P., Zdobnov, E. M., Zody, M. C., and Lander, E. S. (2002) Initial sequencing and comparative analysis of the mouse genome, *Nature* ***420***, 520–562.

5. Venter, J. C., Adams, M. D., Myers, E. W., Li, P. W., Mural, R. J., Sutton, G. G., Smith, H. O., Yandell, M., Evans, C. A., Holt, R. A., Gocayne, J. D., Amanatides, P., Ballew, R. M., Huson, D. H., Wortman, J. R., Zhang, Q., Kodira, C. D., Zheng, X. H., Chen, L., Skupski, M., Subramanian, G., Thomas, P. D., Zhang, J., Gabor Miklos, G. L., Nelson, C., Broder, S., Clark, A. G., Nadeau, J., McKusick, V. A., Zinder, N., Levine, A. J., Roberts, R. J., Simon, M., Slayman, C., Hunkapiller, M., Bolanos, R., Delcher, A., Dew, I., Fasulo, D., Flanigan, M., Florea, L., Halpern, A., Hannenhalli, S., Kravitz, S., Levy, S., Mobarry, C., Reinert, K., Remington, K., Abu-Threideh, J., Beasley, E., Biddick, K., Bonazzi, V., Brandon, R., Cargill, M., Chandramouliswaran, I., Charlab, R., Chaturvedi, K., Deng, Z., Di Francesco, V., Dunn, P., Eilbeck, K., Evangelista, C., Gabrielian, A. E., Gan, W., Ge, W., Gong, F., Gu, Z., Guan, P., Heiman, T. J., Higgins, M. E., Ji, R. R., Ke, Z., Ketchum, K. A., Lai, Z., Lei, Y., Li, Z., Li, J., Liang, Y., Lin, X., Lu, F., Merkulov, G. V., Milshina, N., Moore, H. M., Naik, A. K., Narayan, V. A., Neelam, B., Nusskern, D., Rusch, D. B., Salzberg, S., Shao, W., Shue, B., Sun, J., Wang, Z., Wang, A., Wang, X., Wang, J., Wei, M., Wides, R., Xiao, C., Yan, C., Yao, A., Ye, J., Zhan, M., Zhang, W., Zhang, H., Zhao, Q., Zheng, L., Zhong, F., Zhong, W., Zhu, S., Zhao, S., Gilbert, D., Baumhueter, S., Spier, G., Carter, C., Cravchik, A., Woodage, T., Ali, F., An, H., Awe, A., Baldwin, D., Baden, H., Barnstead, M., Barrow, I., Beeson, K., Busam, D., Carver, A., Center, A., Cheng, M. L., Curry, L., Danaher, S., Davenport, L., Desilets, R., Dietz, S., Dodson, K., Doup, L., Ferriera, S., Garg, N., Gluecksmann, A., Hart, B., Haynes, J., Haynes, C., Heiner, C., Hladun, S., Hostin, D., Houck, J., Howland, T., Ibegwam, C., Johnson, J., Kalush, F., Kline, L., Koduru, S., Love, A., Mann, F., May, D., McCawley, S., McIntosh, T., McMullen, I., Moy, M., Moy, L., Murphy, B., Nelson, K., Pfannkoch, C., Pratts, E., Puri, V., Qureshi, H., Reardon, M., Rodriguez, R., Rogers, Y. H., Romblad, D., Ruhfel, B., Scott, R., Sitter, C., Smallwood, M., Stewart, E., Strong, R., Suh, E., Thomas, R., Tint, N. N., Tse, S., Vech, C., Wang, G., Wetter, J., Williams, S., Williams, M., Windsor, S., Winn-Deen, E., Wolfe, K., Zaveri, J., Zaveri, K., Abril, J. F., Guigo, R., Campbell, M. J., Sjolander, K. V., Karlak, B., Kejariwal, A., Mi, H., Lazareva, B., Hatton, T., Narechania, A., Diemer, K., Muruganujan, A., Guo, N., Sato, S., Bafna, V., Istrail, S., Lippert, R., Schwartz, R., Walenz, B., Yooseph, S., Allen, D., Basu, A., Baxendale, J., Blick, L., Caminha, M., Carnes-Stine, J., Caulk, P., Chiang, Y. H., Coyne, M., Dahlke, C., Mays, A., Dombroski, M., Donnelly, M., Ely, D., Esparham, S., Fosler, C., Gire, H., Glanowski, S., Glasser, K., Glodek, A., Gorokhov, M., Graham, K., Gropman, B., Harris, M., Heil, J., Henderson, S., Hoover, J., Jennings, D., Jordan, C., Jordan, J., Kasha, J., Kagan, L., Kraft, C., Levitsky, A., Lewis, M., Liu, X., Lopez, J., Ma, D., Majoros, W., McDaniel, J., Murphy, S., Newman, M., Nguyen, T., Nguyen, N., Nodell, M., Pan, S., Peck, J., Peterson, M., Rowe, W., Sanders, R., Scott, J., Simpson, M., Smith, T., Sprague, A., Stockwell, T., Turner, R., Venter, E., Wang, M., Wen, M., Wu, D., Wu, M., Xia, A., Zandieh, A., and Zhu, X. (2001) The sequence of the human genome, *Science (New York, N.Y)* ***291***, 1304–1351.

6. Nakabachi, A., Yamashita, A., Toh, H., Ishikawa, H., Dunbar, H. E., Moran, N. A., and Hattori, M. (2006) The 160-kilobase genome of the bacterial endosymbiont Carsonella, *Science (New York, N.Y)* ***314***, 267.

7. Elgar, G., and Vavouri, T. (2008) Tuning in to the signals: noncoding sequence conservation in vertebrate genomes, *Trends Genet* ***24***, 344–352.

8. Ohno, S. (1972) So much "junk" DNA in our genome, *Brookhaven symposia in biology* ***23***, 366–370.

9. Shiraki, T., Kondo, S., Katayama, S., Waki, K., Kasukawa, T., Kawaji, H., Kodzius, R., Watahiki, A., Nakamura, M., Arakawa, T., Fukuda, S., Sasaki, D., Podhajska, A., Harbers, M., Kawai, J., Carninci, P., and Hayashizaki, Y.

(2003) Cap analysis gene expression for high-throughput analysis of transcriptional starting point and identification of promoter usage, *Proceedings of the National Academy of Sciences of the United States of America* ***100***, 15776–15781.

10. Bussemaker, H. J., Li, H., and Siggia, E. D. (2001) Regulatory element detection using correlation with expression, *Nature genetics* *27*, 167–171.
11. Elemento, O., Slonim, N., and Tavazoie, S. (2007) A universal framework for regulatory element discovery across all genomes and data types, *Molecular cell* *28*, 337–350.
12. Pennacchio, L. A., Ahituv, N., Moses, A. M., Prabhakar, S., Nobrega, M. A., Shoukry, M., Minovitsky, S., Dubchak, I., Holt, A., Lewis, K. D., Plajzer-Frick, I., Akiyama, J., De Val, S., Afzal, V., Black, B. L., Couronne, O., Eisen, M. B., Visel, A., and Rubin, E. M. (2006) In vivo enhancer analysis of human conserved non-coding sequences, *Nature* *444*, 499–502.
13. Visel, A., Blow, M. J., Li, Z., Zhang, T., Akiyama, J. A., Holt, A., Plajzer-Frick, I., Shoukry, M., Wright, C., Chen, F., Afzal, V., Ren, B., Rubin, E. M., and Pennacchio, L. A. (2009) ChIP-seq accurately predicts tissue-specific activity of enhancers, *Nature* *457*, 854–858.
14. Lee, T. I., Jenner, R. G., Boyer, L. A., Guenther, M. G., Levine, S. S., Kumar, R. M., Chevalier, B., Johnstone, S. E., Cole, M. F., Isono, K., Koseki, H., Fuchikami, T., Abe, K., Murray, H. L., Zucker, J. P., Yuan, B., Bell, G. W., Herbolsheimer, E., Hannett, N. M., Sun, K., Odom, D. T., Otte, A. P., Volkert, T. L., Bartel, D. P., Melton, D. A., Gifford, D. K., Jaenisch, R., and Young, R. A. (2006) Control of developmental regulators by Polycomb in human embryonic stem cells, *Cell* ***125***, 301–313.
15. Heintzman, N. D., Stuart, R. K., Hon, G., Fu, Y., Ching, C. W., Hawkins, R. D., Barrera, L. O., Van Calcar, S., Qu, C., Ching, K. A., Wang, W., Weng, Z., Green, R. D., Crawford, G. E., and Ren, B. (2007) Distinct and predictive chromatin signatures of transcriptional promoters and enhancers in the human genome, *Nature genetics* *39*, 311–318.
16. Pfeiffer, B. D., Jenett, A., Hammonds, A. S., Ngo, T. T., Misra, S., Murphy, C., Scully, A., Carlson, J. W., Wan, K. H., Laverty, T. R., Mungall, C., Svirskas, R., Kadonaga, J. T., Doe, C. Q., Eisen, M. B., Celniker, S. E., and Rubin, G. M. (2008) Tools for neuroanatomy and neurogenetics in Drosophila, *Proceedings of the National Academy of Sciences of the United States of America* ***105***, 9715–9720.
17. Dupuy, D., Bertin, N., Hidalgo, C. A., Venkatesan, K., Tu, D., Lee, D., Rosenberg, J., Svrzikapa, N., Blanc, A., Carnec, A., Carvunis, A. R., Pulak, R., Shingles, J., Reece-Hoyes, J., Hunt-Newbury, R., Viveiros, R., Mohler, W. A., Tasan, M., Roth, F. P., Le Peuch, C., Hope, I. A., Johnsen, R., Moerman, D. G., Barabasi, A. L., Baillie, D., and Vidal, M. (2007) Genome-scale analysis of in vivo spatiotemporal promoter activity in Caenorhabditis elegans, *Nature biotechnology* *25*, 663–668.
18. Birney, E., Stamatoyannopoulos, J. A., Dutta, A., Guigo, R., Gingeras, T. R., Margulies, E. H., Weng, Z., Snyder, M., Dermitzakis, E. T., Thurman, R. E., Kuehn, M. S., Taylor, C. M., Neph, S., Koch, C. M., Asthana, S., Malhotra, A., Adzhubei, I., Greenbaum, J. A., Andrews, R. M., Flicek, P., Boyle, P. J., Cao, H., Carter, N. P., Clelland, G. K., Davis, S., Day, N., Dhami, P., Dillon, S. C., Dorschner, M. O., Fiegler, H., Giresi, P. G., Goldy, J., Hawrylycz, M., Haydock, A., Humbert, R., James, K. D., Johnson, B. E., Johnson, E. M., Frum, T. T., Rosenzweig, E. R., Karnani, N., Lee, K., Lefebvre, G. C., Navas, P. A., Neri, F., Parker, S. C., Sabo, P. J., Sandstrom, R., Shafer, A., Vetrie, D., Weaver, M., Wilcox, S., Yu, M., Collins, F. S., Dekker, J., Lieb, J. D., Tullius, T. D., Crawford, G. E., Sunyaev, S., Noble, W. S., Dunham, I., Denoeud, F., Reymond, A., Kapranov, P., Rozowsky, J., Zheng, D., Castelo, R., Frankish, A., Harrow, J., Ghosh, S., Sandelin, A., Hofacker, I. L., Baertsch, R., Keefe, D., Dike, S., Cheng, J., Hirsch, H. A., Sekinger, E. A., Lagarde, J., Abril, J. F., Shahab, A., Flamm, C., Fried, C., Hackermuller, J., Hertel, J., Lindemeyer, M., Missal, K., Tanzer, A., Washietl, S., Korbel, J., Emanuelsson, O., Pedersen, J. S., Holroyd, N., Taylor, R., Swarbreck, D., Matthews, N., Dickson, M. C., Thomas, D. J., Weirauch, M. T., Gilbert, J., Drenkow, J., Bell, I., Zhao, X., Srinivasan, K. G., Sung, W. K., Ooi, H. S., Chiu, K. P., Foissac, S., Alioto, T., Brent, M., Pachter, L., Tress, M. L., Valencia, A., Choo, S. W., Choo, C. Y., Ucla, C., Manzano, C., Wyss, C., Cheung, E., Clark, T. G., Brown, J. B., Ganesh, M., Patel, S., Tammana, H., Chrast, J., Henrichsen, C. N., Kai, C., Kawai, J., Nagalakshmi, U., Wu, J., Lian, Z., Lian, J., Newburger, P., Zhang, X., Bickel, P., Mattick, J. S., Carninci, P., Hayashizaki, Y., Weissman, S., Hubbard, T., Myers, R. M., Rogers, J., Stadler, P. F., Lowe, T. M., Wei, C. L., Ruan, Y., Struhl, K., Gerstein, M., Antonarakis, S. E., Fu, Y., Green, E. D., Karaoz, U., Siepel, A., Taylor, J., Liefer, L. A., Wetterstrand, K. A., Good, P. J., Feingold, E. A., Guyer, M. S., Cooper, G. M., Asimenos, G., Dewey, C. N.,

Hou, M., Nikolaev, S., Montoya-Burgos, J. I., Loytynoja, A., Whelan, S., Pardi, F., Massingham, T., Huang, H., Zhang, N. R., Holmes, I., Mullikin, J. C., Ureta-Vidal, A., Paten, B., Seringhaus, M., Church, D., Rosenbloom, K., Kent, W. J., Stone, E. A., Batzoglou, S., Goldman, N., Hardison, R. C., Haussler, D., Miller, W., Sidow, A., Trinklein, N. D., Zhang, Z. D., Barrera, L., Stuart, R., King, D. C., Ameur, A., Enroth, S., Bieda, M. C., Kim, J., Bhinge, A. A., Jiang, N., Liu, J., Yao, F., Vega, V. B., Lee, C. W., Ng, P., Shahab, A., Yang, A., Moqtaderi, Z., Zhu, Z., Xu, X., Squazzo, S., Oberley, M. J., Inman, D., Singer, M. A., Richmond, T. A., Munn, K. J., Rada-Iglesias, A., Wallerman, O., Komorowski, J., Fowler, J. C., Couttet, P., Bruce, A. W., Dovey, O. M., Ellis, P. D., Langford, C. F., Nix, D. A., Euskirchen, G., Hartman, S., Urban, A. E., Kraus, P., Van Calcar, S., Heintzman, N., Kim, T. H., Wang, K., Qu, C., Hon, G., Luna, R., Glass, C. K., Rosenfeld, M. G., Aldred, S. F., Cooper, S. J., Halees, A., Lin, J. M., Shulha, H. P., Zhang, X., Xu, M., Haidar, J. N., Yu, Y., Ruan, Y., Iyer, V. R., Green, R. D., Wadelius, C., Farnham, P. J., Ren, B., Harte, R. A., Hinrichs, A. S., Trumbower, H., Clawson, H., Hillman-Jackson, J., Zweig, A. S., Smith, K., Thakkapallayil, A., Barber, G., Kuhn, R. M., Karolchik, D., Armengol, L., Bird, C. P., de Bakker, P. I., Kern, A. D., Lopez-Bigas, N., Martin, J. D., Stranger, B. E., Woodroffe, A., Davydov, E., Dimas, A., Eyras, E., Hallgrimsdottir, I. B., Huppert, J., Zody, M. C., Abecasis, G. R., Estivill, X., Bouffard, G. G., Guan, X., Hansen, N. F., Idol, J. R., Maduro, V. V., Maskeri, B., McDowell, J. C., Park, M., Thomas, P. J., Young, A. C., Blakesley, R. W., Muzny, D. M., Sodergren, E., Wheeler, D. A., Worley, K. C., Jiang, H., Weinstock, G. M., Gibbs, R. A., Graves, T., Fulton, R., Mardis, E. R., Wilson, R. K., Clamp, M., Cuff, J., Gnerre, S., Jaffe, D. B., Chang, J. L., Lindblad-Toh, K., Lander, E. S., Koriabine, M., Nefedov, M., Osoegawa, K., Yoshinaga, Y., Zhu, B., and de Jong, P. J. (2007) Identification and analysis of functional elements in 1% of the human genome by the ENCODE pilot project, *Nature 447*, 799–816.

19. Halfon, M. S., Gallo, S. M., and Bergman, C. M. (2008) REDfly 2.0: an integrated database of cis-regulatory modules and transcription factor binding sites in Drosophila, *Nucleic acids research 36*, D594–598.
20. Griffith, O. L., Montgomery, S. B., Bernier, B., Chu, B., Kasaian, K., Aerts, S., Mahony, S., Sleumer, M. C., Bilenky, M., Haeussler, M., Griffith, M., Gallo, S. M., Giardine, B., Hooghe, B., Van Loo, P., Blanco, E., Ticoll, A., Lithwick, S., Portales-Casamar, E., Donaldson, I. J., Robertson, G., Wadelius, C., De Bleser, P., Vlieghe, D., Halfon, M. S., Wasserman, W., Hardison, R., Bergman, C. M., and Jones, S. J. (2008) ORegAnno: an open-access community-driven resource for regulatory annotation, *Nucleic acids research 36*, D107–113.
21. Deplancke, B. (2009) Experimental advances in the characterization of metazoan gene regulatory networks, *Briefings in functional genomics & proteomics 8*, 12–27.
22. Fields, S., and Song, O. (1989) A novel genetic system to detect protein-protein interactions, *Nature 340*, 245–246.
23. Wang, M. M., and Reed, R. R. (1993) Molecular cloning of the olfactory neuronal transcription factor Olf-1 by genetic selection in yeast, *Nature 364*, 121–126.
24. Li, J. J., and Herskowitz, I. (1993) Isolation of ORC6, a component of the yeast origin recognition complex by a one-hybrid system, *Science (New York, N.Y) 262*, 1870–1874.
25. Deplancke, B., Dupuy, D., Vidal, M., and Walhout, A. J. (2004) A gateway-compatible yeast one-hybrid system, *Genome research 14*, 2093–2101.
26. Deplancke, B., Mukhopadhyay, A., Ao, W., Elewa, A. M., Grove, C. A., Martinez, N. J., Sequerra, R., Doucette-Stamm, L., Reece-Hoyes, J. S., Hope, I. A., Tissenbaum, H. A., Mango, S. E., and Walhout, A. J. (2006) A gene-centered C. elegans protein-DNA interaction network, *Cell 125*, 1193–1205.
27. Rozen, S., and Skaletsky, H. (2000) Primer3 on the WWW for general users and for biologist programmers, *Methods in molecular biology Clifton, N.J 132*, 365–386.
28. Massouras, A., Hens, K., Gubelmann, C., Uplekar, S., Decouttere, F., Rougemont, J., Cole, S. T., and Deplancke, B. Primer-initiated sequence synthesis to detect and assemble structural variants, *Nature methods 7*, 485–486.
29. Massouras, A., Decouttere, F., Hens, K., and Deplancke, B. WebPrInSeS: automated full-length clone sequence identification and verification using high-throughput sequencing data, *Nucleic acids research 38* Suppl, W378–384.
30. Vermeirssen, V., Deplancke, B., Barrasa, M. I., Reece-Hoyes, J. S., Arda, H. E., Grove, C. A., Martinez, N. J., Sequerra, R., Doucette-Stamm, L., Brent, M. R., and Walhout, A. J. (2007) Matrix and Steiner-triple-system smart pooling assays for high-performance transcription regulatory network mapping, *Nature methods 4*, 659–664.

Part IV

Visualization of GRNs

Chapter 21

BioTapestry: A Tool to Visualize the Dynamic Properties of Gene Regulatory Networks

William J.R. Longabaugh

Abstract

BioTapestry is an open source, freely available software tool that has been developed to handle the challenges of modeling genetic regulatory networks (GRNs). Using BioTapestry, a researcher can construct a network model and use it to visualize and understand the dynamic behavior of a complex, spatially and temporally distributed GRN. Here we provide a step-by-step example of a way to use BioTapestry to build a GRN model and discuss some common issues that can arise during this process.

Key words: Genetic regulatory networks, Network modeling, Network visualization, Systems biology software, Drawing gene networks, Layout of gene networks, Modeling, Computational biology

1. Introduction

Because of the complex, dynamic, and spatially and temporally distributed nature of genetic regulatory networks (GRNs), it is essential to have specially designed software tools to model and visualize the behavior of such networks. BioTapestry (1, 2) (http://www.BioTapestry.org/) is a freely available, open-source software tool that is specifically designed to create, maintain, and share such GRN models.

BioTapestry was originally developed as a tool to support the ongoing effort to model the GRN that controls the development of the endomesoderm up to 30 hours in the purple sea urchin *Strongylocentrotus purpuratus* (3). Because it was designed specifically to deal with the properties of GRNs in developing embryos, it has advantages in this context over more general-purpose tools for handling biological networks such as Cytoscape (4).

Bart Deplancke and Nele Gheldof (eds.), *Gene Regulatory Networks: Methods and Protocols*,
Methods in Molecular Biology, vol. 786, DOI 10.1007/978-1-61779-292-2_21,

BioTapestry has now been used to model a variety of different systems, both as an online interactive tool (3, 5–8) and as static presentations (9–22).

This article provides an overview of how BioTapestry models are structured and presented and then proceeds to a step-by-step description of a possible way to use the program to construct a dynamic GRN model. This description is accompanied by pointers on dealing with some of the important issues that can arise while building the model.

1.1. The Model Hierarchy

A GRN behaves differently in different cell types, spatial domains, environmental conditions, and at different times. To handle this variability, BioTapestry represents a GRN as a *hierarchy of models* and uses this structure to help the user organize these variations in the network state in a coherent fashion.

Figure 1 provides three different screen shots that illustrate how BioTapestry's model hierarchy can be used to illustrate different aspects of a GRN. Each screen shot portrays a view of a different level of the model hierarchy; the user can navigate between the different levels by clicking on elements in the tree view on the left side

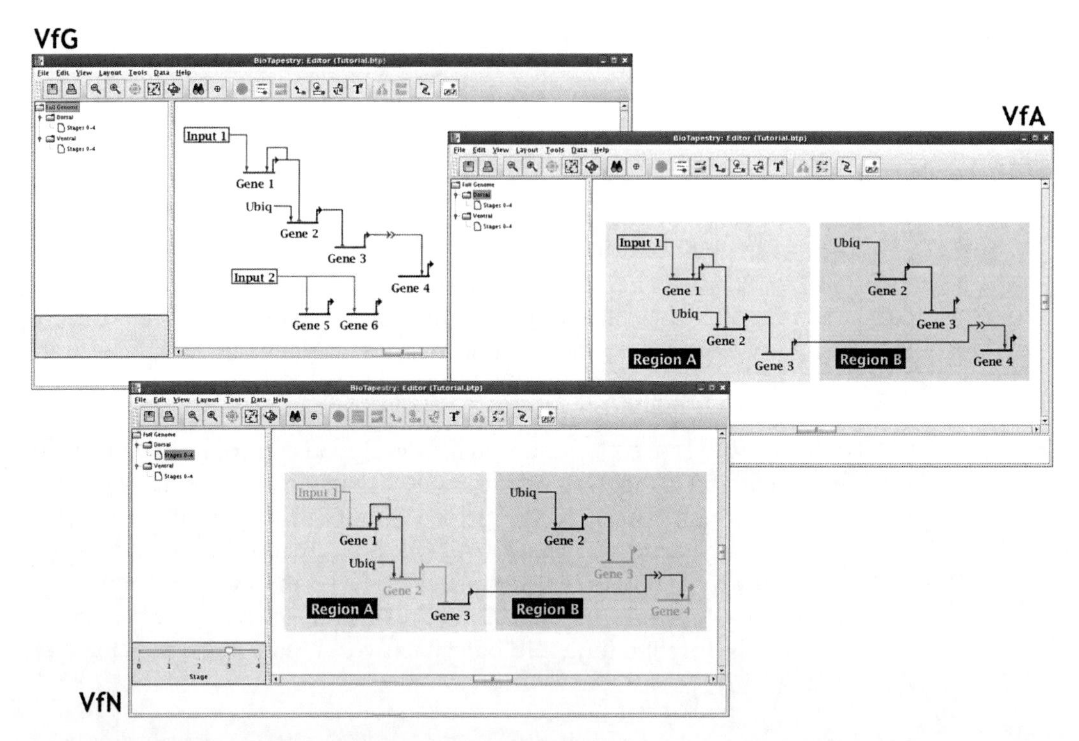

Fig. 1. An example of a BioTapestry model hierarchy. Clockwise from upper left: View from the Genome (VfG), View from All nuclei (VfA), and View from the Nucleus (VfN). See text for discussion.

of the window. Moving clockwise from the upper left screen shot, the three views are as follows:

- The top-level or root model, named *Full Genome*, provides a View from the Genome (VfG) (using terminology discussed in (23)), which is a summary of all inputs into each gene, regardless of when and where those inputs are relevant. One and only one copy of each network element must be shown in this view. The VfG shows the entire regulatory program of each gene in the GRN; it provides the basic network definition for the model hierarchy, but does not address the different states of the GRN.
- The next model down that is depicted, named *Dorsal*, is a View from All nuclei, (VfA) and is derived from the top-level model. This level in the model hierarchy introduces the concept of *regions*. In an abstract sense, each region contains a subset of the top-level VfG network, and depicts a different regulatory state of that top-level network. In concrete terms, each region in the VfA is used to depict a different cell type in a developing embryo, or different cells in an organ. In this particular example, this model is showing an overall summary of the relevant network elements active in two neighboring regions over the entire time period of interest.
- Each of the lowest level models, such as the selected *Stages 0–4*, is a View from the Nucleus (VfN). These views describe a specific state of the network defined in the parent VfA at a particular time and place. Inactive portions of the network are indicated in gray, while the active elements are shown colored.

The following discussion continues to use the VfG/VfA/VfN terminology to refer to the top, middle, and lower levels of the model hierarchy. In fact, the depth of the hierarchy is not restricted to just these three levels, as there can be any number of levels below the VfN, to depict successively finer-grained subsets of the model. When referring to the hierarchy, it is also common to use the terms submodel (for models below a particular model in the hierarchy), parent model (for the model above a model in the hierarchy), and sibling models (for models that share the same parent).

While it might require some extra effort to become comfortable working with BioTapestry's multilevel model hierarchy framework, in contrast to just creating a single top-level model, the payback in terms of expressive power is significant. Except for extremely simple GRNs, it is essential to avoid the tendency to just create a single, top level VfG-like model of a GRN and consider that sufficient to describe and understand the system. Even if accompanied by an extensive prose description of when and where certain portions of the network are active, a single summary GRN representation is insufficient to effectively convey the dynamic variations of network behavior in time and space. Such a representation is also likely to succumb to the dangers of verbal ambiguity.

1.2. Network Representation

BioTapestry is designed to present a GRN so that the essential information of how the transcription of a gene is regulated by other genes in the network can be quickly understood; see Fig. 2. Genes, and the *cis*-regulatory regions of genes, are organized in a structured fashion that makes them stand out. In particular, genes are represented by a symbol that provides an explicit schematic representation of the *cis*-regulatory modules of the gene. At the same time, off-DNA interactions are depicted with a small collection of symbols, such as bubbles or intercellular signaling double chevrons, which are not designed to provide extremely detailed information, but are instead just enough for the user to get an intuitive, high-level understanding. More detailed descriptions of the symbols used in BioTapestry are available at http://www.BioTapestry.org/MiMB. If necessary, additional detailed information on the biochemical process represented by a node can be provided through the pop-up page that can be accessed though a right-click of the mouse on the node.

Figure 2 also shows how BioTapestry represents interactions in the GRN. The outputs of transcription factor genes are typically shown as direct inputs into the regulated gene targets, and so the various posttranscriptional processes are not explicitly represented. However, if certain posttranscriptional activities are an essential part of the regulatory architecture, they can be explicitly added as a set of labeled bubbles and boxes.

The other thing to note about BioTapestry links is that they are bundled together and drawn as a group. This approach is an important strategy for reducing visual clutter in a network with many interactions. Furthermore, link color is used primarily to visually differentiate links from different sources. This is in contrast to using color to apply semantic meaning to groups of links. With BioTapestry, it is appropriate to use link evidence glyphs (shown in Fig. 2) or submodels, instead of color, to provide additional information about various sets of links.

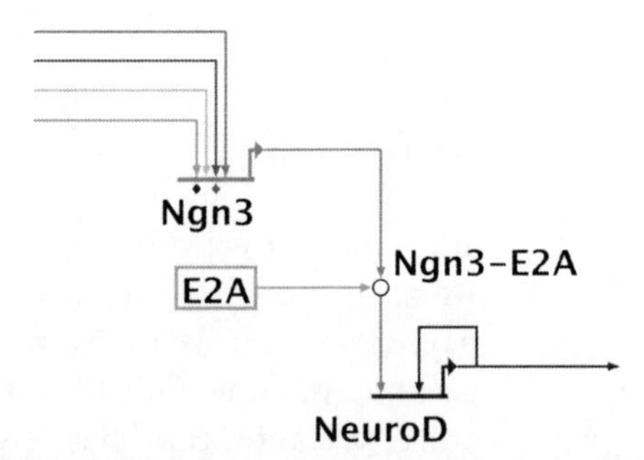

Fig. 2. Detail of a BioTapestry rendering of a GRN. In this example, gene Ngn3 is regulated by four transcription factors. Note how two of the regulatory inputs are tagged with evidence glyphs (*colored diamonds*); these symbols are used instead of link color to convey information about different classes of links, such as the type of experimental evidence supporting the link. *White bubbles* are used to represent off-DNA protein–protein interactions; here it is used to indicate the creation of an Ngn3-E2A protein complex, which then regulates the transcription of the NeuroD gene. NeuroD is also autoregulating.

1.3. Additional Features

Owing to space limitations, only the basic features of BioTapestry can be described here. The software provides a number of useful tools for building and annotating GRN models, and more are being added as part of the ongoing development effort. For example, starting with Version 4.0, the user can create high-level logical abstractions of the network, such as process diagrams, and then use these as a framework for constructing the network. BioTapestry can now also interact with Gaggle (24), which is a framework for sharing data between different software tools and databases. Using Gaggle, for example, users can interactively pass an annotated network definition between Cytoscape and BioTapestry while working simultaneously with both programs on their computer desktop. Also, the ability to run the software as a noninteractive stage of a data processing pipeline on a server allows it to work in concert with other software tools; this feature allows BioTapestry to be used with other software to build a Web service. The tutorials, release notes, and frequently asked questions on the BioTapestry Web site describe these features in more detail.

1.4. Possible Approaches for Building a BioTapestry Model

BioTapestry provides several different ways to create network model hierarchies. First, it is a highly interactive tool with extensive support for manual, fine-grained model drawing and editing. But it can also be used for projects starting from scratch with large sets of interaction data, as it has the ability to create and populate entire network model hierarchies, and to lay them out automatically, from simple lists of interaction data. These interactions can be specified using interactive dialogs, but they can also be imported into the program using a plain-text comma-separated value (CSV) input file that can be created in a spreadsheet or by another computer program.

There are also two different ways to create the lowest-level submodels (VfNs) that show which network elements are active in a particular time and place. One approach creates what are called *static* submodels, while the other creates *dynamic* submodels. A static submodel shows one view of the network, such as a single time point, and is created interactively by clicking on those elements present in the parent model that are to be included in the submodel. In the resulting submodel, included elements are shown colored, while excluded elements are shown as an inactive gray color. The CSV import feature described above is another way to create these static submodels in a noninteractive fashion.

Alternatively, dynamic submodels use underlying tables of experimental data that store the temporal and spatial expression of the network elements, as well as the time spans when the inputs into each are active. These tables are then used to automatically drive which submodel elements are shown as active or inactive. This approach allows the modeled network behavior to be changed by simply updating the tables, and also provides a highly interactive "time slider" tool to dynamically show how the network evolves over time. However, to create dynamic submodels, the user still must first build the top level networks (the VfG and VfAs) via drawing or in an automated fashion such as a CSV import. The dynamic tables only drive what is shown as active and inactive in an already created model hierarchy, and are not used to define the underlying model itself.

There are online tutorials at the BioTapestry home page that deal with a number of important topics, and these are an important resource for learning how to use the software. In particular, the *Quick Start* tutorial provides a good introduction to creating a static model hierarchy through manual drawing, and is the best way to learn the details of drawing a network and manually editing the network layout. There are also tutorials on building a completely static network model hierarchy using a CSV definition file, and on interactively creating dynamic submodels.

The example being worked here steps through the process of building a model hierarchy, with dynamic submodels, that maximizes the use of text file imports. It combines the techniques of building the upper-level models using a CSV file, coupled with the ability to create dynamic submodels by importing extensible markup language (XML) files containing expression and temporal input data. This technique is the most appropriate for building up a network model with the minimum of interaction, and that uses file-based inputs that could be output by a computer program. Note that the example presented here must be somewhat artificial, since due to space requirements and a desire to focus on the essentials, the model in this example is very small. In reality, such very small models are most quickly and easily built using the interactive features of BioTapestry, and not by manually creating separate files beforehand. But for large models and quick visualization of preexisting computerized interaction and expression data, the following technique is ideal.

1.5. Overview

The method that follows involves three major phases:

1. Creating and importing the files describing the network hierarchy.
2. Using some simple but powerful interactive layout manipulation operations that are useful when working with automatically generated networks.
3. Working through the interactive steps that are required to combine the automated creation of a model hierarchy via CSV with the automation afforded by dynamic XML data file imports.

Also, the subsequent Notes section will deal in depth with some of the decisions and issues that should be considered during the process of building a model hierarchy.

2. Materials

BioTapestry is written in Java, and is therefore platform-independent. It runs on any computer system with a Java Runtime implementation, e.g., Windows, Mac OS X, and many flavors of Unix, including Linux.

Since BioTapestry is written in Java, the only prerequisite is that the freely available Java Runtime Environment (JRE) is installed on your computer. For several years, Apple Macintosh computers have shipped with Java preinstalled. For computers that do not already have Java installed, the JRE can be downloaded from http://www.java.com/ by following the instructions found at that link.

BioTapestry is currently configured to run on Java 1.5, and it is still possible to run it using the older Java 1.4 if necessary. The computer requirements are not stringent; it is recommended that the system have at least 512 MB of main RAM, with a minimum screen resolution of 1,024 × 768, though more memory and a larger screen will certainly improve performance and ease of use.

BioTapestry is run as a Java WebStart application, which means that there is no explicit step to download and install the application. Instead, once Java is installed, a user can simply go to the BioTapestry home page at http://www.BioTapestry.org/ and click on the launch link provided there. The code is downloaded onto the computer and maintained locally in the cache of the WebStart system. WebStart will ensure that BioTapestry is kept up-to-date by checking for the latest version with the BioTapestry.org Web server, but also allows the program to be launched like any other program via a desktop icon. Note that although the software is downloaded, maintained, and updated via the Web, your data stays on your machine as locally saved files and is never uploaded or visible to the server.

3. Methods

3.1. Decide on the Structure of the Model Hierarchy

Before actually running BioTapestry, the first thing to decide is what the model hierarchy is going to look like. For this tutorial example, the hierarchy being built is very simple; see Fig. 3. The VfG is shown on the first row. This example then has two separate VfAs, shown in the second row, covering the same period of time, *Stage 0* to *Stage 4*. The first VfA (*Dorsal*), on the left, shows two regions, since they interact during this period via a signal from *Region A* to *Region B*. This juxtaposed presentation also illustrates how the two regions, with the same underlying network architecture, are evolving differently due to slightly different initial conditions.

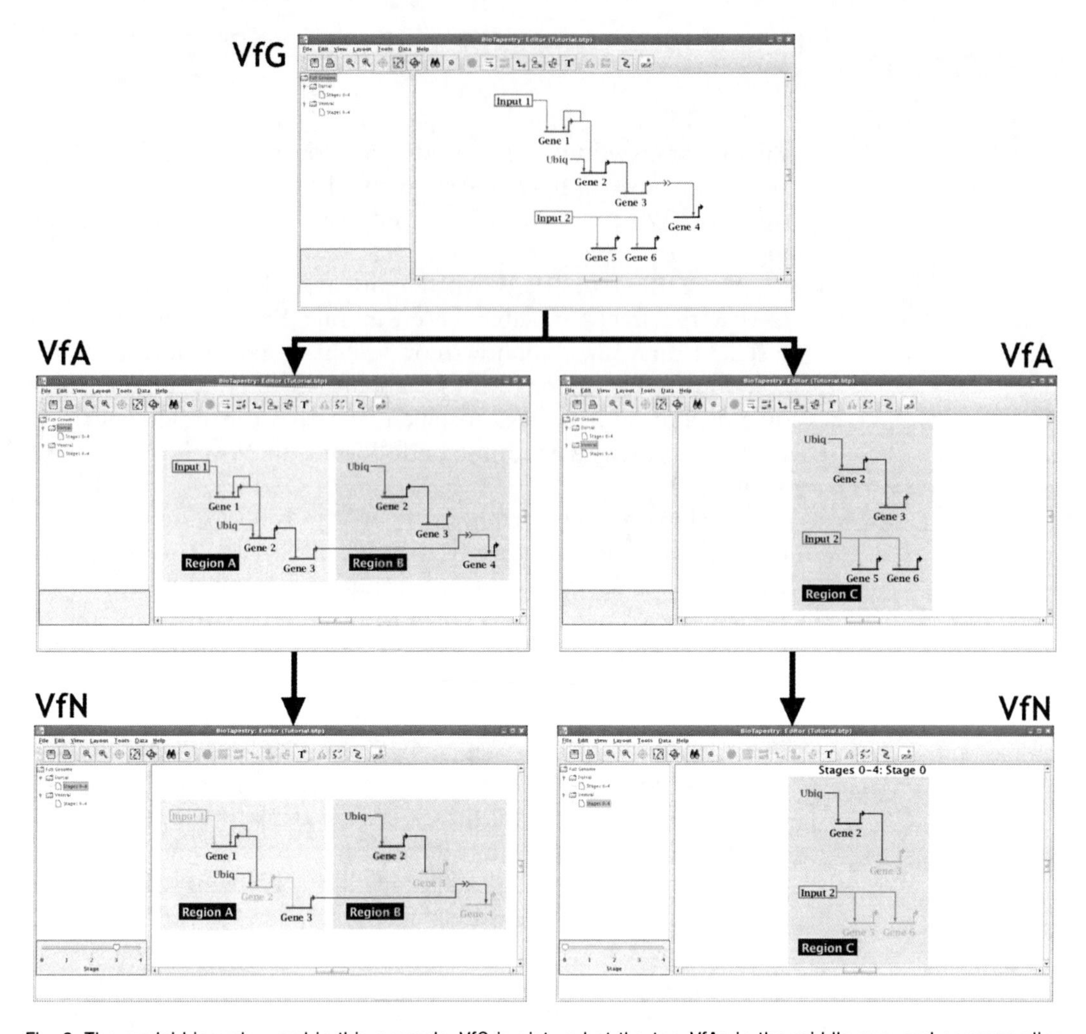

Fig. 3. The model hierarchy used in this example. VfG is pictured at the top, VfAs in the middle row, and corresponding dynamic VfNs in the bottom row.

By contrast, the second VfA (*Ventral*), on the right, depicts a single *Region C* that is not yet interacting with its neighbors. If these models were not so simple, this approach of using multiple VfAs to focus on smaller sets of regions would be an important technique for reducing the complexity of any single view of the network. Finally, the third row of Fig. 3 shows the dynamic VfNs that show the time evolution of their respective VfA parent. To create a clear and informative presentation of a GRN using the model hierarchy, several issues need to be considered (see Notes 1–4).

3.2. Create the Data Files That Will Be Imported into the Program

1. The easiest way to create the comma-separated value (CSV) input file that will define the model hierarchy and populate the top-level VfG and VfA models is to use a spreadsheet program such as Microsoft Excel or OpenOffice Calc, and then save the file as CSV. Start up the spreadsheet program.

2. The model being built is shown in Table 1 as a grid that matches its appearance in a spreadsheet program. Enter these values as shown into the spreadsheet. Alternately, the completed Spreadsheet.csv file is available from: http://www.BioTapestry.org/MiMB. Note that the set of commands is broken up into three sections: one for *model* commands, one for *region* commands, and one for *general* interaction commands. Model commands specify the tree-structured model hierarchy by listing each model and its parent (except the top-level root model has no parent). In this example, there are two child VfA models under the root, *Dorsal* and *Ventral*, which were described in Subheading 3.1. The next command block defines the regions that are present in each model; in this example, the *Dorsal* model contains two regions (*Region A and Region B*), while the *Ventral* model has one region (*Region C*). Note that the last column of the region command provides a

Table 1
CSV input file (shown arranged in spreadsheet cells)

Model	Root							
Model	Dorsal	Root						
Model	Ventral	Root						
Region	Dorsal	Region A	A					
Region	Dorsal	Region B	B					
Region	Ventral	Region C	C					
General	Dorsal	Box	Input 1	Gene	Gene 1	Positive	A	A
General	Dorsal	Gene	Gene 1	Gene	Gene 1	Positive	A	A
General	Dorsal	Gene	Gene 1	Gene	Gene 2	Negative	A	A
General	Dorsal	Gene	Gene 2	Gene	Gene 3	Negative	A	A
General	Dorsal	Bare	Ubiq	Gene	Gene 2	Positive	A	A
General	Dorsal	Gene	Gene 3	Intercel	Signal	Neutral	A	B
General	Dorsal	Intercel	Signal	Gene	Gene 4	Positive	B	B
General	Dorsal	Bare	Ubiq	Gene	Gene 2	Positive	B	B
General	Dorsal	Gene	Gene 2	Gene	Gene 3	Negative	B	B
General	Ventral	Box	Input 2	Gene	Gene 5	Positive	C	C
General	Ventral	Box	Input 2	Gene	Gene 6	Positive	C	C
General	Ventral	Gene	Gene 2	Gene	Gene 3	Negative	C	C
General	Ventral	Bare	Ubiq	Gene	Gene 2	Positive	C	C

unique abbreviation (no more than three characters) for the region; this is what is always used in subsequent commands to refer to the region. Finally, the third block contains one general interaction command for each link in the models. These lines each describe the relevant model, the type (e.g., a *gene*, *bubble*, or *intercel* (signaling) node) and name of the source and target nodes, the sign of the interaction (*positive*, *negative*, or *neutral*), and the source and target region abbreviations. In this example, only the interaction from *Gene 3* in *Region A* to the *Signal* node in *Region B* has different source and target regions; all the other interactions occur in a single region. The complete description of all commands is presented in Table S1, which is available from http://www.BioTapestry.org/MiMB. When reviewing the commands available in the CSV format, it is important to note that there are differences between the features available using CSV input compared to what can be specified interactively (see Note 5).

3. Once the spreadsheet is filled in, it needs to be saved in the plain-text CSV file format that BioTapestry can read (BioTapestry cannot read native .xls Excel files). In Excel, CSV saves are done by choosing *File→ Save As…*; for *Files of type* select *CSV (Comma delimited)*, then click *Save*. A couple of dialog boxes will pop up warning about the limitations of the CSV format, but these can be ignored; click *OK* and *Yes* as needed to complete the save to CSV. Note that in some language locales, a CSV file saved in Excel may actually use semicolons instead (see Note 6).

4. Create an Extensible Markup Language (XML) file that describes the temporal input data. XML files are structured in the same way as the familiar Hypertext Markup Language (HTML) files used by Web browsers, with elements delineated by balanced opening and closing tags, and with the opening tag containing a name-value pair for each attribute. Elements can themselves optionally contain other elements. Table 2 contains an example of an XML file. In a real situation, this file would very likely be created by a computer script, but this example will just create it manually. While there are dedicated XML editing tools available, any plain text editor such as Notepad (on Windows) or TextEdit (on Mac OS X) works well. Note that since the XML file must be saved as plain (ASCII or UTF-8) text, it is best to avoid using a word-processing program such as Microsoft Word for this task unless you are adept at saving files as plain text. Start up the text editor. If you are using TextEdit on a Mac, some preferences then need to be specified before continuing (see Note 7).

5. The entire file that needs to be created is shown in Table 2. Alternately, the completed InputData.xml file is available from

Table 2
Temporal input XML data file

```
<TemporalInputRangeData>
  <temporalRange name="Gene 1">
    <inputTimeRange input="Input 1">
      <range region="Region A" minTime="0" maxTime="1" sign="promote" />
    </inputTimeRange>
    <inputTimeRange input="Gene 1">
      <range region="Region A" minTime="1" maxTime="4" sign="promote" />
    </inputTimeRange>
  </temporalRange>
  <temporalRange name="Gene 2">
    <inputTimeRange input="Gene 1">
      <range region="Region A" minTime="1" maxTime="4" sign="repress" />
    </inputTimeRange>
    <inputTimeRange input="Ubiq">
      <range minTime="0" maxTime="4" sign="promote" />
    </inputTimeRange>
  </temporalRange>
  <temporalRange name="Gene 3">
    <inputTimeRange input="Gene 2">
      <range region="Region A" minTime="0" maxTime="1" sign="repress" />
      <range region="Region B" minTime="0" maxTime="4" sign="repress" />
      <range region="Region C" minTime="0" maxTime="4" sign="repress" />
    </inputTimeRange>
  </temporalRange>
  <temporalRange name="Gene 4">
    <inputTimeRange input="Signal">
      <range region="Region B" minTime="3" maxTime="4" sign="promote" />
    </inputTimeRange>
  </temporalRange>
  <temporalRange name="Gene 5">
    <inputTimeRange input="Input 2">
      <range region="Region C" minTime="0" maxTime="4" sign="promote" />
    </inputTimeRange>
  </temporalRange>
  <temporalRange name="Gene 6">
    <inputTimeRange input="Input 2">
      <range region="Region C" minTime="0" maxTime="4" sign="promote" />
    </inputTimeRange>
  </temporalRange>
  <temporalRange name="Signal">
    <inputTimeRange input="Gene 3">
      <range region="Region B" minTime="3" maxTime="4" sign="promote" />
    </inputTimeRange>
  </temporalRange>
</TemporalInputRangeData>
```

http://www.BioTapestry.org/MiMB. Note the basic structure of this file: within a single pair of (required) bounding `TemporalInputRangeData` tags, each node target is described using a `temporalRange` element, each of which contains a set of `inputTimeRange` elements, one for each input. For each input, a `range` element describes the time bounds of the input, optionally qualified by the `region` where

Table 3
Example fragment of expression XML file

```
<TimeCourseData>
  <timeCourse gene="Input 1" baseConfidence="normal" timeCourse="no">
    <data region="Region A" time="0" expr="yes" />
    <data region="Region B" time="0" expr="no" />
    <data region="Region C" time="0" expr="no" />
    <data region="Region A" time="1" expr="yes" />
    <data region="Region B" time="1" expr="no" />
    <data region="Region C" time="1" expr="no" />
    <data region="Region A" time="2" expr="no" />
    <data region="Region B" time="2" expr="no" />
    <data region="Region C" time="2" expr="no" />
    <data region="Region A" time="3" expr="no" />
    <data region="Region B" time="3" expr="no" />
    <data region="Region C" time="3" expr="no" />
    <data region="Region A" time="4" expr="no" />
    <data region="Region B" time="4" expr="no" />
    <data region="Region C" time="4" expr="no" />
  </timeCourse>
  <timeCourse gene="Input 2" baseConfidence="normal" timeCourse="no">
                              ⋮
</TimeCourseData>
```

this input occurs. If no `region` is specified, the time range applies to all regions in the model. The full syntax for this file is presented in Table S2, which is available from http://www.BioTapestry.org/MiMB. There are several caveats to keep in mind when creating this file (see Note 8). Save this file as plain text.

6. Using the same text editor, create the XML file containing the spatial/temporal expression data for the model. The format of this file is roughly sketched out in Table 3.

Within a single pair of (required) bounding `TimeCourseData` tags, each node is listed using a `timeCourse` element, each of which contains a set of data elements. There is a separate data element to describe the expression level (`noData`, `no`, `weak`, `yes` or `variable`) for every region at every time point. Since this particular format is verbose and lengthy, only the first of ten entries is shown in its entirety in Table 3 (along with the opening tag for the second entry). The other nine entries that follow the illustrated block for *Input 1* will be identical, except for the `gene` name and the `expr` values. Note that the ordering of the `data` elements must be identical across all the `timeCourse` elements. All the

Table 4
Summary of expression values to use in the expression XML data file

Node	Region	Stage 0	Stage 1	Stage 2	Stage 3	Stage 4
Input 1	A	Yes	Yes	No	No	No
	B	No	No	No	No	No
	C	No	No	No	No	No
Input 2	A	No	No	No	No	No
	B	No	No	No	No	No
	C	Yes	Yes	Yes	Yes	Yes
Ubiq	A	Yes	Yes	Yes	Yes	Yes
	B	Yes	Yes	Yes	Yes	Yes
	C	Yes	Yes	Yes	Yes	Yes
Gene 1	A	Yes	Yes	Yes	Yes	Yes
	B	No	No	No	No	No
	C	No	No	No	No	No
Gene 2	A	Yes	Yes	No	No	No
	B	Yes	Yes	Yes	Yes	Yes
	C	Yes	Yes	Yes	Yes	Yes
Gene 3	A	No	No	No	Yes	Yes
	B	No	No	No	No	No
	C	No	No	No	No	No
Gene 4	A	No	No	No	No	No
	B	No	No	No	No	Yes
	C	No	No	No	No	No
Gene 5	A	No	No	No	No	No
	B	No	No	No	No	No
	C	No	Yes	Yes	Yes	Yes
Gene 6	A	No	No	No	No	No
	B	No	No	No	No	No
	C	No	Yes	Yes	Yes	Yes
Signal	A	No	No	No	No	No
	B	No	No	No	Yes	Yes
	C	No	No	No	No	No

values that need to be entered for these two attributes are presented more compactly in Table 4. Alternately, the completed ExpressionData.xml file can be downloaded from http://www.BioTapestry.org/MiMB. The full syntax for this file format is presented in Table S3, which is also available from http://www.BioTapestry.org/MiMB. As with the previous file, there are issues to take into consideration when this file is created (see Note 8). Save the file as plain text.

3.3. Start BioTapestry and Import the CSV File

1. Go to the BioTapestry home page at http://www.BioTapestry.org/ and click on the start link embedded in the prominent sentence "Click HERE to run the BioTapestry Editor." (This assumes that Java has been installed, as described in Subheading 2). If you are running the program for the first time, you will need to accept the security certificate verifying the origin of the program. The BioTapestry Editor window will appear. This is usually a straightforward step, though some issues may occasionally arise (see Note 9).
2. BioTapestry has a few different automatic layout strategies it can use when importing a CSV network. For example, there is a specialized *bipartite strategy* that works well when the network can be partitioned into distinct sets of source and target nodes. The more typically used *general strategy* can itself be instructed to use a couple of different techniques for grouping source nodes. The default *hierarchical* approach works best with larger numbers of genes, while the *single source cluster* option often works best with smaller numbers of nodes. For this small network, the latter technique is preferable. To specify the technique to use, select *Layout→ Set Automatic Layout Options.* In the dialog box that appears, set the *Source Grouping Strategy* in the *General Strategy* section to *Single Source Cluster* and click *OK.*
3. Select *File→ Import→ Import Full Model Hierarchy from CSV...* to load the CSV file into BioTapestry. This will bring up a dialog for the CSV load options. Choose *Completely replace existing network*, and check the box to *Compress child models.* Click *OK*. This will then bring up a standard Open File dialog; navigate to the directory where you saved the CSV file, select the file, and click *Open*. If the file was not saved with a .csv extension, you will need to set *Files of Type* to *All Files* to see the file so it can be selected. Once the file is loaded, you should see the network, as shown in Fig. 4 (though colors of genes, nodes, and links in your network will be different from those depicted). If you get an error message, make sure the syntax of your CSV definition is correct and try importing it again.
4. Once the network is loaded up, get a feel for navigating around. Note how the model hierarchy specified in the CSV file appears in the left-hand navigation panel. You can click on the various models in this panel to see them, use the zoom buttons near the left end of the toolbar to adjust the zoom level, select nodes and links by clicking on them, and so forth. Return to viewing the top-level *Full Genome* model before continuing.
5. Building the model hierarchy from a CSV file creates an internal interaction list in BioTapestry which then forms the formal definition of the network model; this type of network definition

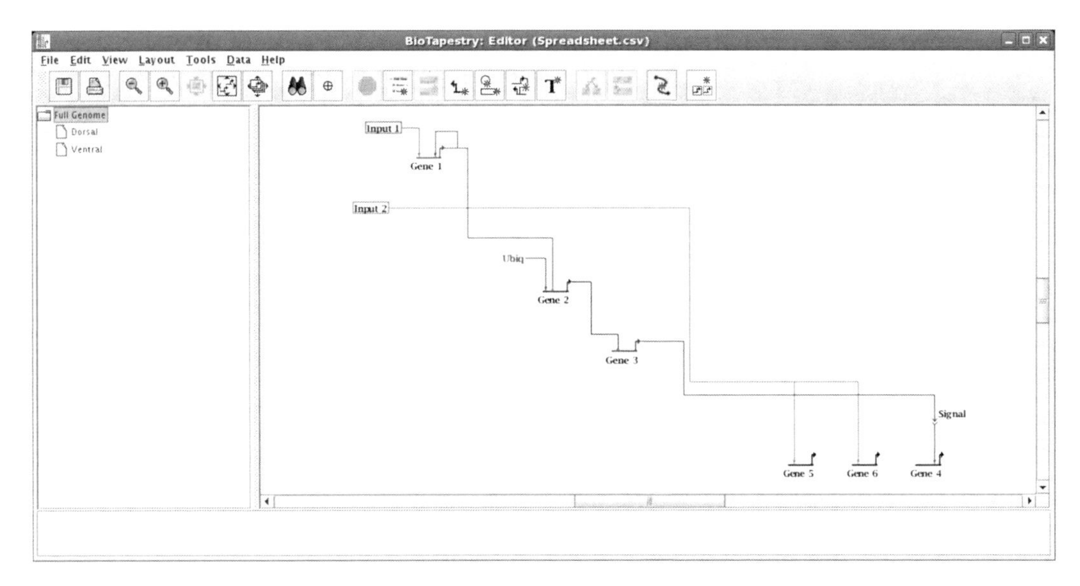

Fig. 4. View of the top-level model in BioTapestry after the CSV file is imported; node and link colors have been modified here for improved clarity in grayscale.

has some implications (see Note 10). If this interaction list is retained, then some of the following steps will cause a warning dialog to appear. To avoid these warnings in the rest of the tutorial, discard those build instructions. Select *Tools→Drop All Interaction Tables Used to Build Networks* and then click *Yes* in the confirmation dialog.

3.4. Steps for Optimizing the Layout

1. The automatic layout algorithm used in a CSV load is designed to handle much larger networks, so the resulting organization of this tiny network can be improved with some manual tweaks that move some nodes and links around. The next few steps will introduce some very basic layout editing procedures to illustrate what can be done to manually edit network layouts. The much more extensive *Quick Start* tutorial covering how to draw and edit network layouts is available on the BioTapestry Web site. The modifications will be done to the top-level network first, so make sure to click on the top-level *Full Genome* model in the navigation tree so that the *Full Genome* model is displayed.
2. Drag the *Input 2* box (click on the *Input 2* box, move the mouse while holding down the button, and then release the mouse button at the destination) down to the left of *Gene 5*. Also, drag the *Signal* node over to the right of *Gene 3*, placing it along the outbound link from *Gene 3*. To then change the orientation of the intercellular symbol to point right, *right-click* (note to Mac users: if you're using a one-button mouse,

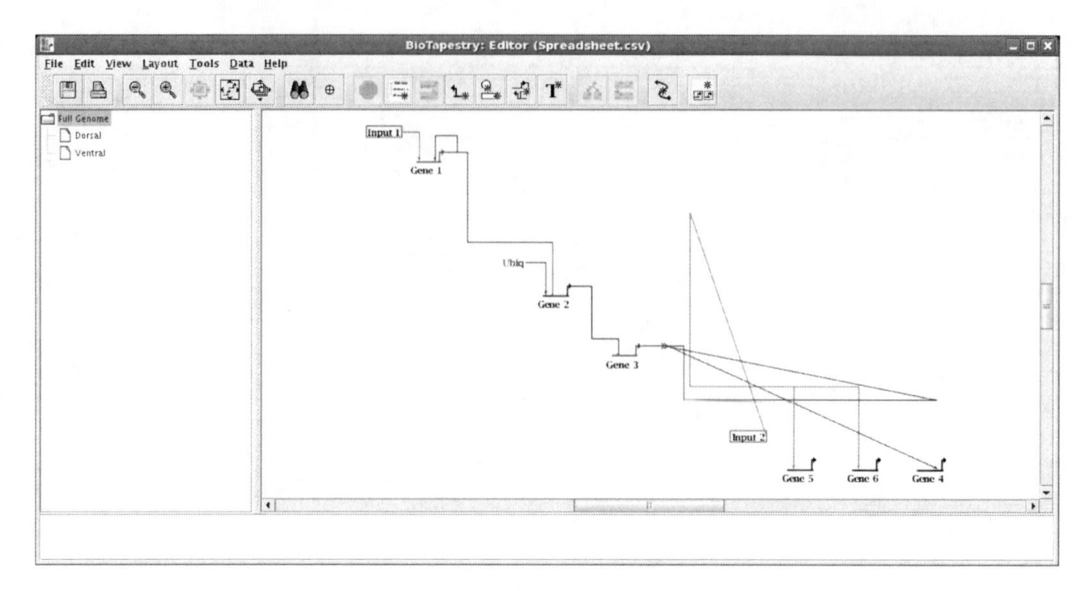

Fig. 5. Appearance of top-level network after nodes have been dragged to new locations.

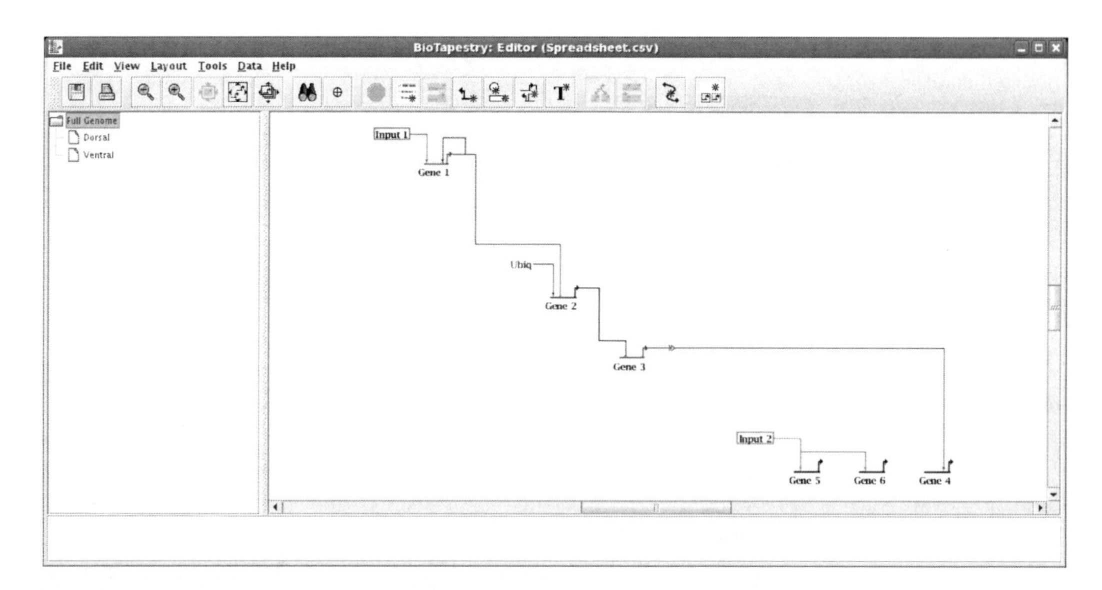

Fig. 6. Top-level network after links have been cleaned up.

press the *Control* (*Ctrl*) key while clicking the mouse to do this) on the *Signal* node, select *Properties…* from the popup menu, and then select the *Presentation Properties* tab. On this tab, set the *Orientation* to *Right*. You can also change the (first) *Color* of the node to match the input link color, and check the *Hide node name* box; finish by clicking *OK*. The result of these steps is shown in Fig. 5. Then, to fix the link layout, select *Layout→ Other Automatic Layout Tools→ Layout Only Irregular Links*. The result should look something like Fig. 6, with all links appearing orthogonal. Keep in mind that any edit step *X* can be reversed by selecting *Edit→ Undo X*.

3. Now is a good time to save your work. Select *File→ Save As...*, then enter the name you wish to use in the dialog box, e.g., *Tutorial*, and click *Save*.
4. The extra bends in the various links are an artifact of the technique the automatic layout algorithm uses to bundle and organize very large sets of links in big networks. In this tiny network, the clearest presentation has these bends removed. To do this quickly, select *Layout→ Other Automatic Layout Tools→ Run Single Link Optimization Path*.
5. This particular layout can also be improved now by compressing the network to eliminate unnecessary space. Select *Layout→ Compress Network...*, make sure the sliders for horizontal and vertical compression are both set to *100* in the dialog that pops up, and click *OK*.
6. The previous compression step removed unused extra space, but is designed to not change the relative positioning of the nodes in the network. An even more compact layout would shift the upper half of the network over to be directly above the bottom row. To do this, move the mouse cursor to somewhere in the empty space to the upper left of the *Input 1* box, press and hold the left mouse button down, drag the mouse over to the lower right of *Gene 3*, and release the button. This dragging operation will show a rubber-band box as you proceed. Once this is complete, all the nodes and link segments inside the box will be selected. Then, you can press on and drag one of the selected items (e.g., *Gene 3*) over to the right, making sure that the unselected link segments stay horizontal; all selected items will follow the drag. Alternately, use the keyboard by holding down the *Shift* key while repeatedly pressing the right arrow key to incrementally shift the selected nodes to the right. The final result is shown in Fig. 7. When complete, click the mouse in an unoccupied space somewhere to unselect everything (or select *Edit→ Select None*).

3.5. Synchronizing the Layouts

1. It is important to keep in mind that the top-level *Full Genome* model and each of its child models in the model hierarchy have independent network layouts (see Note 11). So once layout changes have been made to the *Full Genome*, the other models need to be synchronized with the *Full Genome* layout for those changes to be universally applied. To begin, click on the top-level *Full Genome* model in the navigation tree to make sure the Full Genome model is shown.
2. Select *Layout→ Synchronize All Layouts...*. The first dialog box that pops up allows you to choose which regions in which models you want to synchronize. The default is all regions in all models, so just click *OK*. This brings up a second dialog, which allows you to set several synchronization

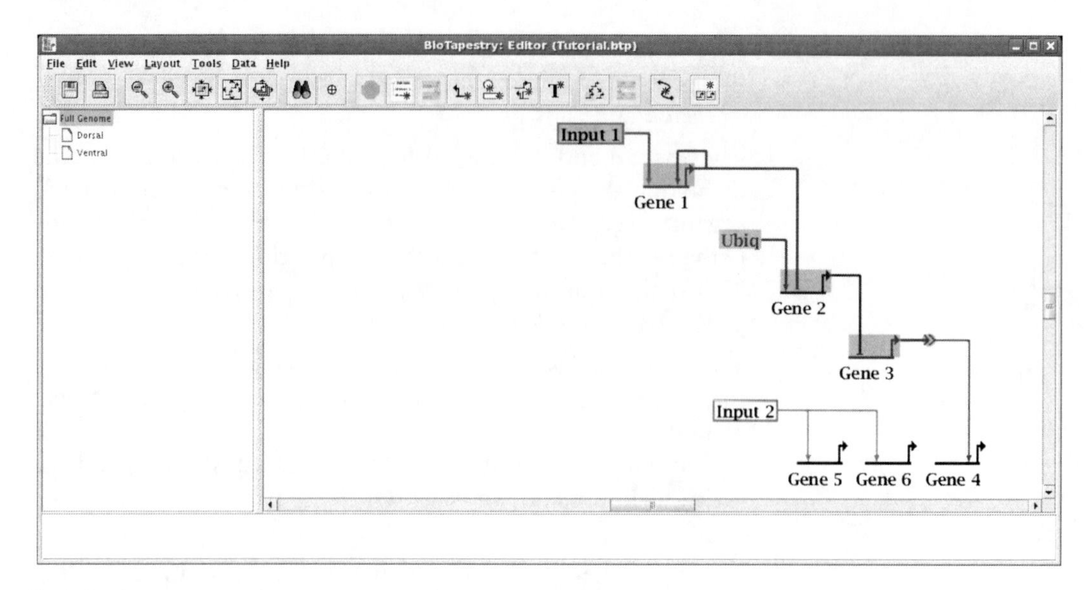

Fig. 7. Final arrangement of top-level network after relocating selected nodes.

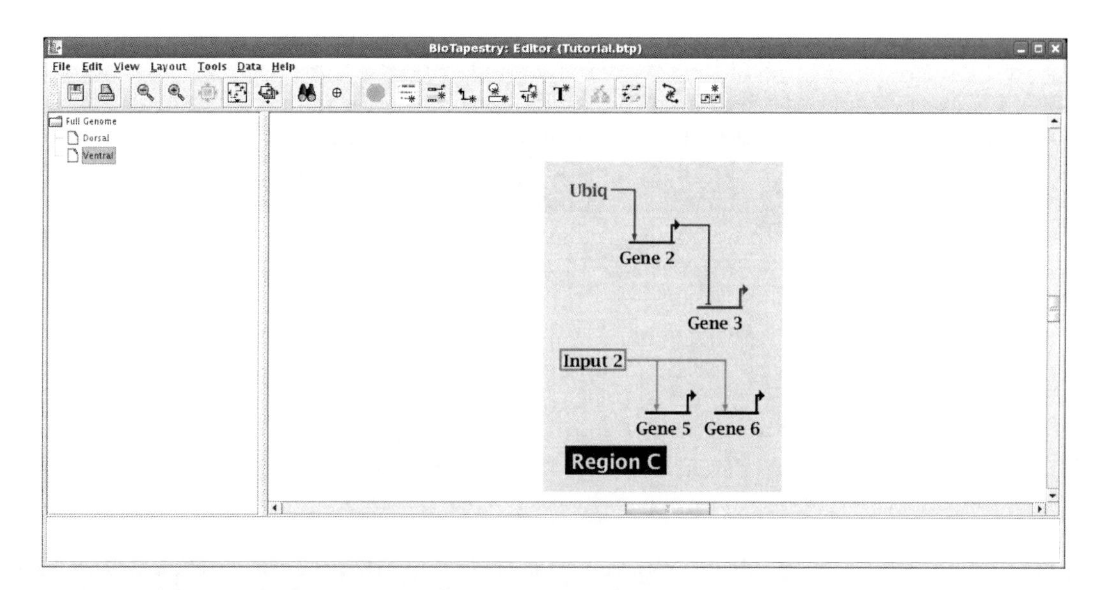

Fig. 8. The *Ventral* VfA model after layout synchronization.

options (see Note 12). For this example, make sure *Compress layout in child models* is checked, *Retain region positions* is unchecked, and *Swap Link Pads* is checked. Since there are no network overlays in this model, the *Network Overlay Strategy* can be left alone. Click *OK*. If you then select the *Ventral* model, you will see that the layout of the network in *Region C* is a compressed subset of the *Full Genome* version. To slightly improve the layout, drag the *Region C* tag down slightly below the network elements; the result is shown in Fig. 8.

3. Another aspect of dealing with a collection of separate model views is to keep them roughly (or in some cases, precisely) aligned with each other so that moving between the different model views is smooth and seamless. To do this, select *Layout→ Layout Centering and Alignment→ Align Centers of All Model Layouts....* In the dialog, choose *Align layout centers* instead of the alternative *Try to overlay matching elements*; the latter option is more appropriate when the VfA layout matches the VfG layout almost exactly. Also, check the box for *Shift all models to workspace center after alignment*. The *Ignore network overlays when centering option* can be ignored, since again there are no overlays in this model. Then click *OK*. After this step, all the models are centered on top of each other in the workspace. The view will be zoomed out completely to show that the current model is at the very center of the entire workspace; you can click on the *Zoom As Needed to Bound All Models in Hierarchy* button in the toolbar (just to the left of the *Search...* binoculars button) to zoom back in again.

3.6. Import Dynamic XML Files

1. Before the dynamic expression and temporal input XML files can be imported, the time scale needs to be specified by selecting *Data→ Set Time Units....* The dialog box that appears allows you to select from the common time scales; you can also specify named or numbered stages. For this tutorial, set *Units* to *Numbered Stages*, set *Custom Unit Name* to *Stage*, set *Custom Unit Abbreviation* to *St*, uncheck the box for *Units appear as a suffix*, and click *OK*. By doing this, the integer values entered for times in the XML files will now represent stage numbers. Note that once time-based data has been entered into BioTapestry, the time scale definition is locked and cannot be changed.
2. Now import the two XML files created at the start of the tutorial. First, select *File→ Import→ Import Time Expression XML Data...*, and navigate to the location where the expression data XML file is stored. Assuming the files were saved as .txt files, it will be necessary to change the *Files of Type* setting to *All Files* to be able to see and select the file. Select the file and click *Open*. To then load the temporal input XML file, select *File→ Import→ Import Temporal Input XML Data...*, and repeat the process of opening the file. Again, it will probably be necessary to change the *Files of Type* setting first.

3.7. Create Dynamic Submodels

1. Now that the dynamic data tables have been loaded, the dynamic submodels for both the *Dorsal* and *Ventral* models can be created. First, the time bounds of the existing *Dorsal* and *Ventral* models must be set before dynamic submodels can be created under them. For the *Dorsal* model, *right-click* on the *Dorsal* entry in the left-hand navigation tree, and select

Edit Model Properties… in the pop-up menu. In the dialog box that appears, check the box for *Specify time bounds*, then set the *Minimum Time (Stage)* to *0*, the *Maximum Time (Stage)* to *4*, and click *OK*. Then, repeat this process to set the time bounds for the *Ventral* model.

2. Now create a dynamic submodel for the *Dorsal* model. Right-click on the *Dorsal* entry in the left-hand navigation tree, and this time select *Create Dynamic Submodel…*in the pop-up menu. In the dialog box that appears, change *Type* to *Every Stage*, set *Min. Stage* to *0*, set *Max. Stage* to *4*, set *Model Name* to *Stages 0–4*, and click OK. Again, repeat this step for the *Ventral* model.

3. The final step when creating dynamic submodels is to specify which regions are going to be included in the model. If you select the *Stages 0–4* submodel of the *Dorsal* model, it should look like Fig. 9: neither *Region A* nor *Region B* have been included in the model. To add regions to the submodel, click on the *Choose Subset of Parent* button in the toolbar (third button from the right). Alternately, select *Edit→ Choose Subset of Parent* from the main menu. The cursor changes to a crosshair; click on the grayed-out *Region A* and then *Region B* to include them in the model, and then click on the stop sign icon (or press the *Esc* key on the keyboard) to end the adding process. In this example, the Signal node between the two regions does not need special attention, but sometimes needs to be handled as an *extra dynamic node* (see Note 13). Complete this step by going to the *Ventral* model and including *Region C*.

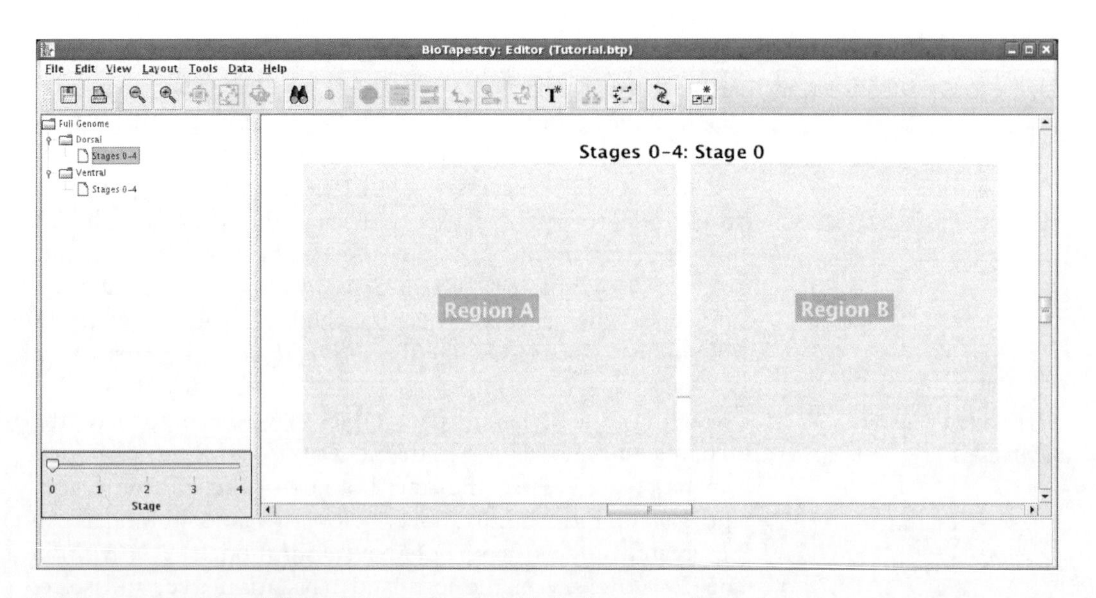

Fig. 9. Appearance of *Dorsal Stages 0–4* dynamic submodel prior to region inclusion.

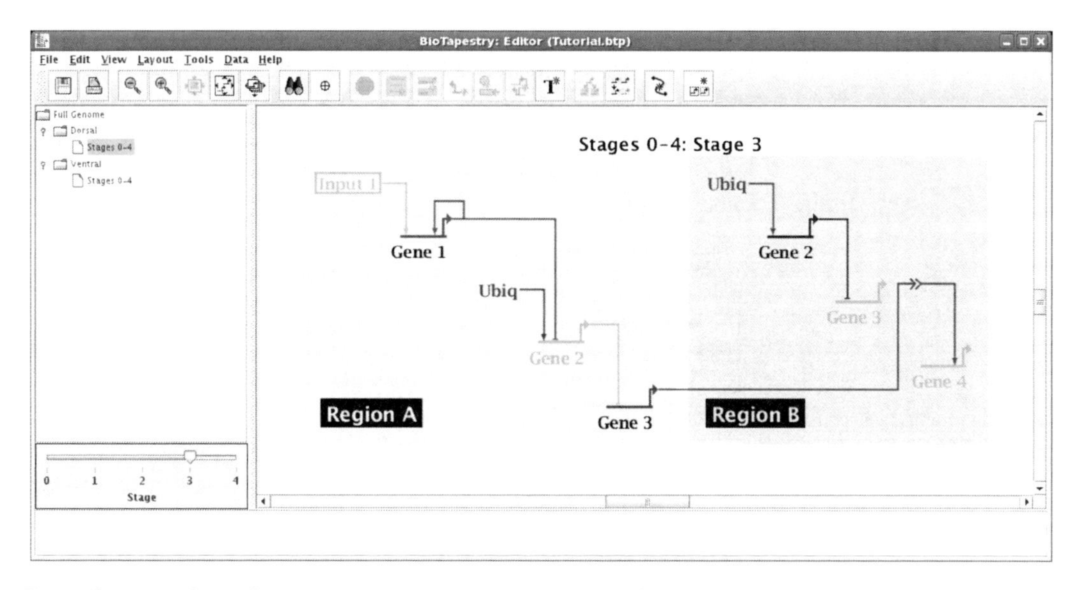

Fig. 10. Completed Dorsal Stages 0–4 model, with time slider selecting Stage 3.

4. Going to the *Dorsal* model, the view should look like Fig. 10. Run the time slider back and forth to see the network behavior over time.

3.8. Set the Startup View and Zoom Behavior

1. The first model displayed when the program starts up has an important influence on the user's appreciation of the whole hierarchy. In this tutorial, make this the *Dorsal* model. *Right-click* on the *Dorsal* entry in the left-hand navigation tree, and select *Make This Model the View on Startup* from the pop-up menu.
2. Some other important parameters are what the zoom level is when a file is loaded, and how the zoom behaves when moving between models. To set these options, select *Edit→ Set Display Options...*to bring up the *Set Display Options* dialog. The two relevant settings are *Set Initial Zoom on Load* and *Zoom When Selecting*. For this exercise, set the former to *Set first zoom to bound all models*, and set the second to *Don't change zoom when selecting models*. Be sure to save the finished model using for example *File→ Save*.

3.9. Adding Experimental Data to the Model

1. One of the important interactive features of BioTapestry is that users can *right-click* on any node and select *Experimental Data* in the popup menu to see a window that displays supporting data for the network. Since the tutorial model already has expression and temporal input tables defined for the nodes, these tables are already included in this display. However, additional information can be added easily. Right-click on *Gene 3*

and select *Properties....* This brings up the *Set Gene Properties* dialog. Click on the *Text Annotations* tab; in the text box, type in a simple text fragment that uses basic HTML markup tags for italics and bold: `<i>Hello</i><b>World</b>`, and then click *OK*. Right-click again on *Gene 3*, select *Experimental Data*, and the text *Hello* **World** will be shown in the window.

2. To provide even more in-depth information, the user can provide a set of Uniform Resource Locators (URLs) that will be displayed in the Experimental Data window for a node. As before, right-click on *Gene 3* and select *Properties...*; this brings up the *Set Gene Properties* dialog. This time, click on the *Data URLs* tab. For *Add URL:* type in http://www.BioTapestry.org/developers/webPluginExample.html and click the adjacent *Add* button. Then click the *OK* button to close the dialog. The Experimental Data window will then display a small Web page with an embedded image. The Web pages that can be specified with this technique have some important restrictions (see Note 14).

3.10. Sharing the Network

1. There are a variety of different ways to disseminate a BioTapestry model. In this step and the next, we outline the two most popular approaches. The preferable and significantly more powerful way to distribute a BioTapestry model is to share a read-only, interactive version of it by hosting a copy of the BioTapestry Viewer for the model on a Web server. The BioTapestry Viewer is a stripped-down, read-only version of the Editor that allows the user to interactively explore the model hierarchy, view supporting data, search for gene targets and input sources, mouse over footnotes for additional information, and perform other powerful operations. There is a ZIP archive on the BioTapestry Web site that contains the necessary files and instructions for setting up a BioTapestry Viewer.
2. A more common and quicker approach to sharing a model is to create static images for publication. Go and view the *Dorsal* model, and then select *File→Export→Export Image for Publication....* In the dialog box, set the *Units to use* to *inches*, set the *Resolution* to *300 dots/in.*, make sure the *Keep original aspect ratio box* is checked, enter the *Width* as *7.5 in.*, keep the *Height* and *Magnification* settings that are set automatically to track the entered width, and set the *Image Format* to *PNG*. Using PNG instead of JPEG is highly recommended (see Note 15). Then click *OK*. In the subsequent *Save* dialog, navigate to the directory to save in, and enter the name of the file. If you do not specify the file extension, it will be based on the file type specified in the previous dialog. Click *Save*.

4. Notes

1. Implications of the Model Hierarchy
   BioTapestry's use of a model hierarchy to represent a network imposes some constraints on what network elements can appear in a given model. These constraints guarantee model consistency; without such constraints, it would be easy for the different model layers to accumulate inconsistencies over time as the models evolve. The most important thing to remember is that a gene or link cannot be added to a lower level model unless it is also present in the models above it in the hierarchy. Thus, if *Gene 1* is added to *Region A* in a VfN, the software will ensure that *Gene 1* is present in *Region A* in the parent VfA, and that *Gene 1* is also present in the VfG. If this is not the case, BioTapestry will add these elements automatically as needed. This guarantees consistency between the model levels. The complement of this rule is also enforced by BioTapestry: deleting a network element from a model automatically removes it from all the models below it (but it still remains in all the models above it). Also, consistent with the definitions of VfGs and VfAs, regions do not exist in the top-level model, and multiple copies of genes cannot be drawn in the top-level model. Thus, if a user wants to create a model that is depicting the interaction between two regions, particularly if the regions share common network elements, they need to be working with a second-level VfA model.
2. Model Hierarchy Organization
   The various BioTapestry models that are currently hosted in online viewers (3, 5–8) can provide good starting points for seeing different ways to organize a model hierarchy. When building the hierarchy, it is important to keep in mind how network layouts are assigned. The top-level VfG model has its own layout, each VfA model below it has its own independent layout, and all the VfN models under a given VfA share the layout of that VfA.

   This framework leads to the requirement that different arrangements of regions will need to be in separate VfAs, as each arrangement requires its own layout. So as a developing embryo differentiates into an increasing number of cell types, this progression will need to be represented by a set of sibling VfA models that track this increase in complexity.

   Another motivation for multiple VfAs arises when there are a large number of different regions to present. While these could all be shown with a single VfA, it is often clearer to reduce complexity by creating manageable subsets of the regions and then distributing them across several VfAs. In the limit, there might be only one region per VfA. The region subsets are ideally chosen to group together regions that are interacting via signals, or to juxtapose regions that illustrate the behavior of the network under different initial conditions. This was the motivation behind the organization of the small example used in this tutorial.

The VfNs below a particular VfA show a subset of the items in that VfA, while sharing the same layout. So those VfNs can be used to isolate and focus on one or more of the regions present in the parent VfA, or to focus on a particular time. Since the model hierarchy is not depth-limited, one useful approach is to create one VfN layer that isolates some subset of regions in the VfA, and then create another VfN layer beneath the first that shows different time points for the isolated set of regions displayed in the parent VfN.

The job of illustrating the behavior of the network over time, with some elements shown as active (colored) and others shown as inactive (gray), is typically handled by the VfNs. Indeed, the interactive technique for creating VfNs is designed to make this approach quick and easy, and the dynamic submodels created in this tutorial are in fact only supported at the VfN level. However, there can be cases where geometry is changing quickly, and it makes sense to then create a series of VfA models, one for each time point, and to show the active and inactive components right at the VfA level; such an approach would not even have a separate VfN level. If desired, this can be done by setting node and link properties in each VfA to *active* or *inactive* as needed (property dialogs are accessed by *right-clicking* on a node or link).

Though VfNs can isolate out and focus on only some regions of the parent VfA, all regions in the VfA still appear in the VfN; if regions are not included, they still are represented as light gray rectangles. To be able to make this approach scale up to support truly large hierarchies, it would be desirable to allow these unused regions to be completely hidden, and to allow VfN elements to be shifted as needed to make for a compact representation. This is a current limitation of BioTapestry that should be addressed in a future release.

Finally, it is useful to study the presentation of the EGRIN model (6) to see a more abstract usage of regions. In this model, each VfA consists of a single abstract "region" that just contains a small subset of a very large top-level network, with the region showing a small related subset of network elements from the top network. In this more abstract sense, "regions" could also be used to, for example, show different network states arising from different experimental conditions.

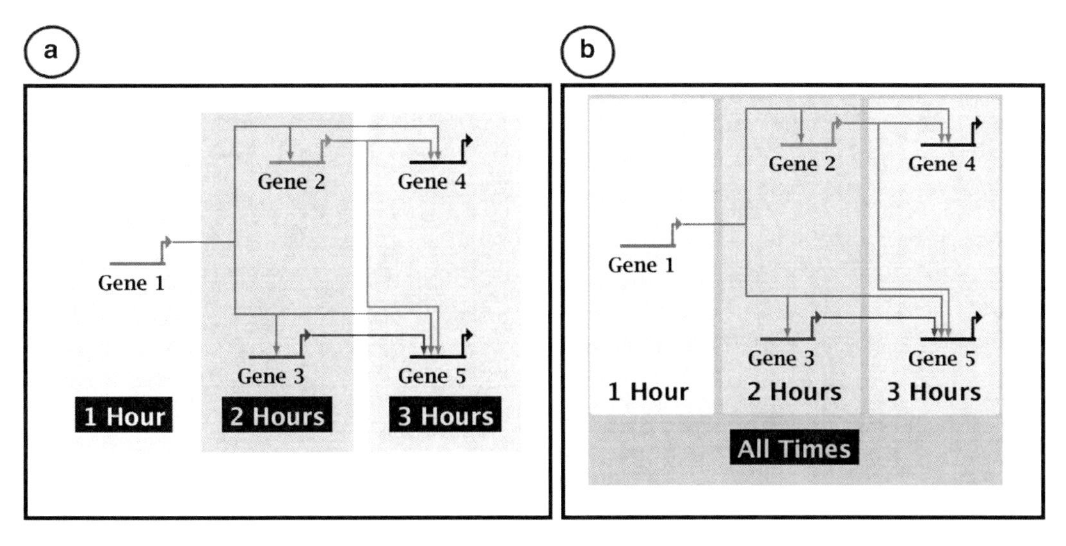

Fig. 11. Comparison of ways to group nodes. (**a**) Using regions; this approach is discouraged. (**b**) Using the BioTapestry Version 4 network overlay feature.

3. Common Issues with Region Organization
   It is best to think of a BioTapestry region as a sufficient subset of the top-level VfG network to describe the regulatory state of, for example, a cell type or particular experimental condition. While regions can be used in other ways, care should be exercised when doing so.

   For example, it is tempting to use regions to group nodes that have some common property; see Fig. 11. In Fig. 11a, separate regions are being used to describe the time of expression of groups of genes within a particular network. But the correct way in BioTapestry to apply descriptive groupings to the network is through the network overlay and module feature introduced in BioTapestry Version 4; Fig. 11b shows how this feature achieves the same results as in Fig. 11a, with the added benefit that any number of different groupings can be overlaid on the network. Note that one issue affecting the use of overlays is that at present the CSV input format does not support them (see Note 6).

   Using overlays and modules instead of regions to provide network annotations also has the advantage of working well with the automatic layout algorithms that BioTapestry uses. If regions are used in the manner shown in Fig. 11a, the model usually ends up with many links going between different regions. But the auto layout engine uses the VfG layout as a template for the layout within each region, and assumes that there are few links among multiple regions, which is the case when such links are just a limited number of, for example, interregion signals. If this assumption is violated, layout quality can suffer.

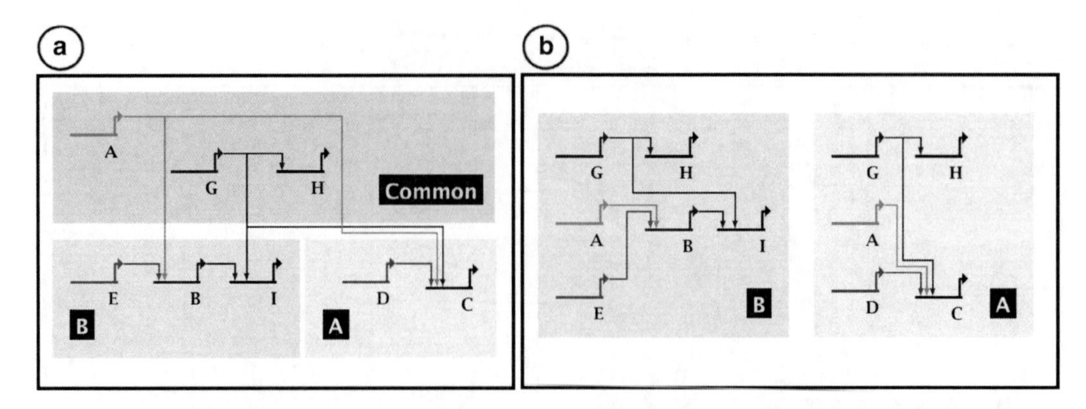

Fig. 12. Illustration of using a common region. (**a**) Shared network elements appear in a single common region. (**b**) Two independent regions that contain all necessary elements.

Another thing to be aware of are the issues involved with the use of "common regions" to create more compact layouts. Figure 12a shows how a common region is used to contain those network elements that exist in two or more regions of a VfA; Fig. 12b shows the same network, except with duplicated elements in two independently defined regions. The common region approach can be used, but it is important to realize that this presentation has an effect on the organization of the model hierarchy. This is because that approach makes it impossible to have a single VfN submodel that shows both regions simultaneously if the common components are behaving differently between the two regions. Instead, there would need to be two VfNs; one showing the behavior of region A, and the other only showing region B. Also, this approach suffers from the same issue that can prevent effective automatic layouts, since this presentation typically creates many links between regions.

Another network arrangement to avoid is the case of a VfG model masquerading as a VfA. A very simple example is shown in Fig. 13: *Gene 1* expresses in *Region 1*, and *Gene 2* expresses in *Region 2*. The reason *Gene 2* does not express in *Region 1* is because the active *Gene 1* is repressing it. In *Region 2*, *Gene 2* can express because *Gene 1* is off (for reasons not depicted here). Figure 13a shows a valid VfA presentation of such a system, where the relationships and causality in the system are perfectly obvious.

Figure 13b shows a VfG that underlies such a system, and the case to avoid for this system is shown in Fig. 13c; the latter shows a VfA that just has the VfG network (i.e., every gene and link appears only once in the network) overlaid over the multiple regions. The rationale for this approach is that it is correctly showing where the various genes express, and it suggests the mechanism for the pattern. However, it is rare that every

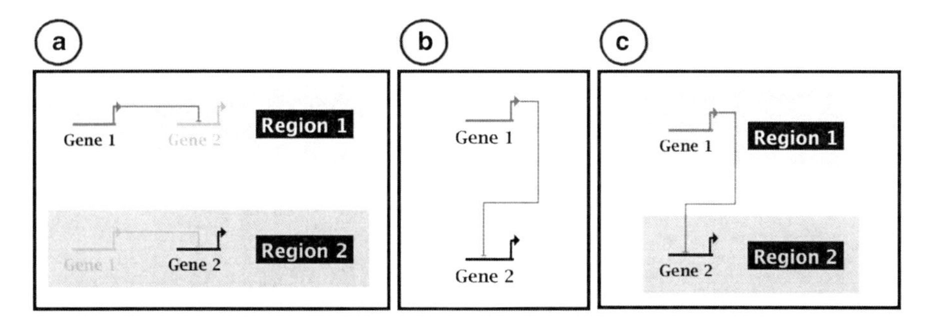

Fig. 13. The "fake VfA." (**a**) A VfA where appropriate duplication of network elements in separate regions provides a clear, unambiguous depiction of the network states in two regions. (**b**) The VfG model for the previous VfA. (**c**) A "fake VfA" where the VfG network is just overlaid on multiple regions. This representation is ambiguous.

gene in the VfG is truly isolated to a single region. With a system of any complexity, this "fake VfA" results in a very confusing presentation, and even with the detailed prose explanation that must accompany it, the result lacks the clarity of a correctly constructed VfA. Note: while this is a recommendation to avoid creating an exact copy of the VfG spread across a *multiple-region* VfA, there are many good reasons to create such a complete copy in a VfA containing just a *single* region.

4. What Elements to Include in a Region

   Regions in a VfA can contain any subset of the top-level network, including the entire network. So the question often arises as to how much of the underlying network should appear in a region. One approach is to include the entire VfG network in each region, with the quiescent portions set to *inactive*, and thus grayed out. The opposite approach is to show just those network elements that are active in the region. The former approach is the most complete, but often includes excessive unused detail. The latter approach is more compact, but has the potential of ignoring the question of *why* the unused sections are active in one context, but inactive in another. The best approach is middle of the road; a region should contain active elements, but also include enough descriptive power and inactive elements to explain why those subnetworks are not running.

5. CSV Support

   While the comma-separated value input format provides the ability to create a full model hierarchy with a single file import, it does not support all of the features that are available via the interactive interface. It is anticipated that as BioTapestry development proceeds, the CSV format will be upgraded to support many of the interactive features. But at this time, the CSV file format has some limitations:

   (a) As described in Subheading 1.2, and shown in Fig. 2, links can be assigned different evidence levels. However, the CSV input format does not currently support the specification of an evidence level for a link.

(b) When a gene has an active repression input, and thus is not expressing in a VfA or static VfN, it is preferable to be able to set the gene's activity level to *inactive*. However, the activity level of a node or linkage cannot currently be specified using the CSV format.

(c) Subheading 1.3 mentioned how BioTapestry now supports the creation of high-level logical abstractions of the network, and the feature was illustrated in Fig. 11b (see Note 3). To implement this feature, the program uses *network overlays*, which contain *network modules*; the latter are sets of genes and nodes. However, this feature is not yet supported by CSV import. Thus, all overlay and module definitions, and node membership in the modules, must be added manually.

(d) Subheading 1.4 described the difference between static and dynamic submodels; the current CSV import format only supports the creation of static submodels.

(e) Subregions, and subregion membership, cannot be specified via CSV input. Subregions are a useful tool for depicting those portions of the network that belong to different lineages emerging from a common territory, and genes in a region can be assigned to one or more subregions contained in a region. The discussion at http://www.BioTapestry.org/faq/FAQ-Modeling.html#subregions provides more details about this feature.

(f) When creating networks interactively, the user can define specific *cis*-regulatory modules for a gene and then draw links terminating in those specific modules. This level of fine-grained specification is not yet available using the CSV import function.

Also, while the CSV import mechanism does support incremental additions to the network while retaining current layouts, making large changes to the network based on the incremental layout algorithm is not recommended.

If certain information cannot be specified using the CSV file format, one workaround to make large sets of changes automatically is to write computer scripts to directly modify BioTapestry's native XML storage (.btp) file, which is human-readable plain text.

6. Microsoft Excel CSV Semicolon Issue

In some configurations, Excel will create a comma-separated value file that doesn't use commas to separate the values. In at least some versions of Windows, Excel chooses to use the regional settings for formatting the CSV file. In some European localizations, this means that semicolons are used instead of commas. If your Windows computer is regionalized to a setting that doesn't use commas in CSV files, use this fix:

(a) Click the Windows *Start* menu.

(b) Click *Control Panel.*

(c) Open the *Regional and Language Options* dialog box.

(d) Click the Regional Options tab.

(e) Click *Customize.*

(f) Type a new separator (a comma) in the *List separator* box.

(g) Click *OK* twice.

7. Using TextEdit to Create Plain Text Files
To use TextEdit on a Mac to create an unformatted plain text file, certain preferences need to be set. With TextEdit running, select *TextEdit→ Preferences…*; on the *New Document* tab, for *Format*, choose the *Plain Text* option, and close the preferences dialog. Then, when selecting *Save As…*, select a plain text encoding of *Unicode (UTF-8)*.

8. Temporal Input Format
One of the key characteristics of both the temporal input XML data and the expression XML data is that just because entries are included for network elements in these files does not mean that these elements get displayed in the network model. For example, the data files may state that *Gene 1* expresses in *Region A*, and that it has two active inputs from *Gene 2* and *Gene 3* during the time period of interest. However, unless *Gene 1* is actually drawn in *Region A* of the VfA parent model, and there are *Gene 2* and *Gene 3* inputs drawn into *Gene 1* in *Region A*, the dynamic information is ignored for the omitted elements.

Other things to note about these files

(a) While the usual XML syntax has separate balanced opening and closing tags (e.g., `<data>…</data>`), note that the shortcut syntax `<data/>` is acceptable when the element does not enclose other elements. This syntax is being used for the `data` and `range` elements in the example.

(b) Unless the optional `internalOnly="yes"` attributes are specified, the dynamic data will be accessible by right-clicking on a node and selecting *Experimental Data* from the popup menu.

(c) Note that the two `baseConfidence` and `timeCourse` attributes are required for the XML `timeCourse` element. The `timeCourse` attribute (which is not currently used) must be set to `no`. The `baseConfidence` attribute has a variety of settings; use `normal` unless you want to take advantage of this feature. The possible `baseConfidence` settings are shown in Table S2, which is available from http://www.BioTapestry.org/MiMB.

(d) The attribute name gene in the `timeCourse` element is something of a misnomer (`name` would be more descriptive) as all network nodes (e.g., genes, bubbles, boxes) need to have a `timeCourse` element defined.

(e) If a node is assigned an expression level `expr="variable"` in a data element, it must also have a value attribute that is assigned a number between 0.0 and 1.0 (inclusive).

(f) The `range` element only specifies two `sign` types: `promote` and `repress`. For neutral links, use the `promote` value. For an example, consider the *Gene 3* input into *Signal* as seen in Fig. 4. The spreadsheet (Table 1) specifies it as `neutral`, while the `sign` of the interaction is specified as `promote` in Table 2.

(g) Note how the *Ubiq* input into *Gene 2* is defined using only a single `range` element, and that element does not include the optional `region` attribute. This is because the time range for this input (0–4) is valid for all regions.

(h) The `range` element also has an optional `source` attribute that can be assigned a region name. This can be useful to differentiate when a node (usually an intercellular node) in a region has inputs from different copies of the same node, each in different regions. The `source` attribute can disambiguate this case.

9. Runtime Issues

In most cases, using Java WebStart to launch BioTapestry is problem-free. However, there are some issues to keep in mind.

If you are trying to run BioTapestry while connected to the Internet through a password-protected wireless router, you will want to make sure to have logged in through your Web browser before proceeding. This will keep the WebStart system from encountering the router login page and incorrectly thinking it is the BioTapestry startup file. The safest approach in these situations is to start BioTapestry without an active Internet connection.

If there are issues getting BioTapestry to run, it is managed through the Java WebStart configuration tool. Accessing this tool depends on the type of computer you have. For Windows, a common path is as follows:

(a) Click the Windows *Start* menu.

(b) Click *Control Panel.*

(c) Open the *Java* dialog box.

(d) Click the *General* tab.

(e) The WebStart system is managed using the *Temporary Internet Files* section.

Clicking on the *Settings...* button takes you to a dialog that allows you to clear the local cache (click *Delete Files...*), which should be tried if BioTapestry is not downloading and starting correctly. Clicking on the *View...* button, and selecting the BioTapestry Editor entry, allows you to do things like create a desktop shortcut for the program. For a Mac, finding the WebStart tool varies due to the installed Java version. The best place to start looking is in the `Applications::Utilities` folder.

10. CSV-Generated Interaction List
When a network model is created using either a CSV import or the *Build Network from Description* dialog, BioTapestry creates an internal interaction list which becomes the fundamental definition of the network model. This is in contrast to models built by drawing, which maintain no such list. If you create a model with an interaction list, then subsequent drawing operations that add or delete network elements have no effect on this underlying definition list, and rebuilding the network from the list will discard any of the drawn changes. Note that this caveat does not apply to layout changes (just moving nodes and links around instead of adding or deleting them), which can be kept when rebuilding the network. So when an interaction list is being used to define the network model, all changes to the network structure (not the layout) should be made through that list, and not via drawing.

It is also important to realize that with CSV imports, any future CSV imports will replace the entire interaction list (though again you can choose to keep the existing layout). Subsequent CSV imports should contain all the original interaction commands (that are not being deleted) in addition to newly added interactions. If you choose to start with a CSV file, but wish to change to drawing, then use the *Tools→Drop All Interaction Tables Used to Build Networks* command. Otherwise, all drawing operations will invoke a warning message about this issue.

11. Layout Independence
It was explained above that the top-level VfG model has its own layout, each VfA model below it has its own independent layout, and all the VfN models under a given VfA share the layout of that VfA. This separation of layout definitions is necessary given the significant variation in network geometry that can occur across the many VfAs, and between the VfAs and the top-level VfG network.

However, there are many situations where sharing of a particular layout across some or all VfAs, or between the VfG and some VfAs, is desirable. Particularly in model hierarchies where

VfAs are designed to contain a single region containing a full copy of the VfG, it is essential to be able to make changes to the VfG layout and propagate them to all the VfAs at once. The synchronization tool for doing this is described in Subheading 3.5.

It is also useful to be able to reorganize the VfG based upon changes that have been made in lower-level models. This is a tougher problem than downward synchronization, since the information needed to lay out the entire VfG may be spread across many fragments of the network in various VfAs. In the typical case, different VfAs will provide conflicting arrangements of common network elements that need to be resolved to generate a single layout for the VfG from the different lower-level versions.

The tool for upwardly propagating layout fragments to create a consistently organized VfG is available by selecting *Layout→ Propagate Layout to Full Genome Model….* The dialog for this operation allows the user to select which VfAs and regions to use to place each network element.

The final type of layout synchronization is between two or more specific regions spread across one, two, or many VfAs. This effect can be achieved when initially adding regions to the models, as multiple regions (with their associated layouts) can be copied within or across VfA models. To copy a single region, right-click within an unoccupied part of the region, and select *Duplicate Region…* from the popup menu. Alternately, select *Edit→ Copy Multiple Regions…*from the main menu.

However, once regions have been created and their contents have potentially diverged, there is no tool to synchronize just these regions; the only current way to do this is by propagating the layout upward to the top-level VfG, then synchronizing the result back down to the desired target region. This should be addressed in a future version of BioTapestry.

12. Layout Synchronization Options

In the example, the options to use for layout synchronization were specified, instead of being explained in detail. Here are the tradeoffs:

(a) *Compress layout in child models*

The basic operation in layout synchronization is to use the global VfG layout as a template for the layout in every region being synchronized. If only a fraction of the full VfG network is present in a region, particularly if the pieces are scattered, then applying the compression step to remove unneeded space will create much more compact regions. The alternative is that the VfG network layout is used directly in the region. If the intention is to create exact VfG copies in a VfA, then compression should be skipped.

(b) *Retain region positions*

If regions have been arranged in a particular order in the VfAs, and it is important to keep them, then this option should be checked. Otherwise, BioTapestry will organize the regions left to right using a hierarchical ordering technique. The downside of retaining region positions is that interregion links often end up being more convoluted.

(c) *Swap link pads*

If this option is checked, the places where links originate and terminate on nodes (link pads) will be copied from the VfG layout. If unchecked, the current assignment used in the VfA is retained. Note that in a VfA, there might be multiple copies of a link inbound to a node, each coming from a different copy of the source node, each in a different region. If the various links have been assigned to different landing pads, then swapping the links will collapse them back to all terminate on the same pad.

13. Signal Nodes

Every node in a VfA belongs to some region. This requirement sometimes raises questions about how to deal with intercellular (i.e., signal) nodes, which might be thought of as existing between regions. The user needs to decide whether to make a signal node a member of the source region, or to add a separate copy to each of the target region(s). This decision is usually dependent on the specifics of a particular modeling situation. Once the signal node is added to the region, a visual presentation trick can be used to make it appear outside of the region: the bounding margin of the containing region can be set to a negative value on the side of the region with the signal.

An issue can arise when creating dynamic submodels that might only include one of several regions, but the user wants to make a signal node in a nonincluded region show up as active, without including the whole region along with it. This can be done by selecting *Edit→Add Extra Dynamic Submodel Node* and then clicking on the signal node while viewing the dynamic submodel in question. This technique assumes that the signal node has been made visible by way of the region boundary padding approach described above.

14. Setting Up Experimental Data Web Pages

The intent of the feature that allows URLs to be assigned to network nodes is that a lab would set up a Web server that hosts a very basic page or set of pages for each gene or node of interest. The URL for each different node would have a unique file name (e.g., *Gene3.html*), and this would be specified on the Data URLs tab. Since the Experimental Data window does not provide an extensive set of HTML support, it is unlikely to produce the desired results if the provided URLs are pointing to arbitrary Web pages. These are the restrictions to keep in mind:

(a) The Web pages should only use basic HTML (italic, bold, paragraph, line break, and simple table tags).

(b) Do not point to Web pages that use Cascading Style Sheets or contain Javascript.

(c) The *bodies* of each separate Web page in the URL list will be excised and concatenated into the body of the single Web page displayed by the Experimental Data window.

(d) Image tags in the pages must be specified using absolute URLs.

(e) Since the Experimental Data display does not currently support even basic Web navigation, there should be no embedded hypertext links in the page.

(f) The Java security model used by the BioTapestry Viewer imposes a restriction that the Web pages and images specified for a node must be located on the same Web server as the one hosting the Viewer. The Editor, which is signed with a security certificate and has enhanced permissions, does not have this restriction. Be sure to check this if a model developed with the Editor is then deployed using the Viewer.

15. Use PNG, not JPG, for Image Exports
Images of BioTapestry models are basically line art, not photographs. The JPG image format is designed for the latter, and using it to export an image of a BioTapestry model (even without compression) introduces artifacts that noticeably reduce the quality of the image. The best way to export BioTapestry images is with PNG. If a publication accepts only JPG or TIFF images, exporting with PNG and then using a program such as Adobe Photoshop or GIMP to convert the image to TIFF will provide superior results.

Acknowledgments

BioTapestry development is a collaboration between the Institute for Systems Biology and the Davidson Lab at the California Institute of Technology; many thanks are due to Dr. Eric Davidson for his vision, leadership, and support. Also, thanks to Dr. Hamid Bolouri for his excellent comments and suggestions on this manuscript. BioTapestry is supported by NIGMS grant GM061005.

References

1. Longabaugh, W. J. R., Davidson, E. H., and Bolouri, H. (2005) Computational representation of developmental genetic regulatory networks, *Developmental biology* ***283***, 1–16.
2. Longabaugh, W. J. R., Davidson, E. H., and Bolouri, H. (2009) Visualization, documentation, analysis, and communication of large-scale gene regulatory networks, *Biochimica et biophysica acta* ***1789***, 363–374.
3. Davidson, E. H., Rast, J. P., Oliveri, P., Ransick, A., Calestani, C., Yuh, C. H., Minokawa, T., Amore, G., Hinman, V., Arenas-Mena, C., Otim, O., Brown, C. T., Livi, C. B., Lee, P. Y., Revilla, R., Rust, A. G., Pan, Z., Schilstra, M. J., Clarke, P. J., Arnone, M. I., Rowen, L., Cameron, R. A., McClay, D. R., Hood, L., and Bolouri, H. (2002) A genomic regulatory network for development, *Science* ***295***, 1669–1678. URL: http://sugp.caltech.edu/endomes/index.html
4. Shannon, P., Markiel, A., Ozier, O., Baliga, N. S., Wang, J. T., Ramage, D., Amin, N., Schwikowski, B., and Ideker, T. (2003) Cytoscape: a software environment for integrated models of biomolecular interaction networks, *Genome research* ***13***, 2498–2504.
5. Vokes, S. A., Ji, H., McCuine, S., Tenzen, T., Giles, S., Zhong, S., Longabaugh, W. J. R., Davidson, E. H., Wong, W. H., and McMahon, A. P. (2007) Genomic characterization of Gli-activator targets in sonic hedgehog-mediated neural patterning, *Development* ***134***, 1977–1989. URL: http://www.mcb.harvard.edu/McMahon/BioTapestry/index.html
6. Bonneau, R., Facciotti, M. T., Reiss, D. J., Schmid, A. K., Pan, M., Kaur, A., Thorsson, V., Shannon, P., Johnson, M. H., Bare, J. C., Longabaugh, W., Vuthoori, M., Whitehead, K., Madar, A., Suzuki, L., Mori, T., Chang, D. E., Diruggiero, J., Johnson, C. H., Hood, L., and Baliga, N. S. (2007) A predictive model for transcriptional control of physiology in a free living cell, *Cell* ***131***, 1354–1365. URL: http://baliga.systemsbiology.net/drupal/content/egrin
7. Georgescu, C., Longabaugh, W. J. R., Scripture-Adams, D. D., David-Fung, E. S., Yui, M. A., Zarnegar, M. A., Bolouri, H., and Rothenberg, E. V. (2008) A gene regulatory network armature for T lymphocyte specification, *Proceedings of the National Academy of Sciences of the United States of America* ***105***, 20100–20105. URL: http://www.its.caltech.edu/~tcellgrn/TCellMap.html
8. Chan, T. M., Longabaugh, W., Bolouri, H., Chen, H. L., Tseng, W. F., Chao, C. H., Jang, T. H., Lin, Y. I., Hung, S. C., Wang, H. D., and Yuh, C. H. (2009) Developmental gene regulatory networks in the zebrafish embryo, *Biochimica et biophysica acta* ***1789***, 279–298. URL: http://www.zebrafishworld.org/
9. Levine, M., and Davidson, E. H. (2005) Gene regulatory networks for development, *Proceedings of the National Academy of Sciences of the United States of America* ***102***, 4936–4942.
10. Stathopoulos, A., and Levine, M. (2005) Genomic regulatory networks and animal development, *Developmental cell* ***9***, 449–462.
11. Wellmer, F., Alves-Ferreira, M., Dubois, A., Riechmann, J. L., and Meyerowitz, E. M. (2006) Genome-wide analysis of gene expression during early Arabidopsis flower development, *PLoS genetics* ***2***, e117.
12. Mori, A. D., Zhu, Y., Vahora, I., Nieman, B., Koshiba-Takeuchi, K., Davidson, L., Pizard, A., Seidman, J. G., Seidman, C. E., Chen, X. J., Henkelman, R. M., and Bruneau, B. G. (2006) Tbx5-dependent rheostatic control of cardiac gene expression and morphogenesis, *Developmental biology* ***297***, 566–586.
13. Hoffman, B. G., Zavaglia, B., Witzsche, J., Ruiz de Algara, T., Beach, M., Hoodless, P. A., Jones, S. J., Marra, M. A., and Helgason, C. D. (2008) Identification of transcripts with enriched expression in the developing and adult pancreas, *Genome biology* ***9***, R99.
14. Kioussi, C., and Gross, M. K. (2008) How to build transcriptional network models of mammalian pattern formation, *PloS one* ***3***, e2179.
15. Smith, J., and Davidson, E. H. (2008) Gene regulatory network subcircuit controlling a dynamic spatial pattern of signaling in the sea urchin embryo, *Proceedings of the National Academy of Sciences of the United States of America* ***105***, 20089–20094.
16. Ririe, T. O., Fernandes, J. S., and Sternberg, P. W. (2008) The Caenorhabditis elegans vulva: a post-embryonic gene regulatory network controlling organogenesis, *Proceedings of the National Academy of Sciences of the United States of America* ***105***, 20095–20099.
17. Materna, S. C., and Oliveri, P. (2008) A protocol for unraveling gene regulatory networks, *Nature protocols* ***3***, 1876–1887.
18. Tamplin, O. J., Kinzel, D., Cox, B. J., Bell, C. E., Rossant, J., and Lickert, H. (2008) Microarray analysis of Foxa2 mutant mouse embryos reveals novel gene expression and inductive roles for the gastrula organizer and its derivatives, *BMC genomics* ***9***, 511.

19. Bolouri, H. (2008) *Computational Modeling of Gene Regulatory Networks: a Primer*, Imperial College Press.
20. Morley, R. H., Lachani, K., Keefe, D., Gilchrist, M. J., Flicek, P., Smith, J. C., and Wardle, F. C. (2009) A gene regulatory network directed by zebrafish No tail accounts for its roles in mesoderm formation, *Proceedings of the National Academy of Sciences of the United States of America* ***106***, 3829–3834.
21. Su, Y. H., Li, E., Geiss, G. K., Longabaugh, W. J. R., Krämer, A., and Davidson, E. H. (2009) A perturbation model of the gene regulatory network for oral and aboral ectoderm specification in the sea urchin embryo, *Developmental biology 329*, 410–421.
22. Sansom, S. N., Griffiths, D. S., Faedo, A., Kleinjan, D. J., Ruan, Y., Smith, J., van Heyningen, V., Rubenstein, J. L., and Livesey, F. J. (2009) The level of the transcription factor Pax6 is essential for controlling the balance between neural stem cell self-renewal and neurogenesis, *PLoS genetics* **5**, e1000511.
23. Davidson, E. H. (2006) *The Regulatory Genome: Gene Regulatory Networks in Development and Evolution*, Elsevier.
24. Shannon, P. T., Reiss, D. J., Bonneau, R., and Baliga, N. S. (2006) The Gaggle: an open-source software system for integrating bioinformatics software and data sources, *BMC bioinformatics 7*, 176.

Part V

Modeling of GRNs

Chapter 22

Implicit Methods for Qualitative Modeling of Gene Regulatory Networks

Abhishek Garg, Kartik Mohanram, Giovanni De Micheli, and Ioannis Xenarios

Abstract

Advancements in high-throughput technologies to measure increasingly complex biological phenomena at the genomic level are rapidly changing the face of biological research from the single-gene single-protein experimental approach to studying the behavior of a gene in the context of the entire genome (and proteome). This shift in research methodologies has resulted in a new field of network biology that deals with modeling cellular behavior in terms of network structures such as signaling pathways and gene regulatory networks. In these networks, different biological entities such as genes, proteins, and metabolites interact with each other, giving rise to a dynamical system. Even though there exists a mature field of dynamical systems theory to model such network structures, some technical challenges are unique to biology such as the inability to measure precise kinetic information on gene–gene or gene–protein interactions and the need to model increasingly large networks comprising thousands of nodes. These challenges have renewed interest in developing new computational techniques for modeling complex biological systems.

This chapter presents a modeling framework based on Boolean algebra and finite-state machines that are reminiscent of the approach used for digital circuit synthesis and simulation in the field of very-large-scale integration (VLSI). The proposed formalism enables a common mathematical framework to develop computational techniques for modeling different aspects of the regulatory networks such as steady-state behavior, stochasticity, and gene perturbation experiments.

Key words: Gene regulatory networks, Signaling pathways, Boolean networks, BDDs, Synchronous, Stochasticity, Cell differentiation, T-helper network

1. Gene Regulatory Networks

A protein that activates (or inhibits) the functionality of another gene/protein is said to interact with the target gene/protein by an activating (or an inhibiting) mechanism. A graphical representation where the nodes of the graph represent the functional form of a protein and directed edges represent the activation (or inhibition)

Bart Deplancke and Nele Gheldof (eds.), *Gene Regulatory Networks: Methods and Protocols*, Methods in Molecular Biology, vol. 786, DOI 10.1007/978-1-61779-292-2_22,

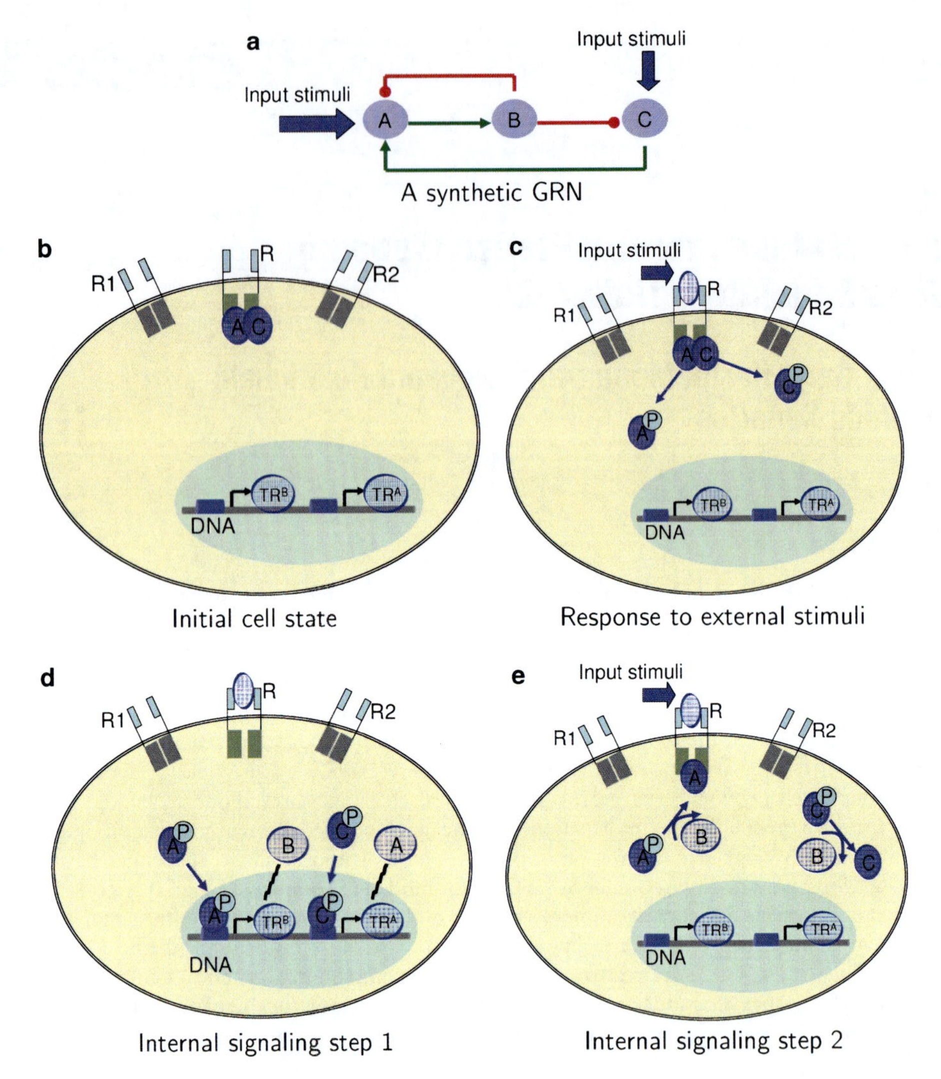

Fig. 1. (**a**) A gene regulatory network (GRN). (**b**) A resting cell state. (**c–e**) Cellular signaling in response to input stimuli. R, R1, and R2 are the cell surface receptors, TR^A and TR^B are the genes encoding proteins A and B. A^P and C^P represent the phosphorylated form of proteins A and C, respectively.

mechanism is commonly referred to as a *gene regulatory network* (GRN). In this chapter, we use the terminology GRNs to refer to combination of both the signaling pathways (a term traditionally used for interactions among the proteins) and GRNs (a term traditionally used for interactions among the proteins such as transcription factors and the genes). A small synthetic GRN is illustrated in Fig. 1a, where activating edges are represented by arrow-headed lines and inhibiting edges are represented by circle-headed lines.

Biological phenomena underlying different interactions in the GRN of Fig. 1a are shown in Fig. 1b–e. Figure 1b represents the cell state in the absence of any input stimuli. When an input stimulus is sensed at a surface receptor (Fig. 1c), proteins A and C dissociate from the membrane and get phosphorylated. In the next time instance (Fig. 1d), phosphorylated A and C can transcriptionally express protein B and A, respectively. This is represented in the GRN by activating edges from A-to-B and C-to-A. Finally (in Fig. 1d), the expression of protein B dephosphorylates A and C, leading to loss in their functionality. This mechanism is represented by inhibiting edges from B-to-A and B-to-C.

The GRNs, such as the one in Fig. 1a, summarize known interactions among a set of proteins that participate in a biological phenomenon. Such a GRN represents a dynamical system where a node changes its value over time depending upon the state of the neighboring nodes in the network. By modeling these GRNs as a dynamical system, one can capture some aspects of cellular dynamics. GRN modeling techniques in the literature can be broadly categorized into continuous and discrete approaches.

In continuous modeling (or more commonly known as kinetic modeling) techniques, methods such as ordinary differential equations (ODEs) and partial differential equations (PDEs) are used for modeling the continuous evolution of the concentration of proteins in a GRN. These models use parameters such as rate constant of chemical reactions, membrane diffusion constants, and protein degradation rates to simulate the kinetics of the system. Kinetic models have been applied successfully in the past for modeling various GRNs (1–3). However, their applications have been restricted to only small well-studied GRNs due to the fact that kinetic parameters are not known for most of the interactions in a GRN. Kinetic parameters for all the gene–gene or gene–protein interactions in a GRN are rarely available in the literature, and measuring these parameters is currently infeasible using the available experimental techniques.

In discrete modeling of GRNs, a node can take only discrete expression values, unlike the continuous range of values that are possible in kinetic models. In the most restrictive case, a node exists in only two expression states corresponding to the active and inactive gene (or protein), represented by Boolean 1 and 0, respectively. In Boolean models, an inhibiting edge can change the expression of the node from 1-to-0 and an activating edge can change the state from 0-to-1 (only when no inhibition is present). If the expression of a gene (or protein) needs to be modeled at more than two expression levels, then Boolean models can be extended to multiple-valued networks with expression states such as low, medium, and high. Various formulations of Boolean networks have been proposed in the literature (4–12).

The decision to use continuous or discrete modeling approaches to a given problem depends upon how much information is available with respect to the given biological problem and what is desired from the modeling effort. Kinetic models are more suitable for modeling the quantitative aspects, e.g., to estimate the

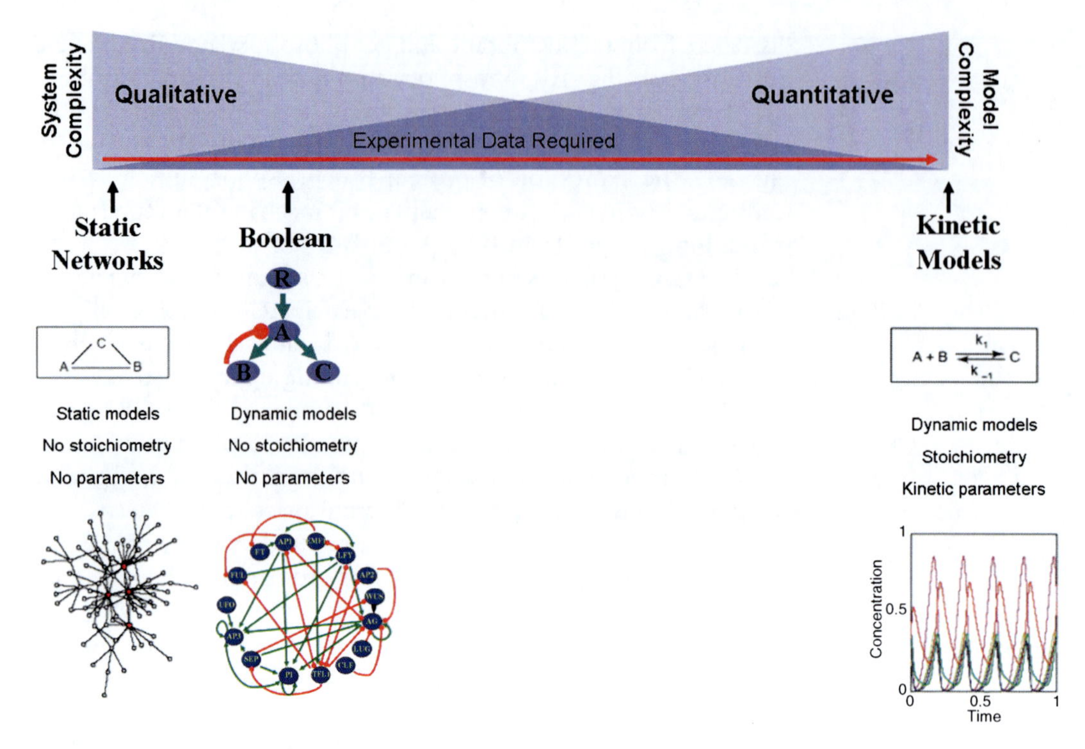

Fig. 2. Different modeling techniques for simulating GRNs categorized on the basis of their complexity and the complexity of the systems they can be modeled (adapted from ref. 13).

concentration of a drug required for a desired cellular response or the amount of protein produced. In contrast, Boolean models are useful in studying the qualitative aspects of biological phenomena, e.g., determining the active proteins in a steady state or understanding the steady-state protein activity changes in the presence of gene knockdown and overexpression.

Figure 2 summarizes a few commonly used modeling techniques for GRNs and where they stand with respect to their complexity and the complexity of systems that can be modeled using these techniques. One needs significantly more detailed information about the gene–gene, gene–protein, or protein–protein interactions in the continuous models as compared to discrete models. This makes the continuous models such as ODEs more complex to build than the discrete models such as Boolean networks. Model complexity along with the availability of experimental data determines the complexity of biological phenomena (or the system complexity) that can be simulated using that model. While the largest systems that have been successfully modeled using ODEs had at most 20 nodes in a GRN, Boolean modeling has been applied on GRNs with over hundreds of nodes.

At the other extreme of the spectrum are the static networks such as protein–protein interaction networks which capture the static interactions independent of the direction of the interaction (or flow of signaling event). Such networks are available for very large biological systems such as the whole yeast proteomic network or even the human protein–protein interaction network. However, only a limited type of analyses such as the critical proteins (defined in terms of hubs), highly correlated genes/proteins (defined by the clusters), and small world network analysis can be performed on such interaction networks.

Variants of Boolean modeling such as multiple-valued networks and combined Boolean-ODE modeling techniques have been proposed to reduce the gap between pure Boolean and pure ODE type of modeling approaches. However, in practice, no single modeling technique can capture all aspects of a biological phenomenon and multiple methods are often used during different phases of hypothesis generation and testing. This chapter describes some of the algorithms for Boolean modeling of GRNs and demonstrates the power of qualitative modeling by applying these algorithms for modeling T-helper cell differentiation GRNs.

2. Boolean Modeling

2.1. Boolean Mapping of GRNs

Edges in a GRN represent biological phenomena underlying the activation (or inhibition) of the functional form of a gene (or protein). Almost all the underlying biological functionalities of a GRN can be represented by a combination of Boolean functions (or logic gates) in the set {AND, OR, IAND, BUFF, NOT}. Figure 3 illustrates a few instances of mapping between biological functionalities and its corresponding GRN and Boolean function representation. For example, in the first row of Fig. 3, the blue protein loses its activity after undergoing degradation by the green protein. This can be represented by a NOT gate, which shows negative influence of the green protein on the blue protein. Similarly, when a protein undergoes phosphorylation or dephosphorylation upon interacting with another protein, it can either lose its activity (first row) or gain its activity (second row). Hence, phosphorylation and dephosphorylation are categorized under both the NOT gate and the BUFF gate (which stands for buffer). Scaffolding and a promoter-assisted gene transcription processes (in the third row) require assembling of two or more proteins into multimolecular complexes. This can be represented by an AND gate requiring the simultaneous presence of two or more inputs. If the presence of a protein has an inhibiting effect on another protein–protein interaction (e.g., phosphorylation blocked by the presence of red protein in the fourth row) or gene–protein interaction (e.g., transcription blocked by the red protein), then it can be represented by an IAND gate. Finally, the Boolean function

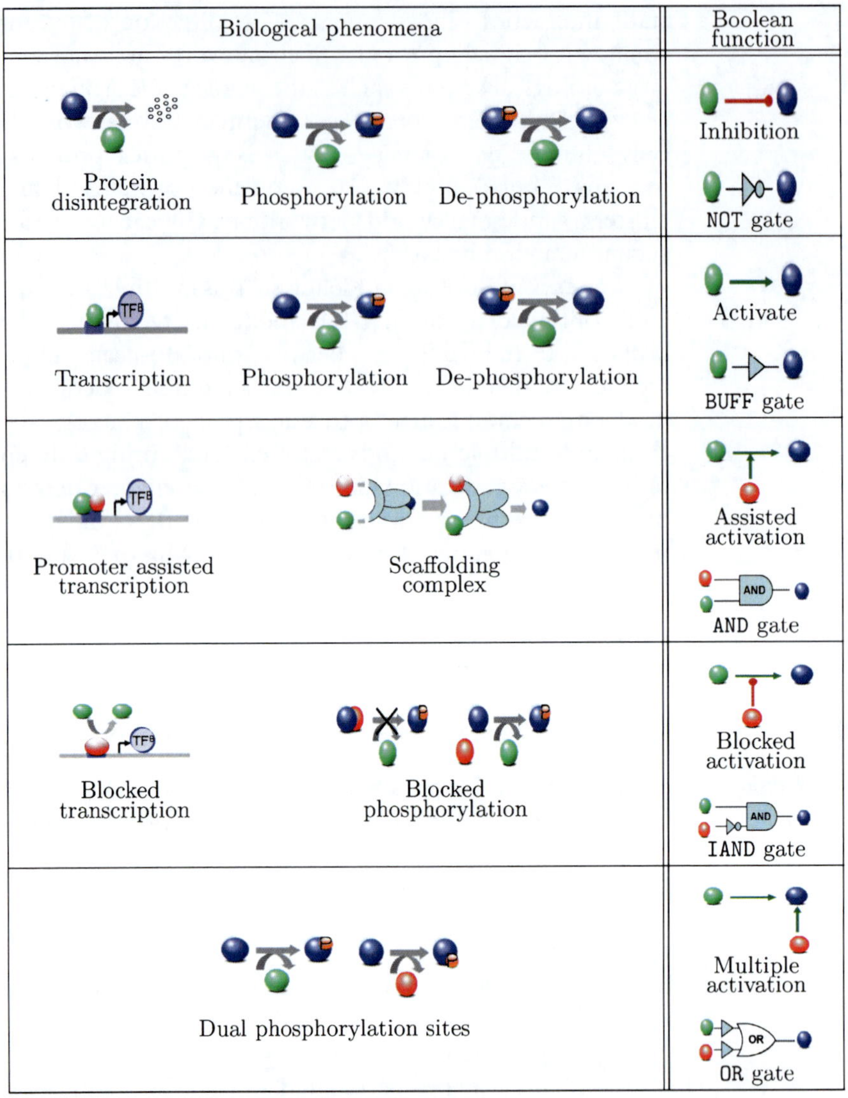

Biological phenomena	Boolean function
Protein disintegration; Phosphorylation; De-phosphorylation	Inhibition; NOT gate
Transcription; Phosphorylation; De-phosphorylation	Activate; BUFF gate
Promoter assisted transcription; Scaffolding complex	Assisted activation; AND gate
Blocked transcription; Blocked phosphorylation	Blocked activation; IAND gate
Dual phosphorylation sites	Multiple activation; OR gate

Fig. 3. Boolean function mapping of biological phenomena in GRNs.

OR represents a choice between alternate biological mechanisms to activate or deactivate a protein. For example, in the fifth row, blue protein can be phosphorylated at two different sites by two different proteins. Phosphorylation at any site can activate the blue protein. This phenomena is captured by the OR gate. However, the OR gate in itself does not have any biological meaning. The dual phosphorylation sites are captured by two BUFF gates in the fifth row of Fig. 3. It should be noted that the logic gates {AND, OR, IAND} can have two or more inputs, while {BUFF, NOT} are only single input functions. With the mapping defined in Fig. 3, a synthetic GRN in Fig. 4a can be translated into a logic gate representation as in Fig. 4b. The shaded logic functions in Fig. 4b correspond to a biological function.

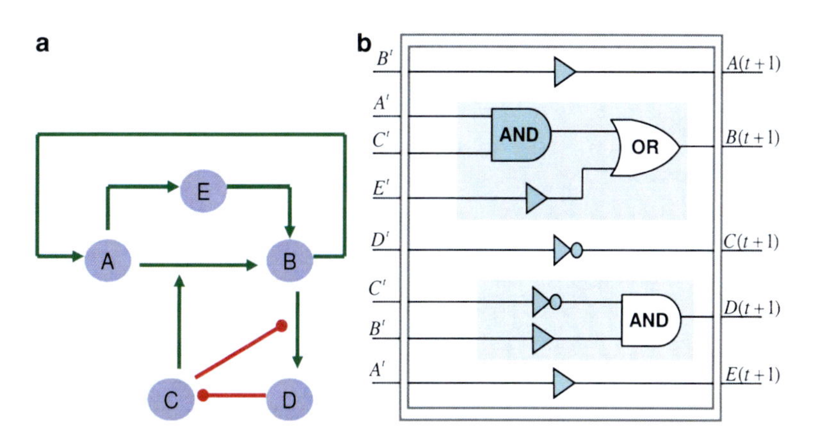

Fig. 4. (**a**) A gene regulatory network (GRN). (**b**) The GRN mapped to Boolean.

2.2. Network States and Steady States

In Boolean modeling of GRNs, a node can have only two expression states: active and inactive. These expression states of a node are represented by 1 and 0, respectively. A snapshot of the activity level of all the nodes in the GRN at time t is called the state of the network. The network can transition from one state to another state as defined by the underlying Boolean functions. Hence, a state of the network evolves over time by making a transition into another state until it stabilizes into an attractor (or the steady state). Such an attractor represents the long-term behavior of the genes/proteins in the regulatory networks. Under the Boolean assumption, if there are N genes in a GRN, the network can be in any of the 2^N possible states. This exponential state space makes the identification of attractors both computationally and memory intensive.

In biology, a state of the network corresponds to the measured activity of all the genes (or proteins) in a cell at a given time instant and can be experimentally measured using techniques such as flow cytometry or microarray. The state transition of a GRN biologically corresponds to the dynamic evolution of the cell. With the existing experimental techniques, it is difficult to monitor all the state transitions inside a cell. However, steady-state behavior, which corresponds to the end point of an experiment when all cells stabilize, is easier to measure experimentally. Furthermore, measurement and comparison of the transient states across multiple experiments are often difficult as the dynamics may vary in each experiment. Steady-state behavior, which corresponds to the end point of an experiment when all cells stabilize, is easier to measure and compare with similar experiments. In the dynamic simulation of GRNs, a state of the network evolves over time and stabilizes in an attractor (or the steady state). Hence, an attractor represents the long-term behavior of the genes/proteins in the regulatory networks. Attractors (or the steady states) of Boolean networks are hypothesized to correspond to the cellular steady states (or phenotypes) (14, 15).

2.3. Problem Formulation

Given a GRN (such as in Fig. 4a), expression of a node (or gene) i at time t is represented by a Boolean variable x^t_i. If $x^t_i = 0$, the node i is inactive and if $x^t_i = 1$, the node i is active. The state of the network at time t is represented by a Boolean vector, $\mathbf{x}^t$, of size N (number of genes in the network) and is referred to as a present state vector. Each bit of this vector (represented by x^t_i) represents whether the gene is active or inactive. Another Boolean vector, $\mathbf{x}^{t+1}$, of size N is used to represent the state of the network in the next step and is called the next state vector. The expression of each gene i at time $t+1$ can be written as a function $x_i(t+1)$ of the state of the genes acting as its input at time t. Equations. 1 and 2 can be used to compute the function $x_i(t+1)$ and can be understood better with the help of Fig. 4 and Example 1. In Eq. 1, an activator (or an inhibitor) function $f^{\text{ac}}_{x_i}(t)$ (or $f^{\text{in}}_{x_i}(t)$) represent the set of genes that have a collective activating (or inhibiting) impact on the gene i.

$$x_i(t+1) = \left(\bigvee_{l=1}^{n} f^{\text{ac}}_{x_{i,1}}(t) \right) \wedge \neg \left(\bigvee_{l=1}^{n} f^{\text{in}}_{x_{i,l}}(t) \right) \tag{1}$$

$$f^{\text{ac,in}}_{x_{i,l}}(t) = \left(\bigwedge_{j=1}^{p} x_j^{t,\text{ac}} \right) \wedge \left(\bigwedge_{j=1}^{p} \neg x_j^{t,\text{in}} \right) \tag{2}$$

$$x_j \in \{0,1\}$$

$f^{\text{ac}}_{x_m}$ are the set of activator functions of x_i,
$f^{\text{in}}_{x_n}$ are the set of inhibitor functions of x_i,
x^{ac}_p are the set of activators of functions f_{x_i},
x^{in}_q are the set of inhibitors of functions
f_{x_i}, and $\wedge$, $\vee$, and $\neg$ represent Boolean AND, OR, and NOT

Example 1

For the synthetic GRN in Fig. 4, Boolean functions describing the expression of nodes are given by the following equations:

$$x_{\text{A}}(t+1) = x^t_{\text{B}}$$
$$x_{\text{B}}(t+1) = \left(x^t_{\text{A}} \wedge x^t_{\text{C}}\right) \vee x^t_{\text{E}}$$
$$x_{\text{C}}(t+1) = \neg x^t_{\text{D}}$$
$$x_{\text{D}}(t+1) = x^t_{\text{B}} \wedge \neg x^t_{\text{C}}$$
$$x_{\text{E}}(t+1) = x^t_{\text{A}}$$

Boolean variables x^t_{A}, x^t_{B}, x^t_{C}, x^t_{D}, and x^t_{E} represent the expression of nodes A, B, C, D, and E, respectively.

Equations 1 and 2 represent the dynamics of individual genes independent of the dynamics of the other genes in the network. To model the dynamics of the complete network, one has to couple the dynamics of these genes. This can be done by defining a transition function, $T(\mathbf{x}^t, \mathbf{x}^{t+1})$, of the state of the network. Function $T(\mathbf{x}^t, \mathbf{x}^{t+1})$ represents the transition from the present state $\mathbf{x}^t$ to the next state $\mathbf{x}^{t+1}$.

2.4. State Transition Function

If the transition function is synchronous, expression of all genes is updated at the same time. A synchronous model can be described by the following set of equations:

$$T_i\left(\mathbf{x}^t, \mathbf{x}^{t+1}\right) = x_i^{t+1} \leftrightarrow x_i(t+1) \quad (3)$$

$$T(\mathbf{x}^t, \mathbf{x}^{t+1}) = T_0(\mathbf{x}^t, \mathbf{x}^{t+1}) \wedge \ldots \wedge T_N(\mathbf{x}^t, \mathbf{x}^{t+1}) \quad (4)$$

Equation 3 gives the transition function for a single gene i. Symbol $\leftrightarrow$ stands for logical equivalence and in Eq. 3 represents that the value of a gene in the next time step, x_i^{t+1} is equal to the value of the function $x_i(t+1)$ (which in turn is defined as in Eq. 1). Equation 4 states that all genes in the network make a simultaneous transition from the present state $\mathbf{x}^t$ to the next state $\mathbf{x}^{t+1}$.

If a state transition graph is constructed using Eqs. 3 and 4, then each state has only one outgoing transition. Hence, assuming that all genes can take "0" or "1" levels of expression, the number of states and the number of transitions in the state transition graph can both equal 2^N, where N is the number of genes in the network. The state transition diagram for the synthetic GRN in Fig. 4a is shown in Fig. 5. The N bit state vector $\mathbf{x}^t$ is packed into an integer counterpart in Fig. 5 for aesthetical reasons.

2.5. Boolean Attractors

Boolean attractors and steady states of a GRN can be described formally by the following set of definitions:

Definition 1 Successor Given a state of the network $\mathbf{x}^t$, all the states $\tilde{\mathbf{x}}^t$ such that $T(\mathbf{x}^t, \tilde{\mathbf{x}}^t) = 1$ are the successor states of the state $\mathbf{x}^t$.

Definition 2 Predecessor Given a state of the network $\mathbf{x}^t$, all the states $\tilde{x}^t$ such that $T(\tilde{\mathbf{x}}^t, \mathbf{x}^t) = 1$ are the predecessor states of the state $\mathbf{x}^t$.

Definition 3 Forward Image Given a set of states $S(\mathbf{x}^t)$, the forward image $I_T^f(S(\mathbf{x}^t))$ is the set of immediate successors of the states in the set $S(\mathbf{x}^t)$ under the state transition graph defined by the transition function T.

Definition 4 Backward Image Given a set of states $S(\mathbf{x}^t)$, the backward image $I_T^b(S(\mathbf{x}^t))$ is the set of immediate predecessors of the states in the set $S(\mathbf{x}^t)$ under the state transition graph defined by the transition function T.

Definition 5 Forward Reachable States Given a set of states S_0, forward reachable states $\text{FR}(S_0)$ are the set of states that can be reached from the states in the set S_0 by iteratively computing the

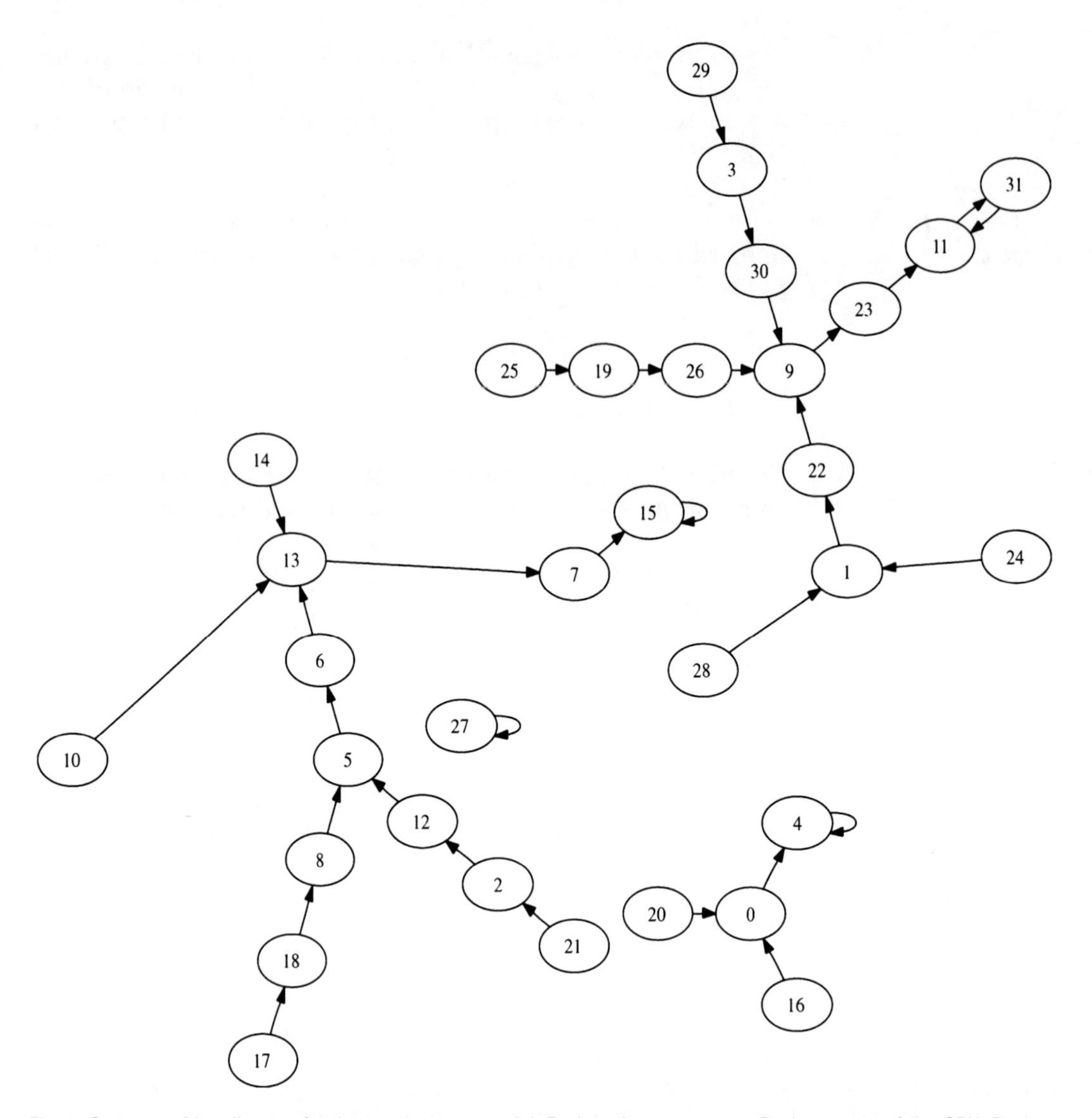

Fig. 5. State transition diagram for the synchronous model. Each node represents a Boolean state of the GRN. Boolean states of the GRN are represented using their decimal counterpart using the following rule: $A.2^0 + B.2^1 + C.2^2 + E.2^3 + D.2^4$. For example, Boolean state {A = 0, B = 0, C = 0, E = 0, D = 0} is represented by the node labeled as 0; and Boolean state {A = 1, B = 0, C = 0, E = 0, D = 0} is represented by the state labeled as 1.

forward image under the transition function T until no new states are reachable.

Definition 6 Backward Reachable States Given a set of states S_0, the backward reachable states BR(S_0) are the set of states $\mathbf{x}^t$ whose forward reachable set contains at least one state in S_0.

Definition 7 Attractor An attractor is a set of states SS($\mathbf{x}^t$) such that for all the states $s \in$ SS($\mathbf{x}^t$), the forward reachable set FR(s) is the same as SS($\mathbf{x}^t$) (i.e., FR(s) = SS($\mathbf{x}^t$) $\forall s \in$ SS($\mathbf{x}^t$)).

Definition 8 A steady state is an attractor that consists of a single state.

Example 2

Let us assume that the system is in a state $S_0 = 5$ in the synchronous state transition diagram of Fig. 5. The successor of state S_0 is the state 6 and the predecessors of S_0 are the states 8 and 12. The set of forward reachable states from S_0 is given by FR(S_0) = {6, 13, 7, 15} and the set of backward reachable states is given by BR(S_0) = {17, 18, 8, 12, 2, 21}. Since, FR(S_0) ≠ S_0, the starting state S_0 does not belong to an attractor.

There are four attractors in Fig. 5, given by $SS_1 = \{4\}$, $SS_2 = \{27\}$, $SS_3 = \{15\}$, and $SS_4 = \{11, 31\}$. Of these four attractors, SS_1, SS_2, and SS_3 are the steady states.

Following Definition 7, a single-state attractor forms a self-loop, otherwise attractor is an oscillating set of states. A self-loop attractor is called a steady state (Definition 8).

Computing attractors (or steady states) of a GRN is of interest to biologists as attractors have a biological correspondence to cell states (or cell phenotypes) (14, 15). Furthermore, since the steady state of a GRN corresponds to the end point of an experiment when all cells stabilize, it is easier to experimentally validate the steady-state behavior than it is to validate the behavior of transient states of a GRN. In the next section, we give algorithms that can be used for identifying all the attractors in a GRN.

2.6. Algorithms for Computing Attractors

As explained in Subheading 2.4, a state transition graph can have an exponential state space, and if this graph is explicitly represented and traversed, then an exponential number of states restricts the computation to small-sized networks. Furthermore, identifying attractors by enumeration becomes difficult as one will have to consider all possible subsets of states that can form an attractor (which can be super-exponential in the worst case). To avoid explicit enumeration of subsets of states, a set of theorems proposed in (15) can be used (stated again in Theorems 1 and 2).

Theorem 1 A state $i \approx S$ is a part of an attractor if and only if $FR(i) \subseteq BR(i)$. State i is transient otherwise.

Theorem 2 If state $i \approx S$ is transient, then states in BR(i) are all transient. If state i is a part of an attractor, then all the states in FR(i) are also part of the same attractor. In the latter case, set $\{BR(i) - FR(i)\}$ has all the transient states.

Example 3

Let us assume that we would like to test whether the state $S_0 = 5$ in the synchronous state transition diagram of Fig. 5 belongs to an attractor. From Example 2, FR(S_0) = {6, 13, 7, 15} and

$BR(S_0) = \{17, 18, 8, 12, 2, 21\}$. As we can clearly see $FR(S_0) = BR(S_0)$. Hence, Theorem 1 implies that the state S_0 does not belong to an attractor. Then Theorem 2 implies that none of the states in $BR(S_0) = \{17, 18, 8, 12, 2, 21\}$ should belong to an attractor either. One can verify from attractors listed in Example 2 that none of the states in the set $BR(S_0)$ indeed belong to any of the four attractors.

Based on Theorems 1 and 2, the procedure for attractor computation is given in Algorithm 1. This algorithm takes as input the transition function $T(\mathbf{x}^t, \mathbf{x}^{t+1})$. In Line 5 of Algorithm 1, a seed state is selected from the state space T^f, and forward and backward reachable states from this seed state are computed in Lines 6 and 7. Then Theorem 1, as implemented in Line 8, checks whether the seed state (from Line 5) is a part of an attractor. If the seed state is indeed a part of an attractor, then using Theorem 2 (as implemented in Lines 9–12) all the states in the forward reachable set are declared to form an attractor in Line 9, and the rest of the states in the backward reachable set are declared transient in Line 10. Otherwise, the seed state and all the other states in the backward reachable set are declared transient in Line 12. In Line 13, the state space is reduced by removing those states that have already been tested for reachability, and the process is repeated to find another attractor on the reduced state space. This process is iterated until the whole state space is explored (i.e., until $T \neq \emptyset$). The states in the backward reachable set are removed from the state space in each iteration, resulting in the continuous reduction in size of the latter. One should note that the number of iterations of Lines 4–13 depends upon how the seed state is selected in Line 5.

Function initial_state() in Algorithm 1 selects a seed state from the given state space T^f. In this function (implemented in Lines 17–25), a random initial state is selected from the transition state space T in Line 17. The forward reachable set from this random initial state is then computed in Lines 19–24. During the forward set computation, when the frontier set evaluates to $\emptyset$ in iteration k, a random state is taken from the frontier set in iteration $k-1$ and returned as the seed state. The motivation behind this function is that a state in the last frontier set is more likely to be a part of an attractor than a random state in the state space T. For synchronous models, it can be proved that the seed state selected in this way is guaranteed to be a part of an attractor.

Algorithm 1 also uses functions forward_set() and backward_set() for computing the forward reachable states $FR(S)$ and backward reachable states $BR(S)$, respectively. These functions are given in Algorithm 2. In Algorithm 2, the *while* loop in Lines 6–10 computes the reachable states iteratively starting from the initial set of states S^0, where kth iteration represents the states reachable in

Algorithm 1: Algorithm for computing Attractors

1 all_attractors(T)
2 **begin**
3 $T' \longleftarrow T$
4 **while** $T' \neq \emptyset$ **do**
5 $s \longleftarrow$ initial_state(T')
6 $FR(s) \longleftarrow$ forward_set(s, T')
7 $BR(s) \longleftarrow$ backward_set(s, T')
8 **if** $FR(s) \wedge \overline{BR(s)} = \emptyset$ **then**
9 report $FR(s)$ as an attractor
10 report $BR(s) \wedge \overline{FR(s)}$ as all transient states
11 **else**
12 report $s \vee BR(s)$ as all transient states
13 $T' \longleftarrow T' \wedge \overline{s \vee BR(s)}$
14 **end**
15 initial_state(T)
16 **begin**
17 $s(V_t)$ = random_state(T)
18 $RS^{(0)} \longleftarrow \emptyset, FS^{(0)} \longleftarrow \{s\}$
19 $k \longleftarrow 0$
20 **while** $FS^{(k)} \neq \emptyset$ **do**
21 $FS^{(k+1)} = I_T^f(FS^{(k)})(\mathbf{x}^{t+1} \leftarrow \mathbf{x}^t) \wedge \overline{RS^{(k)}}$
22 $RS^{(k+1)} = RS^{(k)} \vee FS^{(k+1)}$
23 $k \longleftarrow k+1$
24 $s \longleftarrow$ random_state$(FS^{(k-1)})$
25 **return** s
26 **end**

Algorithm 2: Computing Forward and Backward reachable sets

1 forward_set$(\mathrm{S}^0, \mathrm{T})$
2 /* backward_set$(\mathrm{S}^0, \mathrm{T})$ */
3 **begin**
4 $RS^{(0)} \longleftarrow \emptyset, FS^{(0)} \longleftarrow \{S^0\}$
5 $k \longleftarrow 0$
6 **while** $FS^{(k)} \neq \emptyset$ **do**
7 $FS^{(k+1)} = I_T^f(FS^{(k)})(x^{t+1} \leftarrow x^t) \wedge \overline{RS^{(k)}}$
8 /* $FS^{(k+1)} = I_T^b(FS^{(k)})(x^t \leftarrow x^{t+1}) \wedge \overline{RS^{(k)}}$ */
9 $RS^{(k+1)} = RS^{(k)} \vee FS^{(k+1)}$
10 $k \longleftarrow k+1$
11 **return** $(FR(S^0) \longleftarrow RS^{(k)})$
12 /* **return** $(BR(S^0) \longleftarrow RS^{(k)}(x^{t+1} \leftarrow x^t))$ */
13 **end**

k time steps from S^0. The set of states FS^k and RS^k represents the frontier set and the reachable set, respectively, in the kth iteration of the while loop. The Frontier set in the kth iteration contains the states that have been reached for the first time in the $(k-1)$th iteration of the while loop. The reachable set in the kth iteration contains all reachable states from the initial set S^0 up to k iterations. The Frontier set in iteration $k+1$ is computed by taking the forward image (backward image for backward reachable set computation) of the frontier set in the kth iteration and removing from this image set, the states that have already been explored in previous iterations (which are stored in Reached Set). The Reached Set is updated by adding the new states from the frontier set. This process is iterated until no new states can be added to Reached Set. The final Reached Set represents the forward (backward) reachable set from the set of initial states S^0.

2.7. Algorithm Complexity

Given a graph $G(V, E)$, where V is the set of vertices representing Boolean states of the GRN and E is the set of edges representing transition among those states of the network (e.g., Fig. 5), identifying all the attractors in G can be performed in linear time $O(|V|+|E|)$ using depth-first search algorithms. However, for Boolean functions, there can be an exponential number of vertices (i.e., $|V|=2^N$) in the graph for N Boolean variables. Consequently, the problem of identification of all the attractors in the Boolean state space has exponential space complexity. Binary Decision Diagrams (BDDs) can be used to represent the Boolean state space within reasonable memory requirements and perform efficient reachability analysis on the underlying graph. However, it is difficult to analyze the complexity of the BDD-based algorithm which further depends upon the size of the BDD representation of the problem. BDD representation can be exponential with the problem size in the worst case scenario, but in most practical cases, it has a mild growth with the size of the problem as has been demonstrated in the literature (17–18). Therefore in the worst case scenario, the complexity of the BDD-based Algorithm 2 can be the same as an explicit depth-first search algorithm. But, as we will see in the next section, it runs efficiently for most of the big GRNs.

2.8. Computational Results

Runtimes of Algorithm 1 on some of the benchmark networks are given in Table 1. From the results, it can be seen that the synchronous

Table 1
Benchmarking of the synchronous model using Algorithm 2

Network	Nodes	Edges	Number of attractors	Time taken (seconds)
Mammalian	10	39	3	0.1
T-helper	23	34	3	0.12
Dendritic	114	129	1	0.32
T-cell receptor	40	58	10	3
Network 1	1,263	5,031	1	200

Mammalian Cell Network is taken from (21), T-helper from (22), and T-cell receptor from (23). The Dendritic Cell network was generated by semiautomatic mining of literature evidence. Network 1 is a full literature mined Insulin Growth Factor regulatory network. It has been developed through automatic literature mining tools that build a tentative regulatory network based on the set of keywords such as activation/inhibition

algorithm scales well with the size of the network and can compute all the attractors in reasonable time and memory. The benchmarking was performed on a 1.8 GHz Dual Core Pentium machine with 1 GB of RAM running Linux Fedora Core 5.

3. Modeling Gene Perturbations

Computing attractors on GRNs gives an insight into the cell differentiation process. If the computed Boolean attractors of the GRN have a biological explanation, then the GRN is likely to represent the biological process under investigation. In that case, it would be interesting for biologists to study the results of gene perturbation experiments on the given network.

Gene perturbations (or mutations) can be either in the form of a gene knockout which leads to constant absence of a protein inside the cell or in the form of a constant high expression of a gene leading to overproduction of the corresponding protein. Such mutations may naturally exist in a cell (i.e., inherited from parents) or they could be temporarily induced as a result of a disease or the impact of a drug compound. In both cases, it is interesting for biologists to study the impact of such mutations on the dynamics of the cell. Due to the presence of mutations in a GRN, some sections of the pathways may lose their dynamics completely. Therefore, the consequence of a mutation in a single gene can easily propagate to remotely related genes.

Example 4

A synthetic GRN of Fig. 4a in the presence of the knocked-out and the overexpressed node A is shown in Fig. 6a, b, respectively. The modified pathway can obviously result in a different set of steady states than those in the wild-type (i.e., unperturbed) GRN. In Fig. 6a, when A is knocked-out, node E will always stay at low expression and node C will not have any impact on B. Hence, the node B always stays in a low expression state. This in turn leads to low expression of D, which further causes C to always stay in high expression state. With all the nodes permanently fixed in one expression state, the network loses its ability to produce multiple attractors. A similar explanation can be extended for a scenario in which node A is overexpressed as is shown in Fig. 6b.

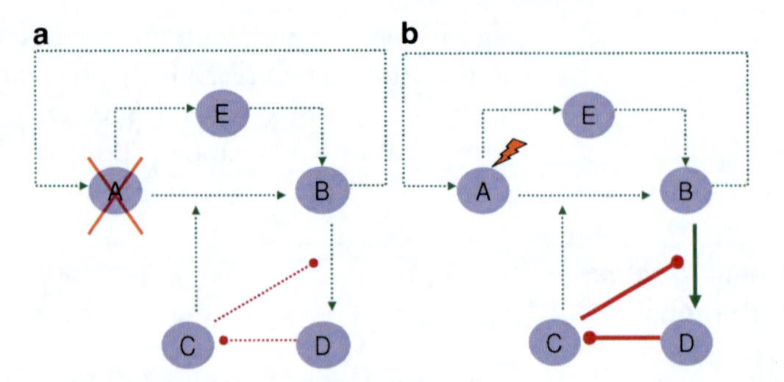

Fig. 6. *Dashed edges* do not form a part of the dynamics of the GRN due to a mutation in one of the participating nodes. (**a**) Modified GRN with the knocked-out node A. (**b**) Modified GRN with the overexpressed node A. In (**b**), when the node A is overexpressed it will keep the node E at high level of activation, which in turn maintains the node B in the high activation state. However, expression of node D can be modulated by the expression of node C; therefore the subpart of the network formed by the edges B-D-C remains functional when the node A is overexpressed. This is in contrast to the situation in (**a**) where knocking down the node A brings all the other nodes in the network into a low activation state.

In the next section, we extend our Boolean model of GRNs to perform in silico perturbation experiments. A gene (or protein) in a GRN can exist in one of the following three states:

1. *Overexpression*. This represents the constant expression of a gene at a high activation level. In Boolean logic, this means that the gene is "ON" or "1" all the time.
2. *Knock-down*. This represents the case when a gene is silenced and it does not participate in the network dynamics. That means gene is "OFF" or at level "0" all the time.
3. *Wild-type*. An unperturbed gene (i.e., neither overexpression nor knockdown) is said to be present in the wild-type condition.

The overexpression and knockdown of genes in GRNs have a similar notion in digital circuits as stuck-at-1 and stuck-at-0 faults, respectively (16).

3.1. Problem Formulation

We modify Boolean Eqs. 1 and 2 to encode knowledge about all possible gene perturbations in the Boolean model and then, during the analysis phase, we select the genes to be perturbed dynamically. Encoding this information in the model itself helps in sharing information between different perturbation experiments. Also, such a modeling approach permits the computation of the set of all minimal gene perturbations that can generate desired steady states. The formulation shown below helps in identifying all such minimal gene perturbation sets without explicitly enumerating and simulating all the possible gene perturbations.

In the presence of perturbations, each node in a GRN can exist in one of the following three states: wild-type, knocked-out, and overexpressed. In addition to Boolean vector $\mathbf{x}^t$, which was used in previous sections for representing the expression state of all the nodes of a GRN, we use two additional N bit Boolean vectors $\mathbf{x}^{\uparrow}$ and $\mathbf{x}^{\downarrow}$ to represent overexpressed and knocked-out nodes, respectively, in the GRN. If a bit i of $\mathbf{x}^{\uparrow}$ evaluates to 1 and the bit i of $\mathbf{x}^{\downarrow}$ evaluates to 0 (i.e., $x^{\uparrow}_i = 1$ and $x^{\downarrow}_i = 0$), then it means that the gene i is overexpressed. Similarly, $x^{\uparrow}_i = 0$ and $x^{\downarrow}_i = 1$ represent a knocked-out node. If both $x^{\uparrow}_i$ and $x^{\downarrow}_i$ evaluate to 0 or 1 for any given i, then the node i is modeled as a wild-type node. This constraint ensures that a node is not both knocked-out and overexpressed at the same time. To encode this information, we use the modified Boolean variables $\tilde{x}_i$ s as in Eq. 5.

$$\tilde{x}_i = \left\{x_i \wedge \left(\neg x_i^{\downarrow} \vee x_i^{-}\right)\right\} \vee \left(\neg x_i^{\downarrow} \wedge x_i^{-}\right) \tag{5}$$

Equation 5 states that if the gene i has been knocked-out (i.e., if $x^{\uparrow}_i = 0$ and $x^{\downarrow}_i = 1$), then $\tilde{x}_i = 0$. If the gene i is overexpressed (i.e., if $x^{\uparrow}_i = 1$ and $x^{\downarrow}_i = 0$), then $\tilde{x}_i = 1$. If the gene is wild type (i.e., if both $x^{\uparrow}_i$ and $x^{\downarrow}_i$ either evaluate to 0 or 1), then $\tilde{x}_i = x_i$. Equation 5 accommodates the perturbation information in the input expression of a node. To specify the perturbation information in the Boolean function that defines the expression of a node, we modify Eqs. 1 and 2 to Eqs. 6 and 8, respectively.

$$\tilde{x}_i(t+1) = \left\{x_i(t+1) \wedge \left(\neg x_i^{\downarrow} \vee x_i^{-}\right)\right\} \vee \left(\neg x_i^{\downarrow} \vee x_i^{-}\right) \tag{6}$$

$$x_i(t+1) = \left(\bigvee_{t=1}^{n} f_{x_{i,l}}^{\mathrm{ac}}(t)\right) \wedge \neg \left(\bigvee_{l=1}^{n} f_{x_{i,l}}^{\mathrm{in}}(t)\right) \tag{7}$$

$$f_{x_{i,l}}^{\mathrm{ac,in}}(t) = \left(\bigwedge_{j=1}^{p} \tilde{x}_j^{\mathrm{ac},t}\right) \wedge \left(\bigwedge_{j=1}^{p} \neg \tilde{x}_j^{\mathrm{in},t}\right) \tag{8}$$

Equation 6 states that if gene i is overexpressed (i.e., if $x^{\uparrow}_i = 1$ and $x^{\downarrow}_i = 0$), then $\tilde{x}_i(t+1) = 1$. If gene i has been knocked-out (i.e., if $x^{\uparrow}_i = 0$ and $x^{\downarrow}_i = 1$), then $\tilde{x}_i(t+1) = 0$. If the gene is wild type (i.e., both $x^{\uparrow}_i$ and $x^{\downarrow}_i$ either evaluate to 0 or 1), then Eq. 6 is the same as Eq. 1. Equation 8 is the counterpart of Eq. 2, with the modified variables $\tilde{x}_i$ s.

Example 5

The synthetic GRN of Fig. 4b mapped with modified Boolean gates is shown in Fig. 7. As one can see from this figure, extra Boolean logic is appended both at the input and at the output to reflect the choice between overexpression and knockdown for each input and output variable.

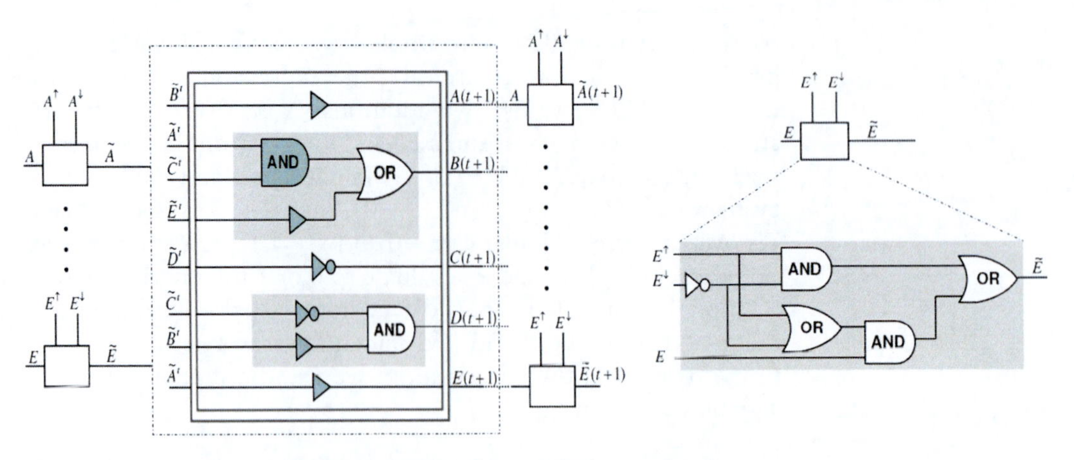

Fig. 7. Modified Boolean gates to accommodate the gene perturbation experiments.

Transition functions for the synchronous model as defined by Eq. 3 are modified to Eq. 9, while Eq. 4 does not change for the perturbation model.

$$T_i(\mathbf{x}^t, \mathbf{x}^{t+1}) = \left\{ x_i^{t+1} \leftrightarrow \tilde{x}_i(t+1) \right\} \tag{9}$$

Using the above formulation, one can model multiple gene perturbations in a GRN. A set of perturbations define a single experiment. Multiple experiments spaced over different time points (also referred to as levels of experiments) can also be performed using the above formulation. The modified transition function $T(\mathbf{x}^t, \mathbf{x}^{t+1})$ represents the relation between the current state and the next state in the presence of any possible gene perturbation. Given a perturbation experiment (which specifies the set of perturbations to be performed), we restrict the transition function state space to only those perturbations which are part of the experiment and compute attractors on that restricted state space. For this, we define three Boolean functions, f^{x^-}, $f^{x^\downarrow}$, and f^p to represent information of the knocked-out, overexpressed, and perturbed genes, respectively. Function f^p is further expressed in terms of f^{x^-} and $f^{x^\downarrow}$. These functions are given in Eqs. 10–12.

$$f^p = f^{x^\downarrow} \wedge f^{x^-} \tag{10}$$

$$f^{x^\downarrow} = \left(\bigwedge_{i:x_i^\downarrow = 1} x_i^\downarrow \right) \wedge \left(\bigwedge_{i:x_i^\downarrow = 0} \neg x_i^\downarrow \right) \tag{11}$$

$$f^{x^-} = \left(\bigwedge_{i:x_i^- = 1} x_i^- \right) \wedge \left(\bigwedge_{i:x_i^- = 0} \neg x_i^- \right) \tag{12}$$

In the next section, we describe the algorithm to perform perturbation experiments using the above formalism.

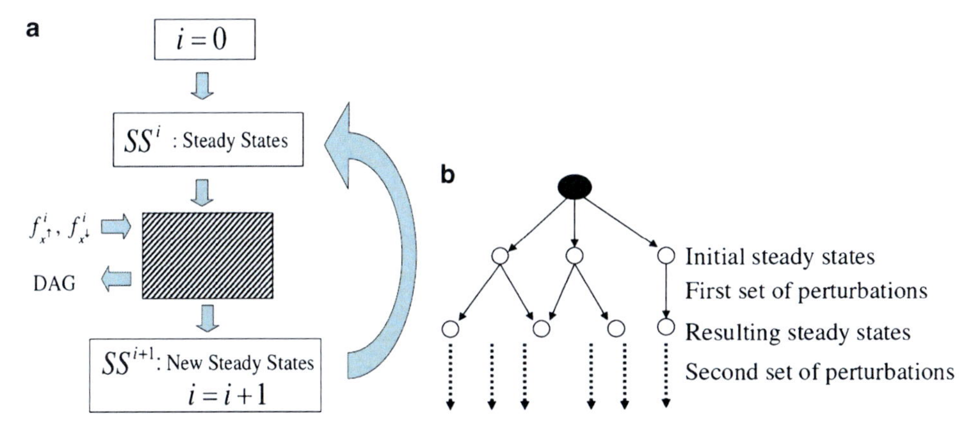

Fig. 8. (**a**) Flowchart of gene perturbation algorithm. (**b**) Directed acyclic graph (DAG) showing the transitions among steady states.

3.2. An Algorithm for Gene Perturbations

A biological experiment often involves more than one perturbation, either concurrently or in a time spaced manner. Let us define the set of concurrent perturbations as a single test and a sequence of tests as an experiment. Tests in an experiment are always performed at different time points defined by their sequence specified in the input experiment file. With this generalized experiment model, the basic idea behind the gene perturbation algorithm can be summarized as in Fig. 8.

In Fig. 8a, the black box computes steady states for every test, following the sequence of tests specified in the experiment. The first test in an experiment is always said to be wild type (with no perturbations). Starting from the second test, the black box in Fig. 8a checks whether the steady states from the previous test can reach the steady states of the current test. The output from this black box is in the form of a directed acyclic graph (DAG) as shown in Fig. 8. In Fig. 8b, starting from the level after the root node of the DAG, steady states results are shown in the sequence in which the tests are specified in the experiment. The first level of the DAG gives all the steady states in the unperturbed network, the second level of DAG gives all the steady states after the first perturbations, and so on. Edges among the nodes represent the possible transitions among steady states in the presence of the perturbation. One should note that while steady states cannot transition among each other in the absence of a perturbation, they can transition among each other when some nodes are perturbed. This change in stable behavior is captured by the absence of edges between nodes at the same level and presence of edges among the nodes at different levels of the DAG in Fig. 8b.

In Fig. 8b, some of the steady states in different levels may be the same or may have some measure of similarity. This similarity measure can be computed by counting the percentage of genes which have the same level of expression in two different steady states.

Algorithm 3: Algorithm for *in-silico* gene perturbation experiments

Input : Genetic Network and perturbation experiments.
Output: DAG representing steady state analysis.

1 **begin**
2 compute T_{sync}
3 **for** $i = 0$ *to* L **do**
4 compute f_k^i, f_e^i and f_{GP}^i
5 $T'_{sync} = T_{sync} \wedge f_{GP}^i$
6 $SS^i[\,] = all\_attractors(T''_{sync})$
7 **if** $i \neq 0$ **then**
8 **for** $k = 0$ *to* $SS^{i-1}.size()$ **do**
9 $FR \longleftarrow forward\_set(SS^{i-1}[k], T_{sync})$
10 **for** $j = 0$ *to* $SS^i.size()$ **do**
11 **if** $SS^i[j] \subseteq FR$ **then**
12 Draw an edge between nodes $SS^{i-1}[k]$ and $SS^i[j]$
13 **end**

Here, we compute the similarity measure between steady states of first level and steady states of all the other levels. This way one can make more sense out of results of the perturbation experiments and can draw more conclusions such as: the system moves from steady state A to steady state B on perturbing gene X, where A and B are the steady states in the original unperturbed network.

Figure 8 is formally described in Algorithm 3. In this algorithm, the main loop in Lines 3–13 is iterated over the sequence of tests (given by a set L) in the experiment. For every test i, corresponding functions $f_i^{x^-}$, $f_i^{x^+}$, and f_i^p are constructed using Eqs. 10–12 in Lines 4 and 5. Then the transition function is restricted to the state space defined by this perturbation experiment (Lines 6 and 7). The attractors are computed on the perturbed network in Line 8 using Algorithm 1. Once the attractors are found, we compare the forward reachability of attractors of the previous test with the attractors of the current level of a perturbation experiment. This is done in Lines 9–14. For every attractor computed in the previous level, we compute the forward reachable states on the new transition function (Line 11). Then we check whether all the attractors in the current test i are contained in this forward reachable set (Line 13). Lines 3–14 can be repeated for different experiments on the same network without having to modify the GRN.

4. Case Study: T-Helper Network

The vertebrate immune system is made of diverse cell populations; some of them are antigen-presenting cells, natural killer cells, and B and T lymphocytes. A subpopulation of T lymphocytes, the T-helper, or Th cells have received a lot of attention from the modeling point of view in the literature. Th cells can be divided into precursor Th0 cells and effector Th1 and Th2 cells, depending on the pattern of secreted molecules (cytokines), which are responsible for the central role of Th cells in cell-mediated immunity (Th1 cells) and humoral responses (Th2 cells). Understanding the molecular mechanisms that regulate the differentiation process from Th0 toward either Th1 or Th2 is very important, since an immune response biased toward the Th1 phenotype results in the appearance of autoimmune diseases, and an enhanced Th2 response can originate allergic reactions (22, 23).

There are several factors at the cellular and molecular levels that determine the differentiation of T-helper cells. Importantly, the cytokines present in the cellular milieu play a key role in directing Th cell polarization. On one hand, IFN-γ, IL-12, IL-28, and IL-27 are the major cytokines that promote Th1 development (24). On the other hand, IL-4 is the major cytokine responsible for driving Th2 responses. Besides the positive role of cytokines in the differentiation process, there also exists a mutual inhibitory mechanism. Specifically, IFN-γ plays a role in inhibiting the development of Th2 cells, whereas IL-4 inhibits the appearance of Th1 cells. This interplay of positive and negative signals, at both the cellular and molecular levels, creates a complexity that is suitable for analysis by modeling the GRN.

Due to its physiological relevance, there are various mathematical models that have been proposed for describing the differentiation, activation, and proliferation of T-helper lymphocytes. Most of these models, however, focus on interactions established among the diverse cell populations that somehow modify the differentiation of Th cells (25, 26). Also, other modeling efforts have been aimed at understanding the mechanism of the generation of antibody and T-cell receptor diversity, as well as the molecular networks of cytokine or immunoglobulin interactions (29, 30). Recently, there have been some publications on the regulatory network that controls the differentiation of Th cells (22, 29). The regulatory network presented (reproduced in Fig. 9) constitutes the most extensive attempt to model the regulatory network controlling the differentiation of Th lymphocytes to date. The topology of the network was derived from published experimental data (22). The network (Fig. 9) is made of 23 nodes and 26 positive and 8 negative interactions.

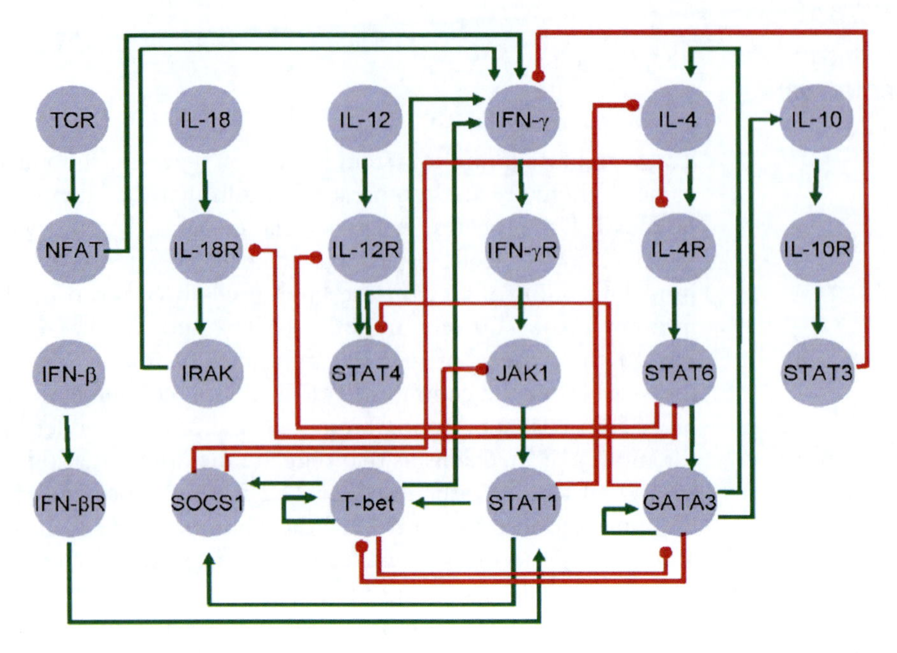

Fig. 9. T-helper gene regulatory network (102). *Green edges* represent the positive interactions and the *red edges* represent the negative interactions.

Table 2
Steady states of the T-helper cell

Perturbed genes	Active genes in steady states								Cell type
Wild type	All the genes are inactive								Th0
	IFN-γ	Tbet	SOCS-1	IFN-γR					Th1
	IL-10	IL-10R	GATA-3	STAT3	STAT6	IL-4	IL-4R		Th2
IL-12$^{+/+}$	IFN-γ	Tbet	SOCS-1	IFN-γR	IL-12	IL-12R	STAT4		Th1
	IL-10	IL-10R	GATA-3	STAT3	IL-12	STAT6	IL-4	IL-4R	Th2
IL-4$^{+/+}$	IFN-γ	Tbet	SOCS-1	IFN-γR	IL-4				Th1
	IL-10	IL-10R	GATA-3	STAT3	STAT6	IL-4	IL-4R		Th2

4.1. Simulation Results

On applying the algorithms developed in previous sections on the T-helper cell network of Fig. 9 (see also Subheading 7), three wild-type steady states as listed in Table 2 are found. These steady states correspond to the molecular profiles observed in Th0, Th1, and Th2 cells, respectively. The first steady state reflects the pattern of Th0 cells, which are precursor cells that do not produce any of the cytokines included in the model (IFN-β, IFN-βR, IL-10, IL-12, IL-18, and IL-4). The second steady state represents Th1 cells with high activation of IFN-γ, IFN-γR, Tbet, and SOCS1. Finally, the third steady state corresponds to the activation observed

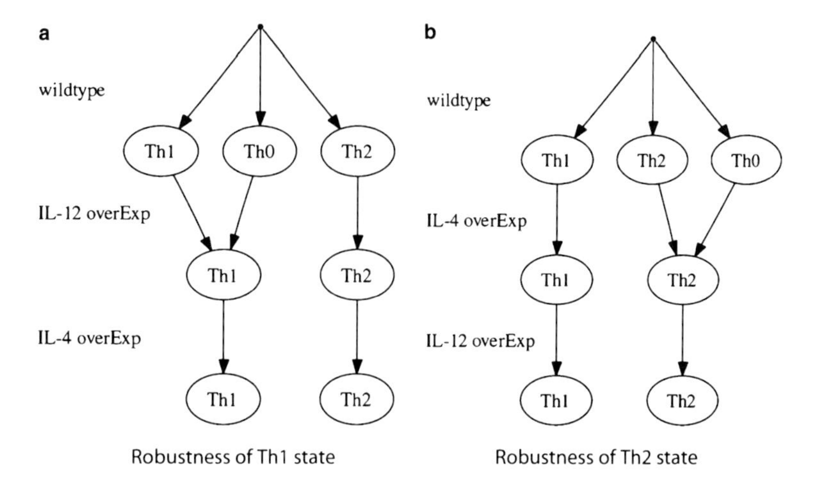

Fig. 10. Results of gene perturbation experiments on the T-helper cell differentiation GRN.

in Th2 cells, with high level of activation of GATA3, IL-10, IL-10R, IL-4, IL-4R, STAT3, and STAT6. These results also match those published in (22).

Next, the response of Th cells was studied to two consecutive stimuli, first a constant saturating concentration of IL-12, and then changing it to a saturating concentration of IL-4. As shown in Fig. 10a, this combination of signals has the result of eliminating the Th0 steady state. If the system is in the Th0 state, the constant activation of IL-12 moves it to the Th1 state, where it stays even after the inactivation of IL-12 and the constant presence of IL-4. In contrast, if the system starts in the Th1 or Th2 states, the two consecutive signals are incapable of moving the system to another attractor. The steady-state profile on overexpression of IL-12 and IL-4 is shown in Table 2. Figure 10b shows the simulation results of applying the same perturbations as described above in the reverse order (i.e., activating IL-4 to its highest level, and then inactivating it and activating IL-12 instead). Results are the same as before, shown in Fig. 10b. The only difference is that in this simulation, the network in the Th0 states receives the IL-4 and moves to Th2, where it stays after the elimination of IL-4 and the activation of IL-12.

These simulations show that Th0 state is unstable under the perturbation of IL-12 or IL-4, which act as differentiation signals to take the system to the Th1 or Th2 states, respectively. By contrast, the Th1 and Th2 states are stable under the perturbation of the IL-4 and IL-12 nodes. These results are in total agreement with experimental data (25) and reported simulations of the Th network using a different mathematical framework (22). In the literature, modeling of Th cell differentiation at the molecular level has been shown to be very useful to bring insight into the origin of the unexpected phenotypes. Using the algorithms proposed in this chapter, one can easily perform such simulations in silico.

5. Non-determinism in GRNs

The process of gene activation (or inhibition) is a complex mechanism that involves binding of one or more transcription factors, transcription of DNA into mRNA, and translation of mRNA into protein. All these stages of gene activation involve some amount of stochasticity. The level of stochasticity induced depends upon the complexity of the binding mechanism, length of the protein coding regions, concentration of transcription factors, and various other factors. Experimentally, gene regulation processes have been shown to be inherently stochastic (32–36). In the presence of stochasticity, a gene may not get expressed even in the presence of required transcription factors, leading to stochastic fluctuations in the states of the GRN. These stochastic fluctuations may cause the dynamics of identical cells to behave differently in the same environmental conditions.

Due to varying levels of complexity, biological functions can show varying levels of stochasticity in their behavior. Although it is experimentally difficult to quantify the measure of stochasticity involved in different biological functions, it is a well-known fact that some functions such as proteasome degradation are least prone to stochasticity, while functions such as scaffolding complexes that integrate signals arising from different pathways are most likely to behave most stochastically. In practice, most biological functions behave somewhere between the above two extremes. Keeping this observation in mind, the probability of stochasticity can be broadly classified into the following three different classes: low probability of error ($\varepsilon \approx 0$), medium probability of error ($\varepsilon \approx 0.5$), and high probability of error ($\varepsilon \approx 1$). Figure 11 gives an example of a few biological functions divided into these different classes of stochasticity.

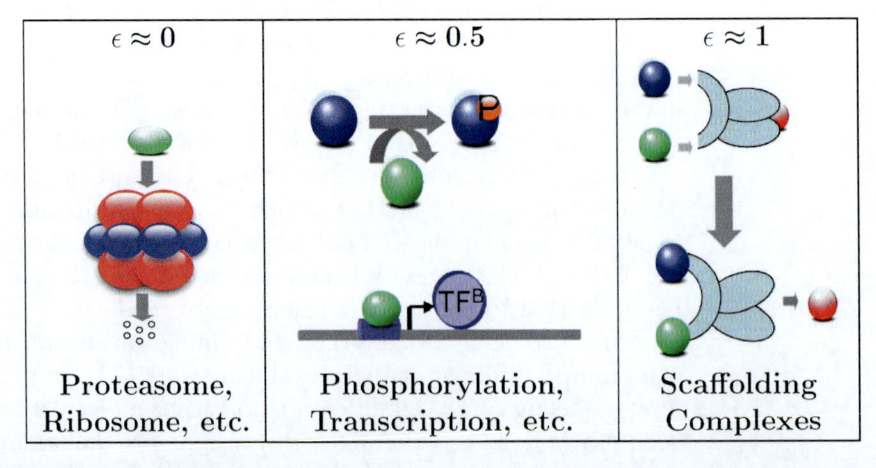

Fig. 11. Biological functions categorized into three different classes of stochasticity and error probability (37). From *left* to *right*, biological processes are classified from very stable structures to highly stochastic systems involving scaffold proteins.

The stochasticity in continuous modeling of GRNs is traditionally modeled as associated with the concentration of the reacting species. At low concentrations of reacting species, the probability of two molecules undergoing a biochemical reaction decreases, thereby adding a stochastic effect on the reaction product concentration. This approach of noisy gene regulation can be efficiently simulated in continuous modeling approaches using chemical master equations (CMEs) and Gillespie's algorithm (38–42). Just like continuous modeling of GRNs, stochastic modeling using Gillespie's algorithms requires that all the kinetic rate constants be known a priori. This requirement restricts its application to only small well-studied networks.

Traditionally, the stochasticity in Boolean models of GRNs is modeled by flipping the expression of nodes in a GRN from 0 to 1 or vice versa with some predefined flip probability (43–46). This model of stochasticity is referred to as stochasticity in node (SIN) in this chapter. However, the SIN model of stochasticity does not take into account the complexity of underlying biological function while flipping the expression of a node.

Another method called the stochasticity in functions (SIF) has been introduced recently for modeling the stochasticity in Boolean models of GRNs (46). In the SIF model, stochasticity is induced at the level of biological functions rather than at the level of expression of a protein/gene. SIF associates a probability of failure with different biological functions and models stochasticity in these functions depending upon the expression of the input nodes. With the above two constraints in the SIF model, the probability of a node in the GRN flipping its expression from 0 to 1 or vice versa at a given time instant depends upon the probability of function failure and the activity of other nodes in the network at that instant in time, thereby making it possible to integrate the stochasticity due to complexity of a biological function with the dynamics of the GRN.

Example 6

The difference between the SIN and SIF models of stochasticity can be better understood from Fig. 12. In Fig. 12, we model a biological function that requires simultaneous activity of inputs A and B to activate the protein C. Under the SIN model, probability of stochastic expression of node C is given by P $(C^{\Delta}|C) = 0.5$. Under the SIF model, however, the probability that node C shows a noisy behavior is given by P $(C^{\Delta}|A, B, f)$, which is a function of the biological phenomena f and its inputs {A, B}. The table next to the figure represents the stochastic expression of C under the SIN and the SIF models. In Fig. 12, the values corresponding to C^{Δ} under the SIF model are the same as the corresponding non-stochastic values in the first three

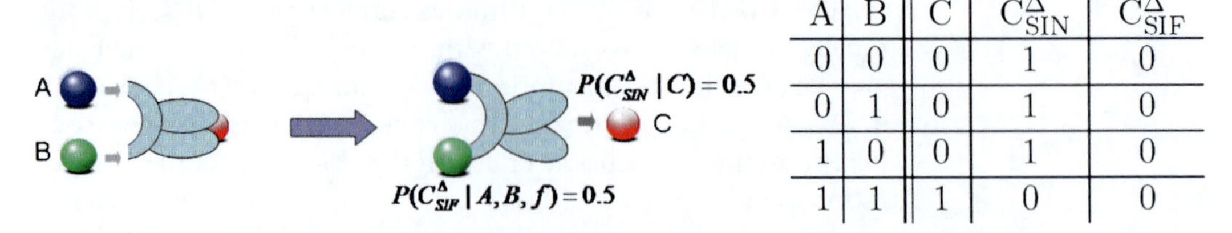

A	B	C	C^{Δ}_{SIN}	C^{Δ}_{SIF}
0	0	0	1	0
0	1	0	1	0
1	0	0	1	0
1	1	1	0	0

Fig. 12. A small example demonstrating the difference between the SIN and SIF in terms of the activity of the output gene (or protein).

rows because the underlying biological function is inactive when only one input is present. However, in the fourth row, the biological function is functionally active and can behave stochastically giving an output C^{Δ} different from its non-stochastic counterpart. Under the SIN model, in the fourth column of the truth table, the stochasticity is independent of the fact whether the underlying biological function is functionally active and is always modeled to be flipped from the true non-stochastic value.

5.1. Impact of Stochasticity

The impact of stochasticity on cellular dynamics can be experimentally monitored by measuring the fraction of different cell phenotypes in an experiment when cells have reached the steady state. Here, we compare the results obtained using SIN and SIF stochasticity models with respect to two properties of steady states: (a) cellular differentiation in response to an external stimulus and (b) robustness of attractors.

5.1.1. Cellular Differentiation

In the absence of stochasticity, all biological functions behave as per their description and an initial state of the network differentiates into a specific steady state. However, in the presence of stochasticity in these functions, a network simulation starting from the same initial state may stabilize into different steady states. The probability of differentiating into one steady state can be different from the probability of differentiating into another steady state. This simulation behavior can be used to explain the well-known biological observation of emergence of phenotypically distinct subgroups within an isogenic cell population in response to an input stimuli (such as on exposure to external ligands) (35).

A sample simulation experiment on a T-helper differentiation network (22), as in Fig. 9, can be effectively used to describe the stochastic differentiation of naive T-helper cells (i.e., Th0) in response to a pulse of IFN-γ, a key cytokine known to play an important role in Th0 to Th1 differentiation. In Fig. 13a, cells are initially in a naive undifferentiated cell state (i.e., Th0). On receiving an input stimulus on IFN-γ, cells must differentiate into Th1

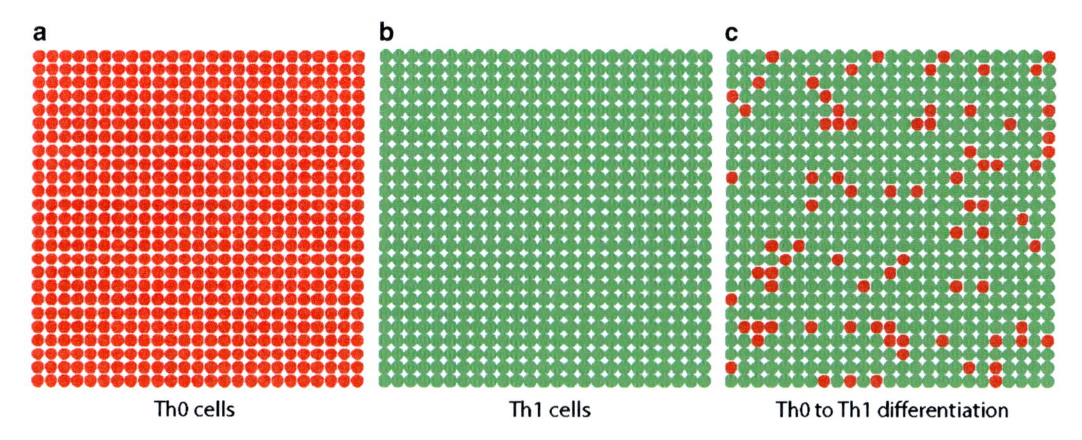

Fig. 13. Simulation results showing the effect of noise on T-helper cell differentiation process with an external stimulus of IFN-γ. Each *small circle* is representative of a T-helper cell and each cell is modeled to behave independent of the neighboring cells. *Red cells* represent the naive undifferentiated Th0 cells, *green cells* represent Th1 cell state. Ratio of number of *red* (or *green*) cells to total number of cells in a panel is representative of the probability of differentiating from Th0 into Th1 cell state. (**a**) Th0 cell state. (**b**) In the absence of any stochasticity, all Th0 cells differentiate to Th1 cell state on receiving IFN-γ. (**c**) In the presence of stochasticity, Th0 cells differentiate into Th1 cells while some cells cannot differentiate on receiving IFN-γ and revert to Th0 cell state.

cell state in the absence of any stochasticity. This is shown in Fig. 13b. Biologically, it is known that while most of the cells should differentiate into Th1 state in response to an IFN-γ dosage, a few cells can revert to the Th0 state (25, 28). This difference in response across the cell population is often attributed to inherent stochasticity in biological functions (Fig. 13c).

5.1.2. Robustness of Attractors

Robustness of attractors of a GRN can be defined as the probability of an attractor reverting back to itself when the expression of one or more nodes is perturbed from its original expression value. In the absence of any stochasticity in the biological functions, there should be no transition among two different attractors. If a perturbation changes the state of an attractor, it is possible that the new perturbed state may transition into a different attractor. The perturbed state may be generated in response to external stimuli such as ligands, inhibitors, or due to internal stochasticity of the cell. Biologically, cellular steady states are highly robust to internal stochasticity due to redundancy of critical biological functions. Redundant alternative biological pathways to control the expression of genes/proteins are nature's solution to the short-term stochastic behavior of subsections of the pathway and are known to exist in abundance in any biological system. To associate high confidence in a GRN, it is imperative that the robustness of cellular steady states is reflected by the robustness of attractors under the stochastic simulations of the corresponding GRNs. Hence, a biologically motivated stochastic model for quantifying the robustness properties of a GRN is essential to compare multiple network configurations for the same biological problem.

6. Modeling Stochasticity in Boolean Networks

We saw in Subheading 2.1 that how different biological functionalities can be well described using a small set of Boolean functions {BUFF, NOT, AND, OR, IAND}. Truth tables defining the characteristic function of these Boolean functions are shown in Table 3. In the presence of stochasticity in Boolean functions, output of these logic gates can be different from those specified by their characteristic functions (represented by C^{Δ} in the Truth tables 3a–e).

As explained earlier in Subheading 2.1, the shaded Boolean functions in Fig. 4b have a corresponding biological functionality and hence should be modeled for stochastic behavior. How this stochasticity is modeled in a GRN varies in the SIN and SIF models and also depends upon the fault model used (as explained in the next few sections).

6.1. Fault Model

A fault in a GRN is defined as the stochastic behavior of a node (i.e., gene expression) or the Boolean function in the GRN. Under the SIN model, faults are modeled in gene expression, and under the SIF model, faults are modeled in Boolean functions. Fault in a node or Boolean function i is represented by a Boolean variable Δ_i. If $\Delta_i = 1$, then the gene (or Boolean function) i takes the faulty value and $\Delta_i = 0$ represents the normal expression value in the absence of any stochasticity. There are multiple nodes (or Boolean functions) in a GRN and more than one of them can be susceptible to faults. Faults in a GRN at a given instant of time are represented by a Boolean vector Δ and is referred to as a fault configuration in the GRN.

Table 3
Truth tables representing the transfer function of different Boolean logic gates. A and B are the input genes, C represents the output gene expression in the absence of stochasticity, and C^{Δ} represents the output gene expression in the presence of stochasticity. The values taken by the variable C^{Δ} can be different depending upon the mathematical model used for capturing the stochasticity in the network

A	C	C^{Δ}
0	0	1
1	1	0

(a) BUFF

A	C	C^{Δ}
0	1	0
1	0	1

(b) INV

A	B	C	C^{Δ}
0	0	0	1
0	1	0	1
1	0	0	1
1	1	1	0

(c) AND

A	B	C	C^{Δ}
0	0	0	1
0	1	1	0
1	0	1	0
1	1	1	0

(d) OR

A	B	C	C^{Δ}
0	0	0	1
0	1	0	1
1	0	1	0
1	1	0	1

(e) IAND

Here, we make an assumption that at most one gene or one function in a GRN can have a fault at a given instant in time, and that multiple faults are spread over different time instants. This implies that $\Delta_i = 1$ for at most one i in a fault configuration vector Δ. A sequence of network states from a given starting state to an attractor is called the trajectory of the state. If n faults exists in the network, then at most n faults can lie on any trajectory. However, multiple faults cannot exist on a trajectory at the same time instant. We refer to this fault injection model as the single fault model. Furthermore, under the single fault model, given a state of the network, all the possible single faults are independent of each other and can exist with equal probability. This leads to multiple outgoing trajectories from a single state. The assumption of a single fault at a time has been widely used in the literature for stochastic Boolean modeling of GRNs under the SIN model (43–46). The single fault model corresponds to a small probability of two distinct biological functions behaving stochastically at the same instant of time.

Next we modify the transition functions introduced in Subheading 2.4, for modeling the stochasticity in GRNs.

6.2. Stochasticity in Nodes

In the SIN model, any node can flip its expression, from 0 to 1 or vice versa, due to internal stochasticity in gene regulation mechanisms. When a node flips from its normal expression value due to stochasticity, a fault is said to be injected in the GRN. Fault in the node i in the GRN is represented by a Boolean variable Δ_i. The transition function for a single gene, $T_i(\mathbf{x}^t, \mathbf{x}^{t+1})$ (Eq. 3) introduced in Subheading 2.4 can be modified such that x_i^{t+1} is equal to the value of the function $x_i(t+1)$ if there is no fault (i.e., $\Delta_i = 0$). Otherwise x_i^{t+1} takes the value opposite to the current value of the function (i.e., $\neg x_i(t+1)$). Equation 13 represents the modified transition function $T_i(\mathbf{x}^t, \mathbf{x}^{t+1}, \Delta)$ in the presence of a fault in the SIN model. Transition function $T(\mathbf{x}^t, \mathbf{x}^{t+1})$ of the GRN is the same as in Eq. 4 and is given again here in Eq. 14.

$$T_i(\mathbf{x}^t, \mathbf{x}^{t+1}, \Delta) = \left[\left(x_i^{t+1} \leftrightarrow x_i(t+1)\right) \wedge \neg\Delta_i\right] \vee \left[\left(x_i^{t+1} \leftrightarrow \neg x_i(t+1)\right) \wedge \Delta_i\right] \quad (13)$$

$$T_i(\mathbf{x}^t, \mathbf{x}^{t+1}, \Delta) = T_0(\mathbf{x}^t, \mathbf{x}^{t+1}, \Delta) \wedge \ldots \wedge T_N(\mathbf{x}^t, \mathbf{x}^{t+1}, \Delta) \quad (14)$$

Since any node can flip its expression in the SIN model, there are exactly N possible faults in the network at a given instance of time, where N is the number of genes in the network. Under the single fault model, if the faults in the network are represented by a Boolean vector Δ of size N, at most one gene x_i has a fault (i.e., $\Delta_i = 1$ for at most one bit). Since all the faults are independent of each other, given the state of the network $\mathbf{x}^t$, a set of independent and equiprobable fault configuration vectors Δs can exist in the network. That is, if we represent the set of possible fault configuration vectors by a set $D = \{\Delta^1, \Delta^2, \ldots\ldots, \Delta^N\}$, the probability of selecting the fault vector Δ^i is given by Eq. 15.

$$P(\Delta = \Delta^i) = \frac{1}{N} \quad (15)$$

If the probability of flipping a node i is given by ε^i, then the probability that the gene i has a fault (i.e., $P(\Delta_i = 1)$) in the fault vector Δ is given by Eq. 16.

$$P(\Delta_i = 1) = \Delta_i \cdot \epsilon^i \quad (16)$$

6.3. Stochasticity in Functions

SIF models the stochasticity in biological functions that are represented using Boolean gates AND, OR, BUFF, NOT, and IAND in Fig. 4b. In the presence of faults, the output of these logic gates can be different from their normal value. For example, the noisy output value of these Boolean functions is given in the last column of Table 3a–e. The noisy output of these Boolean functions can take different values depending upon how the faults are modeled. For example, Table 4 shows another set of transition functions in the presence of noise. Since the above Boolean functions represent an underlying biological functionality, it is important that the way faults are modeled in these functions reflects the biological reasoning behind stochasticity in underlying gene or protein regulation mechanism. Keeping this in mind, Table 4 has a closer correspondence with biological mechanisms than Table 3 as explained below.

In Table 4a–e, noise has an impact on the function only when all the positive inputs are "active" or 1. This constraint corresponds to the fact that a biological function can behave stochastically only when it is functionally active. For example, transcription of a gene can take place only when the transcription factor is present, and there is a natural stochasticity involved in the process of transcription. On the other hand, if the transcription factor is absent, there

Table 4
Truth tables representing the transfer function of different Boolean logic gates. A and B are the input genes, C represents the output gene expression in the absence of stochasticity, and C^Δ represents the output gene expression in the presence of stochasticity under the SIF model

A	C	C^Δ
0	0	0
1	1	0

(a) BUFF

A	C	C^Δ
0	1	1
1	0	1

(b) INV

A	B	C	C^Δ
0	0	0	0
0	1	0	0
1	0	0	0
1	1	1	0

(c) AND

A	B	C	C^Δ
0	0	0	0
0	1	1	1
1	0	1	1
1	1	1	1

(d) OR

A	B	C	C^Δ
0	0	0	0
0	1	0	0
1	0	1	0
1	1	0	1

(e) IAND

can be no stochasticity in the transcription process and the gene would never be expressed. Boolean gates BUFF, NOT, and AND have all the input ports as positive inputs. Boolean gate IAND has some ports which go through the NOT gate and act as negative inputs. Boolean OR gate is modeled to have no stochasticity because it just represents that the two alternate biological functions can have an impact on the same gene/protein. Note that the noise in these alternate biological functions is already modeled with the remaining stochastic gates (i.e., AND, NOT, and BUFF).

In the SIN model, the fault variable Δ represents stochasticity in the expression of a gene, whereas in the SIF model Δ represents stochasticity in Boolean functions. Since not all Boolean functions in a GRN behave stochastically, let us define $G = \{G_1, G_2, \ldots, G_p\}$ as a set of stochastic functions in the mapped GRN of Fig. 4b. If $\Delta_i = 1$, then Boolean function G_i behaves stochastically and take the expression value as defined by the last column of truth tables 4a–e. Otherwise, if $\Delta_i = 0$, Boolean function G_i behaves per its original description. Equations 17–21 formally describe the stochastic Boolean functions.

The expression of each gene i at time $t+1$ is specified by the function $x_i(t+1)$ of the state of the genes acting as its input at time t. The construction of function $x_i(t+1)$ was given in Eq. 1. Alternatively, the function $x_i(t+1)$ can be formed by composing Boolean gates as in Fig. 4b and using the corresponding Eqs. 17–21.

$$\text{BUFF:} f^{\mathrm{B}}(x_a) = [(x_c \leftrightarrow 0) \wedge \Delta] \vee [(x_c \leftrightarrow x_a) \wedge \neg\Delta] \tag{17}$$

$$\text{NOT:} f^{\mathrm{N}}(x_a) = [(x_c \leftrightarrow 1) \wedge \Delta] \vee [(x_c \leftrightarrow \neg x_a) \wedge \neg\Delta] \tag{18}$$

$$\text{OR:} f^{\mathrm{O}}(x_1, \ldots, x_p) = \left(x_c \leftrightarrow \bigvee_{i=1}^{p} x_i \right) \tag{19}$$

$$\text{AND:} f^{\mathrm{A}}(x_1, \ldots, x_p) = [(x_c \leftrightarrow 0) \wedge \Delta] \vee \left[(x_c \leftrightarrow \bigwedge_{i=1}^{p} x_i) \wedge \neg\Delta \right] \tag{20}$$

$$\text{IAND}: f^{\mathrm{IA}}(x_1, \ldots, x_p) = \left[\left\{ x_c \leftrightarrow \left(\bigwedge_{i=1}^{p^{\mathrm{in}}} \neg x_i^{\mathrm{in}} \wedge \bigwedge_{j=1}^{p^{a}} x_j^{a} \right) \right\} \wedge \neg\Delta \right] \vee \left[\left\{ x_c \leftrightarrow \bigwedge_{i=1}^{p^{a}+p^{\mathrm{in}}} x_j \right\} \wedge \Delta \right] \tag{21}$$

Example 7

For the node B in the GRN in Fig. 4, $x_{\mathrm{B}}(t+1)$ is defined in Eq. 22.

$$x_{\mathrm{B}}(t+1) = f^{\mathrm{O}}(f^{\mathrm{B}}(x^{t}_{\mathrm{E}}), f^{\mathrm{A}}(x^{t}_{\mathrm{A}}, x^{t}_{\mathrm{C}})) \tag{22}$$

In the above equation, f^{B}, f^{A}, and f^{O} are defined in Eqs. 17, 19, and 20 respectively, and x^{t}_{C} and x^{t}_{E} represent the expression of nodes C and E at the time instant t.

The transition function $T_i(\mathbf{x}^t, \mathbf{x}^{t+1})$ of a gene i for the SIF model is given by Eqs. 23 and 24 for the synchronous Boolean networks. Note that while the SIN model modifies the transition function of a gene, the description of Boolean functions is modified in the SIF model.

$$T_i(\mathbf{x}^t, \mathbf{x}^{t+1}, \Delta) = \left\{x_i^{t+1} \leftrightarrow x_i(t+1, \Delta)\right\} \tag{23}$$

$$T(\mathbf{x}^t, \mathbf{x}^{t+1}, \Delta) = T_0(\mathbf{x}^t \leftrightarrow \mathbf{x}^{t+1}, \Delta) \wedge \ldots \wedge T_N(\mathbf{x}^t, \mathbf{x}^{t+1}, \Delta) \tag{24}$$

Given a state of the network $\mathbf{x}^t$, not all Boolean functions in the set G behave stochastically. We use a Boolean vector Δ of size $|G|$ to represent the faulty Boolean functions in the network. In a given fault vector Δ, the bit $\Delta_i = 1$ only if all the positive inputs to the function G_i are active or 1. Hence, the number of faults in the network at a given time instant t depends upon the current state of the network. Again, assuming the single fault model, a set $D = \{\Delta^1, \Delta^2, \ldots\ldots, \Delta^{|D|}\}$ of independent fault vectors Δ^i may exist such that in each fault vector, at most one Boolean function G_i has a fault (i.e., $\Delta_i = 1$ for at most one bit). The probability of selecting the fault vector Δ^i is given by Eq. 25.

$$P(\Delta = \Delta^i) = \frac{1}{|D|} \tag{25}$$

Boolean functions in the set G correspond to biological functions and have a probability of failure ε^i associated with each G_i. The probability of failure ε^i is independent of the state of the network and solely depends upon the complexity of the biological function that it represents. The probability that Boolean function i has a fault (i.e., $P(\Delta_i = 1)$) in a given fault vector Δ is given by Eq. 26.

$$P(\Delta_i = 1) = \Delta_i \cdot \in^i \tag{26}$$

While the set D of fault configuration vectors does not depend upon the state of the network and is always the same in the SIN model, both the size of the set D and its elements depend upon the current state of the network $\mathbf{x}^t$ in the SIF model. Furthermore, the size of fault configuration vector Δ is different for the SIN and the SIF models. While the size of the vector Δ is equal to the number of genes in the network in the SIN model, it is equal to the number of stochastic gates in the SIF model of stochasticity.

With these two models of stochasticity in GRNs, we provide, in the next two sections, algorithms to compute the probability of cellular differentiation and robustness.

6.4. Algorithms for Stochastic Cellular Differentiation

Given a state of the network $\mathbf{x}^t$, a fault vector Δ, and the transition function $T_i(\mathbf{x}^t, \mathbf{x}^{t+1})$, the state of the network in the next time step, $\mathbf{x}^{t+1}$, can be computed using Eq. 27 where the symbol $\exists_y$ stands for existential quantification (17) over a variable *y*.

$$x^{t+1} = \exists_{\Delta}\exists_{x^{t+1}} \left\{T(\mathbf{x}^t, \mathbf{x}^{t+1}, \Delta) \wedge \mathbf{x}^t \wedge \Delta\right\} \tag{27}$$

Given a fault vector Δ of length *n* and a current state of the network $\mathbf{x}^t$, the probability that the network would exist in the faulty state $\mathbf{x}^{t+1,\Delta}$ and the fault-free state $\mathbf{x}^{t+1}$ is given by Eqs. 28 and 29, respectively.

$$P\left(\mathbf{x}^{t+1,\Delta} \mid (\mathbf{x}^t, \Delta)\right) = \sum_{i=1}^{n} P(\Delta_i = 1) \tag{28}$$

$$P\left(x^{t+1} \mid (\mathbf{x}^t, \Delta)\right) = \sum_{i=1}^{n} \left(1 - P(\Delta_i = 1)\right) \tag{29}$$

By applying Bayes' rule on Eqs. 28 and 29, the probability of the network being in state $\mathbf{x}^{t+1}$ at the next time instant is given by Eq. 30. In Eq. 30, we marginalize the probability over the set of all possible fault configuration vector Δ*s*. Using Eq. 25 in Eq. 30, we get the probability of generating the faulty next state $\mathbf{x}^{t+1,\Delta}$ from the current state $\mathbf{x}^t$ in Eq. 31. If the network can be in only one starting state $\mathbf{x}^t$, the probability of generating the faulty state $\mathbf{x}^{t+1,\Delta}$ is given by Eq. 32.

$$P(\mathbf{x}^{t+1,\Delta} \mid \mathbf{x}^t) = \sum_{\delta \in D} \left\{P(\mathbf{x}^{t+1,\Delta} \mid (\mathbf{x}^t, \Delta = \delta)) \cdot P(\Delta = \delta)\right\} \tag{30}$$

$$= \frac{1}{|D|} \sum_{\delta \in D} P(\mathbf{x}^{t+1,\Delta} \mid (\mathbf{x}^t, \Delta = \delta)) \tag{31}$$

$$P(\mathbf{x}^{t+1,\Delta}) = P(\mathbf{x}^{t+1,\mathrm{D}} \mid \mathbf{x}^t) \cdot P(\mathbf{x}^t) \tag{32}$$

A similar set of equations exist for the fault-free next state $\mathbf{x}^{t+1}$ (Eqs. 22 and 34).

$$P(\mathbf{x}^{t+1} \mid \mathbf{x}^t) = \frac{1}{|D|} \sum_{\delta \in D} P(\mathbf{x}^{t+1} \mid (\mathbf{x}^t, \Delta = \delta)) \tag{33}$$

$$P(\mathbf{x}^{t+1}) = P(\mathbf{x}^{t+1} \mid \mathbf{x}^t) \cdot P(\mathbf{x}^t) \tag{34}$$

If the network may exist in a set of initial sta tes *S* and the probability of each initial state is specified, then the probability of being in the next states $\mathbf{x}^{t+1,\Delta}$ and $\mathbf{x}^{t+1}$ is given by Eqs. 35 and 36, respectively.

$$P(\mathbf{x}^{t+1,\Delta}) = \sum_{\mathbf{x}_t \in S} P(\mathbf{x}^{t+1,\Delta} \mid \mathbf{x}^t) \cdot P(\mathbf{x}^t) \tag{35}$$

$$P(\mathbf{x}^{t+1}) = \sum_{\mathbf{x}_t \in S} P(\mathbf{x}^{t+1} \mid \mathbf{x}^t) \cdot P(\mathbf{x}^t) \quad (36)$$

Algorithm 4 describes how the probability of the set of next states S_{t+1} is computed from a given set of initial states S_t. In line 5 of Algorithm 4, the possible fault configuration vectors are computed from an initial state S_t^i ($i=1, 2, \ldots, |S_t|$). For each fault configuration vector in the set D, the next states are generated in line 9 and the probability of each next state is computed in lines 10–14 using Eqs. 30–36. In Algorithm 4, P_t and P_{t+1} represent the probability of states in the set S_t and S_{t+1}, respectively.

Algorithm 5 describes how the probability of transition into different steady states can be computed from a given set of initial states. In Algorithm 5, given a set of states S, backward reachable set BR(S) is a set of all the states of the network that can make a transition into the states in S in one or more time steps under the "no stochasticity" condition (i.e., $\Delta = \mathbf{0}$). If SS_{a_i} represents the set of states in an attractor a_i, one can test whether the current state of the network $\mathbf{x}^t$ can differentiate into the attractor a_i given a fault vector Δ by testing whether $BR(SS_{a_i}) \cap \mathbf{x}^{t+1} \neq \emptyset$, where $\mathbf{x}^{t+1}$ is computed using Eq. 27. The probability of making a transition to an attractor a_i is then given by the sum of $P(\mathbf{x}^{t+1})$ for all the states $\mathbf{x}^{t+1}$ that can make a transition to a_i. The function stochastic_differentiation() in lines 15–23 of Algorithm 5 computes the probability of differentiation into all the cellular steady states from an initial set of states S. In line 17, we compute the probability of all the faulty states that may exist by injecting a single fault in the network. Multiple faults may exist in the network that we model using the function stochastic_differentiation_k_faults() in lines 1–14. This function models k sequential faults in the network. However, each fault is injected under the single fault model and multiple faults exist only in consecutive time steps.

6.5. Algorithm for Robustness Computation

In the absence of any stochasticity, attractors in a GRN are first computed using the Algorithm 1 proposed in Subheading 2.6 for the deterministic Boolean modeling of GRNs. By the definition of an attractor (see Subheading 2.5), there can be no path among the attractors in a deterministic model. If SS_{a_i} represents the states in the attractor a_i, then using the function stochastic_differentiation_k_faults(), defined in Algorithm 5, the probability of differentiation into various attractors can be computed for every attractor a_i. The probability to differentiate into various attractors in turn represents the robustness of an attractor a_i. Algorithm 6 formally describes the procedure to compute the robustness of attractors.

6.6. Simulation Results

In this section, the SIN and SIF models of stochasticity are applied on the T-helper network (22) shown in Fig. 9. The network is modeled under increasing number of faults. The results and discussion

Algorithm 4: Algorithm for computing probability of faulty next states.

1 stochastic_next_states(T, S_t, P_t, G)
2 **begin**
3 $S_{t+1} = \emptyset$
4 **for** $i = 0$ *to* $|S_t|$ **do**
5 $D = \text{construct\_fault\_config}(S_t^i, G)$
6 $\mathbf{\Delta} = \mathbf{0}$
7 $s_{tmp} = \exists_{\Delta} \exists_{x^{t+1}} \{T(\mathbf{x}^t, \mathbf{x}^{t+1}, \mathbf{\Delta}) \wedge S_t^i \wedge \mathbf{\Delta}\}$
8 **for** $j = 0$ *to* $|D|$ **do**
9 $s_{tmp}^{\Delta} = \exists_{\Delta} \exists_{x^{t+1}} \{T(\mathbf{x}^t, \mathbf{x}^{t+1}, \mathbf{\Delta}) \wedge S_t^i \wedge D^j\}$
10 $P(s_{tmp}^{\Delta}) = P(s_{tmp}^{\Delta} \mid S_t^i) \cdot P_t^{S_t^i}$
11 $P(s_{tmp}) = P(s_{tmp} \mid S_t^i) \cdot P_t^{S_t^i}$
12 $P_{t+1}^{s_{tmp}^{\Delta}} = P_{t+1}^{s_{tmp}^{\Delta}} + P(s_{tmp}^{\Delta})$
13 $S_{t+1} = S_{t+1} \cup s_{tmp}^{\Delta}$
14 $P_{t+1}^{s_{tmp}} = P_{t+1}^{s_{tmp}} + P(s_{tmp})$
15 $S_{t+1} = S_{t+1} \cup s_{tmp}$
16 **return** (S_{t+1}, P_{t+1})
17 **end**

Algorithm 5: Algorithm for computing probability of differentiation into various attractors in the presence of up to k faults.

1 stochastic_differentiation_k_faults(T, S, G, k, SS)
2 **begin**
3 **for** $i = 1$ *to* $|S|$ **do**
4 $P_t^{S^i} = 1/S.size()$
5 **for** $i = 1$ *to* $|SS|$ **do**
6 $P_{SS_i} = 0$
7 $S_t = S$
8 **for** $i = 1$ *to* k **do**
9 $(S_{t+1}, P_{t+1}, \tilde{P}_{SS}) = \text{stochastic\_differentiation}(T, S_t, G, P_t, SS)$
10 **for** $j = 1$ *to* $|SS|$ **do**
11 $P_{SS_j} = P_{SS_j} + \tilde{P}_{SS_j}/k$
12 $t = t + 1$
13 **return** P_{SS}
14 **end**
15 stochastic_differentiation(T, S, G, P, SS)
16 **begin**
17 $(S_{t+1}, P_{t+1}) = \text{stochastic\_next\_states}(T, S, P, G)$
18 **for** $j = 1$ *to* $|SS|$ **do**
19 **for** $i = 1$ *to* $|S_{t+1}|$ **do**
20 **if** $BR(SS_j) \bigcap S_{t+1}^i \neq \emptyset$ **then**
21 $P_{SS_j} = P_{SS_j} + P_{t+1}^{S_{t+1}^i}$
22 **return** $(S_{t+1}, P_{t+1}, P_{SS})$
23 **end**

Algorithm 6: Algorithm for computing Robustness of Attractors in the presence of up to k faults.

1 robust_attractors_k_faults(T, S, G, k)
2 **begin**
3 $\widetilde{T} = \exists_{\Delta} \left[\{T(\mathbf{x}^t, \mathbf{x}^{t+1}, \mathbf{\Delta})\} \wedge \{\mathbf{\Delta} = \mathbf{0}\}\right]$
4 $SS = \text{all\_attractors}(\widetilde{T})$
5 **for** $i = 1$ *to* $|SS|$ **do**
6 $P_{SS}[i] = \text{stochastic\_differentiation\_k\_faults}(T, SS_i, G, k, SS)$
7 **end**

are organized under the earlier-mentioned two properties of steady states, i.e., cellular differentiation and robustness of attractors.

6.6.1. Cellular Differentiation

On simulating the Th0 to Th1 cellular differentiation in response to external IFN-γ stimulus under the SIN model of stochasticity, we found that an almost equal number of cells differentiate into Th1 and Th2 from the Th0 cell state and a few cells revert to Th0 (Fig. 14c). Biologically it is known that Th0 cells cannot differentiate

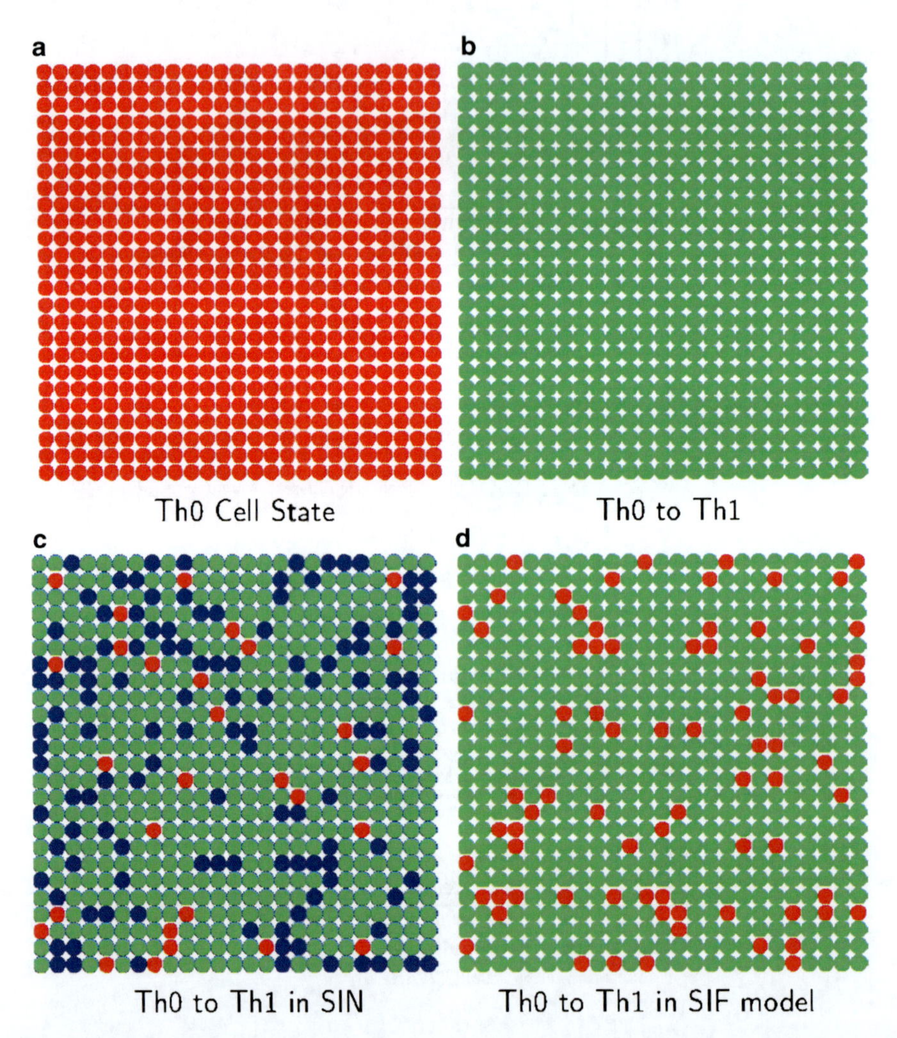

Fig. 14. Simulation results showing the effect of noise on the T-helper cell differentiation process with an external stimulus of IFN-γ. Each *small circle* is representative of a T-helper cell, and each cell is modeled to behave independent of the neighboring cells. *Red cells* represent the naive undifferentiated Th0 cells, *green cells* represent the Th1 cell state, and *blue cells* represent the Th2 cell state. Ratio of number of *red* (*green* or *blue*) cells to total number of cells in a panel is representative of the probability of differentiating into Th0 (Th1 or Th2) cell state. (**a**) Th0 cell state. (**b**) In the absence of any stochasticity, all Th0 cells differentiate to Th1 cell state on receiving IFN-γ. (**c**) Th0 cells differentiate into Th1 and Th2 under the SIN model of stochasticity. Few cells revert to Th0 state as seen by the few patches of *red color*. (**d**) SIF model of stochasticity shows that Th0 cells differentiate into Th1 cells while some cells cannot differentiate on receiving IFN-γ and revert to Th0 cell state. None of the cells differentiate into Th2 cell state. The probability of failure (i.e., Δ_i) is 0.5 for all the nodes (functions) in the SIN model (SIF model). The stochasticity models are simulated with up to four faults in a GRN.

to Th2 state in response to an IFN-γ stimulus (25). The difference in the simulation results from the known biological observation can be a shortcoming of the GRN, of Boolean modeling, or of the model of stochasticity. The equally likely cellular differentiation of steady states under the SIN model of stochasticity has been observed earlier in (48) and was tagged as a shortcoming of Boolean models. However, in our opinion, the SIN model of stochasticity is the main reason behind this discrepancy in simulation results. If we use the more biologically motivated SIF model of stochasticity, where the stochasticity in a biological function is tightly linked with the activity of other nodes in the network, we see that a major subpopulation of Th0 cells differentiate into Th1 cellular state and a few cells revert to Th0 in response to IFN-γ dosage (Fig. 14d). This is consistent with the expected biological behavior of T-helper cells (25) and thereby makes a strong case for the refined SIF model.

6.6.2. Robustness of Attractors

Robustness results of the T-helper network under the SIN and the SIF models of stochasticity are shown in Fig. 15. Since, in the absence of any stochasticity (i.e., $n = 0$), an attractor cannot make

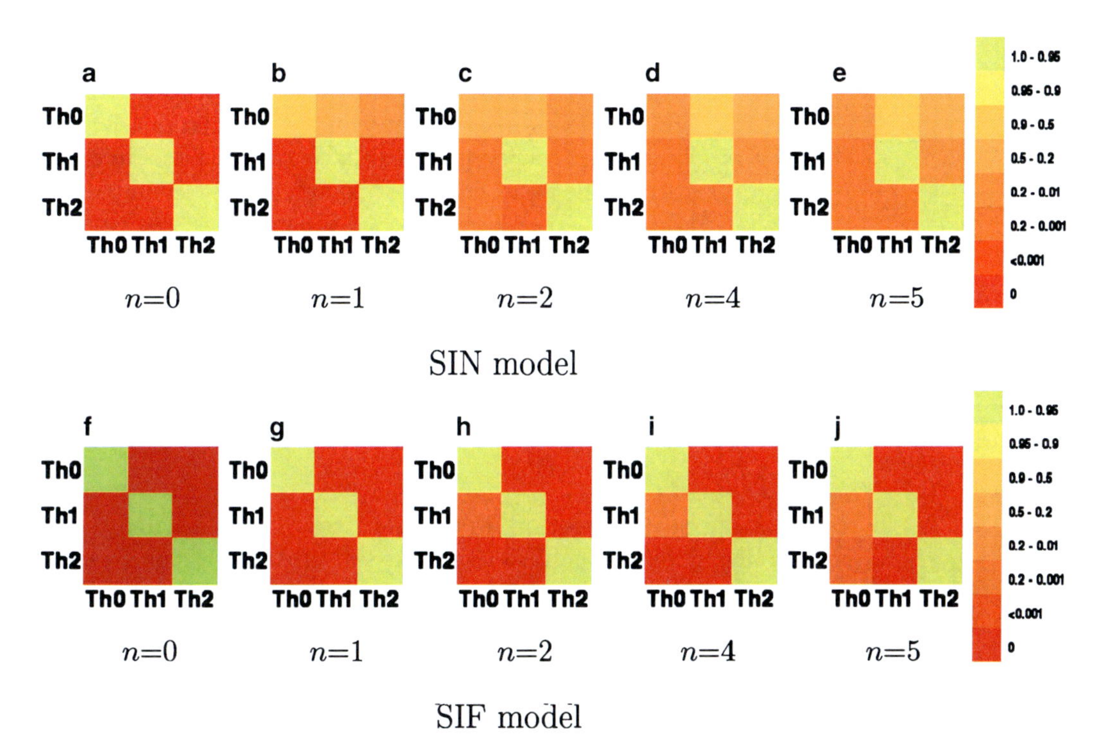

Fig. 15. Simulation results showing the transition probability among Th0, Th1, and Th2 cell states of the T-helper network in the absence of an external stimulus. (**a–e**) Transition probabilities in the SIN model as the number of faults n in the network is increased from $n = 0$ to $n = 5$. (**f–k**) Transition probabilities in the SIF model. In each figure, the intensity of *yellow color* in the entry *i–j* corresponds to the probability of transition from the attractor *i* to the attractor *j*. The *colorbar* in the rightmost column indicates the color-probability encoding. The probability of failure (i.e., Δ_j) is 0.5 for all the nodes (functions) in the SIN model (SIF model).

a transition to another attractor, Fig. 15a and 15f have non-red entries only along the diagonal. In Fig. 15, under the SIN model, all the three attractors (i.e., Th0, Th1, and Th2) are found to make a transition into each other with a significant probability (represented by the intensity of the yellow color in Fig. 15). Robustness of attractors is measured as the number of faults in the network are increased from 0 (i.e., no stochasticity) to 5 faults. One can see from the top row in Fig. 15 that robustness decreases if the number of faults in the network is increased under the SIN model. Under the SIF model, Th0 cell state is found to be robust to stochasticity (bottom row of Fig. 15). Th1 and Th2 cellular states are very robust as most of the cells stay in the original attractor state even with five sequential faults in the network. Moreover, Th1 and Th2 cells do not show transition among each other as the number of faults is increased. This observation further increases our confidence in the SIF model as biologically Th0, Th1, and Th2 cell states are known to be robust and the underlying T-helper network of Fig. 9 is a well-established GRN in the literature. Since the SIF model is closer to the biological phenomenon of inducing faults in biological functions and does not always give low robustness measure, it can provide an effective way to compare the robustness of two different configurations of GRNs in response to internal stochasticity.

From the above results, it is evident that the SIN model of stochasticity often leads to overrepresentation of noise in GRNs by making all the genes/proteins equally likely to flip, independent of the expression of the input genes and complexity of the underlying biological function. With the improved SIF model of stochasticity, it is possible to simulate biological phenomena such as gene perturbation experiments more accurately and to construct GRNs that exhibit strong robustness properties.

7. genYsis Toolbox

The algorithms proposed in this chapter are implemented in our modeling toolbox genYsis (37). The software is written in C++ and makes use of the CUDD software package (49) for the BDD manipulation. Executable binaries of genYsis for both windows and linux have been made available in the public domain and can be downloaded from the following link (50).

7.1. Installing and Running genYsis

User interface of genYsis is based on command line and can be run from terminal window in either Unix or Windows using the following command:

```
Linux:
    ./genYsis -<option> <option_argument>
Windows:
    genYsis.exe -<option> <option_argument>
```

For brevity, only the usage in linux is used for demonstrating the usage of genYsis in this chapter. However, all the commands works in the similar way in the windows installation by replacing "./genysis" with "genYsis.exe" in windows.

Different options available within genYsis can be listed using the following command where –h stands for "help" option:

```
./genYsis -h
```

The output of the above command is listed in Fig. 16. In order to compute the steady states for the small GRN as described in Example 1 and shown in Fig. 4, first one need to construct a file

```
Usage : ./genYsis -<option> <option_argument>

Options list:
-h   Print the usage of genYsis

-p   sets the process id
     Usage: -p<process_id>
          process_id : {1,2 or 3}
               1 : synchronous
               2 : asynchronous
               3 : synchronous-asynchronous combined

-e   sets the experiment file
     Usage: -e <experiment_file_name>

-f   sets the network file
     Usage: -f <network_file_name>

-o   sets the output file
     Usage: -o <output_file_name>

-r   sets the robustness method
     Usage: -r <stochasticity_model>
          stochasticity_model: {SIN or SIF}
                    SIN : Stochasticity in Node
                    SIF : Stochasticity in Function

-c   number of cycles of fault injection for robustness
     Usage: -c <number_of_faults>
     Useful only when -r option is used.
     Required with -r switch.
```

Fig. 16. Different options for modeling GRNs provided in genYsis toolbox.

describing the interactions in the given network. GenYsis uses a text-based network file format. File extension is not important for the software as far as the network description follows the syntax as described below.

```
<input_genes_name> <influence_type> <output_
gene_name>
```

The influence type can be either an activating or an inhibiting interaction and is represented by the following symbols:

```
<influence_type>
   Inhibiting influence is represented by the symbol -|
   Activating influence is represented by the arrow ->
```

There can be multiple genes that together may have an influence on a given output gene. Each input gene may have a positive or negative polarity. For example, if two genes A and B have an influence on a gene C such that the presence of both A and B together activates C, then it can be specified by the following syntax:

```
A&B -> C
```

If the presence of A and absence of B at the same time activate C, then it can be specified by the following syntax:

```
A&^B -> C
```

In the above case, B has negative polarity and is represented by ^B. If the presence of both A and B together deactivates C, then it can be specified by the following syntax:

```
A&B -| C
```

A network file describes the network using the influence syntax as described above. Each line describes only one influence and there is no blank line between two different influences.

Blank spaces are not allowed in `<input_genes_name>` and `<output_gene_name> and at least one` blank space is mandatory in between `<input_genes_name>`, `<influence_type>` and `<output_gene_name>`.

Following the above network description format, the GRN given in Example 1 and displayed in Fig. 4 is described below:

```
B -> A
A&C -> B
E -> B
D -| C
B -> D
C -| D
A -> E
```

7.2. Computing Attractors in genYsis

In order to obtain the attractors in the synchronous modeling of GRNs, the network file as described above is passed as an argument to genYsis with the –f option and the output files to store the attractors is supplied using the –o argument as follows:

```
./genYsis -p1 -f networks/net_sample.net -o results/ss_sample.tsv
```

This command generates the following output giving further details on the number of attractors found and the number of states within each attractor.

```
parsing network file
constructing BDD
BDD construction done
Attractor 1 ::: number of states = 2
Attractor 2 ::: number of states = 1
Attractor 3 ::: number of states = 1
Attractor 4 ::: number of states = 1
States of the attractor 1 are written in the file results/ss_sample_1.tsv
States of the attractor 2 are written in the file results/ss_sample_2.tsv
States of the attractor 3 are written in the file results/ss_sample_3.tsv
States of the attractor 4 are written in the file results/ss_sample_4.tsv
```

A sample output file describing the states in the first attractor is shown below:

```
Gene Name/State No. S_1 S_2
A               1 1
B               1 1
C               0 1
E               1 1
D               0 1
```

7.3. Performing Perturbation Experiments in genYsis

In order to specify the perturbation experiments, the nodes that should be overexpressed or knocked-out are represented using the following file format:

```
<Number of levels of experiments>
/* For each level of experiment one has to
define the following */
<no. of knocked genes> <no. of over expressed
genes> <no. of previously perturbed genes that
return to normal expression>
<names of knocked genes>
<names of over-expressed genes>
<names of previously perturbed genes that
return to normal expression>
```

Following the above file format, the perturbation experiment of Fig. 6b can be described as shown below:

```
2
0 0 0
0 1 0
A
```

The above experiment file can be passed as an argument to genYsis program using the –e option as described below:

```
./genYsis -p1 -f networks/net_sample.net -e experiments/
exp_sample.txt
```

On executing the above command, genYsis generates a file "results/reach_net_sample.txt" describing the transitions among attractors when the node A is overexpressed. The contents of this file for the above command are shown in Fig. 17.

```
###### unperturbed network ##################
States of the attractor 1 are written in the file results/_SS_1_1.txt
States of the attractor 2 are written in the file results/_SS_1_2.txt
States of the attractor 3 are written in the file results/_SS_1_3.txt
States of the attractor 4 are written in the file results/_SS_1_4.txt
###############################################

####### A over-expressed #######
States of the attractor 1 are written in the file results/_SS_2_1.txt
States of the attractor 2 are written in the file results/_SS_2_2.txt
States of the attractor 3 are written in the file results/_SS_2_3.txt
***** Reachability Analysis *****
SS_1_1 ----> SS_2_2
SS_1_2 ----> SS_2_1
SS_1_3 ----> SS_2_2
SS_1_4 ----> SS_2_3
```

Fig. 17. Contents of the reachability file generated by genYsis to model overexpression of gene A in the network of Fig. 3.

```
parsing network file
Constructing Bdd
Computing Steady States
Attractor 1 ::: number of states = 1
Attractor 2 ::: number of states = 1
Attractor 3 ::: number of states = 2
Attractor 4 ::: number of states = 1
States of the attractor 1 are written in the file results/net_sample_SS_1.txt
States of the attractor 2 are written in the file results/net_sample_SS_2.txt
States of the attractor 3 are written in the file results/net_sample_SS_3.txt
States of the attractor 4 are written in the file results/net_sample_SS_4.txt
Computing Robustness
computing robustness of Attractor 1
computing robustness of Attractor 2
computing robustness of Attractor 3
computing robustness of Attractor 4
Robustness results are written in the file `results/outPertFunc_1.txt'
```

Fig. 18. Output of genYsis on modeling the stochasticity in the GRN of Fig. 6a.

7.4. Computing Robustness of Attractors in genYsis

To compute the robustness of attractors in the presence of stochastic behavior of GRNs, one has to specify whether SIN or SIF model of stochasticity is used and how many cycles of faults are injected in the dynamic simulation. These arguments can be specified through the options –r and –c to the genYsis.

For the GRN given in Example 1, the robustness of GRN under the SIF model of stochasticity and two cycles of fault injection can be computed using the following command:

```
./genYsis -p1 -f networks/net_sample.net -r SIF -c 2
```

```
1(1): 1(1)
2(1): 2(0.91)  3(0.085) 4(0.002)
3(1): 2(0.096) 3(0.870) 4(0.033)
4(1): 2(0.022) 3(0.413) 4(0.564)
```

Fig. 19. Probability of transition among different attractors after two cycles of fault injection.

The above command generates the output in Fig. 18. The attractor descriptions are stored in the results subdirectory and the corresponding robustness of attractors in the files "outPertFunc_1.txt" and so on, where 1 represents single fault injection. The contents of one such file after two cycles of fault injection are shown in Fig. 19.

Further details on available options in genYsis can be found in the user manual available on the download webpage of genYsis (49).

8. Conclusions

In this chapter, we have formally introduced the problem of modeling the dynamics in GRNs using Boolean algebra and implicit representation of finite-state machines. The formalism proposed in this chapter enables computation of steady states and dynamical attractors of large GRNs with over 1,000 nodes. Based on these algorithms for computing steady states, further algorithms were developed for modeling multiple gene perturbation experiments. These algorithms can be used for in silico simulation of experimental protocols which involve application of different stimuli on a cell culture over a range of time. The applicability of algorithms was shown by modeling the T-helper cell GRN. It was demonstrated how T-helper cells can differentiate into Th0, Th1, and Th2 cell types when subjected to gene perturbations.

The proposed Boolean formalism was further extended to incorporate inherent stochasticity in biological phenomena. A gene or protein can be normally activated by more than one biological phenomena. This redundancy is inherent in biology to avoid a complete breakdown of the system in the event of malfunctioning in a single gene (or protein) regulation phenomena. However, every gene regulation phenomena has a probability of failure which is normally reflected by the complexity of the regulation mechanism itself. In our Boolean modeling approach, the gene (or protein) regulation mechanisms are represented by Boolean functions. To model stochasticity in gene regulation mechanisms, a fault modeling approach called SIF was introduced in this chapter. In SIF, Boolean functions have a probability of failure reflecting the stochasticity (and hence complexity) of underlying biological functions.

The SIF model was shown to give more biological realistic results as compared to existing approaches for modeling stochasticity in GRNs. By applying SIF on T-helper and T-cell receptor GRN, it was demonstrated that while the two GRNs will be declared non-robust to stochasticity for a small amount of noise under the existing models, SIF can efficiently model the impact of gradually increasing noise on the robustness of GRNs.

In future, as the experimental tools available to biologists will become more advanced, additional insights into gene and protein regulation mechanisms will become available. However, with this improved understanding of biological phenomena, there will be a desire to computationally model increasingly complex biological systems. To deal with the complex biological systems, various abstraction mechanisms will have to be used in the computational tools. Therefore, even though the formalism proposed in this chapter seems to be an abstraction of actual biological mechanisms, computational methods such as the ones proposed in this chapter will form the basis of modeling in biology in the future. Such discrete modeling and analysis techniques will find more applications as the complexity of the systems that one would like to computationally model will grow, and discrete modeling will be the direction of research in systems biology in the coming years.

References

1. Goodwin BC. Temporal organization in cells; A dynamic theory of cellular control processes. Academic Press, New York, 1963.
2. Li S. A quantitative study of the division cycle of caulobacter crescentus stalked cells. PLoS Computational Biology, **4**, 2008.
3. Chen KC. Integrative analysis of cell cycle control in budding yeast. Molecular Biology of Cell, **15**:3841–3862, 2004.
4. Semenov A and Yakovlev A. Verification of asynchronous circuits using time Petri net unfolding. Proceedings of the 33rd annual conference on Design automation, Las Vegas 1996, pages 59–62, 1996.
5. Yakovlev A, Semenov A, Koelmans AM, and Kinniment DJ. Petri nets and asynchronous circuit design. IEEE colloquium on Design and Test Asynchronous Systems, pages 8/1-8/6, 1996.
6. Remy E, Ruet P, Mendoza L, Thieffry D, and Chaouiya C. From logical regulatory graphs to standard petri nets: Dynamical roles and functionality of feedback circuits. Lecture Notes in Computer Science, **4230**:56–72, 2006.
7. Snoussi EH and Thomas R. Logical identification of all steady states: the concept of feedback loop-characteristic states. Bulletin of Mathematical Biology, **55**:973–991, 1993.
8. Steggles LJ, Banks R, Shaw O, and Wipat A. Qualitatively modeling and analysing genetic regulatory networks: a petri net approach. Bioinformatics, **23(3)**:336–343, 2007.
9. Heiner M and Koch I. Petri net based model validation in systems biology. J. Cortadella and W. Reisig (Eds), ICATPN04, LNCS, **3099**: 216–237, 2004.
10. Hofestadt R and Thelen S. Quantitative modeling of biochemical networks. In Silico Biology 1, pages 39–53, 1998.
11. Thomas R. Regulatory networks seen as asynchronous automata: a logical description. Journal of Theoetical. Biology, **153**:1–23, 1991.
12. Reddy VN, Liebman MN, and Mavrovouniotis ML. Qualitative analysis of bio-chemical reaction systems. Computational Biology and Medical Informatics, **26**:9–24, 1996.
13. Kahlem P and Birney E. Dry work in a wet world: computation in systems biology. Molecular Systems Biology, **2**:40, 2006.
14. Huang S, Eichler G, Bar-Yam Y, and Ingber DE. Cell fates as high dimensional attractor states of a complex gene regulatory network. Physics. Review Letters, **94**:128701:1–128701: 4, 2005.

15. Kauffman SA. Metabolic stability and epigenesis in randomly constructed genetic nets. Journal of Theoetical. Biology, **22**:437–467, 1969.
16. Xie A and Beerel PA. Efficient state classification of finite state markov chains. Proceedings of Design Automation Conference, 1998.
17. De Micheli G. Synthesis and optimization of digital circuits. Mc Graw-Hill Higher Education, 2009.
18. Touati HJ, Savoj H, Lin B, Brayton RK, and Sangiovanni-Vincentelli Implicit state enumeration of finite-state machines using BDDs. Proceedings of ICCAD'90, 1990.
19. Roig O, Cortadella J, and Pastor E. Verification of asynchronous circuits by BDD-based model checking of Petri nets. Lecture Notes in Computer Science, Springer Berlin/Heidelberg, **935**:374–391, 1995.
20. Bryant RE. Graph-Based Algorithms for Boolean Function Manipulation. IEEE Transactions on Computers, **35**:677–691, 1986.
21. Fauré A, Naldi A, Chaouiya C, and Thieffry D. Dynamical analysis of a generic Boolean model for the control of the mammalian cell cycle. Bioinformatics, **22**:e124-131, 2006.
22. Mendoza L and Xenarios I. A method for the generation of standardized qualitative dynamical systems of regulatory networks. Theoretical Biology and Medical Modeling, Mar **16**;3:13, 2006.
23. Klamt S, Saez-Rodriguez J, Lindquist JA, Simeoni L, and Gilles ED. A methodology for the structural and functional analysis of signaling and regulatory networks. BMC Bioinformatics, 7, 2006.
24. Agnello D, Lankford CS, Bream J, Morinobu A, Gadina M, O'Shea JJ, and Frucht DM. Cytokines and transcription factors that regulate T helper cell differentiation: new players and new insights. Journal of Clinical Immunology, **23**:147–162, 2003.
25. Murphy KM and Reiner SL. The lineage decisions on helper T cells. Nature Review Immunology, **2**:933–944, 2002.
26. Szabo SJ, Sullivan BM, Peng SL, and Glimcher LH. Molecular mechanisms regulating Th1 immune responses. Annual Reviews Immunology, **21**:713–758, 2003.
27. Yates A, Bergmann C, Van Hemmen JL, Stark J, and Callard R. Cytokine-modulated regulation of helper T cell populations. Journal of Theoretical Biology, **206**:539–560, 2000.
28. Bergmann C, van Hemmen JL, and Segel LA. Th1 or Th2: how an appropriate T helper response can be made. Bulletin of Mathematical Biology, **63**:405–430, 2001.
29. Weisbuch G, DeBoer RJ, and Perelson AS. Localized memories in idiotypic networks. Jour-nal of Theoretical Biology, **146**:483–499, 1990.
30. Krueger GR, Marshall GR, Junker U, Schroeder H, and Buja LM. Growth factors, cytokines, chemokines and neuropeptides in the modeling of T-cells. In Vivo, **17**:105–118, 2002.
31. Mendoza L. A network model for the control of the differentiation process in Th cells. Biosystems, **84**:101–114, 2005.
32. Becskei A and Serrano L. Engineering stability in gene networks by autoregulation. Nature, **405**:590–593, 2000.
33. McAdams HH and Arkin A. Its a noisy business! Genetic regulation at the nanomolar scale. Trends in Genetics, **15**:65–69, 1999.
34. Pedraza JM and Oudenaarden AV. Noise propagation in gene networks. Science, **307**:1965–1969, 2005.
35. Kaern M, Elston TC, Blake WJ, and Collins JJ. Stochasticity in gene expression: From theories to phenotypes. Nature Reviews Genetics, **6**: 451–464, 2005.
36. Losick R and Desplan C. Stochasticity and cell fate. Science, **320**:65–68, 2008.
37. Garg A, Di Cara A, Mendoza L, Xenarios I and De Micheli G. Synchronous vs. Asynchronous modeling of gene regulatory networks. Bioinformatics, **24**:1917–1925, 2008.
38. Rao CV, Wolf DM, and Arkin AP. Control, exploitation and tolerance of intracellular noise. Nature, **421**:231–237, 2002.
39. Gonze D and Goldbeter A. Circadian rhythms and molecular noise. Chaos, **16**:026–110, 2006.
40. Schultz D, Jacob EB, Onuchic JN, and Wolynes PG. Molecular level stochastic model for competence cycles in Bacillus subtilis. Proceedings of National Academy of Science of the USA, **104**:17582–17587, 2007.
41. Gillespie DT. A general method for numerically simulating the stochastic time evolution of coupled chemical reactions. Journal of Computational Physics, **22**:403–434, 1976.
42. Gillespie DT. Exact stochastic simulation of coupled chemical reactions. Journal of Physical Chemistry, **81**:2340–2361, 1977.
43. Ribeiro AS and Kauffman SA. Noisy attractors and ergodic sets in models of gene regulatory networks. Journal of Theoretical Biology, **247**:743–755, 2007.
44. Alvarez-Buylla ER, Chaos A, Aldana M, Bentez M, Cortes-Poza Y, Espinosa-Soto C, Hartasnchez DA, Lotto RB, Malkin D, Escalera Santos GJ, and Padilla-Longoria P. Floral Morphogenesis: Stochastic explorations of a

gene network epigenetic landscape. PLoS ONE, **3**:e3626, 2008.

45. Willadsena K and Wiles J. Robustness and state-space structure of Boolean gene regulatory models. Journal of Theoretical Biology, **249**:749–765, 2007.
46. Davidich MI and Bornholdt S. Boolean network model predicts cell cycle sequence of fission yeast. PLoS ONE, **3**:e1672, 2008.
47. Garg A, Mohanram K, Di Cara A, De Micheli G and Xenarios I. Modeling stochasticity and robustness in gene regulatory networks. Bioinformatics, **25**:i101-i109, 2009.
48. Kadanoff L, Coppersmith S, and Aldana M. Boolean dynamics with random couplings. Springer Applied Mathematical Sciences Series, Special volume:23–89, 2003.
48. Somenzi F. CUDD: CU Decision Diagram Package Release 2.4.1. University of Colorado at Boulder, 2005.
49. GenYsis toolbox. http://lsi.epfl.ch/downloads.

Index

Bart Deplancke and Nele Gheldof (eds.), *Gene Regulatory Networks: Methods and Protocols*,
Methods in Molecular Biology, vol. 786, DOI 10.1007/978-1-61779-292-2,